THE ROUTLEDGE HANDBOOK OF
URBAN DISASTER RESILIENCE

The Routledge Handbook of Urban Disaster Resilience emphasizes the intersection of urban planning and hazard mitigation as critical for community resilience, considering the interaction of social, environmental, and physical systems with disasters. The *Handbook* introduces and discusses the phases of disaster – mitigation, preparedness/response, and recovery – as well as each of the federal, state, and local players that address these phases from a planning and policy perspective.

Part I provides an overview of hazard vulnerability that begins with an explanation of what it means to be vulnerable to hazards, especially for socially vulnerable population segments. Part II discusses the politics of hazard mitigation; the failures of smart growth placed in hazardous areas; the wide range of land development policies and their associated risk; the connection between hazards and climate adaptation; and the role of structural and non-structural mitigation in planning for disasters. Part III covers emergency preparedness and response planning, the unmet needs people experience and community service planning; evacuation planning; and increasing community capacity and emergency response in developing countries. Part IV addresses recovery from and adaption to disasters, with topics such as the National Disaster Recovery Framework, long-term housing recovery; population displacement; business recovery; and designs in disasters. Finally, Part V demonstrates how disaster research is interpreted in practice – how to incorporate mitigation into the comprehensive planning process; how states respond to recovery; how cities undertake recovery planning; and how to effectively engage the whole community in disaster planning.

The Routledge Handbook of Urban Disaster Resilience offers the most authoritative and comprehensive coverage of cutting-edge research at the intersection of urban planning and disasters from a U.S. perspective. This book serves as an invaluable guide for undergraduate and postgraduate students, future professionals, and practitioners interested in urban planning, sustainability, development response planning, emergency planning, recovery planning, hazard mitigation planning, land use planning, housing and community development as well as urban sociology, sociology of the community, public administration, homeland security, climate change, and related fields.

Michael K. Lindell is an Emeritus Professor, Texas A&M University, College Station, and an Affiliate Professor at the University of Washington, Seattle, Boise State University, and Oregon State University. His research interests include organizational emergency preparedness and response, training/exercises, warning systems, evacuation modeling, household disaster preparedness, risk communication, risk perception, household disaster response, disaster impact models, cognitive processing of visual displays, and survey research methods.

THE ROUTLEDGE HANDBOOK OF URBAN DISASTER RESILIENCE

Integrating Mitigation, Preparedness, and Recovery Planning

Edited by Michael K. Lindell

Routledge
Taylor & Francis Group

NEW YORK AND LONDON

First published 2020
by Routledge
605 Third Avenue, New York, NY 10017

and by Routledge
2 Park Square, Milton Park, Abingdon, Oxon, OX14 4RN

First issued in paperback 2022

Routledge is an imprint of the Taylor & Francis Group, an informa business

© 2020 Taylor & Francis

The right of Michael K. Lindell to be identified as the author of the editorial material, and of the authors for their individual chapters, has been asserted in accordance with sections 77 and 78 of the Copyright, Designs and Patents Act 1988.

Publisher's Note
The publisher has gone to great lengths to ensure the quality of this reprint but points out that some imperfections in the original copies may be apparent.

Library of Congress Cataloging-in-Publication Data
Names: Lindell, Michael K., editor.
Title: The Routledge handbook of urban disaster resilience : integrating mitigation, preparedness, and recovery planning / edited by Michael K. Lindell.
Description: New York, NY : Routledge, 2019.
Identifiers: LCCN 2019003631 (print) | LCCN 2019005370 (ebook) | ISBN 9781315714462 (e-book) | ISBN 9781138886957 (hardback)
Subjects: LCSH: City planning. | Hazard mitigation. | Emergency management. | Community development.
Classification: LCC HT166 (ebook) | LCC HT166 .R697 2019 (print) | DDC 307.1/216—dc23
LC record available at https://lccn.loc.gov/2019003631

ISBN: 978-1-03-240128-7 (pbk)
ISBN: 978-1-138-88695-7 (hbk)
ISBN: 978-1-315-71446-2 (ebk)

DOI: 10.4324/9781315714462

Typeset in Bembo
by Swales & Willis Ltd, Exeter, Devon, UK

CONTENTS

FIGURES

CONTRIBUTORS

Sudha Arlikatti is currently an Associate Professor in the Faculty of Resilience at Rabdan Academy, Abu Dhabi, U.A.E. She has over a decade of private sector experience as an architectural-planning consultant and over 16 years of teaching and disaster research experience. Her research interests include disaster warnings and risk communication in multiethnic communities, protective action decision-making, post-disaster sheltering and housing recovery, information and communication technology use by emergency management organizations, and organizational and community resiliency to natural hazards in various international settings.

Sherry I. Bame is Professor Emerita from Texas A&M's Urban Planning program, with an emphasis on health and community services planning as well as environmental health planning and policy. After an almost 20-year career in nursing and public health, she earned her Ph.D. in Health Systems Management and Policy from the University of Michigan. Her ongoing research and publications focus on health and social support needs, services, and access barriers during disasters. Sherry has been honored by the NIH, Texas Governor's Office, International Women's Leadership Association, and International Alliance for I&R Systems for her service, scholarship, and leadership.

Philip R. Berke is a Professor of land use and environmental planning in the Department of Landscape Architecture and Urban Planning at Texas A&M University, and Director of the Institute for Sustainable Communities. His research examines the relationship between community resilience and land use planning with specific focus on methods, theory, and metrics of local planning and implementation.

Thomas Birkland is a Professor of Public Policy in the School of Public and International Affairs at NC State University. His research is in how disasters change public policy and the policy process. He is the author of several articles and two books on disaster policy, including *After Disaster* (1997) and *Lessons of Disaster* (2006).

Samuel D. Brody is a Regents Professor and holder of the George P. Mitchell '40 Chair in Sustainable Coasts in the Departments of Marine Sciences and Landscape Architecture and Urban Planning at Texas A&M University. He is the Director of Center for Texas Beaches and Shores. Dr. Brody's research focuses on coastal environmental planning, spatial analysis, flood mitigation, climate change policy, and natural hazards mitigation. He has published numerous scientific articles on flood risk and mitigation, and recently authored the book, *Rising Waters: The causes and consequences of flooding in the United States*. Dr. Brody teaches graduate courses in environmental planning and coastal resiliency. He has also worked in both the public and private sectors to help local coastal communities to establish environmental and flood mitigation plans. For more information, please visit www.tamug.edu/ctbs.

John T. Cooper, Jr. is Assistant Vice President for Public Partnership & Outreach at Texas A&M University and Director of Texas Target Communities. He previously served as a professor of the practice in Landscape Architecture and Urban Planning, where his work focused on principles related to the design of collaborative planning programs. Before returning to A&M, Dr. Cooper was a program director at MDC, nonprofit based in Durham, North Carolina. At MDC he managed the Emergency Preparedness Demonstration (EPD), a five-year project funded by FEMA ($2.5 million) focused on increasing disaster awareness and preparedness in disadvantaged communities in eight states plus the District of Columbia. Dr. Cooper is president of the Texas Rural Leadership Program and serves on a number of other boards including the US Endowment for Forestry and Communities, the Bill Anderson Fund, and the Coastal Hazards Center (CHC) at the University of North Carolina, Chapel Hill, one of the Department of Homeland Security's Centers of Excellence. Dr. Cooper currently also serves on an advisory panel for the Center for Disaster for Philanthropy's Hurricane Harvey Fund.

William Drake is a doctoral candidate in the Planning, Governance, and Globalization program at Virginia Tech. His research focuses on climate change adaptation planning. He can be reached at wdrake@vt.edu.

Ann-Margaret Esnard is a Distinguished University Professor in the Andrew Young School of Policy Studies at Georgia State University. Her expertise includes urban planning, disaster planning, vulnerability assessment, and GIS/spatial analysis. She has been involved in a number of research initiatives, including NSF-funded projects on topics of population displacement from catastrophic disasters, school recovery after disasters, long-term recovery, and community resilience. She is the coauthor of the 2014 book *Displaced by Disasters: Recovery and Resilience in a Globalizing World*, and co-editor of the 2017 book *Coming Home after Disaster: Multiple Dimensions of Housing Recovery*.

Himanshu Grover is an Assistant Professor in the Department of Urban Design and Planning, University of Washington. He is also the co-director of the Institute for Hazard Mitigation Research and Planning. His research focuses at the intersection of hazard mitigation, land use planning, community resilience, and climate change. In his research, Grover examines inter-linkages between physical development, socio-economic concerns, and the natural environment. He is specifically interested in climate change management (both mitigation and adaptation strategies), urban infrastructure, and hazard mitigation. Grover is also interested in issues related to social vulnerability and environmental justice.

Wesley E. Highfield is an Associate Professor in the Department of Marine Sciences at the Texas A&M University, Galveston Campus. His research and teaching focuses on Geographic Information Systems, Spatial Analysis/Statistics, and Natural Hazard Mitigation & Policy.

Jennifer A. Horney is an Associate Professor of Epidemiology and Biostatistics at the Texas A&M University School of Public Health, a Faculty Fellow of the Hazards Reduction and Recovery Center in the College of Architecture, and the Public Health and Environment Lead of the Institute for Sustainable Communities, part of the University's Environmental Grand Challenge.

Dr. Horney's research focuses on measuring the health impacts of disasters, as well as the linkages between disaster planning and household actions related to preparedness, response, and recovery. Dr. Horney received her Ph.D. and MPH from the University of North Carolina at Chapel Hill, where her research focused on the role of social factors in decision making during disasters. She currently leads research projects funded by the National Science Foundation, National Oceanic and Atmospheric Administration, the National Academies of Sciences, the Department of Homeland Security and other federal, state, and local agencies. Dr. Horney was a member of a team of public health practitioners who responded to Hurricanes Isabel, Charley, Katrina, Wilma, Irene, and Harvey where she conducted rapid assessments of disaster impact on the public health of individuals and communities. She has also provided technical assistance to public health agencies globally around disasters, infectious disease outbreaks, and pandemic influenza planning and response.

Shih-Kai Huang is an Assistant Professor of Emergency Management at Jacksonville State University, the PI of an NSF-sponsored research project, and the co-editor for the *International Journal of Mass Emergencies and Disasters*. His research focuses on warning response issues, hurricane evacuation, protective action decisions, and risk communications.

Fayola Jacobs is a Postdoctoral Associate in the Department of Geography, Environment and Society at the University of Minnesota. She completed her Ph.D. at Texas A&M University in Urban and Regional Sciences. Her research interests include environmental justice, hazards in low-income communities and communities of color, black feminisms and anti-oppression.

Katherine Barbour Jakubcin is a Policy Analyst at the Houston Housing Authority, a Public Housing agency that improves lives by providing quality, affordable housing options and promoting education and economic self-sufficiency to more than 60,000 low-income Houstonians.

Laurie A. Johnson is an internationally recognized Urban Planner based in the San Francisco Bay Area and specializing in disaster recovery and catastrophe risk management. She has been active in research and consulting following many of the world's urban disasters, including the 1995 Kobe Japan earthquake, 2005 Hurricane Katrina, 2011 Tohoku Japan tsunami, and 2010–2011 Canterbury earthquake sequence, all of which are covered in her book, *After Great Disasters: An In-depth Analysis of How Six Countries Managed Community Recovery* (2017). She is also a coauthor of the American Planning Association's guidebook, *Planning for Post-Disaster Recovery: Next Generation* (2014).

Jack D. Kartez is Emeritus Professor of Community Planning and Development in the Edmund S. Muskie School of Public Service and Senior Program Advisor in the EPA Region 1 Environmental Finance Center (EFC), which he co-founded in 2001. An early hazards planning researcher in projects sponsored by the National Science Foundation since 1980, he was a founding Faculty Fellow for the Texas A&M University Hazard Reduction and Recovery Center. He currently works on climate adaptation and stormwater at the EFC with a focus on finance. An environmental issues mediator, his clients include state and local governments and nonprofit organizations.

Yoonjeong Lee is an Assistant Research Scientist in the Center for Texas Beaches and Shores at Texas A&M University Galveston. Her research focuses on resilience, urban flooding, flood risk reduction, and mitigation strategies. She also teaches classes related to resilience and sustainability at Texas A&M.

Michael K. Lindell is an Emeritus Professor, Texas A&M University, College Station and an Affiliate Professor at the University of Washington, Seattle, Boise State University, and Oregon

State University. His research interests include organizational emergency preparedness and response, training/exercises, warning systems, evacuation modeling, household disaster preparedness, risk communication, risk perception, household disaster response, disaster impact models, cognitive processing of visual displays, and survey research methods.

Ward Lyles is Assistant Professor of Urban Planning in the School of Public Affairs and Administration at the University of Kansas. His research interests center on relationships between people, the built environment, and the natural environment, especially in the context of natural hazard mitigation and climate change adaptation. He also examines the role of emotion and compassion in public decision-making.

Jaimie Hicks Masterson is Associate Director of Texas Target Communities at Texas A&M University, a high-impact service learning program that works collaboratively with communities to build resilience. She is author of *Planning for Community Resilience: A Handbook for Reducing Vulnerabilities to Disasters*.

William Merrell is the George P. Mitchell chair of marine sciences at Texas A&M University at Galveston. Following the devastation of Hurricane Ike, Dr. Merrell began the Ike Dike project to provide hurricane surge protection for the Upper Texas Coast including all of Houston and Galveston.

Michelle Annette Meyer is Assistant Professor of Sociology at Louisiana State University. She is a Next Generation of Hazard and Disasters Researchers Fellow and an Early-Career Research Fellow with the National Academies of Sciences Gulf Research Program. Her research interests include social capital, disaster recovery, environmental justice, and nonprofits in disaster. Particularly, she has researched governmental and nonprofit networks in long-term recovery; use of social media across population groups during disasters; recovery variation following technological and natural disasters; social capital and collective efficacy for individual and community resilience; and participatory GIS activities to assess environmental and climate justice in marginalized communities.

Walter Gillis Peacock is Professor of Urban Planning in the Department of Landscape Architecture and Urban Planning and the Director of the Hazard Reduction and Recovery Center at Texas A&M University (TAMU) where he has been a member of the faculty since 2002. He received his Ph.D. from the University of Georgia. He is internationally known for his research on disaster recovery, community resiliency, and social vulnerability. In 2009 he was awarded the *Quarantelli Award for Social Science Disaster Theory*, acknowledging significant theoretical work in disaster and hazards research. Between 2008 and 2012 he was the holder of the *Rodney L. Dockery Endowed Professorship in Housing and the Homeless* and in 2012 he was awarded the *Sandy and Bryan Mitchell Master Builder Endowed Chair at Texas A&M*. In 2014 he received the Distinguished Achievement Award in Research from Texas A&M, an award sponsored by the Association of Former Students.

Carla S. Prater retired in 2014 from Texas A&M University, where she was the Associate Director of the Hazard Reduction & Recovery Center and Senior Lecturer in Landscape Architecture and Urban Planning and Secretary/Treasurer of the International Research Committee on Disasters. Educated at Associação Escola Graduada de São Paulo in Brazil where she grew up, she went on to receive her BA in Modern Languages at Pepperdine University, and later earned her MS in Urban Planning and Ph.D. in Political Science from Texas A&M. Her research has included work on disaster response, hurricane and tsunami evacuation, disaster recovery and mitigation planning. She has taught courses and conducted research in Brazil, Taiwan, Turkey, and India, and has presented papers and participated in workshops in Turkey, Italy, Colombia, Costa Rica, Dominican Republic, Panama, and Taiwan. She has co-authored numerous articles and technical reports, and is co-author of a standard emergency management textbook published by Wiley & Sons.

George Oliver Rogers is a Professor of Landscape Architecture and Urban Planning and a Senior Faculty Fellow of the Hazard Reduction and Recovery Center at Texas A&M University. He has conducted extensive research on human response to risk and hazards and is currently interested in behavioral aspects of sustainability, and the dynamics of risk perception and communication. His work on impact assessment and sustainability focuses on organizational and human behavior.

Garett Sansom is the Associate Director for the Institute for Sustainable Communities at Texas A&M University, part of the University's Environmental Grand Challenge. Dr. Sansom received his DrPH and MPH from the Texas A&M University School of Public Health where his research focused on marginalized communities at risk of health implications from environmental contamination and natural disasters. Dr. Sansom, as part of an interdisciplinary team, has conducted rapid needs assessments following major flooding events in Central Texas as well as the events of Hurricane Harvey along coastal communities. His empirical focus targets marginalized communities that experience environmental justice issues by conducting research that builds local capacity.

Alka Sapat is a Professor in the School of Public Administration and Coordinator, Bachelor of Public Safety Administration program at Florida Atlantic University. Her expertise encompasses disaster management, public policy processes, vulnerability and resilience assessment, and methodology. She has been involved in a number of initiatives, including NSF-funded projects on topics of building code regulations, population displacement after disasters and implications for housing, and the role of the NGOs in disaster recovery. She is the coauthor of the 2014 book *Displaced by Disasters: Recovery and Resilience in a Globalizing World*, and co-editor of the 2017 book *Coming Home after Disaster: Multiple Dimensions of Housing Recovery*.

Gavin Smith is a Professor in the Department of Landscape Architecture at North Carolina State University, where he teaches courses in natural hazards, disasters and climate change adaptation as part of a graduate program he leads in Disaster Resilient Policy, Engineering and Design. His research interests include hazard mitigation, disaster recovery, climate change adaptation, and the translation of research to practice.

Zhenghong Tang is a Professor in the Community and Regional Planning Program, a faculty fellow in the Daugherty Water for Food Global Institute, the Nebraska Water Center, the Center for Great Plains Studies, and the Center for Advanced Land Management Information Technologies at University of Nebraska-Lincoln. His research interests are hazard mitigation planning and environmental planning.

Kristin Taylor is an Assistant Professor of Political Science at Wayne State University. Her research is on the adoption and governance of natural hazard mitigation policies. She studies the ways in which state and local governments learn to mitigate natural hazards after a disaster.

Kenneth C. Topping is Senior Advisor to the California State Hazard Mitigation Plan support team and former lecturer at the City and Regional Planning Department at the California State Polytechnic University-San Luis Obispo. Former Los Angeles city planning director, he has written and consulted on disaster resilience issues in the U.S. and Asia.

Shannon Van Zandt is Professor and Head of the Department of Landscape Architecture & Urban Planning at Texas A&M University. She holds a Ph.D. in City & Regional Planning from the University of North Carolina. She also currently holds the Nicole & Kevin Youngblood Professorship in Residential Land Development in recognition of her scholarship on housing, real estate, and urban development. Dr. Van Zandt's research has created a niche within the housing and disaster fields that focuses on how the spatial distribution of residential land affects exposure, impact, and consequences from natural disasters, particularly for socially vulnerable populations. She serves on the board of the Texas Low-Income Housing Information Service, as well as the advisory committee of Texas Sea Grant, and has recently testified before the Texas State Legislature on issues related to housing recovery after disaster.

Hao-Che Wu is an Associate Professor in the Department of Emergency Management and Disaster Science at University of North Texas and is currently the co-editor of the *International Journal of Mass Emergencies and Disasters*. His research focuses on disaster response, risk analysis, perception of threat, and disaster information use. Within disaster information, he examines the mental model of emergency information use and warning message contents.

Yu Xiao is an Associate Professor in the Toulan School of Urban Studies and Planning at Portland State University. The main area of her research deals with community resilience with a focus on local and regional economic sustainability and resiliency. To fully understand the dynamics of economic adjustment, she has undertaken a mixed-method approach to study economy at various scales ranging from studies of local and regional economies revealing the overall health and prospects of an economy to studies of individual businesses exposing micro-level decision making and interdependency of households and businesses in community post-disaster recovery.

Yang Zhang is an Associate Professor of Urban Affairs and Planning at Virginia Tech. His research interests include hazards mitigation planning, and disaster recovery. Subscribing to the international comparative approach, his work includes case studies in the U.S., China, Haiti, Japan, and Korea. He can be reached at yang08@vt.edu.

PART I

Overview

1

AN OVERVIEW OF HAZARDS, VULNERABILITY, AND DISASTERS

Michael K. Lindell

Introduction

A disaster occurs when an extreme event exceeds a community's ability to cope with that event. Understanding the process by which natural disasters produce community impacts is important for four reasons. First, information from this process is needed to identify the preimpact conditions that make communities vulnerable to disaster impacts. Second, information about the disaster impact process can be used to identify specific segments of each community that will be affected disproportionately (e.g., low-income households, ethnic minorities, or specific types of businesses). Third, information about the disaster impact process can be used to identify the event-specific conditions that determine the level of disaster impact. Fourth, an understanding of disaster impact process allows planners to identify suitable emergency management interventions.

The process by which disasters produce community impacts can be explained in terms of a model originally proposed by Lindell and Prater (2003) and later extended in Lindell et al. (2006) and Lindell (2013a). Specifically, Figure 1.1 indicates the effects of a disaster are determined by three preimpact conditions—hazard exposure, physical vulnerability, and social vulnerability. There also are three event-specific conditions, hazard event characteristics, improvised disaster responses, and improvised disaster recovery. Two of the event-specific conditions, hazard event characteristics and improvised disaster responses, combine with the preimpact conditions to produce a disaster's physical impacts. The physical impacts, in turn, combine with improvised disaster recovery to produce the disaster's social impacts. Communities can engage in three types of emergency management interventions to ameliorate disaster impacts. Physical impacts can be reduced if communities engage in hazard mitigation practices and emergency preparedness practices, whereas social impacts can be reduced by recovery preparedness practices.

The following sections describe the components of the model in greater detail. Specifically, the next section will describe the three preimpact conditions—hazard exposure, physical vulnerability, and social vulnerability. This section will be followed by sections discussing hazard event characteristics and improvised disaster responses. The fourth section will discuss disasters' physical impacts, social impacts, and improvised disaster recovery. The last section will discuss three types of strategic interventions, hazard mitigation practices, emergency preparedness practices, and recovery preparedness practices.

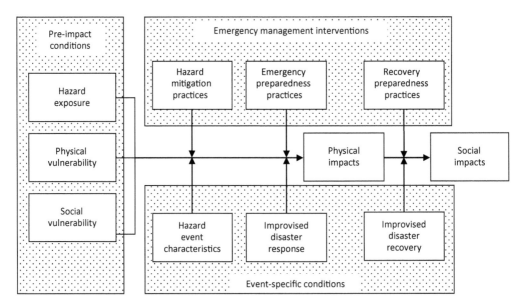

Figure 1.1 Disaster Impact Model (Lindell, 2013a)

Preimpact Conditions

Hazard Exposure

Hazard exposure arises from people's occupancy of geographical areas where they could be affected by specific types of events that threaten their lives and property. For natural hazards, this exposure is caused by living in geographical areas as specific as floodplains that sometimes extend only a few feet beyond the floodway or as broad as the Great Plains of the Midwest where tornadoes can strike anywhere over an area of hundreds of thousands of square miles. For technological hazards, exposure can arise if people move into areas where they could be exposed to events such as explosions or hazardous materials releases. In principle, hazard exposure can be measured by the probability of occurrence of a given event magnitude, but these exceedance probabilities can be difficult to obtain for hazards about which the historical data are insufficient to reliably estimate the probability of very unusual events. For example, many areas of the US have meteorological and hydrological data that are limited to the past century, so the estimation of extreme floods requires extrapolation from a limited data series. Moreover, watershed urbanization causes the boundaries of the 100-year floodplains to change in ways that may be difficult for local emergency managers to anticipate. Even more difficult to estimate are the probabilities of events, such as chemical and nuclear reactor accidents, for which data are limited because each facility is essentially unique. In such cases, techniques of probabilistic safety analysis are used to model these systems, attach probabilities to the failure of system components, and synthesize probabilities of overall system failure by mathematically combining the probabilities of individual component failure.

The greatest difficulties are encountered in attempting to estimate the probabilities of social hazards such as terrorist attacks because the occurrence of these events is defined by social system dynamics that cannot presently be modeled in the same way as physical systems. That is, the elements of social systems are difficult to define and measure. Moreover, the interactions of the system elements have multiple determinants and involve complex lag and feedback effects that are not well understood, let alone precisely measured. Indeed, there are significant social and political constraints that limit the collection of data on individuals and groups. These constraints further inhibit the ability of scientists to make specific predictions of social system behavior.

Physical Vulnerability

Human Vulnerability

Humans are vulnerable to environmental extremes of temperature, pressure, and chemical exposures that can cause death, injury, and illness. For any hazard agent—water, wind, ionizing radiation, toxic chemicals, infectious agents—there often is variability in the physiological response of the affected population. That is, given the same level of exposure, some people will die, others will be severely injured, still others slightly injured, and the rest will survive unscathed. Typically, the most susceptible to any environmental stressor will be the very young, the very old, and those with weakened immune systems.

Agricultural Vulnerability

Like humans, agricultural plants and animals are also vulnerable to environmental extremes of temperature, pressure, chemicals, radiation, and infectious agents. Like humans, there are differences among individuals within each plant and animal population. However, agricultural vulnerability is more complex than human vulnerability because there is a greater number of species to be assessed, each of which has its own characteristic range of responses to each environmental stressor.

Structural Vulnerability

Structural vulnerability arises when buildings are constructed using designs and materials that are incapable of resisting extreme stresses (e.g., high wind, hydraulic pressures of water, seismic shaking) or that allow hazardous materials to infiltrate into the building. The construction of most buildings is governed by building codes that are intended to protect the life safety of building occupants from structural collapse—primarily from the dead load of the building material themselves and the live load of the occupants and furnishings—but do not necessarily provide protection from extreme wind, seismic, or hydraulic loads. Nor do they provide an impermeable barrier to the infiltration of toxic air pollutants. Just as people vary in their physical vulnerability to environmental extremes, so too do buildings. Variation in the designs and construction materials of residential, commercial, and industrial structures—as well as electric power, fuel (e.g., natural gas), water, wastewater, telecommunications, and transportation systems—means that facilities of the same type, subjected to identical environmental stresses, might range from fully functional to completely destroyed.

Social Vulnerability

Social vulnerability has been defined as "the characteristics of a person or group and their situation that influence their capacity to anticipate, cope with, resist, and recover from the impact of a natural hazard" (Wisner et al. 2004: 11). We can define specify people's capacity to "anticipate" in terms of their awareness that their home, work, and other frequent locations are exposed to environmental hazards, as well as their expectations that extreme environmental events could produce casualties, damage, and disruption (Lindell 2013b; Lindell and Perry 2000). We can further define an ability to "cope with, resist, and recover" in terms of their capacity to adopt and implement four types of hazard adjustments—hazard mitigation, emergency preparedness, emergency response, and disaster recovery actions. In addition, we can define their "personal characteristics" in terms of their physical/psychological, material, social/political, and economic resources (Lindell 2018). Demographic categories such as gender, age, ethnicity, education, and income can sometimes serve as predictors of social vulnerability because these population segments can vary systematically in their physical/psychological, material, social/political, and economic resources (Bolin and Kurtz 2018; Enarson et al. 2018).

The central point of the social vulnerability perspective is that, just as people's occupancy of hazard-prone areas and the physical vulnerability of the structures in which they live and work are not randomly distributed, neither is social vulnerability randomly distributed—either geographically or demographically. Thus, just as variations in structural vulnerability can increase or decrease the effect of hazard exposure on physical impacts (property damage and casualties), so too can variations in social vulnerability. Social vulnerability varies across communities and also across households within communities. Variability in vulnerability is a problem for local emergency managers because it requires that they identify the areas within their communities having population segments with the highest levels of social vulnerability.

For example, lower-income households tend to be headed disproportionately by females and racial/ethnic minorities. Such households are more likely to experience destruction of their homes because of preimpact hazard exposure. This is especially true in developing countries such as Guatemala (Peacock et al. 1987), but also has been reported in the US (Peacock and Girard 1997). The homes of these households also are more likely to be destroyed because they were built according to older, less stringent building codes, used lower-quality construction materials and methods, and were less well maintained (Bolin and Bolton 1986). Because lower-income households have fewer resources on which to draw for recovery, they also take longer to transition to permanent housing, sometimes remaining for extended periods of time in severely damaged homes (Girard and Peacock 1997). In other cases, they are forced to accept as permanent what originally was intended as temporary housing (Peacock et al. 1987). Consequently, there may still be low-income households in temporary sheltering and temporary housing even after high-income households all have relocated to permanent housing (Berke et al. 1993; Rubin et al. 1985).

Event-Specific Conditions

Hazard Event Characteristics

Hazard impacts can be difficult to characterize because a given hazard agent can initiate a number of different threats. For example, tropical cyclones (also known as hurricanes or typhoons) can cause casualties and damage through wind, rain, storm surge, and inland flooding (Bryant 1997). Volcanoes can impact human settlements through ash fall, explosive eruptions, lava flows, mudflows and floods, and forest fires (Perry and Lindell 1990; Saarinen and Sell 1985; Warrick et al. 1981). However, once these distinct threats have been distinguished from each other, each can be characterized in terms of six significant characteristics. These are the speed of onset, availability of environmental cues (such as wind, rain, or ground movement), the intensity, scope, and duration of impact, and the probability of occurrence—CDRSS 2006). These characteristics determine people's ability to detect hazard onset, the amount of time they have to respond, the number of affected social units, and thus the event's casualties, damage, and socioeconomic disruption.

A hazard's impact intensity can generally be defined in terms of the physical materials involved and the energy these materials impart. The physical materials involved in disasters differ in terms of their physical state—gas (or vapor), liquid, or solid (or particulate). In most cases, the hazard from a gas arises from its temperature or pressure. Examples include hurricane or tornado wind (note that the atmosphere is a mixture of gases), which is hazardous because of overpressures that can inflict traumatic injuries directly on people. High wind also is hazardous because it can destroy structures and accelerate debris that can itself cause traumatic injuries. Alternatively, the hazard from a gas might arise from its toxicity, as is the case in some volcanic eruptions. Liquids also can be hazardous because of their toxicity, but the most common liquid hazard is water. It is hazardous to structures because of the pressure it can exert and is hazardous to living things when it fills the lungs and prevents respiration. Lava is solid rock that has been liquefied by extreme heat and therefore is hazardous to people and structures because of its thermal energy. Solids also can be hazardous if they take the form of particulates such as airborne volcanic ash or floodborne mud. These are particularly significant because they can leave deposits that have impacts of long duration.

The scope of impact defines the number of affected social units (e.g., individuals, households, and businesses). This is typically defined by the area in which death and destruction occur although, as noted below, this definition is problematic. Impact duration can be short for some hazards, as when hurricane-force wind arrives and departs within hours but indefinitely long for others, as when heavy metals such as lead contaminate an area. The probability of occurrence (per unit of time) is an important characteristic that affects disaster impacts indirectly because more probable hazards are likely to mobilize communities to engage in emergency management interventions to reduce their vulnerability (Prater and Lindell 2000).

Improvised Disaster Response

Disaster myths commonly portray disaster victims as dazed, panicked, or disorganized but people actually respond in a generally adaptive manner when disasters strike (Fischer 2008). Adaptive response is often delayed because *normalcy bias* inhibits people's realization that an improbable event is, in fact, occurring to them. Further delays occur because people have limited information about the situation and, therefore, seek confirmation of any initial indications of an emergency before initiating protective action (*social milling*). However, people also relay warnings to others (Lindell et al. 2016), as well as share rides in evacuating vehicles and offer shelter to friends and relatives (Wu et al. 2012, 2013). In addition, the vast majority of people respond in terms of their customary social units—especially their households and neighborhoods—which usually consumes time in developing social organizations that can cope with the disaster's demands. Contrary to stereotypes of individual selfishness, disaster victims often devote considerable effort to protecting others' persons and property (Drury et al. 2009). In some cases, people even engage in altruistic behaviors that risk their own lives to save the lives of others (Tierney et al. 2001). There is also an increased incidence in other prosocial behaviors such as donating labor and material aid that can produce high levels of social and material convergence in the disaster impact area (Tierney et al. 2001). Finally, emergency responders continue to perform their professional duties despite uncertainties about the safety of their families. Instead, they are more likely to experience burnout from working too many consecutive hours without relief (Quarantelli 1988).

A significant characteristic of improvised emergency response is emergent behavior, which arises when "individuals see needs that are not being met and therefore attempt to address them in an informal manner" (McEntire 2006, p. 175). According to Dynes (1970), *established organizations* perform their normal tasks within normal organizational structures, *extending organizations* perform novel tasks within normal organizational structures, *expanding organizations* perform their normal tasks within novel organizational structures, and *emergent organizations* perform novel tasks within novel organizational structures. Disaster demands that exceed the abilities of individuals acting independently can generate complex emergency response systems whose functioning is difficult to understand without using social network analysis (Petrescu-Prahova and Butts 2008; Uhr et al. 2008). Such systems often produce coordinated responses—"the cooperation of independent units for the purpose of eliminating fragmentation, gaps in service delivery, and unnecessary (as opposed to strategic) duplication of services" (Gillespie 1991, p. 57) through *emergent multiorganizational networks* (EMONs—Drabek et al. 1981). Because of their differences in organizational titles, organizational structures, training, experience, and legal authority, EMONs frequently experience severe difficulties in communicating with each other and coordinating their responses to disasters, especially when they operate according to strict command and control structures or without emergency operations centers (Drabek and McEntire 2003).

Improvised Disaster Recovery

Disaster recovery begins with stabilization of an incident and ends when the community has reestablished normal social, economic, and political routines. It is now generally accepted that disaster recovery encompasses multiple activities, some implemented sequentially and others implemented simultaneously. Immediate tasks in this process include damage assessment, debris clearance, reconstruction of infrastructure (electric power, fuel, water, wastewater, telecommunications, and transportation networks), and reconstruction of buildings in the residential, commercial, and industrial sectors. At any one time, some households or businesses might be engaged in one set of recovery activities while others are engaged in different recovery activities. Thus, attempts to define finely differentiated phases of disaster recovery are inherently limited in their validity so researchers have been less concerned about time phases (e.g., short-term recovery vs. long-term recovery) than about the specific recovery functions that must be performed. The discussion below addresses household and business recovery separately but the recovery of these units is interlinked (Xiao and Van Zandt 2012).

Household Recovery

There are three basic components to household recovery—housing recovery, economic recovery, and psychological recovery (Bolin and Trainer 1978). Many studies of housing recovery, such as Bolin and Stanford (1991, 1998) have adopted Quarantelli's (1982) typology of *emergency shelter* (unplanned and spontaneously sought locations), *temporary shelter* (locations that include food preparation and sleeping facilities), *temporary housing* (which allows victims to reestablish household routines in nonpreferred locations or structures) and *permanent housing* (which reestablishes household routines in preferred locations and structures). There is no single pattern of progression through the stages of housing because households vary in the number and sequence of their moves and the duration of their stays in each type of housing (Cole 2003).

Households' economic recovery is supported primarily by resources obtained from individuals and organizations within the community. The victims themselves might have financial (e.g., savings and insurance) and tangible assets (e.g., property) that are undamaged by hazard impact. As one might expect, low-income victims tend to have lower levels of savings, but they also are more likely to be victims of insurance redlining and, thus, have been forced into contracts with insurance companies that have insufficient reserves and go bankrupt after the disaster. Thus, even those who plan ahead for disaster recovery can find themselves without the financial resources they need (Peacock and Girard 1997). Alternatively, victims can promote their recovery by bringing in additional funds through overtime employment or by freeing up the needed funds by reducing their consumption below preimpact levels. Kinship networks can also contribute to disaster recovery but the significance of this source depends on the physical proximity of other nuclear families in the kin network, the closeness of the psychological ties within the network, the assets of the other families and, of course, the extent to which those families also suffered losses. Friends, neighbors, and coworkers can assist recovery through financial and in-kind contributions but these tend to be less important. Institutional sources of recovery assistance include federal, state, and local government as well as non-governmental organizations and community-based organizations (Phillips 2009). Because the donor-victim relationship is defined by bureaucratic norms, the amount of assistance depends on whether victims meet the qualification standards, usually documented residence in the impact area and proof of loss. Post-disaster loans can be problematic because they involve long-term debt that takes many years to repay (Bolin 1993).

There are significant variations among households in their housing recovery and these are correlated with households' demographic characteristics (Peacock et al. 2006). For example, the percentage of households reporting complete economic recovery after the 1987 Whittier earthquake was 50 percent at the end of the first year but 21 percent reported little or no recovery even at the end of four years (Bolin 1993). Economic recovery was positively related to household income and negatively related to structural damage, household size, and the total number of moves (Bolin 1993). There are also systematic differences in the rate of economic recovery among ethnic groups. For example, Bolin and Bolton (1986) found that Black households (30 percent) lagged behind Whites (51 percent) in their return to preimpact economic conditions eight months after the 1982 Paris, Texas, tornado. Because lower-income households have fewer resources on which to draw for recovery, they also take longer to return to permanent housing, sometimes remaining for extended periods of time in severely damaged homes (Girard and Peacock 1997). Indeed, they sometimes are forced to accept as permanent what originally was intended as temporary housing (Peacock et al. 1987). Conversely, the housing that authorities intend to become permanent is likely to be occupied only temporarily if it is incompatible with victims' preferences for location and features (Arlikatti and Andrew 2012).

Research on psychological recovery indicates that few disaster victims require psychiatric diagnosis and most benefit more from a *crisis counseling* orientation than from a *mental health treatment* orientation, especially if their normal social support networks of friends, relatives, neighbors, and coworkers remain largely intact (Gerrity and Flynn 1997). This is an important finding because some have concluded that the failure to seek formal psychological counseling is a potential threat to the mental health of victims and even first responders.

Business Recovery

Many studies have found that business recovery varies by industry and size (Zhang et al. 2009). Whereas wholesale and retail businesses generally report experiencing significant sales losses, manufacturing and construction companies often show gains following a disaster (Kroll et al. 1990; Webb et al. 2000). Moreover, businesses that serve a large (e.g. regional or international) market tend to recover more rapidly than those that only serve local markets (Webb et al. 2002). Small businesses, in particular, have been found to experience more obstacles than large firms and chains in their attempts to regain their predisaster levels of operations. Compared to their large counterparts, small firms are more likely to depend primarily on neighborhood customers, lack the financial resources needed for recovery, and lack access to governmental recovery programs (Alesch et al. 1993; Kroll et al. 1990).

Disaster Impacts

As noted earlier, disaster impacts comprise physical and social impacts. The physical impacts of disasters include casualties (deaths and injuries) and property damage, and both vary substantially across hazard agents. The physical impacts of a disaster are usually the most obvious, easily measured, and first reported by the news media. Social impacts, which include psychosocial, demographic, economic, and political impacts, can develop over a long period of time and can be difficult to assess when they occur. Despite the difficulty in measuring these social impacts, it is nonetheless important to monitor them, and even to predict them if possible, because they can cause significant problems for the long-term functioning of specific types of households and businesses in an affected community.

Figure 1.2 Disaster Impact Zones (Lindell, 2013a)

Wallace (1956) defined a disaster's concentration in space by a series of impact zones (Figure 1.2), which are simple in concept but problematic to identify in practice because they are not neat circles with well-defined borders. For example, a building's damage from an earthquake depends on its structural resilience and the intensity of earthquake shaking—neither of which has a simple deterministic distribution—so the boundary of the total impact zone can be extremely irregular. Indeed, buildings with seismic-resistant designs and materials can survive in an area that is otherwise completely devastated. Moreover, the infrastructure impacts and social impacts of the fringe impact zone can be equally irregular and extend beyond the boundaries of the resource filter or community aid zones. For example, Zhang et al. (2009) noted that businesses whose buildings and contents are undamaged can fail after a disaster because its customers are now spending their money on repairing their homes.

Physical Impacts

Casualties

The EM-DAT database (www.emdat.be/database) shows that the major disaster-related causes of death from 1900–2018 were drought (11.7 million deaths worldwide) and epidemic (9.6 million). However, Table 1.1 shows that the principal geophysical, hydrological, and meteorological hazards also took a major toll with floods accounting for 7 million deaths, earthquakes for 2.6 million deaths, and severe storms (mostly tropical cyclones) 1.4 million deaths. The death toll varied significantly by region, with Asia accounting for almost 98 percent of deaths from floods, 92 percent of deaths from severe storms, and over 70 percent of deaths from earthquakes.

Of course, there often are major difficulties in determining how many of the deaths and injuries are "caused by" a disaster (Wood and Bourque 2018). In some cases it is impossible to determine how many persons are missing and, if so, whether this is due to death or unrecorded relocation. The size of the error in estimates of disaster death tolls can be seen in the fact that, for many of the most catastrophic events, the number of deaths recorded in the EM-DAT database is rounded to the nearest 1,000 and some even are rounded to the nearest 10,000. Estimates of injuries are similarly

Table 1.1 Worldwide Mortality Data from Environmental Hazards from 1900–2018, by Region

	Flood		Earthquake		Severe Storms		Extreme Temperature		Volcanic Activity		Landslide		Wildfire	
	Incidents	Deaths	Incidents	Deaths	Incidents	Deaths	Incidents	Deaths	Incidents	Deaths	Incidents	Deaths	Incidents	Deaths
Africa	974	28,400	84	21,415	260	6,388	19	364	18	2,218	41	2,604	31	287
Americas	1134	105,464	290	440,489	1,322	105,430	113	9,542	88	67,858	180	20,767	155	1,681
Asia	2026	6,813,628	757	1,833,839	1,719	1,276,751	176	27,052	101	21,890	395	24,689	86	793
Europe	605	9,407	171	278,678	491	3,498	246	145,524	11	735	78	16,829	108	753
Oceania	142	563	56	3,521	316	2,201	7	509	27	3,665	18	546	42	501
Total	4881	6,957,462	1,358	2,577,942	4,108	1,394,268	561	182,991	245	96,366	712	65,435	422	4,015

Source: EM-DAT (www.emdat.be/database)

problematic (see Langness 1994; Peek-Asa et al. 1998; Shoaf et al. 1998, regarding conflicting estimates of deaths and injuries attributable to the Northridge earthquake). Even when bodies can be counted, there are problems because disaster impact may be only a contributing factor to casualties with pre-existing health conditions. Moreover, some casualties are indirect consequences of the hazard agent as, for example, with casualties caused by structural fires following earthquakes (e.g., burns) and destruction of infrastructure (e.g., illnesses from contaminated water supplies).

Damage

The loss of structures, animals, and crops is an important aspect of physical impact, and these losses are rising substantially throughout the world. The EM-DAT database shows that worldwide economic losses were no more than about 20 billion United States dollars (2016 USD) per year during 1960–1975. However, losses ranged from USD 20–120 billion per year during 1976–1994 and from USD 40–400 billion from 1995–2016. As was the case for deaths, Asia accounted for a disproportionate share of the damage. Such losses usually result from physical damage or destruction of property, but they also can be caused by losses of land use to chemical or radiological contamination or loss of the land itself to subsidence or erosion. Damage to the built environment can be classified broadly as affecting residential, commercial, industrial, infrastructure, or community services sectors. Moreover, damage within each of these sectors can be divided into damage to structures and damage to contents. It usually is the case that damage to contents results from collapsing structures (e.g., hurricane winds failing the building envelope and allowing rain to destroy the furniture inside the building). However, some hazard agents can damage building contents without affecting the structure itself (e.g., earthquakes striking seismically resistant buildings whose contents are not securely fastened). Thus, risk area residents may need to adopt additional hazard adjustments to protect contents and occupants even if they already have structural protection.

Perhaps the most significant structural impact of a disaster on a stricken community is the destruction of households' dwellings, which initiates what can be a very long process of housing recovery. Households vary in the progression and duration of each type of housing and the transition from one stage to another can be delayed unpredictably, as when it took nine days for shelter occupancy to peak after the Whittier Narrows earthquake (Bolin 1993). Alternatively, some households might skip one or more stages (e.g., proceeding directly form temporary shelter to permanent housing), stall in a stage (as when "temporary housing" becomes permanent because "permanent housing" is unavailable or unaffordable), or even regress to a previous stage (e.g., moving to temporary shelter after their home is red-tagged).

As is the case with estimates of casualties, estimates of losses to the built environment are uncertain. Damage estimates are most accurate when trained damage assessors enter each building to assess the percent of damage to each of the major structural systems (e.g., roof, walls, floors) and the percentage reduction in market valuation due to the damage. Early approximate estimates are obtained by conducting "windshield surveys" in which trained damage assessors drive through the impact area and estimate the extent of damage that is visible from the street, or by conducting computer analyses using HAZUS (National Institute of Building Sciences 1998). These early approximate estimates are especially important in major disasters because detailed assessments are not needed in the early stages of disaster recovery and the time required to conduct them on a large number of damaged structures using a limited number of qualified inspectors would unnecessarily delay the community recovery process.

Other important physical impacts include damage or contamination to cropland, rangeland, and woodlands. Such impacts may be well understood for some hazard agents but not others. For example, ashfall from the 1980 Mt. St. Helens eruption was initially expected to devastate crops and livestock in downwind areas, but no significant losses materialized (Warrick et al. 1981). There also is concern about damage or contamination to the natural environment (wild lands) because these areas serve valuable functions such as damping the extremes of river discharge and providing habitat for wildlife. In part, concern arises from the potential for indirect consequences such as increased runoff and silting of downstream river beds, but many people also are concerned about the natural environment simply because they value it for its own sake.

Social Impacts

For many years, research on the social impacts of disasters consisted of an accumulation of case studies, but two research teams conducted comprehensive statistical analyses of extensive databases to assess the long-term effects of disasters on stricken communities (Friesma et al. 1979; Wright et al. 1979). The more comprehensive Wright et al. (1979) study used census data from the 1960 (pre-impact) and 1970 (post-impact) censuses to assess the effects of all recorded disasters in the United States. The authors concurred with earlier findings by Friesma et al. (1979) in concluding that no long-term social impact of disasters could be detected at the community level. In discussing their findings, the authors acknowledged their results were dominated by the types of disasters occurring most frequently in the United States—tornadoes, floods, and hurricanes. Moreover, most of the disasters they studied had a relatively small scope of impact and thus caused only minimal disruption to their communities even in the short term. Finally, they noted their findings did not preclude the possibility of significant long-term impacts upon lower levels such as the neighborhood, business, and household.

Nonetheless, their findings called attention to the importance of the *impact ratio*—the amount of damage divided by the amount of community resources—in understanding disaster impacts. They hypothesized that long-term social impacts tend to be minimal in the United States because most hazard agents have a relatively small scope of impact and tend to strike undeveloped areas more frequently than intensely developed areas simply because there are more of the former than the latter. Thus, the numerator of the impact ratio tends to be low and local resources are sufficient to prevent long-term effects from occurring. Even when a disaster has a large scope of impact and strikes a large developed area (causing a large impact ratio in the short term), state and federal agencies and NGOs (e.g., American Red Cross) direct recovery resources to the affected area, thus preventing long-term impacts from occurring. For example, Hurricane Katrina inflicted USD 125 billion in losses to the New Orleans area, but this was only 0.95 percent of the US GDP for that year. Echoing the findings of recovery problems in Hurricane Andrew (Peacock et al. 1997), Comfort et al. (2010: 669) concluded

> that the policy and administrative processes of recovery from catastrophic events have not been well understood, and that the lack of a clear policy design supported by professional administrative practice across jurisdictional levels of authority and action has hindered the recovery of New Orleans.

Psychosocial Impacts

These have been documented in research reviews conducted over a period of 40 years, which have concluded that disasters can cause a wide range of negative psychological responses (Bolin 1985; Gerrity and Flynn 1997; Houts et al. 1988; Perry and Lindell 1978). According to Norris et al. (2002a, 2002b) psychological impacts vary by disaster type (greater in mass violence), victim location (developing countries), victim type (children and middle age, females, and ethnic minorities with existing psychological problems and poor psychosocial resources). These impacts include psycho-physiological effects such as fatigue, gastrointestinal upset, and tics, as well as cognitive signs such as confusion, impaired concentration, and attention deficits. Psychological impacts include emotional signs such as anxiety, depression, and grief. They also include behavioral effects such as sleep and appetite changes, ritualistic behavior, and substance abuse. In most cases, the observed effects are mild and transitory—the result of "normal people, responding normally, to a very abnormal situation" (Gerrity and Flynn 1997, p. 108).

The negative psychological impacts described above arise from perceived threats that elicit *emotion-focused, problem-focused*, and *meaning-focused* coping strategies (Folkman 2013; Lazarus and Folkman 1984). Emotion-focused coping comprises thoughts and actions that are used to manage emotional distress. Adaptive strategies include emotional distancing and social support seeking, whereas maladaptive strategies include daydreaming, blaming others, and substance abuse. Problem-focused coping consists of strategies that focus on eliminating the problem or reducing its impacts. Strategies such as information seeking and protective action are generally adaptive but can be maladaptive if carried to extremes (e.g., seeking absolute certainty about a threat can delay evacuation until it is too late or trying to evacuate across a flooded road). Meaning-focused strategies preserve positive well-being by concentrating on fundamental values (e.g., helping those in need). In general, maladaptive emotion-focused coping disrupts the functioning of only a small portion of the victim population. Instead, most disaster victims engage in adaptive problem-focused coping activities to save their own lives and those of their closest associates.

There also are psychological impacts with long-term adaptive consequences, such as changes in risk perception (beliefs in the likelihood of the occurrence of a disaster and its personal consequences for the individual) and increased hazard intrusiveness (frequency of thought and discussion about a hazard). In turn, these beliefs can affect risk area residents' adoption of household hazard adjustments that reduce their vulnerability to future disasters. However, these cognitive impacts of disaster experience do not appear to be large in aggregate, resulting in modest effects on household hazard adjustment (Lindell and Perry 2000; Solberg, Rossetto and Joffe 2010). One part of the problem is that the psychological effect of experience on later behavior depends on the complex ways that people interpret that experience (Dillon and Tinsley 2016; Dillon et al. 2014). This is compounded by the fact that researchers have attempted to measure people's experience in ways that are too oversimplified to accurately measure the effects of experience (DeMuth 2018; Demuth et al., 2016).

Demographic Impacts

These impacts can be assessed by examining the demographic balancing equation, $P_a - P_b = B - D + IM - OM$, where P_a is the population size after the disaster, P_b is the population size before the disaster, B is the number of births, D is the number of deaths, IM is the number of immigrants, and OM is the number of emigrants (Smith, Tayman and Swanson 2001). In one of the few studies of disaster-related births, Evans, Hu, and Zhao (2010) found that fertility rates in US coastal and slightly inland counties increased with low-severity storm notices and decreased with high-severity storm notices. Unsurprisingly, however, the effect was small—an approximately 2 percent increase or decrease in the number of births.

More frequently studied is the disaster-related death toll which, as noted earlier, can be larger in developing countries than in developed countries. For example, the 2004 Indian Ocean tsunami caused an estimated 227,000 deaths whereas, as noted in the previous section on physical impacts, the number of deaths from recent disasters in the US has been small relative to historical levels. For example, the 6,000 deaths in the 1900 Galveston hurricane were approximately 17 percent of the city's population, whereas the 1400 deaths in New Orleans during Hurricane

Katrina were 0.29 percent of that city's population. Even the death tolls from the 1900 Galveston hurricane and the tsunami caused by the 2011 Tōhoku earthquake, which caused approximately 20,000 deaths in Japan, pale in comparison to those of the 2004 Indian Ocean tsunami (225,000) or the 2010 Haiti earthquake (estimated at over 160,000).

Nonetheless, developed countries are likely to experience significant demographic impacts in the (temporary) post-impact immigration of construction workers and temporary displacement or permanent emigration of population segments that have lost housing. Housing-related emigration is only temporary for communities that are spared disaster impact (e.g., a hurricane threatens but strikes elsewhere), but can be significant in others. For example, about one-sixth of South Dade County households moved out of their homes in the year after Hurricane Andrew, with about half of these moving to North Dade. South Dade regained about half of its population loss in the second year after the hurricane but people were more likely to return if they had not moved far during the interim. Almost three-quarters of those who moved elsewhere in Dade County returned to their original residences, compared to one-tenth of those who left the state altogether (Smith and McCarty 1996).

Some evacuating households will remain in the cities where they sought shelter. This affects reentry planning because of the reduced number of households seeking to return to the evacuation zone. However, there is a limited amount of research on the number of households that migrate permanently from disaster impact areas. Smith and McCarty (1996) studied household relocation after Hurricane Andrew, finding a much higher relocation rate (52 percent or 187,200 households) in heavily damaged South Dade County than in less-damaged North Dade County (10 percent or 166,100 households). In South Dade, structural damage (43 percent) and infrastructure loss (41 percent) were equally important causes of displacement, whereas infrastructure loss (87 percent) accounted for almost all of the departures from North Dade. Infrastructure was restored relatively quickly, so 45 percent of residents in North Dade returned in the following week but this was true for only 6 percent of South Dade residents. There was a ready supply of nearby vacant housing available for temporary occupancy because the impact zone was located in the highly urbanized Miami/Dade County metropolitan area. As a result, most households could relocate inside the county (80 percent for North Dade and 74 percent for South Dade). However, one-third of the displaced population still remained away from their homes after two years. The permanent relocation rate in South Dade was lowest for those who were able to find temporary housing within Dade County (28 percent) and was highest for those forced to leave the state (90 percent).

The most dramatic US disaster-related emigration of recent times occurred in New Orleans, which lost thousands of residents after Hurricane Katrina—dropping from 484,000 in 2000 to 344,000 in 2010 and only recovering to 383,000 in 2016. One study focused on relocation from four Louisiana parishes that were severely damaged by Hurricanes Katrina and Rita. Hori et al. (2009) found that outmigration rates ranged from 33 percent for Orleans to 46 percent for St. Bernard. Temporary housing in nearby communities facilitated temporary reentry ("look and leave"). There was a mixture of immigration and emigration in four buffer parishes (−.43 to +5.1 percent net change in population). Another study of Gulf Coast relocation after these hurricanes found that community characteristics influenced the rates of emigration (Myers, Slack and Singelmann 2008). Communities with higher levels of housing damage, denser development, and greater percentages of disadvantaged populations (lower levels of income per capita, median home valuation, median rent, health insurance; high rates of unemployment, poverty, high school dropouts, and female-headed households) had higher rates of emigration.

The principal cause of households' permanent emigration is their loss of housing. Indeed, even housing that is only moderately damaged takes approximately two years to recover and housing that is extensively damaged takes more than twice as long (Peacock et al. 2014). However, housing reconstruction is sometimes delayed indefinitely—producing "ghost towns" (Comerio 1998).

Households that return after disasters have been found to share a number of characteristics, one of which is homeownership (Elliott and Pais 2006; Kim and Oh 2014; Landry et al. 2007; Paxson and Rouse 2008). Another common characteristic of returning households is minimal storm damage (Elliott and Pais 2006; Landry et al. 2007). Similarly, returning households tended to have lower financial losses (Kim and Oh 2014). Households that had lower expectations of another major hurricane were also more likely to return (Baker et al. 2009; Kim and Oh 2014; Landry et al. 2007; Paxson and Rouse 2008). However, there is some evidence that risk perceptions decrease in importance over time (Shaw and Baker 2010). Ethnicity also predicts post-disaster return, with Whites being more likely to return (Fussell et al. 2010; Groen and Polivka 2010; Kim and Oh 2014; Paxson and Rouse 2008). However, Fussell et al. (2010) reported that race became nonsignificant when controlling for the level of home damage, suggesting that ethnicity might be a proxy for other variables. Similarly, Li et al. (2010) found that Vietnamese Americans had higher rates of return to New Orleans than African Americans, possibly as a result of the local Vietnamese church's support for its community during the early phases of disaster recovery.

Community bondedness also has consistent effects on post-disaster return. Specifically, households with higher levels of community ties (e.g., integration into peer networks and emotional attachments to the community) were more likely to return to New Orleans (Chamlee-Wright and Storr 2009; Fussell et al. 2010; Groen and Polivka 2010; Paxson and Rouse 2008). However, Landry et al. (2007) found that community tenure (the number of years lived in the community) was unrelated to return and living in the community since birth was negatively correlated with return.

Other variables yield less consistent findings. Landry et al. (2007), Fussell et al. (2010), and Groen and Polivka (2010) found that age was positively related to return but Baker et al. (2009), Kim and Oh (2014), and Elliott and Pais (2006) found nonsignificant correlations between these variables. Income has also produced inconsistent correlations with return—positive (Landry et al. 2007), nonsignificant (Baker et al. 2009; Kim and Oh 2014), and negative (Elliott and Pais 2006). Finally, some variables are less well studied; return was more likely among those who are married, less highly educated, and employed before the evacuation in the only study that addressed these variables (Landry et al. 2007).

Economic Impact

One category of economic impact is the property damage whose losses in asset values can be measured by the cost of repair or replacement (CACND 1999). Disaster losses in United States are initially borne by the affected households, businesses, and local government agencies whose property is damaged or destroyed. However, some of these losses are redistributed during the disaster recovery process. There have been many attempts to estimate the magnitude of direct losses from individual disasters and the annual average losses from particular types of hazards (e.g., Mileti 1999). Unfortunately, these losses are difficult to determine precisely because there is no organization that tracks all of the relevant data and some data are not recorded at all (Charvériat 2000; CACND 1999). For insured property, the insurers record the amount of the deductible and the reimbursed loss, but uninsured losses are not recorded so they must be estimated—often with questionable accuracy. Consequently, there have been mixed conclusions about the economic effects of natural disasters. Similar to the conclusions of Friesma et al. (1979) and Wright et al. (1979), Albala-Bertrand's (1993) analysis of six disasters in Latin American countries found that the effects on national economic growth were negligible. Likewise, the Cavallo et al. (2013) analysis of 6,530 disasters in the EM-DAT database from 1970–2008 found that disasters had no significant effect on growth in GDP per capita, which the authors concluded was probably because reconstruction was financed through such methods as reduced consumption, insurance payments, disaster assistance, and increased indebtedness.

By contrast, Strobl's (2012) analysis of 267 hurricanes found that they caused an average 0.8 percent decrease in economic growth rates in Caribbean and Central American countries, but the size of the impact was greater for hurricanes during the hurricane season peak (August and September). Moreover, Strobl (2011) found that hurricanes caused a 0.45 percent decrease in economic growth rates in US coastal counties, compared to a 1.68 percent growth rate under normal conditions. Over a quarter of the loss was due to wealthier people leaving their counties. He also reported that the impacted state experienced an initial negative effect in the quarter in which the hurricane struck, followed by a recovery in the next quarter, and a negligible overall effect for the year. These hurricanes caused no discernible effect at the national level.

Other studies have documented more complex effects. For example, Felbermayr and Gröschl (2014) analyzed the GeoMet database rather than the usual EM-DAT database, finding that the most severe disasters (top 1-percentile) produce a 6.83 percent decrease in GDP per capita whereas top 5-percentile events produce a 0.46 percent decrease and the smallest 25th percentile disasters produce only a 0.01 percent decrease. Moreover, the adverse effects are weakest in countries with higher institutional quality, openness to trade, and financial openness. Finally, Klomp and Valckx (2014) conducted a meta-analysis of 25 disaster economic impact studies and concluded that natural disasters have a significant negative effect on economic growth that is increasing over time. Moreover, the magnitude of the effect is stronger for climatic disasters, such as floods, in developing countries (possibly because they are more dependent on agricultural production). However, the effects are only short-term; GDP per capita ultimately returns to its original trajectory.

Political Impacts

These impacts become manifest in social activism resulting in political disruption, especially during the seemingly interminable period of disaster recovery. The disaster recovery period is a source of many victim grievances and this creates many opportunities for community conflict, both in the US (Bolin 1982, 1993) and abroad (Albala-Bertrand 1993). Many of the relevant studies have reported cross-national comparisons. For example, Olson and Drury's (1997) analysis of disasters in 12 developing countries between 1966–1980 found that prior instability and disaster severity significantly increased civil unrest, whereas level of development and governmental repression significantly decreased it, and disaster aid had a nonsignificant effect. Later, these researchers' reanalysis of those disasters using a slightly different model and analytic procedures found that greater disaster severity, lower levels of development, higher levels of income equality, and lower levels of regime repressiveness were associated with increased political unrest (Drury and Olson 1998).

Other cross-national studies shave also found negative political impacts of disasters. Nel and Righarts' (2008) analysis of data on 187 political entities between 1950–2000 found that rapid-onset natural disasters were significantly associated with civil wars, but the risk of such conflicts was higher in countries that had low to middle income and medium to high inequality, low economic growth rates, and mixed regimes (i.e., neither completely autocratic and thus able to suppress dissent, nor completely democratic and thus able to accommodate dissent). That is, violent civil conflict was more likely when there were increased grievances, increased incentives for challenging existing elites, and weakened elites that were less able to resist.

However, other studies have found neutral or benign political impacts—government change and conflict reduction. Specifically, Chang and Berdiev (2015) analyzed data from 156 countries from 1975–2010, concluding that the occurrence of disasters, their number, and the casualties and damage they caused all increase the probability that the government in power at the time will be replaced. In addition, higher levels of inflation and foreign assistance increased the probability of government change. However, economic growth, trade openness, democratic institutions, and political constraints all inhibited government change. Moreover, Kreutz (2012) analyzed data from 405 disasters in 21 countries from 1990–2004, concluding that natural disasters have positive effects on separatist conflicts by increasing the probability of bilateral talks and ceasefires but had a nonsignificant effect on peace agreements. The effects were stronger in democratic countries and those that experienced more severe disasters, but were strongest when the disaster occurred outside the conflict zone.

The most positive political impacts were documented in Ahlerup's (2013) analysis of post-disaster political change in 157 countries from 1970 to 2008. This study found evidence of increased democratization in mixed regimes, which were more likely to receive external disaster aid and received more aid when they did. This was especially true for mixed regimes that had been in power for fewer than 24 years. The author inferred that this was because the heterogeneous elites in these countries are under more pressure to democratize in order to forestall internal pressures and receive international aid.

Still other studies have provided detailed accounts of intra-community conflicts that take place after disasters. Victims usually attempt to recreate preimpact housing patterns, but it can be problematic for their neighbors if victims attempt to site mobile homes on their own lots while awaiting the reconstruction of permanent housing (Bolin 1982). Conflicts arise because such housing usually is considered to be a blight on the neighborhood and neighbors are afraid the "temporary" housing will become permanent. Neighbors also are pitted against each other when developers attempt to buy up damaged or destroyed properties and build multifamily units on lots previously zoned for single family dwellings. Such rezoning attempts are a major threat to the market value of owner-occupied homes but tend to have less impact on renters because they have less incentive to remain in the neighborhood. There are exceptions to this generalization because some ethnic groups have very close ties to their neighborhoods, even if they rent rather than own.

Attempts to change prevailing patterns of civil governance can arise when individuals sharing a grievance about the handling of the recovery process seek to redress that grievance through collective action. Consistent with Dynes's (1970) typology of organizations, existing community groups with an explicit political agenda can *expand* their membership to increase their strength, whereas community groups without an explicit political agenda can *extend* their domains to include disaster-related grievances. Alternatively, new groups can *emerge* to influence local, state, or federal government agencies and legislators to take actions that they support and to terminate actions that they disapprove. Indeed, such was the case for Latinos in Watsonville, California following the Loma Prieta earthquake (Tierney et al. 2001). Usually, community action groups pressure government to provide additional resources for recovering from disaster impact, but may oppose candidates' re-elections or even seek to recall some politicians from office (Olson and Drury 1997; Prater and Lindell 2000; Shefner 1999). The point here is not that disasters produce political behavior that is different from that encountered in normal life. Rather, disaster impacts might only produce a different set of victims and grievances and, therefore, a minor variation on the prevailing political agenda (Morrow and Peacock 1997).

Emergency Management Interventions

As Figure 1.1 indicates, there are three types of preimpact interventions—known as hazard adjustments (Burton et al. 1978)—that can effect reductions in disaster impacts. Although FEMA defines mitigation, preparedness, response, and recovery as the four "phases" of emergency management, these activities are more properly considered to be functions, because mitigation and preparedness take place concurrently during the preimpact time period; response and recovery take place concurrently trans-impact, and recovery and mitigation take place concurrently post-impact. In general, hazard mitigation and emergency preparedness practices directly reduce a disaster's physical impacts (casualties and damage) and indirectly reduce its social impacts, whereas recovery preparedness practices directly reduce a disaster's social impacts. Although improvised disaster response actions also directly affect disasters' physical impacts, their very nature makes them likely to be much less effective than planned interventions guided by emergency operations plans (EOPs). Similarly, improvised recovery assistance directly affects disasters' social impacts but is likely to be less effective than systematic recovery preparedness practices.

Hazard Mitigation Practices

One way to reduce the physical impacts of disasters is to adopt hazard mitigation practices that protect passively against casualties and damage at the time of hazard impact (as opposed to an active emergency response). Hazard mitigation is often classified as either "structural" or "nonstructural" but this categorization is potentially confusing because engineers design structural protection at many different scales and in structures that have many different functions. The most commonly cited examples of "structural" mitigation are dams, levees, sea walls, and other permanent barriers that prevent floodwater from reaching protected areas. However, engineers also use building designs and construction materials to increase the ability of an individual building's foundation and load-bearing framework to resist environmental extremes. They apply these building construction practices to residential, commercial, and industrial structures as well as to infrastructure facilities such as pipelines for potable water, waste water, and fuel (e.g., oil and natural gas); roads and bridges; radio, television, and cellular telephone towers; and electric transmission lines.

The term "nonstructural mitigation" is also vague because it includes a broad set of mitigation strategies. These include activities as diverse as reducing chemical quantities stored at water treatment plants, purchasing undeveloped floodplains and dedicating them to open space, installing window shutters for buildings located on hurricane-prone coastlines, and bolting water heaters to walls in earthquake zones. In fact, these have little in common other than that they are not designed by engineers. Instead, Lindell et al. (2006, Chapter 7) adapted the system proposed in FEMA (1986)— hazard source control, community protection works, land use practices, building construction practices, and building contents protection. *Hazard source control* acts directly on the hazard agent to reduce its magnitude or duration. For example, patching a hole in a leaking tank truck prevents a gas from being released. *Community protection works*, which limit the impact of a hazard agent on an entire community, include dams and levees that protect against floodwater and sea walls that protect against storm surge. *Land use practices* reduce hazard vulnerability by avoiding construction in areas that are susceptible to hazard impact. The difference between land use practices and land use regulations is important. Landowners can adopt sustainable practices whether or not they are required to do so. Thus, government agencies can encourage the adoption of appropriate land use practices by providing incentives to encourage development in safe locations, establishing sanctions to prevent development in hazardous locations, or engaging in risk communication to inform landowners about the risks and benefits of development in locations throughout the community.

Hazard mitigation can also be achieved through *building construction practices* that make individual structures less vulnerable to natural hazards. Here too, the difference between building construction practices and building codes is important because building owners can adopt hazard resistant designs and construction materials in the absence of government intervention. Government agencies can encourage the adoption of appropriate building construction practices by providing incentives for adopting appropriate designs and materials, establishing code requirements for hazard resistant building designs and materials, or informing building owners about the risks and benefits of different building designs and materials. Finally, hazard mitigation can be achieved by *building contents protection* strategies such as elevating appliances above the base flood elevation or bolting them to walls to resist seismic forces.

Research on hazard mitigation has addressed impediments to implementation such as lack of information about the hazard and suitable mitigation actions (Sadiq and Weible 2010), poor quality of local comprehensive plans (Berke and Godschalk 2009; Tang et al. 2008, 2011), community hazard exposure due to popular urban development strategies such as New Urbanism (Song et al. 2009), and inadequate state hazard mitigation plans (Berke et al. 2012). Other research themes include identifying variables that predict which states have submitted plans (Yoon et al. 2012), ways to increase public support about hazard mitigation (Godschalk, Brody and Burby 2003), ways that individual planners can use their access to the policy process to promote hazard mitigation (Stevens 2010), and local planners' perceptions of different mitigation policies, tools, and strategies (Ge and Lindell 2016).

Disaster researchers have also sought to identify hazard mitigation actions taken voluntarily by individuals, households, and businesses. These studies have examined the influence of hazard proximity, disaster experience, risk perceptions, stakeholder perceptions, and hazard adjustment perceptions on mitigation intentions and actual mitigation actions. A few studies (e.g., Paton et al. 2010) have been cross-national, but similar variables have been used in different countries and across different hazards (Collins 2008, wildfire; Lin et al. 2008, flood and mudslide; Terpstra and Gutteling 2008, flood). One cross-national/cross-hazard result, the Paton et al. (2008) finding of a "false experience" effect for volcano preparedness, replicated an effect previously found after hurricanes. Specifically, Baker (1991) found that people who experienced minor impacts from the fringes of a major hurricane concluded they had survived the worst impacts that a hurricane could inflict. This "false experience" reduces people's risk perceptions, their hazard mitigation actions, and search for further hazard relevant information.

Emergency Preparedness Practices

Emergency preparedness practices are preimpact actions that provide the human and material resources needed to support active responses at the time of hazard impact (Lindell and Perry 2000). An important step in emergency preparedness is to use community hazard/vulnerability analysis to identify the geographic areas and population segments at risk (FEMA 1997). In addition, communities should develop EOPs, the first step of which is to identify the functions that must be addressed. FEMA (2010) identifies 26 functions (in the traditional functional EOP format) and 15 functions (in the emergency support function format), but it is easier to think of organizational performance in disasters as characterized by four basic emergency response functions—emergency assessment, hazard operations, and population protection and incident management (Lindell and Perry 2007). *Emergency assessment* comprises diagnoses of past and present conditions and prognoses of future conditions that guide the emergency response. *Hazard operations* refers to expedient hazard mitigation actions that emergency personnel take to limit the magnitude or duration of disaster impact (e.g., sandbagging a flooding river or patching a leaking railroad tank car). *Population protection* refers to actions—such as sheltering in-place, evacuation, and mass immunization—that protect people from hazard agents. *Incident management* consists of the activities by which the human and physical resources used to respond to the emergency are mobilized and directed to accomplish the goals of the emergency response organization.

After identifying the functions that must be performed, the next step is to determine which community organization will be responsible for accomplishing each function. Once functional responsibilities have been assigned, each organization must develop procedures for accomplishing those functions. Finally, the organizations must acquire response resources (personnel, facilities, and equipment) to implement their plans and they need to maintain preparedness for emergency response by continued planning, conducting emergency response training, acquiring facilities and equipment, and performing emergency drills, exercises, and critiques (Daines 1991; Perry and Lindell 2007).

There are many conditions that determine the effectiveness of local emergency management agencies (LEMAs) and local emergency management committees (LEMCs) in preparing their communities for disasters. Lindell and Brandt (2000) concluded that LEMA effectiveness is determined by individual outcomes and the planning process. The individual outcomes are job satisfaction, organizational commitment, attachment behaviors, and organizational citizenship behaviors. The planning process is defined by preparedness analysis, planning activities, resource development, organizational climate development, and strategic choice. There are five factors that influence the planning process, the first of which are community hazard experience and hazard analyses that indicate the likelihood and expected impacts of future disasters. In addition, hazard experience indirectly affects the planning process by increasing community support from public officials and the news media, as well as different demographic, economic, and political groups in the community. This community support allows policymakers to reallocate community resources such as staff and budget that increase staffing and organization for the LEMA and the LEMC. Finally, communities can access extra-community resources such as professional associations, government agencies, and regional organizations to supplement their own resources (see Lindell and Perry 2007 for a more complete discussion).

Uscher-Pines et al. (2009) reported that LEMAs are poorly prepared for people with disabilities, which Stough and Kelman (2018) attributed to an interaction between individual capabilities and the social environment that results in special needs for communication, medical health, independence, supervision, and transportation (Kailes and Enders 2007). Fox et al. (2007) concluded that LEMAs can't improve because of personnel and funding shortfalls but systematic assessment of the prevalence of different types of disabilities and methods of accommodating those disabilities could guide the retrofit of existing facilities and the design of new facilities. Consequently, increased preparedness is needed for vulnerable population segments (Berke et al. 2010) that will require greater reliance on voluntary organizations such as Community Emergency Response Teams (Flint and Stevenson 2010).

Research on private sector organizations has identified measures that individual firms and community planners need to take to reduce disaster impacts (Chang and Falit-Baiamonte 2002; Zhang et al. 2009). This research has found that size, number of locations, and ownership of its premises are consistently significant predictors of business disaster preparedness across a wide variety of industries ranging from tourism (Bird et al. 2010) to hazardous materials facilities (Cruz and Steinberg 2005).

Many studies have reported that household hazard adjustment adoption is significantly correlated with perceived personal risk, where the latter refers to respondents' judgments of the likelihood that they will be personally affected by specific consequences such as death, injury, property damage, or disruption to daily activities (Lindell 2013b; Lindell and Perry 2000; Solberg et al. 2010). One probable explanation for the studies that have found nonsignificant effects for risk perception is that Weinstein and Nicolich (1993) have shown that the correlation of risk perception (as measured by expected personal consequences) with hazard adjustment adoption must inevitably reach zero over time if people continue to adopt hazard adjustments until they perceive that they have reached the point of diminishing returns. Some studies have reported significant correlations of hazard intrusiveness, defined by frequency of thoughts and discussion about a hazard (Ge et al. 2011; Lindell and Prater 2000). However, expected personal consequences and hazard intrusiveness are highly correlated with each other and with affective responses such as dread, worry, and concern (Wei and Lindell 2017), so all three of these constructs would be expected to require either longitudinal designs that use risk perceptions to predict changes in hazard adjustment or behavioral expectations designs that use risk perceptions to predict hazard adjustment adoption.

There is mixed evidence that personal experience affects responses to hazards—either indirectly (due to its effect on risk perception) or directly (independent of risk perception). There also is conflicting evidence regarding the correlations of hazard proximity with hazard adjustment. Here too, there might be mediating effects of other variables—in this case, the effect of proximity on experience, experience on risk perception, and risk perception on hazard adjustment (Lindell and Hwang 2008). In addition, there is evidence that people's adoption of hazard adjustments is related to the perceived attributes of those adjustments such as efficacy, utility for other purposes, financial cost, knowledge and skill requirements, time and effort requirements, and required social cooperation (Lindell et al. 2009) and that most preparedness actions are motivated by multiple reasons (Bourque et al. 2012).

There is also some evidence that the adoption of hazard adjustments is related to social norms (Solberg et al. 2010) and people's perceptions of stakeholders such as government agencies, elected officials, peers, and self and family (Arlikatti et al. 2007; Lindell and Whitney 2000; Wang et al. 2016; Wei et al. 2018). Some research has emphasized the role of trust, which is sometimes defined in terms of the perceived competence of authorities (Paton et al. 2008) but other research has examined the effects of perceived expertise, trustworthiness (defined as willingness to provide accurate information), and protection responsibility (see Lindell and Perry 2000 and Solberg et al. 2010 for reviews).

Disaster Recovery Preparedness

Household Housing Recovery

Many studies of household housing recovery have concluded that risk area residents should prepare for disaster recovery by purchasing hazard insurance that will provide the funds needed to pay for repairs to their homes and replacement of its contents. However, hazard insurance is often problematic because risk area residents tend to forego it when they consider premiums to be too high and deductibles too large (Palm et al. 1990), as well as recognizing that it is unable to protect persons and specific to a given hazard (Lindell et al. 2009). Hazard insurance varies significantly in its availability and cost—flood, hurricane, and earthquake insurance being particularly problematic (Kunreuther and Roth 1998). Moreover, some ethnic groups cannot afford the rates of high-quality insurance companies or are denied coverage altogether (Peacock and Girard 1997). Finally, a number of studies have reported that policyholders have difficulties obtaining settlements from their insurers that are large enough to pay for the reconstruction of their homes (Lindell et al. 2016).

In addition, local jurisdictions can take a number of steps to avoid delays in displaced residents return to permanent housing. First, they should plan for debris clearance and infrastructure restoration (Phillips 2009). In addition, households seeking to recover from disasters typically encounter delays in obtaining building permits and building inspections (Wu and Lindell 2004) as well as securing financing for reconstruction and agreements with building contractors (Peacock and Girard 1997). Another obstacle in the transition to permanent housing is the length of time required to accomplish the actual construction which, in turn, depends on the original structure's size and complexity, as well as the severity of the damage to that structure (Al-Nammari and Lindell, 2009). Finally, a dramatic increase in the number of construction projects, compared to the time before the disaster, creates severe shortages in equipment and materials that further delay the reconstruction process. In the case of Hurricane Katrina, there was an extraordinary amount of competition for these resources after Katrina that was caused by the landfall of the six other hurricanes that struck that same year (Weiss 2006).

Local, state, and federal planners need to engage in systematic pre-impact recovery planning (Lindell et al. 2006, Chapter 11; Phillips 2009; Schwab et al. 1998) by first estimating the likely number of displaced households and the likely number of local vacancies and then developing plans for the placement of temporary housing and the construction of permanent housing. Planners should develop solutions that avoid the problems experienced after Hurricane Katrina by developing

Recovery and Mitigation Committees that develop procedures for expediting building safety inspection, debris clearance, and utility restoration (Lindell et al. 2006). They should also work to integrate hazard mitigation into disaster recovery by minimizing the siting of either temporary or permanent housing in hazard-prone areas and increase the construction of permanent buildings and infrastructure that can withstand disaster impacts (Evans-Cowley and Gough 2007). Planners also need to develop programs for risk communication that can be used in the aftermath of disasters (Lindell and Perry 2004). In particular, they need to establish crisis communications teams that plan, train, and exercise to prepare themselves for disaster response. Only if it is soundly designed and effectively communicated will a post-disaster housing policy meet the needs of its community. Even a safe and effective temporary housing program can lead to public castigation and lawsuits if it is poorly communicated and confusingly implemented.

Household Economic Recovery

Household economic recovery depends substantially on the ability of businesses to resume normal operation. This depends in part on those businesses' hazard mitigation, emergency preparedness, and disaster recovery actions, but local government agencies can make a vital contribution by articulating a vision of community disaster recovery that strikes a balance between corporate-centered and community-based economic development (Bingham, 2003). According to a *corporate-centered economic development*, usually advocated by the local business community, government provides resources such as land and money to the private sector to invest without any restrictions. This market-based strategy tends to produce results that are good in aggregate but produces an inequitable recovery. By contrast, *community-based economic development* involves active participation by government to ensure that the benefits of recovery will also be shared by economically disadvantaged segments of the community.

The short-term recovery following a major disaster can generate an economic boom as state and federal money flows into the community to reconstruct damaged buildings and infrastructure. These funds are used to pay for construction materials and the construction workforce and, to the extent that the materials and labor are acquired locally, they generate local revenues. In addition, the building suppliers hire additional workers and these, along with the construction workers, spend their wages on places to live, food to eat, and entertainment. Unless there are undamaged communities within commuting distance that can compete for this money, it will all be spent within the community.

Communities must also consider the long-term economic consequences of disaster recovery. What will happen after the reconstruction boom is over? They can attract new businesses if they have a skilled labor pool and good schools—especially colleges whose faculty and students can support knowledge-based industries. Other assets include low crime rates, low cost of living, good housing, and environmental amenities such as mountains, rivers, or lakes (Blakely, 2000). A community can also enhance its economic base if it can attract businesses that are compatible with the ones that are already there. Such firms can be identified by asking existing firms to identify their suppliers and distributors. These new firms might be attracted by the newer buildings and enhanced infrastructure that has been produced during disaster reconstruction.

If a disaster-stricken community does not already have such assets, they can invest in four fundamental components of economic development—locality development, business development, human resources development, and community development. *Locality development* enhances a community's existing physical assets by improving roads or establishing parks on river and lakefronts. *Business development* involves efforts to retain existing businesses or attract new ones. Although it is not easy, this can be accomplished working with businesses to identify their critical needs. In some cases, this might involve establishing a business incubator that allows startup companies to obtain low-cost space and share meetings rooms. *Human resources development* expands the skilled workforce, possibly through customized worker training. Finally, *community development* utilizes NGOs, CBOs, and local firms that will hire current residents of the community whose household incomes are below the poverty level. For example, a comprehensive program for developing small businesses, affordable housing, community health clinics, and inexpensive child care can help to eliminate some of what new businesses might consider to be one of the risks of relocating to the community.

Household Psychological Recovery

Mitchell (1983) developed a system called the *Critical Incident Stress Debriefing*, which involves preincident training, individual crisis support, demobilization (e.g., informational debriefings as personnel rotate off duty), defusing (small group discussions about the emotional significance of the event), family support, and referral to other support services (e.g., psychiatric, psychological, legal, career). Despite its proponents' claims of empirical support for this method, the most rigorous scientific evaluations have found no evidence of its effectiveness (McNally et al. 2003). One problem seems to be that establishing a rigid schedule for victims to discuss traumatic events disrupts their ability to control the alternation between psychological phases of active processing and avoidance (Pennebaker and Harber 1993). A related problem is the requirement for group discussion with their professional peers shortly after the event (usually within 12 hours). In the case of emergency responders, this conflicts with their preference for seeking support from spouses and others outside the workplace (Gist et al. 1999).

Instead, some researchers advocate *Psychological First Aid*, which seeks to accomplish three goals for survivors—provide life's basic necessities (food and water, secure shelter, medical services), help them reduce their immediate stress, and assist them in finding the resources they need to resume their normal lives (Bryant and Litz 2009). Emergency workers also deserve special consideration because they often work long hours without rest, have witnessed horrific sights, and are members of organizations in which discussion of emotional issues may be regarded as a sign of weakness (Rubin 1991). However, there is no evidence that emergency workers need directive therapies either. Nonetheless, there are population segments requiring special attention and active outreach. These include children, frail elderly, people with pre-existing mental illness, racial and ethnic minorities, and families of those who have died in the disaster. These population segments can often benefit from *Cognitive Behavioral Therapy*, which teaches people techniques for more effectively think about their fears and manage their physical reactions to those fears (Bryant and Litz 2009; Norris et al. 2002b).

Community Recovery

Research is beginning to integrate findings on household and business recovery into a coherent theory that is correcting the misconceptions of many researchers and practitioners (Chang and Rose 2012; Johnson and Hayashi 2012; Smith and Birkland 2012; Tierney and Oliver-Smith 2012). Specifically, recovery should not be defined as physical reconstruction and not all households and businesses recover in the same way (or at all), recovery should not be conceived as either purely market-driven or planning-driven, and the goal of recovery should usually be a "new normal" that avoids reproducing previous hazard exposure, physical vulnerability, and social vulnerability. As is the case during emergency response, an urgent need for action conflicts with an equally urgent need for analysis and planning, so preimpact planning is needed to define an organizational structure that can assess the needs and capacities of local households, businesses, and government agencies and develop strategies for achieving recovery (Lindell et al. 2006, Chapter 11).

Indeed, research in this area indicates that communities need to develop preimpact recovery plans just as much as they need preimpact EOPs. There are six important features of a preimpact recovery plan. First, it should define a disaster recovery organization. Second, it should identify the location of temporary housing because resolving this issue can cause conflicts that delay consideration of longer-term issues of permanent housing and distract policymakers altogether from hazard mitigation (Bolin and Trainer 1978; Bolin 1982). Third, the plan should indicate how to accomplish essential tasks such as damage assessment, condemnation, debris removal and disposal, rezoning, infrastructure restoration, temporary repair permits, development moratoria, and permit processing because all of these tasks must be addressed before the reconstruction of permanent housing can begin (Schwab et al. 1998).

Fourth, preimpact recovery plans also should address the licensing and monitoring of contractors and retail price controls to ensure victims are not exploited and also should address the jurisdiction's administrative powers and resources, especially the level of staffing that is available. It is almost inevitable that local government will have insufficient staff to perform critical recovery tasks such as damage assessment and building permit processing, so arrangements should be made to borrow staff from other jurisdictions (via pre-existing Memoranda of Agreement) and to use trained volunteers such as local engineers, architects, and planners. Fifth, these plans also need to address the ways in which recovery tasks will be implemented at historical sites (Spennemann and Look 1998). Finally, preimpact recovery plans should recognize the recovery period as a unique time to enact policies for hazard mitigation and make provision for incorporating this objective into the recovery planning process.

Summary and Conclusions

Over the past seven decades, disaster researchers have made significant progress in addressing the physical (casualties and damage) and social (psychosocial, demographic, economic, and political) aspects of disasters. Their research has addressed the impacts of these events on social units ranging in size from individuals to countries. In addition, these researchers have examined the ways in which populations at risk conduct hazard and vulnerability analyses as well as plan and implement mitigation, preparedness, response, and recovery actions. Ultimately, the incorporation of these research findings into emergency management textbooks will improve the quality of emergency management practice.

References

Ahlerup, P. (2013) "Democratisation in the Aftermath of Natural Disasters," 45–64 in *Globalization and Development*. London: Routledge.

Albala-Bertrand, J.M. (1993) "Natural Disaster Situations and Growth: A Macroeconomic Model for Sudden Disaster Impacts," *World Development* 21(9): 1417–1434.

Alesch, D.J., C. Taylor, S. Ghanty, and R.A. Nagy. (1993) "Earthquake Risk Reduction and Small Business," 133–160 in Committee on Socioeconomic Impacts (eds.) *1993 National Earthquake Conference Monograph 5: Socioeconomic Impacts*. Memphis TN: Central United States Earthquake Consortium.

Al-Nammari, F.M. and M.K. Lindell. (2009) "Earthquake Recovery of Historic Buildings: Exploring Cost and Time Needs," *Disasters* 33(3): 457–481.

Arlikatti, S. and S.A. Andrew. (2012) "Housing Design and Long-Term Recovery Processes in the Aftermath of the 2004 Indian Ocean Tsunami," *Natural Hazards Review* 13(1): 34–44.

Arlikatti, S., M.K. Lindell, and C.S. Prater. (2007) "Perceived Stakeholder Role Relationships and Adoption of Seismic Hazard Adjustments," *International Journal of Mass Emergencies and Disasters* 25(3): 218–256.

Baker, E.J. (1991) "Hurricane Evacuation Behavior," *International Journal of Mass Emergencies and Disasters* 9(2): 287–310.

Baker, J., W.D. Shaw, D. Bell, S. Brody, M. Riddel, R.T. Woodward, and W. Neilson. (2009) "Explaining Subjective Risks of Hurricanes and the Role of Risks in Intended Moving and Location Choice Models," *Natural Hazards Review* 10(3): 102–112.

Berke, P., J. Cooper, D. Salvesen, D. Spurlock, and C. Rausch. (2010) "Disaster Plans: Challenges and Choices to Build the Resiliency of Vulnerable Populations," *International Journal of Mass Emergencies and Disasters* 28(3): 368–394.

Berke, P. and D. Godschalk. (2009) "Searching for the Good Plan: A Meta-Analysis of Plan Quality Studies," *Journal of Planning Literature* 23(3): 227–240.

Berke, P.R., J. Kartez, and D.E. Wenger. (1993) "Recovery After Disaster: Achieving Sustainable Development, Mitigation and Equity," *Disasters* 17(2): 93–109.

Berke, P.R., G. Smith, and W. Lyles. (2012) "Planning for Resiliency: Evaluation of State Hazard Mitigation Plans Under the Disaster Mitigation Act," *Natural Hazards Review* 13(2): 139–150.

Bingham, R. D. (2003) "Economic Development Policies," 237–253 in J.P. Pellisero (ed.) *Cities, Politics, and Policy: A Comparative Analysis.* Washington DC: CQ Press.

Bird, D.K., G. Gisladottir, and D. Dominey-Howes. (2010) "Volcanic Risk and Tourism in Southern Iceland: Implications for Hazard, Risk and Emergency Response Education and Training," *Journal of Volcanology and Geothermal Research* 189(1–2): 33–48.

Blakely, E.J. (2000) "Economic Development," 283–305 in C.J. Hoch, L.C. Dalton, and F.S. So (eds.) *The Practice of Local Government Planning*, 3rd ed. Washington DC: International City/County Management Association.

Bolin, R.C. (1982) *Long-Term Family Recovery from Disaster.* Boulder CO: University of Colorado Institute of Behavioral Science.

Bolin, R.C. (1985) "Disaster Characteristics and Psychosocial Impacts," 3–28 in B.J. Sowder (ed.) *Disasters and Mental Health: Selected Contemporary Perspectives.* Rockville MD: National Institute of Mental Health.

Bolin, R.C. (1993) *Household and Community Recovery After Earthquakes.* Boulder CO: University of Colorado Institute of Behavioral Science.

Bolin, R.C. and P. Bolton. (1986) *Race, Religion, and Ethnicity in Disaster Recovery.* Boulder CO: University of Colorado Institute of Behavioral Science.

Bolin, R.C. and L.C. Kurtz. (2018) "Race, Class, Ethnicity, and Disaster Vulnerability," 181–203 in H. Rodríguez, J. Trainor, and W. Donner (eds.) *Handbook of Disaster Research.* New York: Springer.

Bolin, R.C. and L. Stanford. (1991) "Shelters, Housing and Recovery: A Comparison of U.S. Disasters," *Disasters* 45(1): 25–34.

Bolin, R. C. and L. Stanford. (1998) "Community-Based Approaches to Unmet Recovery Needs," *Disasters* 22(1): 21–38.

Bolin, R. and P.A. Trainer. (1978) "Modes of Family Recovery Following Disaster: A Cross-National Study," 233–247 in E.L. Quarantelli (ed.) *Disasters: Theory and Research.* Beverly Hills CA: Sage.

Bourque, L.B., D.S. Mileti, M. Kano, and M.M. Wood. (2012) "Who Prepares for Terrorism?" *Environment and Behavior* 44(3): 374–409.

Bryant, E.A. (1997) *Natural Hazards.* Cambridge: Cambridge University Press.

Bryant, R.A. and B. Litz. (2009) "Mental Health Treatments in the Wake of Disaster," 321–335 in Y. Neria, S, Galea, and F.H. Norris (eds.) *Mental Health and Disasters.* New York: Cambridge University Press.

Burton, I., R.W. Kates, and G.F. White. (1978) *The Environment as Hazard.* New York: Oxford University Press.

Cavallo, E., S. Galiani, I. Noy, and J. Pantano. (2013) "Catastrophic Natural Disasters and Economic Growth," *Review of Economics and Statistics* 95(5): 1549–1561.

Chamlee-Wright, E. and V.H. Storr. (2009) "There's No Place Like New Orleans: Sense of Place and Community Recovery in the Ninth Ward After Hurricane Katrina," *Journal of Urban Affairs* 31(5), 615–634.

Chang, C. P. and A. N. Berdiev. (2015) "Do Natural Disasters Increase the Likelihood That a Government Is Replaced?" *Applied Economics* 47(17): 1788–1808.

Chang, S.E. and A. Falit-Baiamonte. (2002) "Disaster Vulnerability of Businesses in the 2001 Nisqually Earthquake," *Environmental Hazards* 4(2–3): 59–71.

Chang, S.E. and A.Z. Rose. (2012) "Towards a Theory of Economic Recovery from Disasters," *International Journal of Mass Emergencies and Disasters* 30(2): 171–181.

Charvériat, C. (2000) *Natural Disasters in Latin America and the Caribbean: An Overview of Risk.* Working paper #434. Washington DC: Inter-American Development Bank.

Cole, P.M. (2003) *An Empirical Examination of the Housing Recovery Process Following Disaster.* College Station TX: Texas A&M University.

Comerio, M.C. (1998) *Disaster Hits Home: New Policy for Urban Housing Recovery.* Berkeley CA: University of California Press.

Comfort, L. K., T. A. Birkland, B. A. Cigler, and E. Nance. (2010) "Retrospectives and Prospectives on Hurricane Katrina: Five Years and Counting," *Public Administration Review* 70(5): 669–678.

CACND—Committee on Assessing the Costs of Natural Disasters. (1999) *The Impacts of Natural Disasters: A Framework for Loss Estimation*. Washington DC: National Academy Press.

CDRSS—Committee on Disaster Research in the Social Sciences. (2006) *Facing Hazards and Disasters: Understanding Human Dimensions*. Washington DC: National Academy of Sciences.

Collins, T.W. (2008) "What Influences Hazard Mitigation? Household Decision Making About Wildfire Risks in Arizona's White Mountains," *The Professional Geographer* 60(4): 508–526.

Cruz, A.M. and L.J. Steinberg. (2005) "Industry Preparedness for Earthquakes and Earthquake-Triggered Hazmat Accidents in the 1999 Kocaeli Earthquake," *Earthquake Spectra* 21(2): 285–303.

Daines, G. E. (1991) "Planning, Training and Exercising," 161–200 in T.E. Drabek and G.J. Hoetmer (eds.), *Emergency Management*. Washington DC: International City/County Management Association.

Demuth, J. L. (2018) "Explicating Experience: Development of a Valid Scale of Past Hazard Experience for Tornadoes," *Risk Analysis*, 38(9): 1921–1943.

Demuth, J.L., R.E. Morss, J.K. Lazo, and C. Trumbo. (2016). "The Effects of Past Hurricane Experiences on Evacuation Intentions Through Risk Perception and Efficacy Beliefs: A Mediation Analysis," *Weather, Climate, and Society* 8(4): 327–344.

Dillon, R.L. and C.H. Tinsley. (2016) "Near-Miss Events, Risk Messages, and Decision Making," *Environment Systems and Decisions* 36(1): 34–44.

Dillon, R.L., C.H. Tinsley, and W.J. Burns. (2014) "Near-Misses and Future Disaster Preparedness,"" *Risk Analysis* 34(10): 1907–1922.

Drabek, T.E. and D.A. McEntire. (2003) "Emergent Phenomena and Multiorganizational Coordination in Disasters: Lessons from the Research Literature," *International Journal of Mass Emergencies and Disasters* 20(2): 197–224.

Drabek, T.E., H.L. Tamminga, T.S. Kilijanek, and C.R. Adams. (1981) *Managing Multiorganizational Emergency Responses*. Boulder CO: University of Colorado Institute of Behavioral Science.

Drury, J, C. Cocking, and S. Reicher. (2009) "The Nature of Collective Resilience: Survivor Reactions to the 2005 London Bombings," *International Journal of Mass Emergencies and Disasters* 27(1): 66–95.

Drury, A. C., and R. S. Olson. (1998) "Disasters and Political Unrest: An Empirical Investigation," *Journal of Contingencies and Crisis Management* 6(3): 153–161.

Dynes, R. (1970) *Organized Behavior in Disaster*. Lexington MA: Heath-Lexington Books.

Elliott, J.R. and J. Pais. (2006) "Race, Class, and Hurricane Katrina: Social Differences in Human Responses to Disaster," *Social Science Research* 35(2): 295–321.

Enarson, E., A. Fothergill, and L. Peek. (2018) "Gender and Disaster: Foundations and New Directions for Research and Practice," 205–223 in H. Rodríguez, J. Trainor, and W. Donner (eds.) *Handbook of Disaster Research*. New York: Springer.

Evans, R.W., Y. Hu, and Z. Zhao. (2010) "The Fertility Effect of Catastrophe: US Hurricane Births," *Journal of Population Economics* 23(1): 1–36.

Evans-Cowley, J.S. and M.Z. Gough. (2007) "Is Hazard Mitigation Being Incorporated into Post-Katrina Plans in Mississippi?" *International Journal of Mass Emergencies and Disasters* 25(3): 177–217.

Felbermayr, G. and J. Gröschl. (2014) "Naturally Negative: The Growth Effects of Natural Disasters," *Journal of Development Economics* 111: 92–106.

FEMA—Federal Emergency Management Agency (1986). *Making Mitigation Work: A Handbook for State Officials*. Washington DC: Author.

FEMA—Federal Emergency Management Agency. (1997) *Multihazard Identification and Risk Assessment: A Cornerstone of the National Mitigation Strategy*. Washington DC: Author.

FEMA—Federal Emergency Management Agency. (2010) *Developing and Maintaining Emergency Operation Plans. Comprehensive Preparedness Guide (CPG) 101. Version 2.0*. Washington DC: Author.

Fischer, H.W. III (2008) *Response to Disaster: Fact versus Fiction and Its Perpetuation*, 3rd ed. Lanham MD: University Press of America.

Flint, C.G. and J. Stevenson. (2010) "Building Community Disaster Preparedness with Volunteers: Community Emergency Response Teams in Illinois," *Natural Hazards Review* 11(3): 118–124.

Folkman, S. (2013). "Stress: Appraisal and Coping," 1913–1915 in M. Gellman and J.R. Turner (eds.) *Encyclopedia of Behavioral Medicine*. New York: Springer.

Fox, M.H., G.W. White, C. Rooney, and J.L. Rowland. (2007) "Disaster Preparedness and Response for Persons with Mobility Impairments: Results from the University of Kansas Nobody Left Behind Study," *Journal of Disability Policy Studies* 17(4): 196–205.

Friesma, H.P., J. Caporaso, G. Goldstein, R. Linberry, and R. McCleary. (1979) *Aftermath: Communities After Natural Disasters*. Beverly Hills CA: Sage.

Fussell, E., N. Sastry, and M. VanLandingham. (2010) "Race, Socioeconomic Status, and Return Migration to New Orleans After Hurricane Katrina," *Population and Environment* 31(1–3): 20–42.

Ge, Y. and M.K. Lindell. (2016) "County Planners' Perceptions of Land Use Planning Tools for Environmental Hazard Mitigation: A Survey in the U.S. Pacific States," *Environment and Planning B: Planning and Design* 43(4): 716–736.

Ge, Y., W.G. Peacock, and M.K. Lindell. (2011) "Florida Households' Expected Responses to Hurricane Hazard Mitigation Incentives," *Risk Analysis* 31(10): 1676–1691.

Gerrity, E.T. and B.W. Flynn. (1997) "Mental Health Consequences of Disasters," 101–121 in E.K. Noji (ed.) *The Public Health Consequences of Disasters*. New York: Oxford University Press.

Gillespie, D. (1991) "Coordinating Community Resources," 55–78, in T.S. Drabek and G.J. Hoetmer (eds.), *Emergency Management: Principles and Practice for Local Government*, Washington DC: International City/County Management Association.

Girard, C. and W.G. Peacock. (1997) "Ethnicity and Segregation: Post-Hurricane Relocation," 191–205, in W.G. Peacock, B.H. Morrow, and H. Gladwin (eds.), *Hurricane Andrew: Ethnicity, Gender and the Sociology of Disasters*. New York: Routledge.

Gist, R., B. Lubin, and B.G. Redburn. (1999). "Psychosocial, Community, and Ecological Approaches on Disaster Response," 1–20, in R. Gist and B. Lubin (eds.) *Response to Disaster: Psychosocial, Community, and Ecological Approaches*. Philadelphia PA: Brunner/Mazel.

Godschalk, D.R., S. Brody, and R. Burby. (2003) "Public Participation in Natural Hazard Mitigation Policy Formation: Challenges for Comprehensive Planning," *Journal of Environmental Planning and Management* 46(5): 733–754.

Groen, J. A. and A.E. Polivka. (2010) "Going Home After Hurricane Katrina: Determinants of Return Migration and Changes in the Affected Areas," *Demography* 47(4): 821–844.

Hori, M., M.J. Schafer, and D.J. Bowman. (2009) "Displacement Dynamics in Southern Louisiana after Hurricanes Katrina and Rita," *Population Research and Policy Review* 28(1), 45–65.

Houts, P.S., P.D. Cleary, and T.W. Hu. (1988) *The Three Mile Island Crisis: Psychological, Social and Economic Impacts on the Surrounding Population*. University Park PA: Pennsylvania State University Press.

Johnson, L. and H. Hayashi. (2012) "Synthesis Efforts in Disaster Recovery Research," *International Journal of Mass Emergencies and Disasters* 30(2): 212–238.

Kailes, J.I. and A. Enders. (2007) "Moving Beyond 'Special Needs' A Function-Based Framework for Emergency Management and Planning," *Journal of Disability Policy Studies* 17(4): 230–237.

Kim, J. and S.S. Oh. (2014) "The Virtuous Circle in Disaster Recovery: Who Returns and Stays in Town After Disaster Evacuation?" *Journal of Risk Research* 17(5), 665–682.

Klomp, J. and K. Valckx. (2014) "Natural Disasters and Economic Growth: A Meta-Analysis," *Global Environmental Change* 26: 183–195.

Kreutz, J. (2012) "From Tremors to Talks: Do Natural Disasters Produce Ripe Moments for Resolving Separatist Conflicts?" *International Interactions* 38(4): 482–502.

Kroll, C.A., J.D. Landis, Q. Shen, and S. Stryker. (1990) *The Economic Impacts of the Loma Prieta Earthquake: A Focus on Small Business*. Working Paper 91–187. Berkeley CA: UC Berkeley Institute of Business and Economic Research.

Kunreuther H. and R.J. Roth, Sr. (1998) *Paying the Price: The Status and Role of Insurance Against Natural Disasters in the United States*. Washington DC: Joseph Henry Press.

Landry, C.E., O. Bin, P. Hindsley, J. Whitehead, and K. Wilson. (2007) "Going Home: Evacuation-Migration Decisions of Hurricane Katrina Survivors," *Southern Economic Journal* 74(2): 326–343.

Langness, D. (1994) "The Northridge Earthquake: Planning and Fast Action Minimize Devastation," *California Hospitals* 8(2): 8–13.

Lazarus, R.S. and S. Folkman. (1984) *Stress, Appraisal, and Coping*. New York: Springer.

Li, W., C.A. Airriess, A. Chen, K.J. Leong, and V. Keith. (2010) "Katrina and Migration: Evacuation and Return by African Americans and Vietnamese Americans in an Eastern New Orleans Suburb," *Professional Geographer* 61(1): 103–118.

Lin, S., D. Shaw, and M.-C. Ho. (2008) "Why Are Flood and Landslide Victims Less Willing to Take Mitigation Measures Than the Public?" *Natural Hazards* 44(2): 305–314.

Lindell, M.K. (2013a) "Disaster Studies," *Current Sociology Review* 61(5–6): 797–825.

Lindell, M.K. (2013b) "North American Cities at Risk: Household Responses to Environmental Hazards," 109–130, in T. Rossetto, H. Joffe, and J. Adams (eds.), *Cities at Risk: Living with Perils in the 21st Century*. Dordrecht: Springer.

Lindell, M.K. (2018) "Communicating Imminent Risk," 449–477, in H. Rodríguez, J. Trainor, and W. Donner (eds.) *Handbook of Disaster Research*. New York: Springer.

Lindell, M.K., S. Arlikatti, and C.S. Prater. (2009) "Why People Do What They Do to Protect Against Earthquake Risk: Perceptions of Hazard Adjustment Attributes," *Risk Analysis* 29(8): 1072–1088.

Lindell, M.K. and C.J. Brandt. (2000) "Climate Quality and Climate Consensus as Mediators of the Relationship Between Organizational Antecedents and Outcomes," *Journal of Applied Psychology* 85(3): 331–348

Lindell, M.K., S.D. Brody, and W.E. Highfield. (2016). "Financing Housing Recovery Through Hazard Insurance: The Case of the National Flood Insurance Program," 50–65, in A. Sapat and A.-M. Esnard. *Coming Home After Disaster: Multiple Dimensions of Housing Recovery*. Boca Raton FL: CRC Press.

Lindell, M.K. and S.N. Hwang. (2008) "Households' Perceived Personal Risk and Responses in a Multi-Hazard Environment," *Risk Analysis* 28(2): 539–556.

Lindell, M.K. and R.W. Perry. (2000) "Household Adjustment to Earthquake Hazard: A Review of Research," *Environment and Behavior* 32(4): 590–630.

Lindell, M.K. and R.W. Perry. (2004) *Communicating Environmental Risk in Multiethnic Communities*. Thousand Oaks CA: Sage.

Lindell, M.K. and R.W. Perry. (2007) "Planning and Preparedness," 113–141, in K.J. Tierney and W.F. Waugh, Jr. (eds.) *Emergency Management: Principles and Practice for Local Government*, 2nd ed. Washington DC: International City/County Management Association.

Lindell, M.K. and C.S. Prater. (2000) "Household Adoption of Seismic Hazard Adjustments: A Comparison of Residents in Two States," *International Journal of Mass Emergencies and Disasters* 18(2), 317–338.

Lindell, M.K. and C.S. Prater. (2003) "Assessing Community Impacts of Natural Disasters," *Natural Hazards Review* 4(4): 176–185.

Lindell, M.K., C.S. Prater, and W.G. Peacock. (2007) "Organizational Communication and Decision Making in Hurricane Emergencies," *Natural Hazards Review* 8(3): 50–60.

Lindell, M.K., C.S. Prater, and R.W. Perry. (2006). *Fundamentals of Emergency Management*. Emmitsburg MD: Federal Emergency Management Agency Emergency Management Institute. Available at www.training. fema.gov/hiedu/aemrc/booksdownload/fem/ or hrrc.arch.tamu.edu/publications/books%20and%20 monographs/.

Lindell, M.K. and D.J. Whitney. (2000) "Correlates of Seismic Hazard Adjustment Adoption," *Risk Analysis* 20(1): 13–25.

McEntire, D.A. (2006) "Local Emergency Management Organizations," 168–182, in H. Rodríguez, E.L. Quarantelli, and R.R. Dynes (eds.) *Handbook of Disaster Research*. New York: Springer.

McNally, R.J., R.A. Bryant, and A. Ehlers. (2003). "Does Early Psychological Intervention Promote Recovery from Posttraumatic Stress?" *Psychological Science in the Public Interest* 4(2): 45–79.

Mileti, D.S. (1999) *Disasters by Design: A Reassessment of Natural Hazards in the United States*. Washington DC: Joseph Henry Press.

Mitchell, J.T. (1983). "When Disaster Strikes . . . The Critical Incident Stress Debriefing Process," *Journal of Emergency Medical Services* 8(1): 36–39.

Morrow, B.H. and W.G. Peacock. (1997) "Disasters and Social Change: Hurricane Andrew and the Reshaping of Miami," 226–242, in W.G. Peacock, B.H. Morrow and H. Gladwin, *Hurricane Andrew: Ethnicity, Gender and the Sociology of Disaster*. London: Routledge.

Myers, C.A., T. Slack, and J. Singelmann. (2008) "Social Vulnerability and Migration in the Wake of Disaster: The Case of Hurricanes Katrina and Rita," *Population and Environment* 29(6): 271–291.

National Institute of Building Sciences. (1998) *HAZUS*. Washington DC: Author.

Nel, P. and M. Righarts. (2008) "Natural Disasters and the Risk of Violent Civil Conflict," *International Studies Quarterly* 52(1): 159–185.

Norris, F.H., M.J. Friedman, P.J. Watson, C.M. Byrne, E. Diaz, and K. Kaniasty. (2002a) "60,000 Disaster Victims Speak: Part I. An Empirical Review of the Empirical Literature, 1981–2001," *Psychiatry* 65(3): 207–239.

Norris, F.H., P.J. Friedman, and P.J. Watson. (2002b) "60,000 Disaster Victims Speak: Part II. Summary and Implications of the Disaster Mental Health Research," *Psychiatry* 65(3): 240–260.

Olson, R.S. and A.C. Drury. (1997) "Untherapeutic Communities: A Cross-National Analysis of Post-Disaster Political Unrest," *International Journal of Mass Emergencies and Disasters* 15(2): 221–238.

Palm, R., M. Hodgson, R.D. Blanchard, and D. Lyons. (1990) *Earthquake Insurance in California*. Boulder CO: Westview Press.

Paton, D., S. Sagala, N. Okada, L-J. Jang, P.T. Bürgelt, and C.E. Gregg. (2010) "Making Sense of Natural Hazard Mitigation: Personal, Social and Cultural Influences," *Environmental Hazards* 9(2): 183–196.

Paton D., L. Smith, M. Daly, and D.M. Johnston. (2008) "Risk Perception and Volcanic Hazard Mitigation: Individual and Social Perspectives," *Journal of Volcanology and Geothermal Research* 172(3–4): 179–188.

Paxson, C. and C.E. Rouse. (2008) "Returning to New Orleans after Hurricane Katrina," *American Economic Review* 98(2): 38–42.

Peacock, W.G., N. Dash, and Y. Zhang. (2006) "Sheltering and Housing Recovery Following Disaster," 258–274, in H. Rodríguez, E.L. Quarantelli, and R.R. Dynes. (eds.) *Handbook of Disaster Research*. New York: Springer.

Peacock, W.G and C. Girard. (1997) "Ethnic and Racial Inequalities in Disaster Damage and Insurance Settlements." 171–190 in W.G. Peacock, B. H. Morrow and H. Gladwin (eds.). *Hurricane Andrew: Ethnicity, Gender and the Sociology of Disaster*. London: Routledge.

Peacock, W.G., L.M. Killian, and F.L. Bates. (1987) "The Effects of Disaster Damage and Housing on Household Recovery Following the 1976 Guatemala Earthquake," *International Journal of Mass Emergencies and Disasters* 5(1): 63–88.

Peacock, W.G., B.H. Morrow, and H. Gladwin. (1997) *Hurricane Andrew: Ethnicity, Gender and the Sociology of Disaster*. London: Routledge.

Peacock, W.G., S. Van Zandt, Y. Zhang, and W.E. Highfield. (2014) "Inequities in Long-Term Housing Recovery After Disasters," *Journal of the American Planning Association* 80(4): 356–371.

Peek-Asa, C., J.F. Kraus, L.B. Bourque, D. Vimalachandra, J. Yu, and J. Abrams. (1998) "Fatal and Hospitalized Injuries Resulting From the 1994 Northridge Earthquake," *International Journal of Epidemiology* 27(3): 459–465.

Pennebaker, J.W. and K.D. Harber. (1993). "A Social Stage Model for Collective Coping: The Loma Prieta Earthquake and the Persian Gulf War," *Journal of Social Issues* 49(4): 125–146.

Perry, R.W. and M.K. Lindell. (1978) "The Psychological Consequences of Natural Disaster: A Review of Research on American Communities," *Mass Emergencies* 3(2–3): 105–115.

Perry, R.W. and M.K. Lindell. (1990) *Living with Mt. St. Helens: Human Adjustment to Volcano Hazards*. Pullman WA: Washington State University Press.

Perry, R.W. and M.K. Lindell. (2007) *Emergency Planning*. Hoboken NJ: John Wiley.

Petrescu-Prahova, M. and C.T. Butts. (2008) "Emergent Coordinators in the World Trade Center Disaster," *International Journal of Mass Emergencies and Disasters* 26(3): 133–168. Available at www.ijmed.org.

Phillips, B.D. (2009) *Disaster Recovery*. Boca Raton FL: CRC Press.

Prater, C.S. and M.K. Lindell. (2000) "The Politics of Hazard Mitigation," *Natural Hazards Review* 1(2): 73–82.

Quarantelli, E.L. (1982) *Sheltering and Housing After Major Community Disasters*. Columbus OH: Ohio State University Disaster Research Center.

Quarantelli, E.L. (1988) "Disaster Crisis Management: A Summary of Research Findings," *Journal of Management Studies* 25(4): 373–385.

Rubin, C.B. (1991) "Recovery from Disaster," 224–259, in T.E. Drabek and G.J. Hoetmer (eds.) *Emergency Management: Principles and Practice for Local Government*. Washington DC: International City Management Association.

Rubin, C.B., M.D. Saperstein, and D.G. Barbee. (1985) *Community Recovery from a Major Natural Disaster*. Monograph # 41. Boulder CO: University of Colorado, Institute of Behavioral Science

Saarinen, T. and J. Sell. (1985) *Warning and Response to the Mt. St. Helens Eruption*. Albany NY: State University of New York Press.

Sadiq, A.-A. and C.M. Weible. (2010) "Obstacles and Disaster Risk Reduction: Survey of Memphis Organizations," *Natural Hazards Review* 11(3): 110–117.

Schwab, J., K.C. Topping, C.C. Eadie, R.E. Deyle, and R.A. Smith. (1998) *Planning for Post-Disaster Recovery and Reconstruction*, PAS Report 483/484. Chicago IL: American Planning Association.

Shaw, W.D. and J. Baker. (2010) "Models of Location Choice and Willingness to Pay to Avoid Hurricane Risks for Hurricane Katrina Evacuees," *International Journal of Mass Emergencies and Disasters* 28(1): 87–114.

Shefner, J. (1999) "Pre- and Post-Disaster Political Instability and Contentious Supporters: A Case Study of Political Ferment," *International Journal of Mass Emergencies and Disasters* 17(2): 37–160.

Shoaf, K.I., H.S. Sareen, L.H. Nguyen, and L.B. Bourque. (1998) "Injuries as a Result of California Earthquakes in the Past Decade," *Disasters* 22(3): 218–235.

Smith G. and T. Birkland. (2012) "Building a Theory of Recovery: Institutional Dimensions," *International Journal of Mass Emergencies and Disasters* 30(2): 147–170.

Smith S.K. and C. McCarty. (1996) "Demographic Effects of Natural Disasters: A Case Study of Hurricane Andrew," *Demography* 33(2): 265–275.

Smith, S.K., J. Tayman, and D.A. Swanson. (2001) *State and Local Population Projections: Methodology and Analysis*. New York: Kluwer.

Solberg, C., T. Rossetto, and H. Joffe. (2010) "The Social Psychology of Seismic Hazard Adjustment: Re-Evaluating the International Literature," *Natural Hazards and Earth System Sciences* 10(8): 1663–1677.

Song Y., P. Berke, and M. Stevens. (2009) "Smart Developments in Dangerous Locations: A Reality Check of Existing New Urbanist Developments," *International Journal of Mass Emergencies and Disasters* 27(1): 1–24.

Spennemann, D.H.R and D.W. Look. (1998) *Disaster Management Programs for Historic Sites*. San Francisco CA: Association for Preservation Technology.

Stough, L.M. and I. Kelman. (2018) "People with Disabilities and Disasters," 225–242 in H. Rodríguez, W. Donner and J.E Trainor (eds.) *Handbook of Disaster Research*, 2nd ed. New York: Springer.

Stevens, M. (2010) "Implementing Natural Hazard Mitigation Provisions: Exploring the Role That Individual Land Use Planners Can Play," *Journal of Planning Literature* 24(4): 362–371.

Strobl, E. (2011) "The Economic Growth Impact of Hurricanes: Evidence from US Coastal Counties," *Review of Economics and Statistics* 93(2): 575–589.

Strobl, E. (2012) "The Economic Growth Impact of Natural Disasters in Developing Countries: Evidence from Hurricane Strikes in the Central American and Caribbean Regions," *Journal of Development Economics* 97(1): 130–141.

Tang, Z., M.K. Lindell, C.S. Prater, and S.D. Brody. (2008) "Measuring Tsunami Planning Capacity on the U.S. Pacific Coast," *Natural Hazards Review* 9(2): 91–100.

Tang, Z., M.K. Lindell, C.S. Prater, T. Wei, and C.M. Hussey. (2011) "Examining local Coastal Zone Management Capacity in U.S. Pacific Coastal Counties," *Coastal Management* 39(2): 105–132.

Terpstra, T. and J.M. Gutteling. (2008) "Households' Perceived Responsibilities in Flood Risk Management in the Netherlands," *International Journal of Water Resource Development* 24(4): 555–565.

Tierney, K., M.K. Lindell, and R.W. Perry. (2001) *Facing the Unexpected: Disaster Preparedness and Response in the United States*. Washington DC: Joseph Henry Press.

Tierney, K. and A. Oliver-Smith. (2012) "Social Dimensions of Disaster Recovery," *International Journal of Mass Emergencies and Disasters* 30(2): 123–146.

Uscher-Pines, L., A.J. Hausman, S. Powell, P. DeMara, G. Heake, and M.G. Hagen. (2009) "Disaster Preparedness of Households with Special Needs in Southeastern Pennsylvania," *American Journal of Preventive Medicine* 37(3): 227–230.

Uhr, C., H. Johansson, and L. Fredholm. (2008) "Analysing Emergency Response Systems," *Journal of Contingencies and Crisis Management* 16(2): 80–90.

Wang, F., J.-C. Wei, S.-K. Huang, M.K. Lindell, Y. Ge, and H.-L. Wei. (2016) "Public Reactions to the 2013 Chinese H7N9 Influenza Outbreak: Perceptions of Risk, Stakeholders, and Protective Actions," *Journal of Risk Research*. DOI: 10.1080/13669877.2016.1247377.

Wallace, A.F.C. (1956) *Tornado in Worcester: An Exploratory Study of Individual and Community Behavior in an Extreme Situation*. Washington DC: National Academy of Sciences-National Research Council

Warrick, R.A., J. Anderson, T. Downing, J. Lyons, J. Ressler, M. Warrick, and T. Warrick. (1981) *Four Communities Under Ash after Mount St. Helens*. Boulder, CO: University of Colorado Institute of Behavioral Science.

Webb, G.R., K.J. Tierney, and J.M. Dahlhamer. (2000) "Business and Disasters: Empirical Patterns and Unanswered Questions," *Natural Hazards Review* 1(2): 83–90.

Webb, G.R., K.J. Tierney, and J.M. Dahlhamer. (2002) "Predicting Long-Term Business Recovery from Disasters: A Comparison of the Loma Prieta Earthquake and Hurricane Andrew," *Environmental Hazards* 4(2–3): 45–58.

Wei, H.-L. and M.K. Lindell. (2017) "Washington Households' Expected Responses to Lahar Threat from Mt. Rainier," *International Journal of Disaster Risk Reduction* 22, 77–94.

Wei, H-L., M.K. Lindell, C.S. Prater, F. Wang, J-C. Wei, and Y. Ge. (2018) "Perceived Stakeholder Characteristics and Protective Action for Influenza Emergencies: A Comparative Study of Respondents in the United States and China," *International Journal of Mass Emergencies and Disasters* 36(1): 52–70.

Weinstein, N. D. and M. Nicolich. (1993) "Correct and Incorrect Interpretations of Correlations Between Risk Perceptions and Risk Behaviors," *Health Psychology* 12(3): 235–245.

Weiss, N.E. (2006) *Rebuilding Housing after Hurricane Katrina: Lessons Learned and Unresolved Issues*. Washington DC: Congressional Research Service.

Wisner, B., P. Blaikie, T. Cannon, and I. Davis. (2004) *At Risk: Natural Hazards, People's Vulnerability and Disasters*, 2nd ed. London: Routledge.

Wood, M. and L. Bourque. (2018) "Morbidity and Mortality Associated with Disasters," 357–383, in H. Rodríguez, W. Donner and J.E Trainor (eds.), *Handbook of Disaster Research*, 2nd ed. New York: Springer.

Wright, J.D., P.H. Rossi, S.R. Wright, and E. Weber-Burdin. (1979) *After the Clean-Up: Long-Range Effects of Natural Disasters*. Beverly Hills CA: Sage.

Wu, J.Y. and M.K. Lindell. (2004) "Housing Reconstruction After Two Major Earthquakes: The 1994 Northridge Earthquake in the United States and the 1999 Chi-Chi Earthquake in Taiwan," *Disasters* 28(1): 63–81.

Wu, H.C., M.K. Lindell, and C.S. Prater. (2012) "Logistics of Hurricane Evacuation in Hurricanes Katrina and Rita," *Transportation Research Part F: Traffic Psychology and Behaviour* 15(4): 445–461.

Wu, H.C., M.K. Lindell, C.S. Prater, and S-K. Huang. (2013) "Logistics of Hurricane Evacuation in Hurricane Ike," 127–140 in J. Cheung and H. Song (eds.) *Logistics: Perspectives, Approaches and Challenges*. Hauppauge NY: Nova Science Publishers.

Xiao, Y. and S. Van Zandt. (2012) "Building Community Resiliency: Spatial Links Between Household and Business Post-Disaster Return," *Urban Studies* 49(11): 2523–2542.

Yoon, D.K., G.A. Youngs, Jr., and D. Abe. (2012) "Examining Factors Contributing to the Development of FEMA-Approved Hazard Mitigation Plans," *Journal of Homeland Security and Emergency Management* 9(2) Article 14. DOI: 10.1515/1547-7355.201.

Zhang, Y., M.K. Lindell, and C.S. Prater. (2009) "Vulnerability of Community Businesses to Environmental Disasters," *Disasters* 33(1): 38–57.

2

IMPACTS ON SOCIALLY VULNERABLE POPULATIONS

Shannon Van Zandt

Introduction

Communities are only as strong as their weakest links. They cannot be resilient unless all members are able to withstand and bounce back from an economic, social, or physical disaster. Yet a community's social vulnerabilities are often neglected. When disaster strikes, its impact is not just a function of its magnitude and location. Development patterns characterized by sprawl, concentrated poverty, and segregation shape urban environments in ways that isolate vulnerable populations so that low-income and high-income, White and African-American, owners and renters, primary residents and vacationers, are not just socially separated from one another, but also physically separated in clusters and pockets across the community. In many if not most communities, these development patterns interact with the physical geography to expose vulnerable populations to greater risk. At every stage, socially vulnerable populations are at a disadvantage. These disadvantages compound one another through each phase of disasters, aggravating impacts, slowing recovery, and exacerbating pre-existing inequities. Further, they accumulate over time, undermining community efforts to become more resilient. This chapter describes the processes by which socially vulnerable populations are placed at higher risk of exposure and the impacts that result from it.

Low-Income Households Live in Low-Quality Homes in Low-Lying Areas

Disasters occur when natural or technological hazards interact with socio-ecological systems (Mileti 1999). Disaster impacts are due to interactions between hazard exposure (e.g., low-lying areas), physical vulnerability (e.g., low-quality homes), and social vulnerability (e.g., low-income households). Hazard vulnerability is generally characterized as a function of hazard exposure and the characteristics of the geographic area, built environment, and resident populations that may lead to particular vulnerabilities (Cutter 1996; Manyena 2006). Hazard exposure is the probability that extreme events (e.g., flooding, wind, surge, etc.) will occur, while physical vulnerability refers to the potential damage to the built environment, especially housing (NRC 2006). More recent perspectives have expanded the vulnerability construct to include social vulnerability, which refers to characteristics of a subpopulation that create variability in vulnerability to disasters (e.g., poverty, race, disability, language barriers, or age) (Cutter 1996; Cutter and Finch 2008; Flanagan et al. 2011; Lindell and Prater 2003; NRC 2006; Van Zandt et al. 2012; Wisner et al. 1994).

In the disaster literature, the focus on social vulnerability has typically been on household and individual differences in capacity, power, resources, and access to information, and how this translates into differences in individual and household actions related to preparedness, warning, and evacuation. Accordingly, some of the most robust findings in the social vulnerability literature indicate that minority and low-income households are more likely to experience disproportionate impacts of disasters in terms of damage, casualties, displacement, and recovery. The strength of these findings suggests a role for locational (i.e., neighborhood and community) influences and differences in physical vulnerability and hazard exposure that compound these same social vulnerabilities. In other words, it's not just that people and households vary in their capacity to anticipate, cope with, respond to, and recover from disasters. It's that these socially vulnerable households are not randomly distributed in space, but rather are concentrated in fairly predictable spatial patterns according to these same characteristics. As a result, they experience increased exposure and physical vulnerability to hazards, which has both short- and long-term consequences.

Figure 2.1 captures how pre-existing inequities among sub-populations may become compounded and exaggerated throughout the phases of disaster. Throughout this chapter, the reader may refer back to this figure to better understand how these pre-existing differences are refracted through what the literature refers to as the "lens of vulnerability" (French et al. 2008; Peacock et al. 2015). In the main part of the figure, the graphic shows that differences exist prior to a disaster. Underneath the main part, the phases of disaster are shown, along with bars indicating the approximate length of time of the phase as portrayed in the figure. Time is shown without units since these phases may last from a few hours to a few (or many) years. The lines represent varying trajectories that different sub-populations may experience.

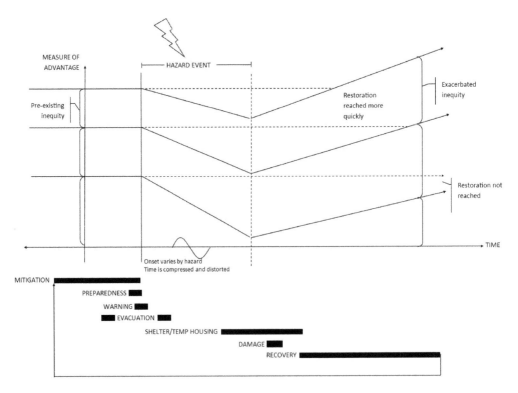

Figure 2.1 Conceptual Model of Recovery Trajectories for Socially Vulnerable Populations (Adapted from Peacock, Van Zandt, Zhang, and Highfield 2014)

In the figure, differences in individual or household characteristics that the social science literature has associated with social advantage or disadvantage (such as income, race/ethnicity, or gender) will matter in terms of how households are able to choose housing that is less exposed to disasters, as well as their ability to make changes to their homes (e.g., elevating it, installing hurricane shutters or safe rooms) that will make them more physically resilient. In the event of disaster, these differences will matter in terms of people's ability to prepare by stockpiling supplies (e.g., food, water) as well as accessing information about the impending disaster and its likely impact (e.g., the use of standard modes of communication by emergency management and government personnel). Socially vulnerable households' limited housing options and constrained ability to act mean that they are more likely to be impacted by disasters, and thus experience greater risk of casualties and property loss. This is shown in the figure by a more acute decline during or after the hazard event. Finally, these differences may manifest themselves in terms of limited resources, both financial (in terms of nonexistent or inadequate insurance and less savings) as well as social (in terms of fewer or poorer connections), to help them overcome obstacles to recovery. Consequently, as the figure shows, recovery trajectories are shallower and longer, and some groups may never achieve restoration (defined in the literature as reaching pre-disaster levels). In the end, the lens of vulnerability has refracted damage and recovery trajectories so that pre-existing gaps in advantage have been widened rather than attenuated.

The phases of disaster are actually more cyclical than linear. Recovery from one disaster is an opportunity—possibly foregone—to mitigate the next one. So, moving forward in time from the graph, these exacerbated inequities will circle back to the mitigation phase, and accumulate over time, continuing to widen differences among sub-populations within the community, undermining resilience. For the increasing number of communities that regularly experience disasters, this can and should be particularly concerning. For planners, it is important to understand how local land use policies, codes and ordinances, and capital investment expenditures can contribute to or detract from community resilience by influencing the exposure of socially vulnerable populations. The opportunities for enhancing positive outcomes for vulnerable populations and thus creating more resilient communities come primarily during the mitigation and recovery phases. Working to increase housing choice and access to opportunities that will maintain and build human capital, along with targeting mitigation and recovery resources to more vulnerable areas of the community, can strengthen weak links within a community's networks to enhance community resilience.

What Is Social Vulnerability?

Policies and practices related to different stages of disasters typically expect that all residents of an area receive, process, and act on information in the same way. Further, they assume that all residents have the same resources and capacity to act in the desired manner, along with equal access to supports and resources for recovery. The term "social vulnerability" has evolved over the past 20 years or so as an alternative explanation for the variation observed in disaster impacts. It refers to differences in a person's or household's capacity to anticipate, cope with, respond to, and recover from disasters (Wisner et al. 1994). Although the literature continues to grow, it has identified the following dimensions of social vulnerability: race/ethnicity (Bolin 1986; Bolin and Bolton 1986; Bolin and Stanford 1998b; Fothergill et al. 1999; Lindell and Perry 2004; Peacock et al. 1997; Perry and Mushkatel 1986), income and poverty (Dash et al. 1997; Fothergill and Peek 2004; Peacock et al. 1997), gender (Enarson and Morrow 1997; Enarson and Morrow 1998; Fothergill 1999), as well as a host of other factors such as age, education, religion, social isolation, housing tenure, etc. (Van Zandt et al. 2012). Frequently, these factors exist in combinations (both poor and Black, for example), which may compound vulnerability (Morrow 1999).

These vulnerability factors shape and influence access to and knowledge of resources (physical, financial, and social), control of these resources, as well as perceived or real power within the larger community or society (Logan and Molotch 1987). They may also influence the capacity of the individual or household to act. For example, although Whites more often rely on media or government to obtain information about threats or hazards, African-Americans more often rely on social connections such as friends or church members (Morrow 1997; Perry and Lindell 1991). Even if one resident has the same information as others, he or she may not have the capacity to react in the desired manner. For example, low-income or elderly residents may have not have cars, preventing them from evacuating in a timely manner or to a location of their choice (Fernandez et al. 2002). Renters are typically more mobile or transient than owners and may not have local family connections to facilitate evacuation or sheltering, whereas owners are more likely to have such resources (Lee et al. 2014). These are but a few examples of how such differences may result in both short-term and long-term disparities in disaster response and recovery.

In the sections that follow, this chapter will walk through the phases of disaster—mitigation, preparedness and response (which are separate phases but will be treated together here), and recovery. Although this chapter focuses on the recovery of socially vulnerable populations, it is not possible to understand the impacts on socially vulnerable populations without discussing all the inequities that lead to increased damage and longer recovery trajectories. The evidence draws heavily from two excellent reviews of the disaster and hazards literature related to race/ethnicity (Fothergill et al. 1999) and poverty (Fothergill and Peek 2004), and builds on them with more recent literature reviewed in Van Zandt et al. (2012); Lee et al. (2014), and Peacock et al. (2014).

Mitigation

Hazard mitigation refers to efforts to reduce or eliminate the risks from natural hazards undertaken prior to an event. Thus, mitigation strives to prevent hazards from becoming disasters or to reduce the severity of their impact. Mitigation strategies are often categorized as structural, including dams, levees, drainage culverts, safe rooms, hardening of structures, etc., or non-structural, which includes land use planning efforts designed to prohibit or limit development in high hazard areas as well as environmental protection efforts designed to shield the natural environment's ability to absorb and deflect hazard impacts. For socially vulnerable populations, mitigation has the potential to keep housing units out of low-lying or otherwise hazard-exposed areas or protect those that are already in such areas. Unfortunately, long-standing development patterns and traditions of discrimination mean that many socially vulnerable households tend to be located in any community's least desirable areas, which often means that they are exposed to hazardous conditions (see for example these excellent reviews of environmental justice research including Bryant and Mohai 1992; Bullard 1993; Pastor et al. 2006).

People and households are not randomly distributed in space. Instead, they are concentrated in fairly predictable spatial patterns based on household characteristics, especially race/ethnicity and income. These patterns aggravate existing economic and social inequalities by creating concentrations of disadvantage contrasted against concentrations of affluence and privilege (Squires and Kubrin 2005). These concentrations of disadvantage mean that although each individual or household may retain household-level disadvantages based on their race, ethnicity, income, education, etc., they also acquire neighborhood-level disadvantage related to access to opportunities and resources. And for disasters, this may often be compounded by proximity and exposure to nearby hazards, both natural and technological (Pastor et al. 2006).

Although both poverty and minority populations have moved from central cities to suburbs over the past 40 years, jurisdiction-level segregation still persists (Pendall 2000; Schneider and Phelan

1993). Remaining pockets of poverty are associated with low-performing schools, social isolation, crime, unemployment, and low levels of educational attainment, whereas privileged areas are associated with high-quality schools, low crime, proximity to services such as health care and banking, and healthier physical and social conditions (for a review, see Squires and Kubrin 2005). Although such public goods may appear at first glance to be unrelated to hazard vulnerability, their availability builds capacity and resilience among community members. Those without access to these amenities, or with poorer access, are less likely to have the necessary resources (social, civic, and economic) available to them in the event of a disaster.

The availability of such public goods in affluent neighborhoods is based—at least in part—on the absence of low-income households (Crane and Manville 2008). Class, as well as racial and ethnic uniformity, appears to facilitate the willingness to pay for neighborhood goods and services (Fischel 2001). Consequently, private actors in the real estate market are motivated to maintain such homogeneity to protect the level of amenities available. Within the real estate industry, different actors have used a variety of techniques to maintain segregation. Real estate agents can use steering, blockbusting, and other forms of differential treatment (Denton 2006; Galster and Godfrey 2005; Yinger 2006). Mortgage lenders have used redlining in their property appraisal techniques (Dane 1993; Jackson 1987; LaCour-Little 1999) as well as discriminatory practices in their underwriting of loans (Apgar and Calder 2005). These practices may include high or even predatory interest rates, denials unrelated to creditworthiness, or differential loan terms (Boehm et al. 2006). In the early twentieth century, developers used restrictive covenants to explicitly exclude owners or occupants of specified racial groups (Dean 1947; LaCour-Little 1999). Although these practices are illegal, and some have fallen out of use, many are still practiced subversively and can be difficult to detect and prosecute.

More recently, attention has turned to the ways that unscrupulous mortgage lending practices have targeted and affected "less desirable areas," first by creating uniformly low- to moderate-income homeowner communities, then by concentrating foreclosure activities there (Immergluck 2009). Racial segregation has both facilitated and exacerbated the 2010 foreclosure crisis (Rugh and Massey 2010). As a result, the foreclosure crisis hit minority communities particularly hard, stripping vulnerable homeowners of financial stability and devastating many low-income minority neighborhoods and communities (Lucy 2010).

Jurisdictions themselves also tend to employ exclusionary practices. Beyond their racist history of discriminatory zoning practices that concentrated racial minorities in undesirable (and frequently low-lying) areas (Hanchett 1997; Silver 1991), public sector actors such as cities and counties continue to act in ways that perpetuate these historical patterns. Cities are similarly motivated to maintain homogeneity to protect neighborhood stability and the local tax base. Responding to political pressure from affluent stakeholders, cities act to maintain neighborhood homogeneity, support affluent and fast-growing areas of communities, and neglect poor or minority areas both within and adjacent to their boundaries. Their actions may come in the form of land use planning, annexation, or capital investment.

Under pressure from the "Not-In-My-Back-Yard" (NIMBY) attitudes of affluent home and property owners, cities use large-lot or low-density zoning and building permit caps to limit the availability of affordable housing options (Pendall 2000). The result is homogeneous developments of single-family detached residences, while land available for multi-family, affordable, or subsidized housing is limited (HUD 2005; Pendall et al. 2006). What affordable housing is available is typically isolated in areas where large proportions of low-income and minority populations are already located. Evidence suggests that these practices exacerbate income and racial segregation (Dawkins 2005; Pendall 2000; Pendall and Carruthers 2003; Talen 2005).

Cities may also react to market forces by disinvesting or refusing to invest in poor neighborhoods. The proliferation of impact fees as a way to fund infrastructure has been seen to shift capital investment to fast-growing areas and away from already-developed areas (Levine 2005).

Poor neighborhoods are more likely to lack access to jobs, transit, health-care facilities, and other public services (Pavel 2008). They are also more likely to be located near undesirable community facilities such as water treatment plants, waste incinerators, or landfills, as well as private sector facilities such as refineries (Bullard 1993). Capital investment programs result in unequal provision of municipal services and infrastructure, such as lighting, sidewalks, as well as water and sewer provisions (Arnold 2007; Blackwell and Fox 2006). Such inequitable provisions have far-reaching consequences for the low-income and minority individuals who live in these underserved neighborhoods (Squires and Kubrin 2005). Without adequate infrastructure and services, residents' health risks increase and property values decrease, perpetuating health and wealth inequalities (Marsh et al. 2010).

Annexation practices also aggravate segregation. By selectively excluding minority communities (also known as municipal underbounding), local governments undermine the political standing of minorities and their ability to advocate for their needs (Johnson et al. 2004). Even more than disinvestment in poor communities within city limits, underbounded minority communities are denied access to the same levels of infrastructure (which may aggravate area flooding and damage from other disasters) as well as access to services including police and fire protection, trash/debris collection, and code enforcement. Although local governments may claim that they are trying to reduce the tax burden on minority or poor communities, evidence shows that the individual provision of water and sewer through septic systems is a greater financial burden than the tax increase would be (Marsh et al. 2010).

Such actions result in cities and communities with settlement patterns that have the potential to increase rather than mitigate exposure and vulnerability in several important ways. First, concentration of households by poverty and race, as well as by status as renters, means that individual- and household-level vulnerabilities are compounded by neighborhood-level disadvantage. Neighborhood-level disadvantage reduces social, economic, and civic resources resulting in lower levels of human capital, capacity, information, engagement, and power, ultimately undermining community resilience and the ability of households to resist, absorb, and recover from disasters. Second, these spatial patterns are likely associated with lower levels of housing quality as well as neighborhood infrastructure. Low-quality structures and poorly maintained infrastructure increase physical vulnerability so that when a neighborhood is hit by a disaster, the damage is likely to be greater, both to individual homes as well as to the neighborhood as a whole. Finally, the locational outcomes for poor, minority, and renting communities are associated with greater levels exposure to both natural and technological hazards (Bullard 1993; Pastor et al. 2006).

Preparedness and Response

Preparedness and response activities take place before and during a disaster. They include preparations for impact, receiving warnings about actions needed to prepare for impact, and decisions to take protective actions such as evacuation. For the most part, these stages of disaster happen outside of urban planning decision-making. These are more typically handled by emergency management personnel and other civic authorities (county judges and perhaps state government officials, depending on the nature of the disaster). However, for planners, it is important to attend to how decisions made prior to a disaster (during the mitigation stage) can influence the ability of residents to act in the desired manner during this phase. Disasters both magnify and accelerate processes already occurring in communities (Bates and Peacock 1987; Kates et al. 2006; Morrow and Peacock 1997; Olshansky et al. 2012; Prince 1920). Consequently, these pre-existing conditions are the most important predictor of impacts. In other words, by facilitating the clustering of disadvantaged populations, land use planning and capital investment approaches may further limit the ability of vulnerable populations to prepare for and react to disasters.

Preparedness

Preparedness refers to "preimpact activities that establish a state of readiness to respond to extreme events that could affect the community" (Lindell et al. 2006, p. 244). At the household level, such actions include stocking up on supplies of food, water, and other daily essentials; checking battery supplies and securing a weather radio; or securing the home and contents by covering windows to protect against wind and projectiles. Although the literature generally suggests that minorities and lower-income populations have higher perceptions of risk for natural and technological hazards (Flynn, et al. 1994; Lindell and Prater 2000; Peacock et al. 2005; Turner et al. 1986; Vaughan and Nordenstam 1991; Vaughan and Seifert 1992), the empirical evidence finds that minorities, low-income households, and renters more often display lower levels of preparedness (e.g., Eisenman et al. 2009).

These findings appear across multiple types of disasters, but are not wholly consistent. Although some researchers found no racial/ethnic variations with respect to flood preparation (Ives and Furseth 1983) or with hurricane preparedness in Miami (Gladwin and Peacock 1997), more often researchers find differences in the levels of preparedness of socially vulnerable populations. Several research-ers have found preparation for earthquakes less common among minorities than Whites (Edwards 1993; Farley 1998; Mileti and Darlington 1997; Turner et al. 1986). Similar findings are reported for African-American households preparing for hurricanes (Norris et al. 1999). Morrow and Enarson (1996) noted that prior to Hurricane Andrew poor women in public housing heard warnings and wanted to prepare, but simply lacked the economic resources for supplies. Even among homeowners in Florida, both low-income and Black households were less likely to have code-compliant hurri-cane shutters to protect their homes from hurricanes.

Housing tenure itself may also matter. Turner and his colleagues (1986) found that even con-trolling for other differences, owners were more likely than renters to have emergency supplies on hand in an earthquake, whereas Burby et al. (2003) found similar results in a case of joint natural-technological disasters. They suggested that renters are less connected to their community and thus less likely to receive warnings or have the resources to prepare.

Warning

Disaster warnings are meant to initiate action to protect life and property. For disaster warning pro-cesses to work, the recipient must receive, heed, and believe the warning. Source credibility and confirmation are critical, and should lead to protective action being taken (Lindell 2017; Lindell and Perry 2004). Although findings are not always consistent, the general pattern suggests that race/eth-nicity, income, and other social vulnerability factors are relevant. Researchers have found that among Hispanics (and Mexican-Americans in particular) and African-Americans, social networks and relatives are more important than more formal channels such as government officials and the media for relaying warning and disaster information (Blanchard-Boehm 1998; Morrow 1997; Perry and Mushkatel 1986; Perry and Nelson 1991; Phillips and Ephraim 1992). Interestingly, Perry and Lindell (1991) found that Whites were somewhat less likely to require message confirmation, which is consistent with conclu-sions by Perry and Mushkatel (1986) that Whites more strongly believe warnings than do either Blacks or Hispanics. Consequently, minorities may experience potential delays in receiving and confirming warning messages since they display greater dependence on informal social and familial networks. Such delays have the potential to delay evacuation and lead to increased injuries or even death.

Evacuation

Research on evacuation is somewhat inconsistent, but generally suggests that minorities, lower-income groups, and the aged are less likely to evacuate when ordered. Early research found that minorities and lower-income populations fail to obey warnings (Moore 1958; Sims and Bauman 1972). Examining flooding response, Perry, Lindell, and Greene (1982) found that Hispanics were less likely to evacuate and Drabek and Boggs (1968) found that Hispanic households were more dependent on extended family ties to facilitate evacuation. On the other hand, Perry and Lindell (1991), examining flooding and hazards material spills, report limited to non-significant ethnic variations in evacuation. A recent meta-analysis of 38 hurricane evacuation decision studies (Huang et al. 2016) indicates that homeownership had significant correlation with evacuation but other sociodemographic variables had nonsignificant correlations.

Individual studies, however, shed light on the conditions under which minorities and low-income populations may delay evacuation. Gladwin and Peacock (1997) found that low-income and Black households were less likely to evacuate prior to Hurricane Andrew. These findings are consistent with the failures of many poorer and minority households to evacuate New Orleans in response to Katrina (Pastor et al. 2006). Gladwin and Peacock (1997) speculate that this is due in part to a lack of resources (particularly private vehicles), ineffective public transportation options, and few refuge options outside evacuation zones. Morrow and Enarson (1996) and Morrow (1997) found that, prior to Hurricane Andrew, poor women and others in public housing lacked transportation, forcing many to walk or hitchhike in order to evacuate. Similarly, Enarson (1999) found that the homeless, unemployed and lower-income women were less able to evacuate in response to Red River Valley flood warnings. Others have found that the elderly are also less able to evacuate due to transportation dependence (Fernandez et al. 2002). Lindell and Perry (2004: 90) suggest that income and education might have consequences for evacuation in response to warning "due to restricted material resources, knowledge, and skill."

Findings regarding preparedness, warning, and evacuation suggest that a wide variety of vulnerability characteristics may affect an individual or household's receptiveness to advice from local authorities and media; however, these findings are not terribly consistent from disaster to disaster. Given that preparedness, warning, and evacuation are essentially individual- or household-level decisions, it is perhaps not surprising that these results are less consistent. Although neighborhood- or community-level influences may impact how warnings are received (through neighbors or social contacts, for example) and the ability of households or individuals to evacuate (transportation-dependent populations are often clustered), these may be less important for preparedness actions than they were for mitigation.

Recovery

The recovery phase of disaster considers factors that affect an individual's or household's ability to "bounce back" from the impact of the disaster. For our purposes, we will consider damage and casualties to be a part of the recovery stage, although these constitute major parts of the actual impact. The extent of this impact is perhaps the most important determinant of recovery.

Casualties

Research examining variations in casualties finds that minorities and low-income populations are much more likely to be disproportionately impacted by disasters. In one of the earliest studies examining disaster casualties, Bates and his colleagues (1962) found significantly higher death rates among African-Americans (322 per 1000) compared to Whites (38 per 1000) in Southern Louisiana and Texas due to Hurricane Audrey. Rossi et al. (1983) examined injuries due to various disasters from 1970 through 1980, and found that lower-income areas experienced significantly higher injuries, particularly in connection with floods and earthquakes. Bolin and Bolton (1986) reported that following the Paris Texas tornado, African-American respondents were significantly more likely to report friends being injured (19.6 percent to 9.9 percent) and killed (31.1 percent vs. 17.5 percent) when compared to Whites. Aguirre (1988) similarly found that the poor had higher injury and deaths following the Saragosa Texas tornado in 1987. Fothergill and Peek (2004) found that nearly 40 percent of all tornado fatalities occur among mobile home residents, who are more likely to be low-income. More recently, Zahran and his colleagues (2008) found that Texas counties with higher concentrations of socially vulnerable populations had higher flood casualty rates from 1997 to 2001. It is clear from these findings that the interactions among race/ethnicity, income, housing type, and housing location place these households at greater risk of bodily injury and death.

Damage

One of the most robust findings in the disaster literature is that low-income and minority house-holds tend to suffer disproportionately higher levels of damage (Bates et al. 1962; Bates 1982; Bates and Peacock 1987; Bolin 1982; Bolin 1986; Bolin 1993; Bolin and Bolton 1986; Dash et al. 1997; Drabek and Key 1984; Haas et al. 1977; Highfield et al. 2014; Peacock and Girard 1997; Quarantelli 1982; Wisner et al. 1994).

As discussed in the section on mitigation earlier in this chapter, low-income and minority households experience both private-sector and public-sector obstacles to locating in less risky and more desirable neighborhoods. Furthermore, filtering, a process by which lower-income house-holds successively occupy homes and neighborhoods as they deteriorate physically (Grigsby 1963; Myers 1975), allocates the poor and minorities to older and poorer quality homes (Bolin 1986; Bolin and Bolton 1986; Charles 2003; Foley 1980; Logan and Molotch 1987; Massey and Denton 1993; Peacock et al. 2006; Peacock and Girard 1997; Phillips 1993; Phillips and Ephraim 1992; Van Zandt 2007). These homes were typically built to older, less stringent building codes, used lower-quality designs and construction materials, and were less well maintained (Bolin 1994; Bolin and Bolton 1983; Bolin and Stanford 1998a; Girard and Peacock 1997). Recent research, however, suggests that much older homes (historic homes) may have been built to better standards and with a better understanding of siting on higher, safer ground (Highfield et al. 2014). However, these homes are not as subject to the same kind of filtering and thus are less likely to be occupied by lower-income households.

In a recent study of Hurricane Ike specifically designed to tease out the independent effects of social vulnerability on damage relative to physical vulnerability and hazard exposure, Highfield and his colleagues (2014) found that even after controlling for a host of geographic and structural characteristics including age of home and proximity to water, homes located in areas with higher proportions of both Hispanic and Blacks were found to have experienced more damage. Their findings also suggest that lower-value homes suffered disproportionately. The compounding effects of social vulnerability assessed in terms of race/ethnicity and socioeconomic status were clearly evident—and potentially even underestimated—given that the analysis excluded multi-family units, which are more likely to house minority and low-income households.

Recovery

The robustness of the finding that low-income and minority households tend to suffer dispropor-tionately higher levels of damage affirms the suggestion that processes leading to inequitable development patterns during the mitigation phase result in similarly inequitable impacts from disasters. Looking back to Figure 2.1, the reader will recognize the more acute drop in advantage that takes place in response to the hazard event for the least advantaged populations. The recovery trajectory, then, starts off from a worse position for socially vulnerable populations, giving them an even wider gap than more advantaged populations to overcome to reach restoration (their starting position).

This immediate post-disaster deficit is compounded by the fact that housing recovery policies in the US have structurally favored middle-class homeowners. Renters and low-income homeowners' recovery needs have been insufficiently addressed and are often overlooked completely (Fothergill and Peek 2004; Kamel and Loukaitou-Sideris, 2004; Mueller et al. 2011; Prince 1920). As noted earlier, disasters simply magnify and accelerate processes already taking place in communities (Bates and Peacock 1987; Kates et al. 2006; Morrow and Peacock 1997; Olshansky et al. 2012). Therefore, many of the same mechanisms that were at work during the mitigation phase will still be relevant during recovery. In other words, households' attempts at recovery may be thwarted by many of the same processes that result in deficits in capacity, information, power, and resources.

Socially vulnerable populations often lack insurance and access to financial resources that can aid in recovery. Minorities, and particularly African-Americans, continue to face racial discrimination in the buying, selling, and renting of housing due to racial steering, redlining, White attitudes, and lender discrimination (Feagin and Sikes 1994; Guy et al. 1982; Horton 1992; Oliver and Shapiro 1995; Sagalyn 1983). African-American households are more likely to be denied a mortgage, must make larger down payments, and when accepted, often pay higher interest rates; after purchasing, their homes are likely to be located in lower-quality neighborhoods where homes appreciate at lower rates (Flippen 2004; Oliver and Shapiro 1995; Van Zandt 2007). Minorities often face similar problems obtaining homeowner's insurance, which makes procuring a home mortgage impossible (Squires 1998; Squires and Velez 1987; Squires et al. 2001). These inequalities mean that minorities and low-income populations are less likely to meet the financial burden of disaster recovery (Bolin and Stanford 1998a, 1998b; Peacock and Girard 1997). Quite simply, they will face a greater struggle identifying and qualifying for assistance, and are more likely to pay more for it.

Early research showed that low-income and minority homeowners were much less likely to qualify for government-backed Small Business Administration (SBA) loans (Bolin 1982; Bolin 1986; Bolin 1993; Bolin and Bolton 1986; Bolin and Stanford 1998a, 1998b; Drabek and Key 1984; Quarantelli 1982), although more recent research suggests that ethnic/racial variations may no longer be significant (Galindo 2007). Early research also tended to find that low-income and minority households were less likely to have homeowner's insurance (Cochrane 1975; Drabek and Key 1984; Moore et al. 1963, Moore and Bates 1964), yet later research found more parity in holding insurance policies, but also that poor and minority households were more likely to report settlements failing to meet repair and reconstruction costs (Bolin 1982; Bolin and Bolton 1986). Peacock and Girard (1997) found a similar pattern in Miami-Dade County following Hurricane Andrew where both Black and Hispanic households were more likely to report insufficient insurance settlements for repairs and reconstruction. Further analysis however, suggested that large national insurance companies that were more likely to provide adequate settlements had systematically failed to underwrite insurance in minority, and particularly African-American, neighborhoods (Peacock and Girard 1997).

In addition to being inadequately insured and/or unable to access that insurance, research further suggests that poorer households and neighborhoods often fall far short in receipt of public assistance to jump-start the recovery process (Berke et al. 1993; Bolin and Stanford 1991; Dash et al. 1997; Phillips 1993; Rubin 1985). The literature suggests that minorities, low-income households, and even female-headed households can be at a disadvantage in part because of low language skills and education when it comes to qualifying for and negotiating the process of obtaining public financial resources such as SBA loans or minimum housing assistance (Bolin 1985; Bolin and Stanford 1990; Morrow 1997; Morrow and Enarson 1997; Phillips 1993).

The literature also suggests that rental housing is slower to recover, which makes it more difficult for minority and low-income households to find post-disaster housing and return to their pre-disaster communities, often extending the recovery process (Bolin 1986, 1993; Bolin and Stanford 1998a, 1998b; Comerio 1998; Comerio et al. 1994; Morrow and Peacock 1997; Quarantelli 1982). Past research has paid fairly little attention to rental housing, but after Hurricane Katrina, several researchers addressed the recovery of this housing sector, given the large proportion of rental housing in New Orleans and its impact on the African-American population. Laska and Morrow (2006) found that rental housing was disproportionately impacted by Hurricane Katrina, and that renters were slower to recover. Quigley (2007) found that rental prices shot up after the hurricane in response to dramatically increased demand, which put even a further crunch on lower-income households.

Although these studies have examined short-term housing shortages and impacts on recovery, rarely have longitudinal data been collected to allow an assessment of housing recovery to determine long-term impacts on socially vulnerable populations. Yet recent work by Peacock and his collaborators (2016) compared two such longitudinal studies—one of recovery after 1992's Hurricane Andrew in Miami-Dade County, Florida, and the other after 2008's Hurricane Ike in Galveston, Texas. Using parcel-level data on both single-family and multi-family housing types, the researchers tracked housing value recovery for four years following the hurricane, comparing it to values in the year of the storm (prior to impact). The data clearly indicate that housing recovery is a highly uneven process. As expected, damage has major consequences for recovery; even after four years, the consequences of damage were evident in the rebuilding process. Whether restoration levels are ever reached is highly dependent upon the initial levels of damage. They found that factors related to social vulnerability, particularly race/ethnicity and income, may affect recovery in different ways in different settings. In Miami, they found that both income and race/ethnicity were critical factors that were associated with higher losses and much slower recovery rates. Yet in Galveston, income was the more critical factor in determining which neighborhoods suffered more damage and lagged significantly in the recovery process. Furthermore, in both settings rental housing in general, and multi-family and duplexes more specifically, suffered higher levels of damage and have been much slower to reach restoration values.

These long-term trajectories provide evidence that inequalities at each stage of disaster may accumulate to exacerbate inequities that can undermine resilience. Particularly in Miami-Dade, where Zhang and Peacock (2009) had an opportunity to observe changes over two decades, they saw that the differences in these damage and recovery trajectories in terms of tenure, housing type, and neighborhood socioeconomic factors led to very different housing opportunities within a community. They found that high levels of damage were correlated with both sales and abandonments, but sales were more often associated with high damage and higher proportions of non-Hispanic Whites whereas abandonments were associated with areas of high damage and higher levels of African-Americans within the neighborhood. These dynamics may be particularly concerning for planners because concentrations of abandonments, more common in minority neighborhoods with low-value homes, are associated with declines in housing values, neighborhood destabilization, and redevelopment (Accordino and Johnson 2000; Greenberg et al. 1990). The turnover that is likely to ensue is also likely to lead to population shifts, a loss of affordable housing, and undesirable changes in land use.

Differences in recovery for rental housing, and housing that generally is associated with rental housing, may contribute to population turnover. Because disasters are grounds for terminating leases in many states, rarely do original renters return to their units after a disaster (Hori and Schafer 2010; Levine et al. 2007). Even when attention is paid to the recovery of rental housing, its return does not guarantee the return of original renters (Quigley 2007). Renters are perhaps most at risk of permanent displacement. They are difficult to track after a disaster, and almost nothing is known about their long-term outcomes.

Conclusion: Social Vulnerability Is the Essence of Resilience

When disasters strike, pundits often wax poetic about how disasters are equal opportunity events, how they affect all: rich and poor, Black and White, even as they wonder at sights like post-Katrina New Orleans. It is true that hazards themselves are fairly indiscriminate. Yet the ways in which they interact with social and ecological systems, as well as the built environment, are anything but indiscriminate. Our communities are built to place the most vulnerable populations in the most vulnerable locations, compounding their individual- and household-level vulnerabilities. Given that a system put under stress will break at its weakest point, the ability of the most vulnerable among us to absorb, deflect, resist, and recover from impact is the essence of resilience.

The research reviewed in this chapter leads us to draw several important conclusions that together make an important statement about how to create more resilient communities. The research findings tell us that pre-existing conditions (not just individual or household characteristics, but also neighborhood- or community-level characteristics) are important predictors of damage, and also that damage is the most important predictor of recovery. Such inequitable recovery may then be expected to exacerbate pre-existing differences going into the next disaster. This cycle compounds and exaggerates negative impacts on socially vulnerable populations and erodes community resilience. It suggests that focusing efforts on a community's most vulnerable populations—reducing spatial inequities through better land use planning, increasing housing choice, fighting discrimination in the lending, real estate, and insurance industries, strengthening safety nets and community social services, and targeting capital investments and recovery resources—will have payoffs not just for the vulnerable populations themselves, but will strengthen communities and enhance overall community resilience (and will do so not just for natural hazards, but also for economic shocks and other threats).

The fundamental conclusion from this review is that to become more resilient, communities must make equity a priority in all planning efforts. While this may seem obvious, it is rarely explicitly done. There are several approaches communities may take to make equity an explicit value.

Assess Equity Impacts of Any and All Proposed Development, Plans, and Capital Investment Decisions

Cities have a powerful role in guiding private development. Yet they often abdicate their responsibility to hold developers accountable for adherence to equity goals and fail to recognize their own role in promoting or undermining such goals. Private-sector impediments are difficult to overcome, yet cities can build in incentives, remedial actions, and rewards for the compliance of private developers and landlords with equity goals. They can also use zoning and other land use planning tools to guide appropriate development in appropriate places. They can ensure that investment in infrastructure and capital improvements is equitable and overcomes historic injustices.

Equity assessments are not a common part of plan evaluation activities, nor are Capital Investment Programs routinely assessed for their impacts on vulnerable populations and neighborhoods. Even in housing plans, equity assessments are expected, but not routinely or appropriately done. Yet some popular federal grant programs, such as the Sustainable Communities grant program co-sponsored by the U.S. Department of Housing & Urban Development, the Environmental Protection Agency, and the Department of Transportation, are now requiring "fair housing and equity assessments" as a part of the reporting associated with grant outcomes. Although tools for such assessments are not widely available, techniques such as "opportunity mapping," developed by Ohio State's Kerwin Institute for Race and Ethnicity, offer potentially useful approaches when done in conjunction with robust public engagement processes that engage vulnerable groups in defining the local context for understanding equity. Yet more standardized tools are needed to allow all planning documents to be assessed with little training or expertise needed.

Involve Minorities and Low-Income Residents in the Planning Process

One powerful way to increase awareness of and demand for more equitable development practices is to actually ask vulnerable groups what they need and want. Racial/ethnic groups are often excluded from community post-disaster planning and recovery activities because they have less economic power and political representation (Bolin and Bolton 1983; Morrow 1997; Morrow and Peacock 1997; Phillips 1993; Prater and Lindell 2000; Quarantelli 1982; Tierney 1989). This is also certainly true in comprehensive or land use planning. Plan makers and governmental officials respond to the loudest and most insistent voices in a community (Forester 1999). A community striving to become more resilient would begin by inviting disadvantaged populations into a process that allows their voices to be heard.

Incorporate Hazard Mitigation and Disaster Recovery into Comprehensive Planning

Along with equity, everyday planning efforts such as comprehensive plans should explicitly incorporate hazard assessments and disaster plans. Proper land use planning can help limit development in low-lying and high-hazard areas and can guide the adoption of building codes that strengthen structures in those areas as they are rebuilt. Importantly, comprehensive planning also provides a vision for change in the community. If equity goals are strong in the plan, they may well provide the possibility of using the recovery and rebuilding period to lessen inequities for socially vulnerable populations rather than exacerbating them (Berke et al. 2010).

Direct Recovery Assistance to Those Who Need It Most

When a disaster occurs, it opens a window of opportunity and infuses the community with disaster recovery funds. Communities that have failed to plan ahead—that do not have a comprehensive plan or a recovery plan—will find themselves unprepared to make the best use of this infusion of funds. In the absence of such a plan, the likelihood is that recovery funds will go towards projects that further exacerbate rather than mitigate community inequities. The presence of a comprehensive plan can help the community to guide recovery in a way that is consistent with the vision laid out in the plan. A strong plan can provide protection against the assertion of special interests (usually those backed by power and money). It is incumbent on local planning offices to assess and monitor differentials in disaster impact and recovery trajectories associated with housing type, tenure, income, and race/ethnicity factors to identify and help target resources to areas that were hardest hit and are lagging.

Attention to disparities in the recovery phase serves as mitigation for the next disaster. The cumulative nature of impacts on socially vulnerable populations suggests that impacts during recovery cannot be separated from impacts at other phases of disaster. At every stage, opportunities are available to diminish inequities to build more resilient communities.

References

Accordino, J. and G. T. Johnson. (2000) "Addressing the Vacant and Abandoned Property Problem," *Journal of Urban Affairs* 22(3): 301–315.

Aguirre, B. E. (1988) "The Lack of Warnings Before the Saragosa Tornado," *International Journal of Mass Emergencies and Disasters* 6(1): 65–74.

Apgar, W. and A. Calder. (2005) "The Dual Mortgage Market: The Persistence of Discrimination in Mortgage Lending," 101–123, in X. de Souza Briggs (ed.), *The Geography of Opportunity: Race and Housing Choice in Metropolitan America*, Washington DC: Brookings Institution Press.

Arnold, C. A. (2007) *Fair and Healthy Land Use: Environmental Justice and Planning*, Planning Advisory Service Report 549/550, Washington DC: American Planning Association.

Bates, F. L (ed.). (1982) *Recovery, Change and Development: A Longitudinal Study of the Guatemalan Earthquake*, Athens GA: Department of Sociology.

Bates, F. L., C. W. Fogleman, V. J. Parenton, R. H. Pittman, and G. S. Travy. (1962) *The Social and Psychological Consequences of a Natural Disaster: A Longitudinal Study of Hurricane Audrey*, Washington DC: National Academy of Sciences – National Research Council, Publication 1081.

Bates, F. L. and W. G. Peacock. (1987) "Disasters and Social Change," 291–330, in R. R. Dynes, B. De Marchi, and C. Pelanda (eds.), *The Sociology of Disasters*, Milan, Italy: Franco Angeli Press.

Berke, P. R., J. Cooper, D. Salvesen, D. Spurlock, and C. Rausch. (2010) "Disaster Plans: Challenges and Choices to Build the Resiliency of Vulnerable Populations," *International Journal of Mass Emergencies and Disasters* 28(3): 368–394.

Berke, P. R., J. Kartez, and D. Wenger. (1993) "Recovery After Disaster: Achieving Sustainable Development, Mitigation and Equity," *Disasters* 17(2): 93–109.

Blackwell, A. G. and R. K. Fox. (2006) "Regional Equity and Smart Growth: Opportunities for Advancing Social and Economic Justice in America," 407–427, in D. C. Soule (ed.), *Urban Sprawl: A Comprehensive Reference Guide*, Westport CT: Greenwood Press.

Blanchard-Boehm, R. D. (1998) "Understanding Public Response to Increased Risk from Natural Hazards: Application of the Hazards Risk Communication Framework," *International Journal of Mass Emergencies and Disasters* 16(3): 247–278.

Boehm, T. P., P. D. Thistle, and A. Schlottmann. (2006) "Rates and Race: An Analysis of Racial Disparities in Mortgage Rates," *Housing Policy Debate* 17(1): 109–149.

Bolin, R. (1982) *Long-Term Family Recovery from Disaster*, Boulder CO: University of Colorado, Institute of Behavioral Science, Program on Environment and Behavior, Monograph #36.

Bolin, R. (1985) "Disasters and Long-Term Recovery Policy: A Focus on Housing and Families," *Policy Studies Review* 4(4): 709–715.

Bolin, R. (1986) "Disaster Impact and Recovery: A Comparison of Black and White Victims," *International Journal of Mass Emergencies and Disasters* 4(1): 35–50.

Bolin, R. (1993) "Post-Earthquake Shelter and Housing: Research Findings and Policy Implications," 107–131, in K. J. Tierney and J. M. Nigg (eds.), Monograph 5: *Socioeconomic Impacts*, prepared for the 1993 National Earthquake Conference, Memphis TN: Central U.S. Earthquake Consortium.

Bolin, R. (1994) *Household and Community Recovery after Earthquakes*, Boulder CO: University of Colorado, Institute of Behavioral Science, Program on Environment and Behavior.

Bolin, R. and P. Bolton. (1983) "Recovery in Nicaragua and the U.S.A," *The International Journal of Mass Emergencies and Disasters* 1(1): 125–144.

Bolin, R. and P. Bolton. (1986) *Race, Religion, and Ethnicity in Disaster Recovery*, Boulder CO: University of Colorado, Institute of Behavioral Science, Program on Environment and Behavior, Monograph # 42.

Bolin, R. and L. Stanford. (1990) "Shelter and Housing Issues in Santa Cruz County," 99–108, in R. Bolin (ed.), *The Loma Prieta Earthquake: Studies of Short-term Impacts*, Boulder, CO: University of Colorado, Institute of Behavioral Science, Program on Environment and Behavior, Monograph #50.

Bolin, R. and L. Stanford. (1991) "Shelter, Housing and Recovery: A Comparison of U.S. Disasters," *Disasters* 15(1): 24–34.

Bolin, R. and L. Stanford. (1998a) *The Northridge Earthquake: Vulnerability and Disaster*, New York: Routledge.

Bolin, R., and L. Stanford. (1998b). "The Northridge Earthquake: Community-Based Approaches to Unmet Recovery Needs," *Disasters* 22(1): 21–38.

Bryant, P. and B. Mohai (1992) "Race, Poverty, and the Environment: The Disadvantaged Face Greater Risks," *EPA Journal* 18(1), 2–5.

Bullard, R. D. (1993) *Confronting Environmental Racism: Voices from the Grassroots*, Boston: South End Press.

Burby, R. J. (2005) "Have State Comprehensive Planning Mandates Reduced Insured Losses from Natural Disasters?" *Natural Hazards Review* 6(2): 67–81.

Burby, R. J., L. J. Steinberg, and V. Basolo. (2003) "The Tenure Trap the Vulnerability of Renters to Joint Natural and Technological Disasters," *Urban Affairs Review* 39(1): 32–58.

Charles, C. Z. (2003) "The Dynamics of Racial Residential Segregation," *Annual Review of Sociology* 29: 167–207.

Cochrane, H. C. (1975). *Natural Hazards and their Distributive Effects*, Boulder CO: University of Colorado, Institute of Behavioral Science, Program on Environment and Behavior.

Comerio, M. C. (1998) *Disaster Hits Home: New Policy for Urban Housing Recovery*, Berkeley CA: University of California Press.

Comerio, M. C., J. D. Landis, and Y. Rofe. (1994) *Post-Disaster Residential Rebuilding*, Working paper 608, Berkeley CA: University of California, Institute of Urban and Regional Development.

Crane, R. and M. Manville. (2008) "People or place? Revisiting the Who Versus the Where of Urban Development," *Land Lines* 20(3): 2–7.

Cutter, S. L. (1996) "Vulnerability to Environmental Hazards," *Progress in Human Geography*, 20(4): 529–539.

Cutter, S. L. and C. Finch. (2008) "Temporal and Spatial Changes in Social Vulnerability to Natural Hazards," *Proceedings of the National Academy of Sciences* 105(7): 2301–2306.

Dane, S. (1993) "A History of Mortgage Lending Discrimination in the United States," *Journal of Intergroup Relations* 20(1): 6–28.

Dash, N., W. G. Peacock, and B. Morrow. (1997) "And the Poor Get Poorer: A Neglected Black Community," 206–225, in W. G. Peacock, B. H. Morrow, and H. Gladwin (eds.), *Hurricane Andrew: Ethnicity, Gender and the Sociology of Disaster*, London: Routledge.

Dawkins, C. J. (2005) "Tiebout Choice and Residential Segregation by Race in U.S. Metropolitan Areas, 1980–2000," *Regional Science and Urban Economics* 35(6): 734–55.

Dean, J. (1947) "Only CaucAsian: A Study of Race Covenants," *Journal of Land and Public Utility Economics* 23(4): 428–432.

Denton, N. A. (2006) "Segregation and Discrimination in Housing," 61–81, in R. Bratt, M. Stone, and C. Hartman (eds.), *A Right to Housing: Foundation for a New Social Agenda*, Philadelphia PA: Temple University Press.

Drabek, T. E. and K. Boggs. (1968) "Families in Disaster: Reactions and Relatives," *Journal of Marriage and the Family* 30(3): 443–451.

Drabek, T. E. and W. H. Key. (1984) *Conquering Disaster: Family Recovery and Long-term Consequences*, New York: Irvington Publishers.

Edwards, M. L. (1993) "Social Location and Self Protective Behavior," *International Journal of Mass Emergencies and Disasters* 11(3): 293–304.

Eisenman, D. P., D. Glik, R. Maranon, L. Gonzales, & S. Asch. (2009). "Developing a Disaster Preparedness Campaign Targeting Low-Income Latino Immigrants: Focus Group Results for Project PREP," *Journal of Health Care for the Poor and Underserved* 20(2): 330–345.

Enarson, E. (1999) "Violence Against Women in Disasters: A Study of Domestic Violence Programs in the US and Canada," *Violence Against Women* 5(7): 742–768.

Enarson, E. and B. H. Morrow. (1997) "A Gendered Perspective: The Voices of Women," 116–140, in W. G. Peacock, B. H. Morrow, and H. Gladwin (eds.), *Hurricane Andrew: Ethnicity, Gender and the Sociology of Disaster*, London: Routledge.

Enarson, E. and B. H. Morrow. (1998) *The Gendered Terrain of Disaster*, Westport CT: Praeger.

Farley, R. (1998) "Comments on Racial and Ethnic Discrimination in American Economic Life," Presentation at the Urban Institute, Washington, DC.

Feagin, J. R. and M. P. Sikes. (1994). *Living with Racism: The Black Middle Class Experience*, Boston MA: Beacon.

Fernandez, L. S., D. Byard, C. C. Lin, S. Benson, and J. A. Barbera. (2002) "Frail Elderly as Disaster Victims: Emergency Management Strategies," *Prehospital and Disaster Medicine* 17(2): 67–74.

Fischel, W. A. (2001) "Why are there NIMBYs?" *Land Economics* 77(1): 144–152.

Flanagan, B. E., E. W. Gregory, E. J. Hallisey, J. L. Heitgerd, and B. Lewis. (2011) "A Social Vulnerability Index for Disaster Management," *Journal of Homeland Security and Emergency Management* 8(1): 1–22.

Flippen, C. (2004) "Unequal Returns to Housing Investments? A Study of Real Housing Appreciation among Black, White, and Hispanic Households," *Social Forces* 82(4): 1523–1551.

Flynn, J., P. Slovic, and C. K. Mertz. (1994) "Gender, Race, and Perception of Environmental Health Risks," *Risk Analysis* 14(6): 1101–1108.

Foley, D. L. (1980) "The Sociology of Housing," *Annual Review of Sociology* 6: 457–78.

Forester, J. (1999) *The Deliberative Practitioner: Encouraging Participatory Planning Processes*, Cambridge MA: MIT Press.

Fothergill, A. (1999) "An Exploratory Study of Woman Battering in the Grand Forks Flood Disaster: Implications for Community Responses and Policies," *International Journal of Mass Emergencies and Disasters* 17(1): 79–98.

Fothergill, A., E. G. Maestas, and J. D. Darlington. (1999) "Race, Ethnicity and Disasters in the United States: A Review of the Literature," *Disasters* 23(2): 156–173.

Fothergill, A. and L. A. Peek. (2004) "Poverty and Disasters in the United States: A Review of Recent Sociological Findings," *Natural Hazards* 32(1): 89–110.

French, S. P., E. Feser, and W. G. Peacock. (2008) "Quantitative Models of the Social and Economic Consequences of Earthquakes and Other Natural Hazards," *Final Report, Mid-America Earthquake Center Project SE-2*, Atlanta GA: Georgia Institute of Technology.

Galindo, K. B. (2007) "Variation in Disaster Aid Acquisition among Ethnic Groups in a Rural Community," PhD Dissertation, College Station TX: Texas A&M University.

Galster, G. C. and E. Godfrey. (2005) "By Words and Deeds: Racial Steering by Real Estate Agents in the U.S. in 2000," *Journal of the American Planning Association* 71(3): 251–268.

Girard, C. and W. G. Peacock. (1997) "Ethnicity and Segregation: Post Hurricane Relocation," 191–205, in W. G. Peacock, B. H. Morrow, and H. Gladwin (eds.), *Hurricane Andrew: Ethnicity, Gender and the Sociology of Disaster*, London: Routledge.

Gladwin, H. and W. G. Peacock. (1997) "Warning and Evacuation: A Night for Hard Houses," 52–74, in W. G. Peacock, B. H. Morrow, and H. Gladwin (eds.), *Hurricane Andrew: Ethnicity, Gender, and the Sociology of Disasters*, London: Routledge.

Greenberg, M. R., F. J. Popper, and B. M. West. (1990) "The TOADS A New American Urban Epidemic," *Urban Affairs Review* 25(3): 435–454.

Grigsby, W. (1963) *Housing Markets and Public Policy*, Philadelphia PA: University of Pennsylvania Press.

Guy, R. F., L. G. Pol, and R. Ryker. (1982) "Discrimination in Mortgage Lending: The Mortgage Disclosure Act," *Population Research and Policy Review* 1(3): 283–296.

Haas, J. E., R. W. Kates, and M. J. Bowden. (1977) *Reconstruction Following Disaster*, Cambridge MA: The MIT Press.

Hanchett, T. (1997) *Sorting Out the New South City*, Chapel Hill NC: University of North Carolina Press.

Highfield, W., W. G. Peacock, and S. Van Zandt. (2014) "Mitigation Planning: Why Hazard Exposure, Structural Vulnerability, AND Social Vulnerability Matter," *Journal of Planning Education & Research* 34(3): 287–300.

Hori, M. and M. J. Schafer. (2010) "Social Costs of Displacement in Louisiana after Hurricanes Katrina and Rita," *Population and Environment* 31(1–3): 64–86.

Horton, H. D. (1992) "Race and Wealth: A Demographic Analysis of Black Homeownership," *Sociological Inquiry* 62(4): 480–489.

Huang, S-K., M.K. Lindell, and C.S. Prater. (2016) "Who Leaves and Who Stays? A Review and Statistical Meta-Analysis of Hurricane Evacuation Studies," *Environment and Behavior*, 48(8): 991–1029.

Immergluck, D. (2009) *Foreclosed: High-risk Lending, Deregulation, and the Undermining of America's Mortgage Market*. Ithaca NY: Cornell University Press.

Ives, S. M. and O. J. Furseth. (1983) "Immediate Response to Headwater Flooding in Charlotte, North Carolina," *Environment and Behavior* 15(4): 512–525.

Jackson, K. T. (1987) *Crabgrass Frontier: The Suburbanization of the United States*. New York: Oxford University Press.

Johnson, J. H., A. Parnell, A. M. Joyner, C. J. Christman, and B. Marsh. (2004) "Racial Apartheid in a Small North Carolina Town," *The Review of Black Political Economy* 31(4): 89–107.

Kamel, N. M. O. and A. Loukaitou-Sideris. (2004) "Residential Assistance and Recovery Following the Northridge Earthquake," *Urban Studies* 41(3): 533–562.

Kates, R. W., C. E. Colten, S. Laska, and S. P. Leatherman. (2006) "Reconstruction of New Orleans after Hurricane Katrina: A Research Perspective," *Proceedings of the National Academy of Sciences* 103(40): 14653–14660.

LaCour-Little, M. (1999) "Discrimination in Mortgage Lending: A Critical Review of the Literature," *Journal of Real Estate Literature* 7(1): 15–49.

Laska, S. and B. H. Morrow. (2006) "Social Vulnerabilities and Hurricane Katrina: an Unnatural Disaster in New Orleans," *Marine Technology Society Journal* 40(4): 16–26.

Lee, J. Y., S. Bame, and S. Van Zandt. (2014) "Differences between Renters' and Owners' Unmet Housing Needs during Disasters," presented at 2014 Association of Collegiate Schools of Planning (ACSP) Conference, Philadelphia PA: 10/30/14–11/2/14.

Levine, J. C. (2005) "Equity in Infrastructure Finance: When are Impact Fees Justified?" *Land Economics* 70(2): 210–222.

Levine, J. N., A. M. Esnard, and A. Sapat. (2007) "Population Displacement and Housing Dilemmas Due to Catastrophic Disasters," *Journal of Planning Literature* 22(1): 3–15.

Lindell, M. K. (2017). "Communicating Imminent Risk," 449–477, in H. Rodríguez, J. Trainor, and W. Donner (eds.), *Handbook of Disaster Research*, New York: Springer.

Lindell, M. K. and R. W. Perry. (2004) *Communicating Environmental Risk in Multiethnic Communities*, Thousand Oaks CA: Sage.

Lindell, M. K. and C. S. Prater. (2000) "Household Adoption of Seismic Hazard Adjustments: A Comparison of Residents in Two States," *International Journal of Mass Emergencies and Disasters* 18(2): 317–339.

Lindell, M. K. and C. S. Prater. (2003) "Assessing Community Impacts of Natural Disasters," *Natural Hazards Review* 4(4): 176–185.

Lindell, M. K., C. S. Prater, and R. W. Perry. (2006) *Introduction to Emergency Management*, Hoboken NJ: John Wiley.

Logan, J. R. and H. L. Molotch. (1987) *Urban Fortunes: The Political Economy of Place*, Berkeley CA: University of California Press.

Lucy, W. (2010) *Foreclosing the American Dream*, Chicago IL: American Planning Association Press.

Manyena, S. B. (2006) "The Concept of Resilience Revisited," *Disasters* 30(4): 434–450.

Marsh, B., A. M. Parnell, and A. M. Joyner. (2010) "Institutionalization of Racial Inequality in Local Political Geographies," *Urban Geography* 31(5): 691–709.

Massey, D. D. and N. A. Denton. (1993) *American Apartheid: Segregation and the Making of the Underclass*, Cambridge MA: Harvard University Press.

Mileti, D. (1999) *Disasters by Design*, New York: Joseph Henry Press.

Mileti, D. S. and J. D. Darlington. (1997) "The Role of Searching in Shaping Reactions to Earthquake Risk Information," *Social Problems* 44(1): 89–103.

Moore, H. E. (1958) *Tornadoes Over Texas*, Austin TX: University of Texas Press.

Moore, H. E., F. L. Bates, M. V. Layman, and V. J. Parenton. (1963) *Before the Wind: A Study of the Response to Hurricane Carla*, Washington DC: National Academy of Sciences/ National Research Council.

Moore, H. E. and F. L. Bates (1964) *. . .and the Winds Blew*, Austin TX: The Hogg Foundation for Mental Health.

Morrow, B. H. (1997) "Stretching the Bonds: The Families of Andrew," 141–170, in W. G. Peacock, B. H. Morrow, and H. Gladwin (eds.), *Hurricane Andrew: Ethnicity, Gender and the Sociology of Disaster*, London: Routledge.

Morrow, B. H. (1999) "Identifying and Mapping Vulnerability," *Disasters* 23(1): 1–18.

Morrow, B. H. and E. Enarson. (1996) "Hurricane Andrew through Women's Eyes: Issues and Recommendation," *International Journal of Mass Emergencies and Disasters* 14(1): 1–22.

Morrow, B. H. and E. Enarson. (1997) "A Gendered Perceptive: The Voices of Women," 116–140, in W. G. Peacock, B. H. Morrow, and H. Gladwin (eds.), *Hurricane Andrew: Ethnicity, Gender and the Sociology of Disasters*, London: Routledge.

Morrow, B. H. and W. G. Peacock. (1997) "Disasters and Social Change: Hurricane Andrew and The Reshaping of Miami?" 226–242, in W. G. Peacock, B. H. Morrow, and H. Gladwin (eds.), *Hurricane Andrew: Ethnicity, Gender and the Sociology of Disaster*, London: Routledge.

Mueller, E. J., H. Bell, B. B. Chang, and J. Henneberger. (2011) "Looking for Home after Katrina: Post-Disaster Housing Policy and Low-Income Survivors," *Journal of Planning Education and Research* 31(3): 291–307.

Myers, D. (1975) "Housing Allowances, Submarket Relationships and the Filtering Process," *Urban Affairs Quarterly* 11(2): 215–240.

NRC—National Research Council. (2006). *Facing Hazards and Disasters: Understanding Human Dimensions*, Washington DC: National Academy of Sciences/National Research Council.

Norris, F. H., J. L. Perilla, J. K. Riad, K. Kaniasty, and E. A Lavisso. (1999) "Stability and Change in Stress, Resources, and Psychological Distress Following Natural Disasters: Findings from Hurricane Andrew," *Anxiety, Stress, and Coping* 12(4): 363–396.

Oliver, M. L. and T. M. Shapiro. (1995) *Black Wealth / White Wealth: A New Perspective on Racial Inequality*, New York: Routledge.

Olshansky, R. B., L. D. Hopkins, and L. A. Johnson. (2012) "Disaster and Recovery: Processes Compressed in Time," *Natural Hazards Review* 13(3): 173–178.

Pastor, M., R. D. Bullard, J. K. Boyce, A. Fothergill, R. Morello-Frosch, and B. Wright. (2006) *In the Wake of the Storm: Environment, Disaster and Race after Katrina*, New York: Russell Sage Foundation.

Pavel, M. P. (2008) "Breaking Through to Regional Equity," *Race, Poverty, and the Environment* 15(2): 29–32.

Peacock, W. G., S. D. Brody, and W. Highfield. (2005) "Hurricane Risk Perceptions among Florida's Single Family Homeowners," *Landscape and Urban Planning* 73(2): 120–135.

Peacock, W. G. and C. Girard. (1997) "Ethnic and Racial Inequalities in Hurricane Damage and Insurance Settlements," 171–190, in W. G. Peacock, B. H. Morrow, and H. Gladwin (eds.), *Hurricane Andrew: Ethnicity, Gender and the Sociology of Disaster*, London: Routledge.

Peacock, W. G., N. Dash, and Y. Zhang. (2006) "Shelter and Housing Recovery Following Disaster," 258–274, in H. Rodriguez, E. L. Quarantelli, and R. Dynes (eds.), *The Handbook of Disaster Research*, New York: Springer.

Peacock, W. G., B. H. Morrow, and H. Gladwin (eds.). (1997) *Hurricane Andrew: Ethnicity, Gender and the Sociology of Disasters*. London: Routledge.

Peacock, W. G. Van Zandt, S., Y. Zhang, and W. Highfield. (2015) "Inequities in Long-Term Housing After Disaster," *Journal of the American Planning Association* 80(4): 356–371.

Pendall, R. (2000) "Local Land Use Regulation and the Chain of Exclusion," *Journal of the American Planning Association* 66(2): 125–142.

Pendall, R. and J. I. Carruthers. (2003) "Does Density Exacerbate Income Segregation? Evidence from US Metropolitan Areas, 1980 to 2000," *Housing Policy Debate* 14(4): 541–589.

Perry, R. W. and M. K. Lindell. (1991) "The Effects of Ethnicity on Evacuation Decision-Making," *International Journal of Mass Emergencies and Disasters* 9(1): 47–68.

Perry, R.W., M. K. Lindell, and M. R. Greene. (1982) "Crisis Communications: Ethnic Differentials in Interpreting and Responding to Disaster Warnings," *Social Behavior and Personality* 10(1): 97–104.

Pendall, R., J. Martin, and R. Puentes. (2006) "From Traditional to Reformed: A Review of the Land Use Regulations in the Nation's 50 Largest Metropolitan Areas," Research Brief, Washington DC: Brookings Press.

Perry, R. W. and A. H. Mushkatel. (1986) *Minority Citizens in Disasters*, Athens GA: University of Georgia Press.

Perry, R. W. and L. Nelson. (1991) "Ethnicity and Hazard Information Dissemination," *Environmental Management* 15(4): 581–587.

Phillips, B. D. (1993) "Culture Diversity in Disasters: Sheltering, Housing and Long-Term Recovery," *International Journal of Mass Emergencies and Disasters* 11(1): 99–110.

Phillips, B. and M. Ephraim. (1992) *Living in the Aftermath: Blaming Processes in the Loma Prieta Earthquake*, Working Paper No. 80, Boulder, CO: University of Colorado, Institute of Behavioral Science, Natural Hazards Research Applications Information Center.

Prater, C. S. and M. K. Lindell. (2000) "Politics of Hazard Mitigation," *Natural Hazards Review* 1(2): 73–82.

Prince, S. H. (1920) "Catastrophe and Social Change: Based upon a Sociological Study of the Halifax Disaster," PhD dissertation, New York: Columbian University, Department of Political Science.

Quarantelli, E. L. (1982) "General and Particular Observations on Sheltering and Housing in American Disasters," *Disasters* 6(4): 277–281.

Quigley, W. P. (2007) "Obstacle to Opportunity: Housing that Working and Poor People Can Afford in New Orleans Since Katrina," *Wake Forest Law Review* 42: 393.

Rossi, P. H., J. D. Wright, E. Weber-Burdin, and J. Pereira. (1983) *Victims of the Environment: Loss from Natural Hazards in the United States, 1970–1980*, New York: Plenum Press.

Rubin, C. B. (1985) "The Community Recovery Process in the United States after a Major Disaster," *International Journal of Mass Emergencies and Disasters* 3(2): 9–28.

Rugh, J. S. and D. S. Massey. (2010) "Racial Segregation and the American Foreclosure Crisis," *American Sociological Review* 75(5): 629–651.

Sagalyn, L. B. (1983) "Mortgage Lending in Older Urban Neighborhoods: Lessons from Past Experiences," *Annals of American Academy* 465: 98–108.

Schneider, M. and T. Phelan. (1993) "Black Suburbanization in the 1980s," *Demography* 30(2): 269–279.

Silver, C. (1991) "The Racial Origins of Zoning: Southern Cities from 1910–40," *Planning Perspective* 6(2): 189–205.

Sims, J. and D. Bauman. (1972) "The Tornado Threat: Coping Styles of the North and the South," *Science* 176: 1386–1392.

Squires, G. D. (1998) "Why an Insurance Regulation to Prohibit Redlining?" *John Marshall Law Review* 31(2): 489–511.

Squires, G. D. and C. E. Kubrin. (2005) "Privileged Places: Race, Uneven Development and the Geography of Opportunity in Urban America," *Urban Studies* 42(1): 47–68.

Squires, G. D., S. O'Connor, and J. Silver. (2001) "The Unavailability of Information on Insurance Unavailability: Insurance Redlining and the Absence of Geocoded Disclosure Data," *Housing Policy Debate* 12(2): 347–372.

Squires, G. D. and W. Velez (1987) "Neighborhood Racial Composition and Mortgage Lending: City and Suburban Differences," *Journal of Urban Affairs* 9(3): 217–232.

Talen, E. (2005) "Land Use Zoning and Human Diversity: Exploring the Connection," *Journal of Urban Planning and Development* 131(4): 214–232.

Tierney, K. J. (1989) "Improving Theory and Research in Hazard Mitigation: Political Economy and Organizational Perspectives," *International Journal of Mass Emergencies and Disasters* 7(3): 367–396.

Turner, R. H., J. M. Nigg, and D. H. Paz. (1986) *Waiting for Disaster: Earthquake Watch in California*, Berkeley CA: University of California Press.

U.S. Department of Housing and Urban Development (HUD) (2005) *Why Not in Our Community? Removing Barriers to Affordable Housing*, U.S. Department of Housing and Urban Development Office of Policy Development & Research. Washington, DC. www.huduser.gov/portal/Publications/wnioc.pdf.

Van Zandt, S. (2007) "Racial/Ethnic Differences in Housing Outcomes for First-Time, Low-Income Home Buyers," *Housing Policy Debate* 18(2): 431–474.

Van Zandt, S., W. G. Peacock, D. Henry, H. Grover, W. Highfield, and S. Brody. (2012) "Mapping Social Vulnerability to Enhance Housing and Neighborhood Resilience," *Housing Policy Debate* 22(1): 29–55.

Vaughan, E. and B. Nordenstam. (1991) "The Perception of Environmental Risks Among Ethnically Diverse Groups," *Journal of Cross-Cultural Psychology* 22(1): 29–60.

Vaughan, E., and M. Seifert. (1992) "Variability in the Framing of Risk Issues," *Journal of Social Issues* 48(4): 119–135.

Wisner, B., P. Blaikie, T. Cannon, and I. Davis. (1994) *At Risk: Natural Hazards, People's Vulnerability and Disasters*, London: Routledge.

Yinger (2006) R. J. "The Promise of Education," *Journal of Education for Teaching* 31(4): 307–310.

Zahran, S., S. D. Brody, W. G. Peacock, A. Vedlitz, and H. Grover. (2008) "Social Vulnerability and the Natural and Built Environment: A Model of Flood Casualties in Texas 1997–2001," *Disasters* 32(4): 537–560.

Zhang, Y. and W. G. Peacock. (2009) "Planning for Housing Recovery? Lessons Learned from Hurricane Andrew," *Journal of the American Planning Association* 76(1): 5–24.

3

RISK COMMUNICATION

A Review and Peek Ahead

George Oliver Rogers

Introduction

The risk communication literature is both vast and varied. Although the stated goals of many studies are similar, the approaches vary from focusing on the message, to emphasis on the process and people, to focusing on the organizations doing the communicating. This chapter reviews the risk communication literature and summarizes some of the critical barriers to risk communication in general. It ends by briefly highlighting the daunting challenges posed by climate change.

Risk communication and hazard communication share many aspects of communication, including informing people of potential dangers, predictions of future events, and technical information on how the danger is likely to materialize and what to do about it. Yet they are also distinct in important ways. Hazard communication is communication about well-understood hazards or emergencies, often as they are unfolding. In hazard communication, prior experience is rich with lessons learned from sufficiently frequent occurrences to provide the necessary context for the communication. Prior experience, including outcomes, is often the basis of determining which behaviors are most appropriate under similar circumstances, and may provide touchstones for guiding behavior.

These communications can be, and have been, succinctly adapted to posters to help people respond appropriately in the early phases of emergencies. For example, a poster from the University of California at Irvine campus has entries for evacuations, fire, criminal activity, hazardous materials, earthquakes, medical emergencies, and shooting. The entry for earthquake splits the behaviors into two groups depending on whether the person is in an inside location or an outside location. The poster advises people what actions are possible that can help protect them from harm or injury during the first few moments of the event. It presumes that emergency personnel will be arriving with instructions for later activities that will be of similar quality. These posters, and there seem to be at least a dozen or so examples just for university communities, are posted prior to the event to serve as succinct reminders of potential hazards and appropriate behaviors should one occur. The one described above was posted in a restroom on a "wall of interest" for at least a few moments each day.

Risk communication is communication about risks that are themselves often fraught with uncertainty, complexity, and incompleteness. Not only are the outcomes often difficult to link directly to specific behaviors, actions, and decisions, there is frequently (sometimes considerable) disagreement even among experts about the links between the risks and the outcomes. For example, even in 2014 when more than nine of ten climate scientists agree that human actions are at the root of climate change, there are still many that find climate change a complete hoax foisted upon the public by proponents of increased government involvement and spending. Similar to so many public health risks, climate change has been subject to bias, including selective reporting by the media, imperfect human information processing, and subjective human decisions required to communicate these complex, multi-faceted problems with uncertain outcomes. For example, nuclear power proponents thought it would be sufficient to communicate the official estimate of the risk of death among the public (2×10^{-6}, or about 2 chances in a million over a lifetime) and compare it to other risks people routinely accept at those levels, like death in a commercial airline crash (in the 1×10^{-5} range) or an automobile accident (in the 1×10^{-4} range, with about 40,000 deaths per year in the US alone). But research soon showed that a variety of other attributes, like dread or common, known or unknown, voluntary or involuntary characteristics of the risks were just as important as risk estimates in producing human response (Slovic et al. 1981). In fact, the research showed that human response was not completely rational in that even risk experts used the bias-inducing cognitive heuristics as laypeople when asked to make risk judgments in areas outside their expertise.

The Goals of Risk Communication

In an effort to help risk communicators learn the art of risk communication without repeating the mistakes of the past, Fischhoff (1995: 138) summarizes the stages of risk communication. His seven stages of risk communication have been widely cited and discussed in the literature, which makes them a "must-read" for any risk communication effort. Abbreviated here for the sake of completeness, the stages suggest that we need to:

- get the numbers right,
- tell the public the numbers,
- explain what the numbers mean,
- show them they accept similar risks,
- show them it's a good deal for them,
- treat them well,
- make them partners.
- all of the above!

These stages range from the public relations of persuasion, to the information transfer of content, to the interpretation of content, to a focus on the process and partnering.

Covello and Sandman (2001: 171–2) summarize the four eras of risk communication. In the first era the strategy was "simply to ignore the public" to the extent possible; protect them certainly, but don't allow them to participate in public policy concerning risk. This was the basic strategy until the mid-1980s when environmental activists began to exert power over environmental policy and, perhaps more importantly, risk communicators found that when these activists were ignored, it exacerbated existing controversies. This gave rise the second era of risk communication, which focused on "how to explain the data better." While risk communicators learned to communicate the risk of small numbers (e.g., parts per billion, deaths per million) in material designed for seventh-to-ninth graders, they also found that the receivers still needed to be motivated to absorb the material. The third era of risk communication was a profound shift in that it addressed public outrage, which in turn required "building a dialogue with the public." To do that the risk communicators found that empathy requires not just awareness of the outrage but recognition that the public was entitled to be outraged—no matter how negligible the risk. But this policy leads to some unsatisfactory situations where, because of the outrage, public funds are used to deal with negligible risks while other more serious concerns are left unaddressed. The fourth era of risk communication came when the dialogue led to "treating the public as a partner." But this profound change in risk communication has faced significant barriers, including the habit and inertia of old behavior, the propensity of the technical people (who are needed to provide credibility) to fall back on clear boundaries with logical unemotional solutions, and the commitment of some activists to ideological positions that leave little room for compromise.

Chess (2001) points out that the range of choices available to individuals is constrained by organizations. She goes on to place risk communication in the context of the loss of confidence in the chemical industry after the catastrophic Bhopal accident, and a similar release in Institute, West Virginia, in 1985. Even though the Chemical Manufacturers Association implemented the Community Awareness and Emergency Response Program on a voluntary basis, Congress passed the Emergency Planning and Community Right to Know Act in 1986. The goals of the Act include providing for (1) emergency planning for chemical emergencies, (2) emergency notification in case of releases, and (3) community right-to-know about toxic and hazardous materials. The inherent goal of risk communication became clear: to provide people what they need to know to make reasonably informed decisions about the risks they face.

Historical Summary

In 1988 the U. S. Environmental Protection Agency released The Seven Cardinal Rules of Risk Communication—a pamphlet drafted by an academic and a government administrator (Covello and Allen 1988). While this may not have been the first government agency to attempt to communicate risk more effectively, it certainly has been the most widely circulated. The seven cardinal rules put forth seem obvious, even sophomoric, but they are often violated in the practice of risk communication. The seven cardinal rules of risk communication are (see also Covello et al. 1988):

1 Accept and involve the public as a legitimate partner.
2 Plan carefully and evaluate your efforts.
3 Listen to the public's specific concerns.
4 Be honest, frank, and open.
5 Coordinate and collaborate with other credible sources.
6 Meet the needs of the media.
7 Speak clearly and with compassion.

The first rule is driven by the idea that in a democracy people have the right to participate in decisions that affect them by impacting their lives, property, or other things they may value. In a democracy, the goal of risk communication should be to produce a well-informed public capable of reasonably informed decisions, rather than to defuse public concerns. The second rule suggests that risk communication is more likely to be successful if it is carefully planned. The public audience is rarely homogeneous enough to be reached by a single message. The multiple audiences are best reached by various strategies focused on specific interests, needs, concerns, and preferences. Hence, planning the risk communication is essential to effective communication.

Rule three suggests that risk communication is a dialogue, and that by listening to the public risk communication can focus on the aspects of the issue that are most important to the audience, like trust, competence, credibility, fairness, and voluntariness. This avoids the miscommunication that often accompanies preconceived notions about what the people know or think, or which actions to take to deal with risk. The fourth rule suggests that trust and credibility are a risk communicator's most precious assets—they are the hardest to obtain, the easiest to lose, and once lost almost impossible to regain. This rule means maintaining transparency in the process.

The fifth rule suggests that credible voices are powerful motivators. If these voices speak in concert with the message, they provide independent confirmation—few things can undermine the credibility of the message more than disagreement from a credible source. Rule six recognizes the media as the primary transmitter of risk information: they play critical roles in setting policy agendas and determining processes and outcome. Unfortunately, journalists' perspectives are quite different from those of risk communicators. They are more interested in politics and simple messages than the complexities so often required to characterize a risk, and danger makes a better story than safety. Finally, rule seven is a reminder that technical language can be a useful shorthand among professionals with shared background, but it creates significant barriers to a general understanding among people that do not share the professional jargon. Moreover, it reminds the risk communicator that it is okay to share empathy with people over the tragedy of an injury or death.

The mental models approach to risk communication starts from the basic assumption that communicating risk to people needs to be placed in the context of their basic understanding of risk. To achieve the desired outcome of providing people with what they need to know about risk in sufficient detail to allow them to make reasonably informed risk decisions, the mental models approach rests on the idea that existing beliefs about risk affect the interpretation of the risk message. Given this emphasis on the context of what the people know and believe about risk and its circumstances, the mental models approach begins with interviews of key stakeholders (Atman et al. 1994; Morgan et al. 2001). These interviews are followed by confirmatory surveys that provide a broader understanding and consolidation of what is known about the risk. The mental models approach engages technical expertise to develop influence diagrams that can be used to represent the risk during risk communication. Influence diagrams help organize information about the risk, inform people as to what actions are most appropriate, and generally inform decisions about risk. The mental models approach is an integrated set of activities that allows designers of risk communication to select risk communication content, structure messages to best communicate about risk, and organize materials to identify and fill existing gaps in local knowledge about risk. Most importantly, because the mental models approach treats each risk communication situation as a unique case, it precludes the use of *en masse* approaches that treat risk communication as a one-way communication from those that know to those that remain uninformed because the risk communication message failed to address their concerns.

The organization model of risk communication argues that organizations shape the communication of risk in a variety of ways. Beyond the crafting of the message itself, organizations set up the framework in which risk communication takes place. From the outset Chess (2001) points out that organizations constrain risk choices by limiting the range and character of available alternatives. This is similar to Ackoff's (1974) argument that planning problems do not simply appear fully formed out of nowhere. When planning problems are identified, the nature of the circumstances themselves must first be deemed problematic. Then withstanding the scrutiny against a value system, which may or may not be explicit, a decision that this problem needs to be addressed can be made. Risk communication problems are similar. The organization has to determine the appropriate amount of risk information to communicate, which is a function of a variety of risk estimates. This means that not only does the organization have to determine which risk estimate best represents the distribution of possibilities (e.g., best-case, most-likely, worst-case, mean-risk), but whether (and how) to communicate the variation and uncertainty. It also means that the view selected by the organization to be communicated is unlikely to be universally shared by all members of the organization. Chess (2001) argues that because organizations, viewed in an open systems framework, are largely shaped by their environments, understanding organizational behavior, including risk communication, is contingent upon understanding the environmental context of that behavior. Hence, with the loss of legitimacy after the accident at Bhopal in December of 1984, and the subsequent passage of the Emergency Planning and Community Right to Know Act in October of 1986, the Chemical Manufacturers

Association found themselves in an increasingly hostile environment. A major accident had served as a catalyst for intense public hostility and scrutiny, which led to the passage of new legal requirements. The resulting turbulence and uncertainty created a substantial threat to the chemical manufacturers. In this environment Chess (2001: 182) suggests that, "risk communication became a corporate survival mechanism." All social organizations "instinctively" fight for survival. Hence to survive, they had to adapt—namely, to communicate about the risks they impose upon a community.

Barriers to Risk Communication

A clear understanding of the level of risk provides the foundation for effective, ethical risk communication. The barriers to effective risk communication are usually summarized in terms of issues surrounding the exchange of information (cf. Covello et al. 2001), like magnitude and significance, situational contexts such as conflict, lack of coordination, and lack of planning, and issues of human irrationality and trust. Issues of differing perspectives, data presentation and manipulation, and associative language and culture are briefly discussed below.

Even when organizations are trying to communicate the risks involved in a given set of activities, the "language of risk" can be a distinct problem. Such was the case when the US Army attempted to communicate the risks involved in the disposal of the unitary chemical weapons stockpile at Johnston Atoll. When Congress mandated the destruction of the unitary chemical weapons stockpile in December 1985, the US Army engaged in an environmental impact assessment process that ended with the production of an Environmental Impact Statement (US Army 1988). Subsequent decisions called for the on-site disposal of the stockpile stored in the continental United States and the removal of the chemical weapons from Germany by 1992 and subsequent disposal at Johnston Island (Ambrose 1989; GAO 1991). Rogers (1992) examined the risk communication between the Army and native Polynesian cultures in terms of the written record of proceedings, including the Environmental Impact Statement and the transcript from the public hearing in Honolulu. While the Army made great efforts to better understand the risks involved in the disposal process, in part to meet the Congressionally mandated requirement for "maximum protection" of the public, the language used to communicate about risk prevented a meaningful dialogue.

Rogers (1992) identified twelve characteristics where the Army and the native Polynesian cultures failed to share a common perspective on the risks (Table 3.1). To begin with, they failed to agree on the name of the place, its location, or its history. In addition, the native culture viewed the secrecy of national security as simply being used to hide environmental and ecological risks, and viewed the extensive planning as suspect. For example, ships full of munitions to be incinerated arriving at Johnston Atoll during hurricane season was repeatedly mentioned by the public as a serious concern. The approaches of the two cultures were diametrically opposed—the Army used a reductionist approach that examined the detail of individual parts and sub-parts, while the native culture viewed the situation much more holistically, with the whole being more than the sum of the intimately interconnected parts. The Army used deductive logic in a linear, often two-dimensional and bounded space, while the native culture was more inductive, nonlinear, multidimensional, and expansive. But perhaps most important was that these differences in perspective pitted the Army's belief and trust in the system against the native culture's long history of perceived insult and egregious deeds committed by outsiders. In short, the two cultures involved failed to communicate in that "[t]hey failed to establish risk communication dialogue, and never established a common framework for effective risk communication" (Rogers 1992: 437). There was no shared meaning and no dialogue to clarify meaning. Unfortunately, this served to undermine the trust and credibility of both perspectives.

Communicating technical risk information is always a challenge. In the Chemical Weapons Demilitarization program, the US Army spent a great deal of time, energy, and resources determining

Table 3.1 Comparing Perspectives

Characteristic	U. S. Army	Native/Polynesian Cultures
Place Name	Johnston Atoll, JA or Johnston Island, JI	Kalama Island
Location	800 miles southwest of Hawaii	Part of the Hawaiian Island chain
History	Military presence since 1940s; weapons stored since 1970s	Stolen from the Kingdom of Hawaii in late 1800s
Secrecy	Necessary for national security	Used to hide environmental & ecological risks
Planning	Reduces risk by decreasing chances & consequences of accidents	Arrive during hurricane season
Perceived Risk	Credible accidents w/probabilities greater than 10^{-8} per program	Accidents happen
Acceptable Risk	Minimal risk to dispose of chemical weapons—e.g., plane crash into barge	No risk—the Titanic sank, the Valdez spilled oil
Perspective	Deductive, linear, two dimensional, bounded	Inductive, nonlinear, multi-dimensional, expansive
Approach	Reductionist—examines detail, sum of segmented parts equals whole	Holistic—whole more than sum of parts
Environmental/Ecological	Non-significant individual impacts	Significant combined impacts disturbances
Belief/Trust	Confidence in system safety	History of perceived insult
Spatial/Temporal Limits	Narrowed around performance period and location of proposed action(s)	Broadened to consider proposed actions as part of long history area wide

Source: Rogers, G. O. (1992) "Aspects of Risk Communication in Two Cultures," *International Journal of Mass Emergencies and Disasters* 10(3): 458, Table 4. Reprinted by permission of the International Research Committee on Disasters.

risks and developing the best ways to portray them to the public (US Army 1988). For example, inasmuch as Congress mandated the "maximum protection" of the public and the prevailing standard developed for commercial nuclear power plants was "reasonable protection" of the public, a new standard had to be developed. Since the "reasonable protection" standard had emerged over several decades to be the development of worst-case scenario emergency plans for risks in the range of 10^{-6} or greater, the "maximum protection" standard was determined to be the 10^{-8} range. These risks were then communicated in graphic form for distributional accidents—not individual accidents but rather accidents that represent a set of similar accidents. For example, a vehicle crash that represents all the potential ways a vehicle can be involved in an accident that could potentially release of chemical agent

More recently the San Jacinto River Authority wanted to communicate the impacts associated with the removal and use of water from Lake Conroe for drinking water. They proposed four scenarios to withdraw 25, 50, 75, and 100 thousand acre-feet of water in four phases of the Groundwater Reduction Plan beginning in 2015, 2025, 2035, and 2045 respectively. Because of the size of Lake Conroe at full pool, these diversions amount to roughly 1 ft/yr, 2 ft/yr, 3 ft/yr, and 4 ft/yr respectively. At the initial public meetings, these were presented as "not much impact" and graphically represented as a plot with the historical lake levels similar to the one presented in Figure 3.1a. But note that if the y-axis of the chart is modified to begin at a lower number, the apparent "impact" depicted appears reduced. So the impact can be manipulated to look greater by raising the point at which the x-axis crosses the y-axis to as much as 189, or reduced to say 100 or even zero would depict the impacts as nearly imperceptible. They are both small in size by a trick of graphics, and spread throughout the entire time of operation (1974 to 2010). A simple reorganization of the information from historical-order to the frequency-of-occurrence-order yields the graphic representation in Figure 3.1b.

Now it can be seen that lake levels that have not occurred in the 30-plus years of operation occur under the groundwater reduction plan scenarios. Table 3.2 presents the chance of falling below a given lake level under each scenario compared to what has occurred in history. This allows the user to see that lake levels that had a limited chance of occurrence in the past would have an increased chance of occurrence in the future. For example, a lake level of four feet below full pool—which local residents know dry-docks many private docks and severely impacts local marinas—has historically only happened 1.0 percent of the time, but under the proposed scenarios it would be likely to occur 2.6 percent, 6.0 percent, 11.0 percent, and 20.5 percent of the time. Having developed these graphics, the first time we showed them to members of lake-front communities, one resident quickly pointed out that our charts were "all wrong," because they assumed that the reduced lake levels were a risk, when "residents know" reduced lake levels are a certainty—not a risk.

In another case the Federal Emergency Management Agency (FEMA) became concerned—partly because of Congressional inquiries—that communities of color and poverty were responding to hazard warnings less effectively than other communities. FEMA launched the Emergency Preparedness Demonstration program of community-based participatory planning to enhance resiliency in six communities (Berke et al. 2011). In conjunction with that effort, Rogers and Burns (2010) were asked to investigate the subjective meaning of words being used to communicate risk and hazard among typical community residents and the emergency planning and response community serving them. As a consequence, a free-response investigation of subconscious use of words in two communities: Washington, DC, and Port Arthur, Texas was undertaken. In Washington, DC, a group

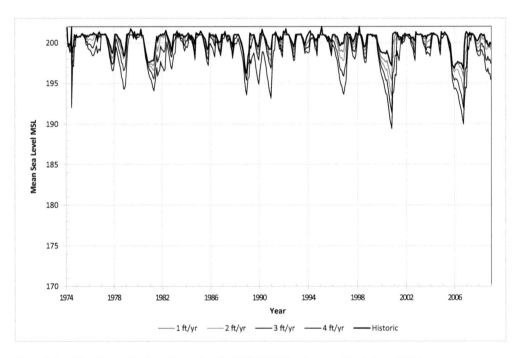

Figure 3.1a Historic and Projected Lake Levels 1974–2009 Emphasizing Depiction of Risk

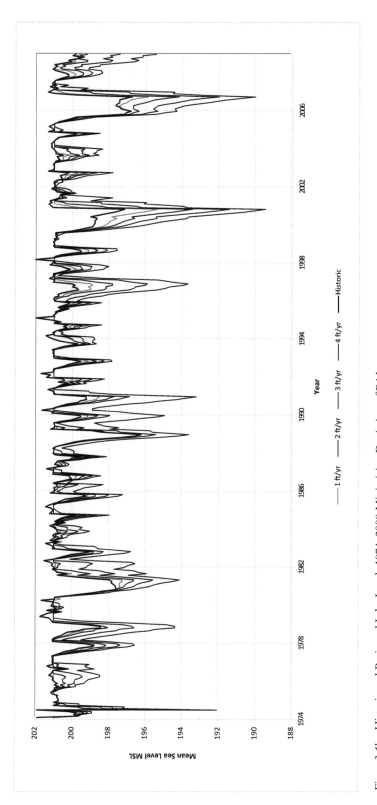

Figure 3.1b Historic and Projected Lake Levels 1974–2009 Minimizing Depiction of Risk

Table 3.2 The Chance of Lake Conroe Falling Below Given MSL Under SJRA GRP Scenarios

Lake level Below (ft)		History	To annually divert			
MSL	Full Pool		25,000 ac ft	50,000 ac ft	75,000 ac ft	100,000 ac ft
200	−1	21.9%	30.0%	36.7%	46.2%	60.5%
199	−2	13.1%	13.3%	19.3%	27.6%	45.5%
198	−3	6.0%	7.4%	9.5%	17.1%	29.0%
197	−4	1.0%	2.6%	6.0%	11.0%	20.5%
196	−5	0.0%	0.7%	1.9%	6.0%	14.3%
195	−6	0.0%	0.0%	1.2%	2.6%	10.0%
194	−7	0.0%	0.0%	0.5%	1.9%	5.2%
193	−8	0.0%	0.0%	0.0%	1.2%	2.6%

of disadvantaged senior citizens was compared with a group of first responders, while in Port Arthur, Texas, the community group (including members of local parent–teacher groups, churches, and community centers) was compared with people working for the local health department, police department, or Community Emergency Response Teams (Rogers and Burns 2010). A list of words was generated from two questions: What ". . .are the most important things that matter in your life?" and ". . .do you feel is most important for the safety and well-being of your community?" Words like family, children, community, friends, health, job, and home topped the list of things most important in your life, while words like police, health, community, education, knowledge, safety, preparedness, and school were most often mentioned as most important for the safety and well-being of the community. A total of 109 unique words were subsequently investigated as stimulus words for the associative group analysis (Szalay and Brent 1967), where participants are asked to respond with as many unique single words as possible to a given stimulus word in 30 seconds.

For example, the simple stimulus word, home, generated nearly 171 and 163 unique words or concepts in Port Arthur, Texas, and Washington, DC, respectively. Some of these response words were only mentioned by one person; others were shared among two or more people. Inasmuch as culture is shared and our emphasis was on shared meaning, we selected 37 response words that were shared by more than one group (Figure 3.2). Fourteen of the words (or concepts) were shared among responders and community members in both communities, including words like safe(ty), family, house, comfort(able), shelter, peace(ful), secure(ity), love, happy(iness), work, life, kids, live, food, and taxes. Eleven response words were shared primarily among responders from both communities, including base, heart, good, castle, nice, dwelling, dog(s), apartment, mortgage, and investment. Another eleven response words come mainly from community members, including owner(ship), warm, mother(mom), children, money, abode, relax(ing), haven, rest, fun, sanctuary, and repair. Some of the response words used in this latter group are shared among responders and community members but not from the same community. Even among the shared response words, there are significant differences between community members and responders. Words like family, comfortable, shelter, and love carry more community members than responders, while words like house are use twice as often among responders compared to community members. Finally, there is also a series of response words that are only used among responders, including base, heart, good, castle, nice, dwelling, dog(s), apartment, mortgage and investment and some other that are mostly used by responders, including haven, rest, and sanctuary. It is interesting to note that responders used response words like mortgage and investment, whereas community members used money and repairs. This may be a reflection of the disadvantaged communities, where money suggests poverty and the lack of resources, and investment suggests the positive asset of a home.

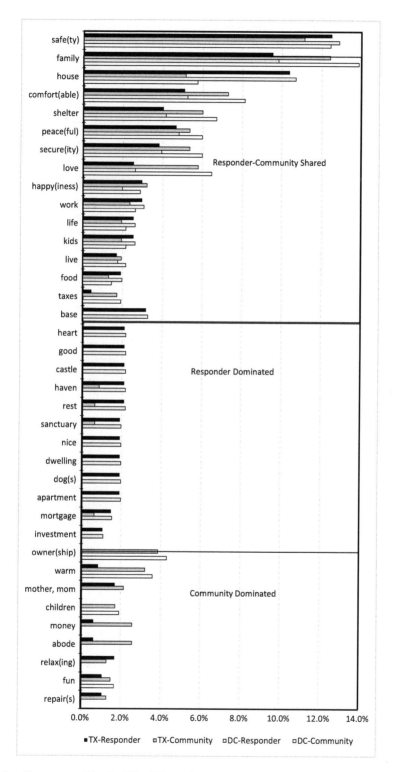

Figure 3.2 Free Response to Stimulus Word Home Among Responders and Community in Washington, DC and Port Arthur, TX

Examples of some of the many potential barriers to risk communication are discussed above, including differing perspectives, data presentation and manipulation, and associative language. These examples underscore some conditions that are more likely to allow risk communication to be successful than when they do not exist. Risk communication efforts are more likely to be successful when

- they rely on *credible information* that remains credible when informed by local scrutiny/knowledge;
- *local participants* are encouraged to actively engage in the process, and *share responsibility* with authorities for potential outcomes;
- risk *outcomes are directly linked to risk events* so that the links between risks and their consequences are well understood and transparent;
- risk *outcomes are relatively uniform and consistent* across a variety of settings;
- the receiver and the sender *share a culture and sub-culture*—a language and its shared usage;
- the *trust between receiver and the sender* is demonstrably genuine, having shared many common experiences, interpretations of the events, and shared responsibility for outcomes;
- the people receiving the risk message perceive themselves to be *treated fairly*;
- the risk communication *message is consistent* with local knowledge and understanding; and
- a *transparent relationship* between risks and desired actions is identified.

Challenges for Climate Change

On March 31, 2014, the Intergovernmental Panel on Climate Change released their Fifth Assessment Report, Phase I Report on Climate Change (IPCC 2014). The Summary for Policy Makers "assesses needs, options, opportunities, constraints, resilience, limits, and other aspects associated with adaptation" to climate change (p. 3). In the first 25 pages of the Summary for Policy Makers the authors use the word risk 181 times—this is before they get to the region-by-region detailed assessment of risks that is the heart of the report. The assessment summarizes the three "key" risks for each region. The regions are Africa, Europe, Asia, Australasia, North American, Central and South America, Polar Regions, Small Islands, and The Ocean. The uncertainty associated with each risk is rated based on the "type, amount, quality and consistency of evidence" (p. 5). The evidence is classified as limited, medium, or robust, while the amount of agreement is classed as low, medium, or high. For example, the first risk discussed in Africa is described as "[c]ompound stress on water resources facing significant strain from overexploitation and degradation at present and increased demand in the future, with drought stress exacerbate in drought-prone regions of Africa (High Confidence)" (p. 27). Each identified risk is assessed as to its climatic drivers and likelihood is assessed for the present, near term (2030–2040), and long term (as either 2°C or 4°C change in the 2080–2100). Each risk is assessed in terms of current conditions, and with high levels of adaptive action to cope with the risk. A momentous undertaking to be sure, but the assessment is done on a five-point scale: Very Low-Low-Medium-High-Very High (Low and High are not listed, but implied by the blank area between Very Low and Medium, and Medium and Very High). The risks listed for North America include "Wildfire induced loss of ecosystem integrity" (p. 28), "Heat-related human mortality" (p. 29), and "Urban floods in riverine and coastal areas" (p. 29); the confidence in each of these is high.

The report goes on to assess the confidence in observations and contribution of climate change on an area-by-area basis for each of four issues: (1) snow & ice, rivers and lakes, floods and drought, (2) terrestrial ecosystems, (3) coastal erosion and marine ecosystems, and (4) food production and livelihoods. For example, the retreat of highland tropical glaciers in Africa, and shrinkage of glaciers in western and northern North America are both high-confidence and reflect a major contribution of climate change. Major contributions of climate change in terrestrial systems include decreases in tree density in Sahel and semi-arid Morocco in Africa, and more frequent wildfires in subarctic conifer forests and tundra in North America. In marine systems, the decline in coral reefs in tropical African waters, and changes in salmon migration and survival in the Pacific Northwest North America are both experiencing major contributions from climate change. Climate change is also assessed as a major contributor to decline in fruit-bearing trees in Sahel, and impacts on the livelihoods of indigenous people in Canada. The report identifies eight key risks with high confidence that contribute to one or more reason for concern, including the risk of

1   death, injury, ill-health, and disrupted livelihoods in low-lying coastal areas;
2   severe ill-health and disrupted livelihoods in large urban areas due to flooding;
3   extreme weather event impacts on infrastructure and critical services like electricity, water supply, and health and emergency services;
4   mortality and morbidity during periods of extreme heat;
5   food insecurity associated with drought and variability of precipitation;
6   loss of livelihoods and income in rural areas due to insufficient drinking and irrigation water;
7   loss of marine and coastal ecosystems; and
8   loss of terrestrial and inland water ecosystems.

(p. 10)

While it is clear that this undertaking is a massive, exhaustive attempt to identify risks associated with climate change throughout the world, it is also clear that the actions needed to address these risks, both in terms of preventing their occurrence and of adapting as they occur, are mostly locally focused. Clearly, the climate change committee is attempting to communicate about the risks associated with climate change, and they have done an extraordinary job of communicating both the certainty and the uncertainty associated with the risks identified. But to be effective in stimulating appropriate responses, these efforts will have to turn their attention to local risks to local communities. Risk communication will have to consider the unique impacts on local populations, creating local resilience through unique local action. Engendering participation and shared responsibility is unlikely to be effective without direct transparent application of the risks to the local area. The risk communication literature suggests a number of conditions summarized in the previous section that are unlikely to be met on a global scale. The risk communication for climate change faces extreme challenges of communicating a global comprehensive risk in which the outcomes are separated from the initiating events by decades, with unknown uncertainties. And they have to be multicultural, multilingual, and fair in face of the variety of actions that will be required and the extant inequalities among those expected to take the actions. The climate change risk communication requirements are almost antithetic to the principles of risk communication from prior experience. Overcoming the associated barriers will undoubtedly require innovation. Risk communication efforts that build from the local community building blocks are more likely to successfully stimulate appropriate adaptive actions.

References

Ackoff, R. L. (1974) "The Systems Revolution," *Long Range Planning* 7(6): 2–20.

Ambrose, J. R. (1989) "Record of Decision: Chemical Stockpile Disposal Program," Washington DC: Department of the Army, Under Secretary of the Army.

Atman, C., A. Bostrom, B. Fischhoff, and M. G. Morgan. (1994) "Designing Risk Communications: Completing and Correcting Mental Models of Hazardous Processes, Part I," *Risk Analysis* 14(5): 779–788.

Berke, P. R., J. Cooper, D. Salvesen, D. Spurlock, and C. Rausch. (2011) "Building Capacity for Disaster Resiliency in Six Disadvantaged Communities," *Sustainability* 3(1): 1–20.

Chess, C. (2001) "Organizational Theory and the Stages of Risk Communication," *Risk Analysis* 21(1): 179–188.

Covello, V. T. and F. Allen. (1988) *Seven Cardinal Rules of Risk Communication*, Washington DC: US Environmental Protection Agency.

Covello, V. T. and P. M. Sandman. (2001) "Risk Communication: Evolution and Revolution," 164–178, in A. Wolbarst (ed.), *Solutions to an Environmental Peril*, Baltimore MD: John Hopkins University Press.

Covello, V. T., R. G. Peters, J. G. Wojtecki and R. C. Hyde. (2001) "Risk Communication, the West Nile Virus Epidemic, and Bioterrorism: Responding to the Intentional or Unintentional Release of Pathogen in an Urban Setting," *Journal of Urban Health: Bulletin of the New York Academy of Medicine* 78(2): 382–391.

Covello, V. T., P. M. Sandman, and P. Slovic. (1988) *Risk Communication, Risk Statistics, and Risk Comparisons: A Manual for Plant Managers*, Washington DC: Chemical Manufacturers Association.

Fischhoff, B. (1995) "Risk Perception and Communication Unplugged: Twenty Years of Process," *Risk Analysis* 15(2): 137–145.

General Accounting Office (GAO). (1991) *Chemical Warfare: DOD's Successful Effort to Remove U.S. Chemical Weapons from Germany*, No. GAO/NSIAD-91-105, Report to Congressional Requesters, Washington DC: US Government Accounting Office.

IPCC. (2014) "Summary for Policymakers," in *Climate Change 2014: Impacts, Adaptation, and Vulnerability. Part A: Global and Sectoral Aspects. Contribution of Working Group II to the Fifth Assessment Report of the Intergovernmental Panel on Climate Change*, Cambridge: Cambridge University Press.

Morgan, M. G., B. Fischhoff, A. Bostrom, and C. J. Atman (eds.). (2001) *Risk Communication: A Mental Models Approach*, New York: Cambridge University Press.

Rogers, G. O. (1992) "Aspects of Risk Communication in Two Cultures," *International Journal of Mass Emergencies and Disasters* 10(3): 37–464.

Rogers, G. O. and G. R. Burns. (2010) *Disaster Preparedness and Response in Disadvantaged Communities: Exploring the Subconscious Foundations of Emergency Behavior*, College Station TX: Texas A&M University Hazard Reduction and Recovery Center.

Slovic, P., B. Fischhoff, and S. Lichtenstein. (1981) "Facts and Fears: Societal Perception of Risk," 497–502, in K. B. Monroe (ed.), *Advances in Consumer Research* 8, Ann Arbor MI: Association for Consumer Research.

Szalay, L. B. and J. E. Brent. (1967) "The Analysis of Cultural Meaning Through Free Verbal Associations," *The Journal of Social Psychology* 72(2): 161–187.

US Army. (1988) "Chemical Stockpile Disposal Program Final Programmatic Environmental Impact Statement," U. S. Army Program Executive Officer, Program Manager for Chemical Demilitarization, Aberdeen Proving Ground, Aberdeen, MD.

PART II

Contributions of Hazard Mitigation Planning to Community Resilience

4

NEXT GENERATION MITIGATION IN A CHANGING WORLD

Jack D. Kartez

Introduction

The next five chapters of this symposium-based handbook appropriately start with the issue of how well we have been mitigating the negative impacts of natural hazards on lives, property, ecological systems, and the continuity of social and economic life in our human communities. Even two generations ago, the focus would have been on emergency response with the tacit view that natural disasters are celestial acts—regrettable events but beyond control and thus subject merely to coping and relief, not anticipatory action and human agency and responsibility. Such coping as was carried out mainly occurred in physical, structural efforts to control the impact of natural events through engineering of waterways and the construction of dams, dikes, and levees, and certainly not by simply avoiding population growth in vulnerable areas, or other nonstructural means. Societal response largely took the form of post-disaster relief actions for immediate medical and subsistence needs and eventual rebuilding—often in the same path of hazard for the future.

Much has changed since then, not only in terms of the reach and scale of disaster impacts on humanity—in the United States and elsewhere—but also the global environmental and human conditions creating vulnerability to natural hazards of all kinds. Coastal population growth worldwide is colliding with climate change effects. The cascading effects of natural hazards on human technological and infrastructure systems in urban regions have made such impacts more complex and the mitigation of future hazards even more challenging. The policy stance towards hazards in the United States has changed over these more than four decades as well—but how much and to what effect?

The chapters from the Mitigation Panel that follow this introduction all starkly question whether our understanding of threats to resilience—if not survival—is truly complete enough and especially whether we have acted on our knowledge sufficiently. Several themes emerge that in part are ones that the disaster policy research community has been ever more urgently voicing over recent decades. The needs for pre-disaster reduction and avoidance of hazards and for achieving equitable burdens of risk, are still not fulfilled by our societal institutions. New themes and imperative challenges have emerged as well, as will be discussed here and in the next five chapters.

Mitigation is the pivotal phase in the disaster cycle because if we adequately achieved such protection from and avoidance of nature's dynamics, the demands for disaster response and recovery would be far less than the growing and costly burdens they currently represent. Yet mitigation remains a societal failure in the United States in many lights, though it is notably more successful in a few small nations such as the Netherlands (as discussed in a September 14, 2011 seminar at Resources for the Future, Inc. in the United States, see www.rff.org/FloodRisk).

Delivering a recent annual Gilbert F. White Lecture in Geographical Sciences at the National Academy of Sciences on December 4, 2014, hazards researcher Susan L. Cutter—lead scientist of the widely referred-to Social Vulnerability Index—baldly declared that mitigation has failed to meet needs in part because our knowledge is not being adequately applied to policy. This is a paradox, she further argues, because our losses continue to mount at the same time our knowledge is greater than ever before (see also Cutter and Emrich 2005; White et al. 2001).

Scope of Chapters 5 through 9

An overview of the calls for action on mitigation and their results over time will help put the contributions of Chapters 5 through 9 of this volume in perspective. Social science fields concerned with place and location, particularly geography and urban and regional land use planning, have been prominent in the history of evaluation of American approaches to hazard mitigation and the consequent calls to action. These will be recapped, especially with respect to flooding and coastal hazards from hurricanes. Chapters on mitigation from those fields include those by Philip Berke and Ward Lyles, and by Sam Brody, Wes Highfield, William Merrell, and Yoonjeong Lee. Further perspectives on the mitigation phase are examined with a focus on the mounting threats and repeated losses from coastal hazards by an interdisciplinary group of authors including Walter Peacock, Michelle Meyer, Fayola Jacobs, Shannon Van Zandt, and Himanshu Grover. Political scientists Kristin O'Donovan and Thomas Birkland examine the broader institutional issues underlying mitigation policy action. Planner Himanshu Grover concludes this section with an analysis of how disaster reduction and climate change adaptation will relate to each other moving forward. In the hazards research community—one of the most interdisciplinary of all knowledge-linking networks—many fields have engaged with the issue of mitigation.

In the following discussion the trajectory of flood and hurricane mitigation policy over time is briefly examined. What shortcomings as well as achievements have brought us to our current situation? What changes are afoot for the future context in which our human and ecological systems will be subject to threats and how we will respond to those threats? Fundamental issues apply to other hazards such as earthquakes, although there are differences in the nature of seismic vulnerabilities and the institutional and market responses to mitigation. In any case, flood and storm hazards represent the most ubiquitous vulnerabilities in the United States and many areas of the world, and ones continuing to be exacerbated by the collision of human settlement patterns and global change such as sea level rise.

The thrusts of the succeeding chapters are then highlighted and expanded in each of those chapters. The result is an agenda for evidence-based action on mitigation and the pressing need to pursue a new generation of policy efforts and societal organization and adjustment given the threatening mixture of trends in global environmental and population settlement conditions.

The Slow Trajectory of US Mitigation Policy

Representing a new generation of scientists of place and human settlement, the late Gilbert F. White sounded a call at the dawn of the New Deal in the 1930s to reconsider the efficacy of trying to hold back nature's forces by dint of brute structures alone (e.g., dams), and the potential role that location decisions (i.e., land use planning) could play in reducing human vulnerability (White 1936). This was just as the nation was embarking on new and unprecedented uses of technological power to manage capricious nature, such as the Tennessee Valley Authority's multi-state flood control projects. Such calls as White's for caution in how the nation deployed new technologies, and the call for more intentional settlement patterns, were also echoed by the Regional Plan Association of America, but they remained limited to a very small community of thought.

By the time the nation had weathered the economic emergency of the Great Depression, the struggle of World War II, and the first Cold War conflict in Korea, a new era of rapid growth and dispersed settlement began the process that would eventually bring flood and storm vulnerabilities to greater national attention. The march of the population to coastlines and new encroachments on natural drainage systems accelerated in pace, as it has to this day. Yet it was the phenomenal, first-time multi-state losses of more than $1 billion and 81 fatalities caused by Hurricane Betsy in 1965 that mobilized legislative will to respond on a national scale in a new way. Federal policy at the time had taken the form of Presidential disaster relief, essentially amounting to a direct transfer of dollars from taxpayers across the nation to those in harm's way whose luck had run out.

The resulting new federal intervention in 1968 creating the National Flood Insurance Program (NFIP) mainly took the form of indemnifying landowners against future losses through the establishment of a risk pool—one that required national subsidy for decades—for those who made personal and policy decisions to locate in vulnerable areas. Without elaborating all details of the NFIP as it will be covered further in various chapters of this book, the national strategy for mitigation stopped fatefully short of simply steering a burgeoning population away from hazards through locational planning.

A key aspect of the NFIP was the nation-spanning effort to map flood hazards based on hydraulic study and likely storm probabilities. The new incentive-based approach required that local governments create overlay zoning for the special flood hazard areas delineated in those maps—consisting of the floodway and 100-year return probability floodplain. This in itself, albeit a phenomenal step towards understanding where to manage vulnerabilities to floods and storm surges, embodied future undesired consequences. Insurance was still grounded in the hold-harmless notion that disaster and hazard risks are beyond our own agency and responsibility. This has created contradictions. The requirement to elevate structures above the 100-year-return "base flood"—rather than prohibit floodplain encroachment—has only led to more development in hazardous locations. Insurance claims for multiple, repetitive losses to the same communities, locations, and landowners have failed to reduce future losses, but have increasingly strained the very fiscal sustainability of the insurance fund. The system has been aimed at supporting failure, not safety, as multiple authors here will argue.

Over time since the NFIP's creation, there has been a slow legislative march towards addressing these contradictions. But due to global change of both climate and population, such reforms are increasingly recognized by hazards researchers and professionals as inadequate. The reforms point to, but do not achieve, the degree of application of hazards knowledge to achieve the sufficient mitigation actions called for by authors of the following chapters.

It is useful to briefly frame what that march of policy change has looked like as a prelude to envisioning the next generation of needed mitigation action, again with the focus on flood and storm surge hazards. The concept of mitigation emerged in the 1974 Federal Disaster Relief Act (PL 93-288) with a directive that the mitigation needs of state and local governments receiving Presidential grant and loan relief "shall be evaluated and appropriate action shall be taken to mitigate such hazards, including safe land-use and construction practices . . ." (PL 93-488, sec. 406). By 1988, the Stafford Act (Stafford Disaster Relief and Emergency Assistance Act, PL 100-707) went a little further by requiring mitigation planning as part of using recovery assistance and also by initiating pre-disaster hazard mitigation planning incentives for states and their local governments. The Stafford Act nonetheless was judged too limited to significantly reduce both repetitive and new exposures (Burby 2006; Godschalk et al. 1998). In 2000, the Disaster Mitigation Act of 2000 (DMA 2000, PL 106-390) strengthened this approach by making pre-disaster state and local mitigation plans a precondition of receiving post-disaster federal assistance in presidentially declared events. Likewise, the Flood Insurance Reform Acts of 1994 and 2004 attempted to rein in the role of the NFIP in promoting indifference to flood hazards and accommodating the moral hazard of repetitive loss communities (PL 103-325 and PL 108-264 respectively). But writing in 1999 at the conclusion of the

Second National Assessment of Research on Natural Hazards (see Mileti 1999), Ray Burby and the other members of the Assessment's Land Use Committee argued that federal policies that make it appear safer to develop hazard areas have only increased vulnerability, not to mention moral hazard, in the nation's use of scarce disaster management resources (Burby et al. 1999).

Multiple chapters that follow in this entire book revisit this fateful path of policy development in detail as well as examine the framework of additional federal and state legislation and agency programs that make up the US system for societal mitigation of natural hazards, including the development of seismic building codes, efforts to manage the sensitive but hazardous coastal zone, and other arenas. But a half-century after the creation of the NFIP, with floods and storm surge damages remaining the most ubiquitous and in aggregate most damaging of threats, the United States still maintains a light hand in the needs for mitigation.

Over time the hazards research community has noted paradoxes in how we approach the mitigation challenge, starting with Burby and French (1981) who identified the land use management paradox: the factors which promote adoption of modest floodplain mitigation measures also stimulate new growth and encroachment in those same hazard areas, thus increasing societal losses—a paradox long raised in the hazards research community—that mounting disaster losses are occurring even as our knowledge of hazards science has greatly improved—is only the latest of such observations. Global change now complicates the challenges in new ways.

New Challenges: Stationarity Is Dead

Climate change now "undermines a basic assumption" that historically has facilitated management of flood hazards as well as water and other natural resources (Milly et al. 2008). The assumption that climate conditions are static—the mean behavior of processes such as precipitation and distributions around those means—has been the foundation for decades of the everyday parameters used to predict water runoff for the purposes of design and construction of roads, bridges, drainage systems, soil conservation, water supplies, and other vital applications supporting human communities. Additionally, those historical conditions have defined the spatial scope and recurrence probabilities of the so-called 100-year flood or "project flood" that is the basis for the US approach to floodplain management, insurance, and regulation, as well as major infrastructure design.

Changing fundamental parameters—a phenomenon called nonstationarity—means that 100-year return floods (i.e., with one percent annual probability) have become more frequent and floodplain inundation extents have grown beyond assumed regulatory boundaries with the greater volumes of precipitation in single storms that occur. While major extreme events like Hurricanes Katrina and Sandy have grabbed public attention in the United States, it is the less visible, incremental impacts of change that affect the entire nation. In a widely cited argument in *Science*, Milly et al. (2008, p. 573) simply argue:

> Planners have tools to adjust their analyses for known human disturbances within river basins, and justifiably or not, they generally have considered natural change and variability to be sufficiently small to allow stationarity-based design. In view of the magnitude and ubiquity of the hydroclimatic change apparently now under way, however, we assert that stationarity is dead and should no longer serve as a central, default assumption in water-resource risk assessment and planning. Finding a suitable successor is crucial for human adaptation to changing climate.

Blown-out culverts, collapsed roadbeds, overflowing sewer systems and other forms of ubiquitous wear and tear are the growing and cumulatively expensive consequences of nonstationarity (Stiles 2014). Furthermore, along the coastlines where population concentrates, sea level rise has amplified the effects of hurricanes and thus another effect of nonstationarity: storm surges and upriver flooding that exceed historically based averages and likelihoods.

Yet another consequence of nonstationarity is not physical but one that vexes and confuses societal decision making. It is difficult to gain understanding among non-specialists, who often include institutional leaders, of the probabilistic nature of our knowledge about climate and the hazards driven by these unseen systems that we must use modeled information to truly grasp. This has long created uncertainty and difficulties for decision makers and their constituents. To borrow from an apocryphal example in the toxic chemical hazards arena, when experts tell laypersons there is a 1 in 100,000 chance that there will be one cancer in 20 years from exposures to a substance or facility, the potentially affected people simply want to know what experts cannot tell them: "am I dead or not?" Likewise, the probabilistic classification of regulatory floods in the United States as 100-year events obscures the fact that events of that magnitude simply have an average probability, not a schedule. Much the same reality of provisional and stochastic knowledge applies to other natural hazards as well.

Nonstationarity now means that standards are no longer accurate because the fundamental distributions are changing. The 100-year flood now may be a 50-year or 25-year flood. Put another way, the magnitudes and frequencies of events are changing. Since 2001, water has reached flood levels four times as frequently each year than before in coastal cities such as Annapolis, Wilmington, Washington, DC, Atlantic City, and Charleston (McNeill et al. 2014). The historical parameters used in civil and transportation engineering, for example—which are codified into legally consequential standards of everyday practice at state and federal levels by institutions such as the American Association of State Highway and Transportation Officials—are simply wrong under emergent conditions.

This author has worked with an interdisciplinary group of both climate scientists and civil engineering researchers and institutional practitioners for several years under a National Science Foundation (NSF)-funded Infrastructure and Climate Network research coordination network project (www.theICNet.org) to explore how to adapt infrastructure standards to the science. The way forward is complex. The added element of uncertainty has barely been coped with as yet in the world of policy and design practice. Where a high level of investment and incentives for mitigation have a closer rational nexus with the benefits, for example with respect to seismic hazards, action has occurred. Pacific states in the US and their underlying local governments have moved to adopt earthquake building codes for commercial and institutional structures that may help reduce catastrophic damage levels and loss of life. Nonetheless, avoiding seismic hazard areas and even unstable slopes is not a well-adopted policy in competition with hardening through engineering solutions and—where available—earthquake insurance.

Next Generation Mitigation

The US policy frame for mitigation may rely on the principles of insuring those who locate in harm's way and attempting to harden against nature's forces. But large-scale climate change is forcing other countries with very different policy traditions to also rethink their fundamental approach to mitigation. In Great Britain, the 1949 Coastal Protection Act excludes national government aid to relieve property owners from impacts of coastal hazards. But storm erosion amplified by rising sea levels now threatens the sustainability of entire small communities along the coasts of the British Isles to the extent that local authorities for the first time are pressed to step in with measures such as buy-outs to facilitate retreat from climate change. In the face of this, Malcolm Kerby of the British NGO National Voice of Coastal Communities has lamented that "we are trying to fix a problem with a 70-year-old toolkit" (Doyle and McNeill 2014). Although the UK government has decided to commit greatly expanded resources to coastal flood defense through retreat from rising seas, it has set up an existential conflict for the future between mostly smaller coastal communities and large upriver cities as far as who has priority for limited adaptation funding and assistance (Doyle and McNeill 2014).

Climate change adaptation is increasing the complexity of hazard mitigation not only in terms of the technical forecasting of the characteristics of the hazards but also in confronting new tensions over resource priorities and basic policy doctrines. The needs for a new generation of approaches to mitigation will likely require what the National Academy of Sciences members who have promoted sustainability science have been arguing for since the late 1990s (Kates et al. 2001; Clark and Dickson 2003) Sustainability science principles include the need for problem-focused interdisciplinary approaches across the sciences which intentionally link knowledge to action and co-produce such knowledge with science's clients in society who must take action. The goal of this approach is to acknowledge and then bridge the science-to-practice gap.

Perhaps ironically, such principles have been hallmarks of the collaborative hazards research-practice community fostered by Gilbert F. White along with his colleague, hazard science pioneer Robert W. Kates, and others for some decades, albeit without all the hoped-for societal results as yet. While it has been difficult to make needed progress in dealing with hazard mitigation under stationarity assumptions, global change is further complicating our situation in multiple ways: with increased uncertainty about the future; with often diversionary debates about blame (such as whether climate change is attributable to human activity or not—an irrelevant issue to mitigating the hazards); and with greater issues of deciding how to apportion increasingly overburdened societal resources.

Basic assumptions about what mitigation should encompass or achieve are being challenged by necessity—for example, whether mitigation means protecting the status quo or adjusting to nature's forces through flexible design or even retreat. Such approaches have been only slowly attempted over time so far, mainly in the *post facto* recovery from catastrophic events such as the Great Midwest Floods of 1993 and Hurricane Sandy in 2012. Adaptation and resilience are only recently emerging concepts that confront the need to rethink the relative roles hazard resistance versus hazard adjustment should play in determining our place in nature.

An example of emerging approaches for coping with the paralyzing patchwork of barriers to making local decisions about hazard adaptation is the Coastal Adaptation to Sea level rise Tool (COAST) planning process for long-term sea level rise and storm surge mitigation (Heberle et al. 2014; Kirshen et al. 2012; Merrill et al. 2010). COAST involves engaging local decision makers in deciding what parameters for future coastal storm risk to analyze and visualize—such as which sea level rise target, what storm return rate and surge category, and what time period to use—rather than arguing about whether climate change is real, human-caused, and consequential. This co-produced basis for futures analysis (related to joint fact-finding in environmental dispute resolution and consensus building, cf. Susskind et al. 1999) then becomes an accepted context for analyzing future losses of property and other assets. This in turn provides an often more acceptable and pragmatic basis for developing and choosing among alternative hazard adaptation actions in order to decide whether to accept future losses, mitigate them now at present value, or choose a staged and/or contingent set of actions over time. The painfully obvious but frequently failed aim of this process is to move from pointless arguments over climate change attribution and uncertainty to action on the hazards. Or as William "Skip" Stiles of Wetlands Watch put it recently in arguing that "all adaptation is local": "The trick is to move more rapidly toward the inevitability of action and minimize the greater expense and disruption that comes from having started too late" (Stiles 2014: 64).

This illustrates the added complexity of decision-making and mitigation politics (as well as its urgency) and the necessity of principles from sustainability science like coproduction to overcome the science-to-society gap:

In fairness, much of the challenge lies not with climate scientists narrowing the range of estimated effects, or even in communicating those risks more effectively. The challenge lies in getting policymakers adapted to living in the new reality of climate change, when they never even adapted to the old reality of gradual sea level rise. In this new reality, decisions must be made on the basis of estimated effects with wider ranges of variability or even using scenarios rather than quantifiable projections. . . . [Policymakers] must learn to live in a world in which the past is no longer prologue and . . . no longer provide[s] guidance for the future.

(Stiles 2014: 63)

The following chapters reflect both a history of learning and a stark awareness of unmet needs both for societal adjustment and for a new generation of mitigation strategy to meet emergent realities. The themes highlighted in the preceding sections are all expanded on in depth. These include the wider and changing conceptions of vulnerability, the nature of resilience as a new frame for action, the institutional context in which action has been often frustrated but in which it must happen (with a key local government role), and specific and more ambitious frameworks for mitigation going forward, learning from past shortfalls in using our tools, especially spatial planning.

In Chapter 5, Kristin Taylor and Thomas Birkland analyze the intergovernmental system of roles and policies that has developed since Hurricane Betsy a half-century ago. Diagnosing the way in which the US system functions, O'Donovan and Birkland explain that the governance of hazard mitigation occurs at the local government level even while operating within a complex multi-level system of federal and state funding mandates and incentives. But action must be sieved through the local level of understanding and the politics of consent. Commitment to mitigation has varied somewhat from one federal administration to another, and while it has grown over time, O'Donovan and Birkland observe that the US policy system still operates more like an entitlement program of post-disaster spending than a capacity-building effort to avoid future disaster losses and increase safety. A renewed effort to promote local level physical land use planning is a crucial link in moving mitigation to needed levels.

In Chapter 6, Philip Berke and Ward Lyles cite the long and continued arguments that spatial planning at the local level is the lynchpin of boosting mitigation and adaptation to needed levels. Berke and Lyles then bring together the quarter-century of research on improving local planning effectiveness through plan quality assessment (PQA) conducted by members of the Second Assessment's Land Use Committee with the newly emergent challenges of climate change and the new perspectives from sustainability science on social-ecological systems (SES). They outline a framework that is a learning-planning system for achieving the "resilient America" called for in the National Research Council's 2012 report *Disaster Resilience: A National Imperative*. This includes resiliency planning indicators to more concretely assess the "untested assumptions" of the effectiveness of planning as called for in the NRC report in the abstract, building on lessons from the PQA research.

In Chapter 7, Walter Peacock, Michelle Meyer, Fayola Jacobs, Shannon Van Zandt, and Himanshu Grover begin with a recognition of the increasing, but nonetheless uncertain, impacts of climate change on community vulnerability through hurricane frequency, hurricane severity, and sea level rise. They note that these threats to the increasing population, housing, and infrastructure of Atlantic and Gulf coastal jurisdictions is widely acknowledged. However, despite extensive advocacy for policies, programs, and action to address hazard mitigation and climate change adaptation, the research literature has largely neglected the processes of local government adoption and implementation of these policies. The chapter begins to address this deficiency by presenting a comprehensive review of the wide variety of hazard mitigation and climate change policies that are available to coastal jurisdictions. The chapter then turns to the presentation of results from a 2013–2014 survey on the adoption of these strategies by a large sample of Atlantic and Gulf coastal jurisdictions. Peacock and his colleagues provide insight into the historical evolution of coastal hazard mitigation by comparing their data with results from a 1989 study of similar jurisdictions. Finally, their discussion of factors shaping the adoption of hazard mitigation and climate change adaptation policies concludes with an integrated model of this process suited to greater future success.

In Chapter 8, Samuel Brody, Wesley Highfield, William Merrell, and Yoonjeong Lee argue that current US policy is rooted in a recovery approach to flood risk reduction that accepts failure and only prepares for losses instead of avoiding them. The system is broken and tinkering, such as through insurance, is not enough. Furthermore, these authors argue, the dichotomy of structural versus non-structural mitigation, which has been the focus of critiques for some time, is not as important in the future that is now emerging as a shift to a focus on proactive, protective strategies. Avoided costs must be made real to decision makers so that there are usable and convincing metrics for robust and pragmatic choice in the face of uncertainty—whether of structural and/or non-structural means—that will increase protection and reduce the most likely future losses.

Finally, in Chapter 9, Himanshu Grover focuses in on the planning processes for hazard mitigation and climate change adaptation in the United States. His chapter systematically compares key concepts underlying contemporary hazard mitigation planning and climate change adaptation. Specifically, he highlights the shared goals of natural hazard risk reduction in both hazard mitigation and more recently emerging climate change adaptation policy making, and identifies vulnerability reduction as their common agenda. The chapter continues with a careful analysis of different conceptualizations of vulnerability that leads to the identification of community resilience as the bridging framework for linking rather than separating these two domains for promoting coordinated and effective local environmental vulnerability reduction. An integrated model for reducing community vulnerability to environmental hazards is presented to confront the escalation of threats due to climate change. A central feature of this model is a community's ability to break the cycle of disaster impacts by learning how to develop its adaptive capacities. The chapter concludes with a call for sustained longitudinal research to assess the dynamics of communities' hazard adjustments as they react to previous disaster impacts, anticipate future disasters, and support needed action.

In sum, these five chapters outline a holistic agenda for designing, advocating for, and testing the effectiveness of a next generation of natural hazard mitigation policy and planning approaches. Where have we been in mitigation policy and why? What are the trends and dynamics of vulnerability that we must address? How does mitigation fit within a more systematic and broad framework of resiliency in a changing world subject to a nonstationary climate, with growing societal strains and uncertainty? And, how shall we better use the mitigation planning and policy toolbox with new understanding gained from costly lessons and emergent knowledge?

References

Burby, R. J. (2006) "Hurricane Katrina and the Paradoxes of Government Disaster Policy: Bringing About Wise Governmental Decisions for Hazardous Areas," *The ANNALS of the American Academy of Political and Social Science* 604(1): 171–191.

Burby, R. J., T. Beatley, P. R. Berke, R. E. Deyle, S. P. French, D. R. Godschalk, E. J. Kaiser, J. D. Kartez, P. J. May, R. Olshansky, R. G. Paterson, and R. H. Platt. (1999) "Unleashing the Power of Planning to Create Disaster-Resistant Communities," *Journal of the American Planning Association* 65(3): 247–258.

Burby, R. J. and S. P. French. (1981) "Coping with Floods: The Land Use Management Paradox," *Journal of the American Planning Association* 47(3): 289–300.

Clark, W. C. and N. M. Dickson. (2003) "Sustainability Science: The Emerging Research Program," *Proceedings of the National Academy of Sciences* 100(14): 8059–8061.

Cutter, S. and C. Emrich. (2005) "Are Natural Hazards and Disaster Losses in the US Increasing?" *EOS, Transactions of the American Geophysical Union* 86(41): 381, 388–389.

Doyle, A. and R. McNeill. (2014) "Why Britain is Flirting with Retreat from its Battered Shores," Reuters News Service. Filed Dec. 12, 2014. Retrieved from the World Wide Web on December 12, 2014 from URL: www.reuters.com/investigates/special-report/waters-edge-the-crisis-of-rising-sea-levels/

Godschalk, D., T. Beatley, P. Berke, D. Brower, E. Kaiser, C. Bohl and R. M. Goebel. (1998) *Natural Hazard Mitigation: Recasting Disaster Policy and Planning*, Washington DC: Island Press.

Heberle, L., S. Merrill, C. Keeley, and S. Lloyd. (2014) "Local Knowledge and Participatory Climate Change Planning in the Northeastern US," 239–252, in F. Leal, F. Alves, S. Caeiro, and U. Azeiteiro (eds.), *International Perspectives on Climate Change: Latin America and Beyond*, London: Springer Publishing.

Kates, R., W. Clark, R. Corell, J. Hall, et al. (2001) "Sustainability Science of Local Communities," *Science* 292(5517): 641–642.

Kirshen, P., S. Merrill, P. Slovinsky, and N. Richardson. (2012) "Simplified Method for Scenario-Based Risk Assessment Adaptation Planning in the Coastal Zone," *Climatic Change* 113(3–4): 919–931.

McNeill, R., D. Nelson, and D. Wilson. (2014) "As Seas Rise, a Slow-motion Disaster Gnaws at America's Shores" Reuters News Service, Posted Sept. 4, 2014. Retrieved from the World Wide Web on December 12, 2014 from URL: http://www.reuters.com/investigates/special-report/waters-edge-the-crisis-of-rising-sea-levels/.

Merrill, S., D. Yakovleff, D. Holman, J. Cooper, and P. Kirshen. (2010) "Valuing Mitigation Strategies: a GIS-Based approach for Climate Adaptation Analysis," *ArcUser*, Fall 2010.

Mileti, D. (1999) *Disasters by Design: A Reassessment of Natural Hazards in the United States*, Washington DC: Joseph Henry Press.

Milly, P. C. D., J. Betancourt, M. Falkenmark, R. M. Hirsch, Z. W. Kundzewicz, D. P. Lettermaier, and R. J. Stouffer. (2008) "Stationarity is Dead: Whither Water Management?" *Science* 319(5863): 573–574.

Stiles, W. A., Jr. (2014) "All Adaptation Is Local," *Issues in Science and Technology* 30(2): 57–64.

Susskind, L., S. McKearnan, and J. Thomas-Larmer. (1999) *The Consensus Building Handbook: A Comprehensive Guide to Reaching Agreement*, Thousand Oaks, CA: Sage Publications.

White, G.F. (1936) "Notes on Flood Protection and Land Use Planning," *Planners Journal* 3(3): 57–61.

White, G. F., R. W. Kates, and I. Burton. (2001) "Knowing Better and Losing Even More: The Use of Knowledge in Hazards Management." *Global Environmental Change Part B: Environmental Hazards* 3(3): 81–92.

5

THE POLITICS AND GOVERNANCE OF MITIGATION

Considerations for Planning

Kristin Taylor and Thomas Birkland

Introduction

It is well known that professional planners and the land use planning process are effective in mitigating disaster damage and improving community readiness for disasters (Burby and Dalton 1994), thereby making communities more resilient to the impacts of disasters (Berke et al. 2012). Land use planning is an important government function that goes beyond preserving community character and quality of life. It also promotes public safety and welfare and therefore supports the central mission of government at all levels. Indeed, Mileti (1999: 155–156) argues that "[n]o single approach to bringing sustainable hazard mitigation into existence shows more promise at this time than increased use of sound and equitable land-use management". This is no less true today. But while planning for disasters—both in the organizational sense and in the land-use sense—seems self-evidently valuable or important to planners, important features of politics and governance work against the adoption of effective land-use planning for public safety. While the system of policy and governance addressing natural hazards performs reasonably well for smaller disasters, for significant disasters—or catastrophic disasters like Hurricane Katrina—the intergovernmental system of disaster planning, preparedness, and mitigation does not always work well. In a system where planning mandates are weak and growth continues unchecked in the most hazardous places in the nation—coastal areas and floodplains, in particular—our existing policies and systems of governance work to aggravate, not mitigate, hazards.

The central premise of this chapter, and its inherent complication, is that while hazard mitigation politics and policymaking occur in the federal, state and local government, the *governance* of mitigation and the implementation of policies primarily centers on local governments. This fact presents two important challenges for hazard mitigation and ultimately disaster resilience. First, local governments implement hazard mitigation policies in a governance system in which federal policy priorities can shift and policies can often work at cross-purposes, undermining what intergovernmental incentives exist to mitigate hazards. Second, the capacity of local governments to engage in effective mitigation varies widely. Local governments are often conflicted as to what they can or should do to balance growth and land use, and they tend to take a short-run view of the benefits of development without accounting for long-term costs to the community in the future. Indeed, if we consider intergenerational equity to be important in the development of sustainable communities, current policies and practices are often unsustainable, as later discussion will show.

This chapter is divided into three parts. First, we describe the political and policy structures for disaster policy governance, including the major federal policies and the intergovernmental challenges of such policies. Second, we describe conflicting visions of the different techniques for mitigation within the context of local government and politics. Third, we conclude with examples of how states have overcome the intergovernmental problems to encourage better disaster management and planning.

The Intergovernmental Policy and Governance System in Natural Hazards

For most of American history, disaster planning, relief, recovery, and mitigation were the responsibilities of state and local governments, with episodic involvement from the federal government. Local self-help was the dominant mode of organization, and some communities sought to avoid the appearance of needing outside assistance, out of fear that such apparent vulnerability would be bad for local development. A significant deviation from this doctrine was the enactment of the Flood Control Act of 1928, which gave the U.S. Army Corps of Engineers significant powers to undertake engineering projects to prevent or mitigate flooding along the Mississippi and Sacramento Rivers.

During the Depression, the federal government became more involved in local disaster relief, often under the auspices of existing New Deal agencies (Dauber 2013). As the federal government was called upon to provide more assistance, Congress realized that some sort of regular and predictable process was necessary to describe—and circumscribe—the federal role in disaster relief. Thus, since the first Disaster Relief Act was enacted in 1950, American disaster policy doctrine has been that emergency management is a state and local responsibility. Under such a system, support was available from the federal government but was not assumed to be automatic, nor was such assistance considered a first resort. Indeed, this 1950 act simply sought to codify long-standing practice under what May and Williams (1986) call "shared governance."

The federal government sets its priorities for disaster preparation, response, recovery, and mitigation via the Stafford Act and its amendments. Godschalk and colleagues (1998: 15) refer to it as an "intergovernmental mitigation system" where governance is shared. Federal Emergency Management Agency (FEMA) regional offices facilitate state-level implementation based upon the individual state's commitment to mitigation and its own mitigation plans, and the state then implements the plan to reduce risks for local governments. Under "shared governance," the federal government not only supports the states when they lack the resources to respond to major disasters, but also encourages effective disaster preparedness and mitigation to reduce damage, thereby making disasters more manageable and creating less demand for disaster relief.

Although the role of the federal government's influence on mitigation policies should not be discounted, the so-called "intergovernmental mitigation system" may be more of an ideal type. This system relies on the states being willing and able to undertake initiatives to plan for and mitigate the effects of disasters through various risk reduction measures. This capacity is highly variable, however, across and within states (Birkland and Waterman 2008; Comfort 2007; Cutter et al. 2008; Green et al. 2007; Vale and Campanella 2005). And as we can see since the 1950 act, subsequent enactments of federal disaster law have expanded, not limited, the federal role in disaster relief. These enactments are shown in Table 5.1. While the federal government is not the first-responder, popular belief notwithstanding, the federal government has become a first financial responder. Indeed, some smaller events that decades ago would not have qualified as "presidentially declared disasters" worthy of federal assistance saw federal funding as states lobbied hard for federal funds as a form of redistributive or "pork" spending.

Table 5.1 Selected Federal Legislation on Natural Hazards, 1950–2010

Year	Legislation	Summary	Mitigation Enactments
1956	Federal Flood Insurance Act, PL 84–1016		Flood insurance program that never started because the House rejected funding for it
1950	Disaster Relief Act of 1950, PL 81–875	Formalized existing practice allowing for funding to repair local public facilities	
1966	Disaster Relief Act of 1966, PL 89–769	Amended 1950 act to allow rural communities to participate, aid for damaged higher education facilities; repair of public facilities under construction	
1968	National Flood Insurance Act of 1968, PL 90–448.		The National Flood Insurance Act of 1968, Title XII of the Housing and Urban Development Act of 1968 (PL 90–448) is passed creating the National Flood Insurance Program
1969	Disaster Relief Act of 1969, PL-91-79	Debris removal, food aid, unemployment benefits, loan programs revised; duration limited to 15 months	
1970	Disaster Assistance Act of 1970, PL 91-606	Continues most provisions of the 1969 law, plus grants for temporary housing or relocation, funding for legal services	
1973	Flood Disaster Protection Act of 1973, PL 93–234.		Amending the 1968 act. Expanded coverage, imposed sanctions on communities in flood zones that failed to participate in flood insurance
1974	Disaster Relief Amendments of 1974, PL 93–288	Defined "major disasters" and "emergencies," broadened categories of allowable expenditures. Served as template for most policy until the Stafford Act. In 1977, this act was reauthorized through 1980 (PL 95-51). Again reauthorized in 1980 (PL 96-568)	Amendments are the first congressional mandate for hazard mitigation as a condition for receiving disaster assistance
1977	National Earthquake Hazards Reduction Act, PL 95-124	Bill enacted to address concerns raised by Alaska and San Fernando earthquakes, among other events	Included provisions to support research on prediction and mitigation. As reauthorized, supports research and other efforts to mitigate earthquake losses
1979			Land acquisition program for flood-damaged properties from Section 1362 of the National Flood Insurance Act of 1968 is funded. (Carter's Water Policy Initiative recommends funding buyout program in June 1978)
1982	Coastal Barrier Resources Act, PL 97-348		Restricted or prohibited use of federal funds to create or improve infrastructure in sensitive coastal areas (see also Platt 1999, 80–81)

Year	Law	Description	Details
1988	Robert T. Stafford Disaster Relief and Emergency Assistance Act, PL 100–707	Amends Disaster Relief Amendments of 1974	Emphasizes acquisition or buyouts of properties, reduce risk through construction in non-hazardous areas
1993	Stafford Act Amendments, 103–181	Enhanced 1988 law to emphasize mitigation	Increased funding for buyouts of disaster-prone land; increased federal share of buyouts from 50% to 75% of costs
1994	The Community Development and Regulatory Improvement Act (PL 103–325), the National Flood Insurance Reform Act of 1994		Most comprehensive flood insurance changes since 1974. Creates the Mitigation Assistance Program, creates the National Mitigation Fund, provides additional coverage for compliance with land use and control measures
1999	Consolidated Appropriations Act (PL 106–113)		Directs FEMA to study the feasibility and justification for reducing buyout assistance to property owners who choose not to buy flood insurance. Authorizes $215 million for buyout/relocation
2000	Disaster Mitigation Act and Cost Recovery Act, PL 106–390	Encourages state and local hazard mitigation, requires enhanced state and local mitigation planning	Amends the Stafford Act, provides technical and financial assistance to state and local governments to assist in the implementation of pre-disaster hazard mitigation measures. Requires states to prepare a comprehensive state program for emergency and disaster mitigation prior to receiving funds from FEMA
2002	Homeland Security Act, PL 107–296	Made FEMA a part of the new Department of Homeland Security	
2004	Bunning-Bereuter-Blumenauer Flood Insurance Reform Act of 2004, PL 108–264		Provisions to encourage owners of repetitively flooded properties to accept buyouts or lose eligibility for flood insurance
2004	National Windstorm Impact Reduction Act of 2004, Title II of PL 108–360	Creates a National Windstorm Impact Reduction Program patterned after the Earthquake Program. This law is part of the Earthquake program reauthorization; the program is far smaller than the earthquake program	Some R&D spending on wind hazards
2006	Post-Katrina Emergency Management Reform Act of 2006	Returned preparedness functions to FEMA, made FEMA a stand-alone agency within the DHS, but limited DHS's ability to greatly change its management. Requires the FEMA administrator to be a professional emergency manager	

Source: Based on May 1985, Tables 2.2, 2.3, and 2.4, with updates.

The primary federal law governing the federal government's role in disasters is the Stafford Act (PL 101-707, 42 USC 5121 et seq.), as amended in 1993, 2000, 2003, 2005, and then again in 2006. The Stafford Act represents 60 years of growth in federal disaster aid programs, including individual assistance programs to people and public assistance to local governments whose government-owned infrastructure has been damaged. Most disasters are small, on a local scale, and are handled reasonably efficiently by local and state governments with some federal assistance—if any is needed at all. However, when a large-scale event occurs and federal response is viewed as inadequate or failed, these failures come under official scrutiny and the harsh glare of the public. In this way the most visible and well-known, but also the most anomalous, disasters tend to attract the greatest attention and become the most important drivers of reappraisal and reform, even when they do not represent the "typical" disaster.

Of particular importance to planners was the so-called Volkmer amendment, a section of the 1993 Hazard Mitigation and Assistance Act that amended the Stafford Act to increase the funds available to states for hazard mitigation (including land buyouts) (Sylves 2008). This reaction to the damage done by the 1993 Midwest floods was a highly visible commitment to mitigation.

The Federal Emergency Management Agency (FEMA) is primarily responsible for Stafford Act implementation. FEMA's poor record of preparedness and response before and during Hurricane Hugo and the Loma Prieta earthquake in 1989 and in South Florida's Hurricane Andrew in 1992 nearly led to the abolition of the agency and the reassignment of its programs to other agencies. Instead of abolishing FEMA, in 1993 President Clinton appointed James Lee Witt, who had served as his state emergency management director in Arkansas, to serve as its director.

Witt and the Clinton administration knew that rapid and effective federal relief would yield substantial public and political support (Cooper and Block 2006: 54; Platt 1999), and the number of disasters that were presidentially declared disasters (under the meaning outlined by the Stafford Act) grew. Witt also created innovative federal programs to improve local and state capacity in disaster planning and response and also to promote hazard mitigation. Emergency managers and students of disaster policy lauded these efforts. At the same time, the Clinton administration was much more willing than previous administrations to issue disaster declarations for a much wider range of events—even "large snowstorms in regions well accustomed to snow" (Allen 1997; Miskel 2006: 110). FEMA thereby stoked demand for rapid and generous federal support, with less emphasis on state and local efforts. Perhaps paradoxically, the Clinton-Witt era was progressive in that it heavily promoted hazard preparedness and mitigation, but it also understood disaster relief to be a political tool and behaved accordingly with generous relief, thereby stoking public expectations further.

Despite this increased federal activism, the "federal support/local responsibility" doctrine was at least rhetorically maintained until September 11, 2001, after which the federal government's top priority became counterterrorism and attack response. While the emergency management system worked very well on September 11, the federal government sought to centralize a great deal of local emergency management decision-making in the federal government (Posner 2007), thereby undermining notions of shared responsibility or shared governance. This centralization was manifest in the formation of the Department of Homeland Security (DHS). The FEMA "brand" was maintained as the public name of the Emergency Preparedness and Response Directorate that, in practice, denied the FEMA head the role of effectively managing and leading federal disaster mitigation, relief, and recovery efforts. These major policy changes reflected "opportunistic federalism" (Conlan 2006), in which the September 11 attacks provided the opportunity to centralize more power in the federal government (Birkland and Waterman 2008).

This centralization undermined the more cooperative form of federalism developed during the Clinton administration, under which the federal government had begun to induce states and local governments to take action to prepare for disasters and mitigate their effects. Of course, this does not deny that federal government disaster relief was much more readily available than it had been in prior administrations. In any case, Hurricane Katrina showed that the attempted centralization of emergency management was a failure, and it also revealed how unprepared the nation was for even a well-known and predictable hazard (Tierney 2005). One of the major outcomes of the storm was the enactment of the Post-Katrina Emergency Management Reform Act (PKEMRA) of 2006, which, among other things, restored FEMA as a quasi-independent organization located within the DHS but with far greater autonomy and improved access to the president during emergencies.

Despite the well-known shortcomings in FEMA's response to Katrina, Americans came to believe that FEMA has powers and capabilities that are far greater than those specified in the Stafford Act. To the extent that criticisms of the federal government's future efforts will sound similar to those criticisms leveled after Hurricane Katrina, these reproaches share a belief that the federal government should be the prime mover in response to all disasters of this magnitude.

Under this shared system of governance, communities are much better served when their local, regional, and state governments understand the limits of federal action and the limited policy tools that the federal government has at its disposal to promote disaster preparedness and mitigation. There are many examples of communities and states having done so and having seized the initiative in preparing for and mitigating disasters.

Conflicting Visions of Implementing Natural Hazard Mitigation Policies

The Promise and Peril of Structural Mitigation

At least since the originator of comprehensive flood plain management, Gilbert White, asked flood managers to "work with nature" to manage flooding and to mitigate its effects, disaster mitigation has been an important tool for protecting lives and property in extreme events. White's notion of working with nature is generally what is considered non-structural hazard mitigation. If working with nature is the preferred way of mitigating disaster losses, planners will need to continue to be at the forefront of community disaster planning and preparedness (see White 1945, 1958, 1975, 1977; White and Kates 1968). But White's approach to *working with nature* has proved to be inconsistent with traditional notions of *taming nature* by building dams, levees, floodwalls, and other structures meant to contain floods or storm surge. While this structural mitigation approach has prevented significant losses, particularly flood losses, it has also encouraged development in hazardous areas.

For example, following Hurricane Betsy in 1965, the federal government supported the construction of a modern levee system in New Orleans built to withstand a 100-year storm (Burby 2006). In the short term the city was able to expand development and protect life and property from regular and minor floods. However, the long-term effect of the levee system was to encourage development in the natural floodplain, subjecting people to catastrophic flood risks who otherwise would have lived in less risky areas. This example represents the careful balance between structural and non-structural mitigation techniques. While in some cases the short-term benefits of some structural mitigation techniques can be a useful—and sometimes the only—option, the long-term effects can lead to unintended consequences that exceed the original risks.

It is important to note that the shared governance system in the United States is inherently biased toward structural mitigation techniques. "Hazard mitigation" as an important element of disaster policy and emergency management is relatively recent. The term gained prominence only in the early 1990s, concurrent with President Clinton's appointment of James Lee Witt to lead the Federal Emergency Management Agency and with Witt's creation of the Mitigation Directorate in 1993. Indeed, "when the Mitigation Directorate was established on November 29, 1993, mitigation became the cornerstone of emergency management for the first time in the history of Federal disaster assistance" (National Flood Insurance Program 2014). The presence of hazard mitigation as a federal policy priority has waxed and waned, often sending unclear messages to state and local officials about both the importance and value of mitigation.

Moreover, the goal of the federal disaster relief and recovery policy, from congressional and administrative perspectives, has been to restore communities to pre-disaster conditions, not to reduce risks in the future (Birkland 1997: 47–73; Birkland 1998). While we noted that disaster mitigation did experience some greater attention during the Clinton administration, this focus on mitigation was an anomaly in the history of FEMA because, after all, the Witt years at FEMA were themselves anomalous—a time in which the agency focused on natural disaster mitigation and was led by professionals (Tierney 2005). More typically, FEMA, and disaster policy in general, has tended to focus on response and recovery, rather than mitigation.

The rationale for this post-disaster focus is simple; disaster policy in the United States is a form of distributive spending and thus has developed into something of an entitlement program (Platt 1999: 11–46) in which the public, supported by local governments and news media, has come to expect intensive federal involvement in and funding of disaster relief as a matter of course. Such spending is, of course, politically popular among elected officials as well (Roberts 2013). This is not meant to suggest that federal officials ignore mitigation so as to create opportunities for pork. Rather, it is clear that disaster relief spending carries with it far greater benefits and fewer political risks than improved mitigation policy affords.

The implications of the mitigation goal mean that communities are rebuilt to be structurally stronger but not necessarily more resilient to the next event. Serious consideration about whether to relocate people and property out of risky areas or to forego development in the name of public safety is rare.

Through the lens of public policy scholarship, we see the challenge of hazard mitigation governance as rooted in the process of policy implementation. Policy implementation essentially means putting laws, regulations, executive orders, programs, and plans into practice. While many policies that promote mitigation are made at the federal level, *implementing* effective disaster policies has been undertaken by state and local governments through, for example, land use planning (including in some places mandated comprehensive planning and disaster elements), zoning, and building codes. Land use planning also implements key requirements of the National Flood Insurance Program (NFIP). So, while state governments and federal agencies play an increasingly important role in all aspects of the disaster cycle, including mitigation, it is state and local governments who codify the rules under which mitigation efforts should proceed.

In the 1970s through early 1980s, understanding policy implementation was a major concern among policy scholars. A particularly useful and prominent model of policy implementation is what Malcolm Goggin and his colleagues call the "Communications Model of Intergovernmental Implementation" (Goggin et al. 1990). This model states that credible messages sent by authoritative senders to willing receivers will be more quickly and faithfully acted upon than will less credible messages sent to unwilling receivers. The question for students of policy, then, is whether the message from top-level governments is credible, whether the sender is considered authoritative, and whether local governments are willing to act upon the information being sent. In this model, state governments often act as intermediaries and interpreters between national and local governments (see also Hill and Hupe 2002: 65–67).

For planners, the important question is whether the messages received from the federal government are perceived to be credible and the extent to which those messages are considered to be authoritative as they are sent from federal to state to local officials responsible for mitigation. The relative authoritativeness of federal government with respect to mitigation has been uneven. For example, federal incentives to participate in the Community Rating System (CRS) have been relatively ineffective at influencing local planning towards mitigation (Berke et al. 2014).

In disaster mitigation, a great deal of responsibility rests with state governments. The role that states play is important because planning mandates, particularly for land use and hazard mitigation, are influential on the capacity and quality of local land use plans (Berke et al. 2014). This is because the federal government shares most resources and creates obligations for states in many cases. State governments are therefore in the middle of the communications chain between the top-level policy designers and local implementers. Crucially, "state-level implementers form the nexus for the communication channels and these implementers are the target of the implementation-related messages transmitted from both federal- and local-level senders" (Lester and Goggin 1998: 4). When the federal government sends implementation messages that are considered unclear (like the incentives under the CRS), states play an important role in sending more authoritative messages to local governments that mitigation—particularly mitigation through land use planning—is important. As recipients, "state-level implementers must interpret a barrage of messages. Structuring the interpretation process are the form and content of the messages and the legitimacy and reputation of the sender. Therein lies the key to implementation's variability" (Lester and Goggin 1998: 4).

This discussion has focused primarily on the role of planning in hazard mitigation. While land use planners and emergency managers aren't often directly concerned with hazard mitigation, planners should be concerned when disaster relief programs create a form of "moral hazard." A moral hazard refers to instances when individuals or organizations are willing to accept risks if they are not responsible for the consequences of an actual disaster. Kousky and Shabman argue that "there is no compelling evidence for a moral hazard in disaster relief programs for households" (2013: 12), but that increasingly generous relief, such as after Hurricane Sandy, may induce moral hazard. Furthermore, this individual moral hazard is a function of relief provided under what is known as the individual assistance program. A parallel program, the public assistance program, provides relief to local governments whose facilities and infrastructure are damaged. When combined with other policies (such as support for structural mitigation or rebuilding projects using Community Development Block Grant [CDBG] funds), such programs may further induce communities to develop in riskier areas. This interaction between potential relief funds for communities and the incentives for mitigation should raise concerns among planners that these policies work at cross-purposes, creating additional risk.

Plans and plan quality can vary widely from one community to another or one state to another. The differences range from comprehensive land use planning that includes hazard mitigation to stand alone hazard mitigation plans to no comprehensive land use planning. There are three potential explanations for such variation in the type of planning and the quality of the plans: capacity of local governments to plan, the commitment to plan, and the effectiveness of the communications associated with planning. In states with comprehensive planning mandates, the quality of local plans can be bolstered if the mandate compels local government officials and citizens to be actively engaged in the planning process (Berke and French 1994). However, the effect of state mandates on local comprehensive land use planning may be limited to hazard mitigation planning and not affect other community plans for transportation, housing, and the like (Lyles et al. 2014).

Land use planning has been found to be an effective non-structural mitigation tool because it channels growth away from hazardous areas, particularly flood-prone areas (Brody et al. 2007). Some states, like North Carolina, require communities in hazardous areas, particularly coastal zones, to engage in hazard-specific land use planning in addition to standard land use planning. But in other states, like New York, there is no requirement for local land use planning and if communities decide not to plan, the existing land use patterns form the overall plan. Such plans tend to be based on the accretion of local practice and experience rather than on considered planning. For this reason, states that require local land use planning compel local governments to adopt such plans when they otherwise would not. These mandates can help communities overcome resistance to hazard mitigation that stems from pro-growth political interests or community affluence, but the overall quality of the plans suggests that the net result might be one of rote compliance (Berke et al. 1996). However, political tensions emerge because mitigation land use planning may reduce the availability of desirable and attractive parcels of land that would otherwise be ripe for local economic development. Moreover, land use planning has been found to be a low priority for local governments as a preventative measure for reducing risk in hazardous areas (Berke et al. 2014)

But these planning mandates—or even local responses to federal funding requirements that call for mitigation planning—do not always yield high quality plans. Local commitment to planning and mitigation can either contribute to the overall quality of the plan or hinder it. The so-called "commitment conundrum" suggests that mandates can only go so far to bolster plan quality—local political commitment to mitigation and planning are also necessary (Burby and May 1989). Plans can be poorly designed and in communities that have standalone mitigation plans, they do not incorporate natural hazards into their comprehensive land use plans (Burby and French 1985; Burby and Kaiser 1987; Godschalk et al. 1998; Olshansky 2001). And, indeed, Burby and May (1989) note that some local government officials are entirely indifferent to mitigation planning, either because they don't understand it or because they fear it will harm economic development prospects in the short term.

Federal attempts to engage in mitigation policy leadership are most evident in the NFIP. While the NFIP's website claims that mitigation is the cornerstone of emergency management, this is not reflected as a priority in national policy, nor has it been fully successful in mitigating hazards. We know federal policies have little influence on local governments to mitigate by means of land use planning (Berke et al. 2014). One explanation is the shifting federal attention towards and commitment to mitigation on the policy agenda over the last 15 years. Indeed, in 2001 the prominence of disaster mitigation declined as a result both of the change of direction that occurred with the Bush Administration's decision to staff FEMA leadership with inexperienced political appointees rather than disaster professionals, thereby largely driving hazard mitigation lower on the agenda (Tierney 2005), and also as a result of the termination of "Project Impact," a FEMA initiative designed to induce greater community participation in hazards mitigation. The advent of the Bush Administration also marked the beginning of substantial cuts to federal mitigation funding.

Further impeding progress on mitigation was the creation of the Department of Homeland Security and FEMA's being subsumed under it, further driving down FEMA's leadership on disasters and emergency management in general. Mitigation, even after FEMA was returned to some autonomy after Hurricane Katrina (Post-Katrina Emergency Management Reform Act 2006), has not returned to its former prominence at FEMA, having been subsumed primarily under the flood insurance program.

This is not to say that reform has not been attempted. Congress passed, and President Obama signed, the Biggert-Waters Flood Insurance Reform Act of 2012. The Biggert-Waters Act was a serious attempt to move the NFIP towards financial solvency. Among other things, this new law would have raised flood insurance rates to reflect actual risks of flooding, thereby, for the first time, making the NFIP a risk-based insurance program, rather than a development subsidy. Furthermore, the Biggert-Waters Act was a way to limit the subsidized risk that had been a long-criticized feature of the NFIP. The law gradually increased flood insurance premiums to reflect the full cost of the flood risk to owners, limited properties that could be grandfathered in for old insurance rates, and reduced subsidized flood insurance rates for vacation homes or properties subjected to repetitive flood losses (Beer 2014).

However, after major provisions of the Biggert-Waters Act were implemented in October 2013, there was significant public concern about the increasing premiums and the relative fairness of the increases (Alvarez and Robertson 2013; Beer 2014). While affected property owners along the Atlantic coast had already experienced an increase in flood insurance rates after Hurricane Sandy, property owners in states like New York, Florida, Louisiana, and Mississippi claimed that they were unduly bearing the burden of the premium increase. Congress subsequently acted to reverse major provisions of the Biggert-Waters Act by passing the Homeowner Flood Insurance Affordability Act (HFIAA) of 2013 (replacing Biggert-Waters as PL 113-89). The legislation was signed into law in March 2014 and it capped premium increases at 18 percent per year, reinstated grandfathering provisions for properties that would be subject to increased premiums when new flood maps are drawn, and refunded money to owners who had already paid increased rates under the NFIP.

The Biggert-Waters reform saga helps explain how difficult it is to design and implement effective hazard mitigation policies, regardless of the manifest value of such policies; in one oft-cited report, analysts claimed that for every dollar spent on hazard mitigation, four dollars in future losses would be saved (Multihazard Mitigation Council 2005). The problem with this calculation is that the dollar being spent, and the dollars being saved, may not be coming from or flowing to the same people or place. Furthermore, the uncertainties of when and where a disaster might strike compete with the current realities of many communities, which, as the urban politics literature has revealed, create great pressures for land development even when risks are present (Logan and Molotch 1988; Peterson 1981). This reflects what Ray Burby (2006) calls the local development paradox, and the dilemma facing policy makers addressing mitigation in a federal system: who should be responsible for mitigation, and who should pay for it?

Overcoming Challenges to Implementing Mitigation Policies

While the federal government, primarily through FEMA, has come to be viewed as the center for natural hazards policy in the United States, the intergovernmental system still, even after all the changes to policies since 1950, focuses responsibility for disaster response, mitigation, and recovery on state and local governments. The federal government creates incentives for states and localities to adopt certain policies. With respect to planning, the federal government can influence states to adopt comprehensive plan mandates in about the same manner as it can influence building code adoption. Under the Disaster Mitigation Act of 2000 (DMA), states were required to include hazards for local land use planning as a condition of receiving disaster recovery assistance, but plan quality as mandated by the DMA was poor (Berke et al. 2012).

But Goggin's Communications Model, when viewed as a classic question of top-down policy formulation and innovation, doesn't account for what scholars understand to be "network governance" of complex problems (Bogason and Musso 2006; Jones et al. 1997). Such a conception of policy design and implementation places much less emphasis on hierarchy, whether organizational or constitutional, and instead understands governance to be a series of interactions between various actors, none of whom are assumed to be at the "top" of a system, but some of whom may, in social network terms, be considered "nodes." In this conception, state governments might be considered nodal, but they would be a part of a larger network that responds to various incentives from other actors to choose to adopt, or not adopt, improved disaster mitigation practices. This network governance approach is also consistent with the norms of the planning profession, which highly values partnerships and collaborations instead of edicts and mandates.

If we understand mitigation to be a policy arena characterized by networked governance, we can understand the many individual success stories that have become exemplars of effective hazard mitigation. We can also better understand why local and state governments innovate.

Scholars of policy innovation note that an important feature of the federal system is the emergence of certain states and localities as leaders or innovators, either taking initiative without significant federal support, or leveraging that federal support to achieve better than typical outcomes. For example, the 1993 Midwest Flood affected nearly every state in the Midwest and resulted in $15 billion in

Table 5.2 Examples of State and Local Mitigation Initiatives

Year	Location	Innovation	Result	Source
1972	California	Alquist–Priolo Act	Prevents building on active fault traces—first earthquake mitigation legislation with a major land use component in California	www.conservation.ca.gov/cgs/rghm/ap/Pages/main.aspx
1984	Tulsa, OK	Food management innovations	After a major flood, Tulsa adopted a series of land use innovations avoid building in flood-prone land. "Today, Tulsa's floodplain and stormwater program is based on respect for the natural systems. . . . We're building parks in the floodplains, sports fields in stormwater detention basins, and greenway trails on creek banks. We are forging strong partnerships with federal and state agencies. And we've stopped creating new problems. Since the City adopted comprehensive drainage regulations 15 years ago, we have no record of flooding in any structure built in accord with those regulations . . . . Because the federal government gave our program its highest ranking, Tulsans enjoy the lowest flood insurance rates in the country."	www.cityoftulsa.org/city-services/flood-control.aspx
1993	Petersburg, IL	Land Buyouts	Following multiple repetitive flood losses, the city of Petersburg, IL sought funding from FEMA under the Hazard Mitigation Grant Program and other source to relocate housing and other structures out of the floodplain The project incorporates "re-use planning" for lots identified for the project and the surrounding sites.	www.illinois.gov/ready/press/pages/071913.aspx www.floods.org/PDF/MSS_IV_Final.pdf
1990	California	Seismic Hazards Mapping Act	"Provide for a statewide seismic hazard mapping and technical advisory program to assist cities and counties in fulfilling their responsibilities for protecting the public health and safety from the effects of strong ground shaking, liquefaction, landslides, or other ground failure and other seismic hazards caused by earthquakes."(California Public Resources Code Sec. 2692)	www.conservation.ca.gov/cgs/shzp/Pages/shmpact.aspx

Date	Location	Name	Description	Source
Late 1990s	Seattle, WA	Seattle Project Impact	"Seattle Project Impact was a public–private partnership whose overall goal was to make regional communities more resistant to the damaging effects of disasters. The Project encouraged people to take action before a disaster occurred through initiatives promoting safer homes, schools, businesses, and better earthquake and landslide hazard mapping." Was one of the most active Project Impact locations in the nation	http://mitigation.eeri.org/resource-library/homeowners/regional-us/seattle-project-impact
2000	North Carolina	North Carolina Flood Mapping	State and federal program to map and update all 100 NC counties for flood hazards; much more active than typical flood mapping and revision programs	www.ncdps.gov/Index2.cfm?a=000003,000010,000176
2014	Boulder County, CO	Post-disaster redevelopment review	Hazard mitigation review process for repairs to damaged homes and construction of homes that had been destroyed	www.bouldercounty.org/flood/pages/default.aspx
1997 through 2000s	Grand Forks, ND	Red River Flood Management	After the devastating Red River floods of 1997, federal, state, and local leaders in the public and private sector worked on a comprehensive flood management and mitigation plan that included the usual engineered works and, crucially, a buy-out of 800 homes and 50 businesses. As one FEMA official noted "Some of the areas that were low-lying — that they just knew it would be difficult to protect, even with a real robust structural mitigation project—they went with the acquisition."	/www.businessinsurance.com/article/20130324/NEWS06/303249977

damage and 50 deaths, but Iowa, Kansas, Minnesota, Missouri, and Nebraska bore the brunt of the flooding (Sylves 2008). Among these states, Iowa was the only affected state to press for land buyouts to mitigate the hazard following the flood. As a local-level example, in spite of having levees in place, Valmeyer, Illinois, was so severely damaged by the 1993 flooding that it relocated itself completely out of the floodplain (Association of State Floodplain Managers 2000). In addition, we know that states were mandating land use planning requirements long before FEMA established the mitigation directorate and before Congress adopted the DMA.

In 1985 Florida adopted the Growth Management Act, requiring local governments to include hurricane and flood mitigation tools in their comprehensive plans (Deyle et al. 2008). In 1986, in the case of seismic safety, the State of California required that earthquake-specific land use planning and development requirements be included in local governments' comprehensive plans in response to the 1971 San Fernando Earthquake (Olshansky 2001). Other examples are provided in Table 5.2. These examples suggest that in the "shared governance" or "networked governance" system of mitigation, some governments or other actors will be leaders in promoting mitigation. This variability is a function of local government capacity and commitment, risk perception, local support for land-use practices that may reduce the amount of developable land, and the dilemmas posed by structural mitigation, which include local economic stimulus and some protection from typical disasters (but not the most significant ones).

The successes listed in Table 5.2—and the much larger collection of such stories published by FEMA (2011) in their comprehensive review of local best practices in mitigation—have a number of common elements that planners will find familiar:

- Local engagement in public and decision-maker education about the value of planning and flood hazard mitigation,
- Partnerships with local and state emergency planners and state economic development officials to promote actions that would support better mitigation decisions,
- Local efforts to promote state mandates or support for local comprehensive planning with a distinct hazard planning and mitigation element,
- State efforts at local capacity building that focus both on the technical aspects of mitigation planning and the political concerns for local plan acceptance, and
- Partnerships with the environmental policy community, including state agencies and interest groups, to ensure that hazard mitigation projects are ecologically sensible and thereby sustainable as effective policies.

As Dennis Mileti argues, these policies will be successful when they are

> integrated into the considerations of the daily activities of everyone who has an influence on disaster losses. This, in turn, will not be possible until hazards mitigation is housed within a redesigned national culture that favors sustainable development and people are reorganized to support that cultural shift.
>
> *(1999: 267)*

Conclusion

Effective planning and governance require, at the outset, that planners and policy makers know both the constraints and opportunities afforded by the substance and organization of disaster policy. Mitigation is not an impossible task or a far-off goal; it is something in which people, communities, and planners must engage every day. This chapter outlines those constraints and opportunities to offer encouragement for better policy.

In a nation governed under a federal system, successful hazard mitigation is very challenging. This challenge derives not only from the nature of federalism, but also from the very reason the founders proclaimed the superiority of federalism. The vast size of the United States and the physical and demographic diversity of its states provide a strong case for federalism. After all, the concerns faced by people living in Alaska, Arizona, or Alabama will likely be quite different. At the same time, the federal government can, using the vast legal and material resources at its disposal, influence state and local decision making so as to promote improved disaster mitigation. Studies of federalism, policy making, and implementation in an intergovernmental system focus on these challenges. We do the same in this chapter, but we also acknowledge that the federal system, as it has evolved into a system of shared and networked governance, also contains opportunities for state and local policy makers to leverage existing policies to create superior mitigation practices that become exemplars of best practice. While one might be pessimistic about the slow progress made in hazard mitigation, a review of the last 75 years of policy history suggests that significant improvements in mitigation policy have been made at all levels. At the same time, the challenge of mitigation in a system of shared governance is unlikely to remain the same. As climate change poses potential new hazards and as people continue to move to and develop areas near coasts, rivers, and other hazards—some of the riskiest areas of the nation—the challenges of mitigation will continue as well.

References

Allen, J. T. (1997) "Defining Disaster Down," *Slate*, January 18. www.slate.com/id/1883/. Accessed December 12, 2014.

Alvarez, L. and C. Robertson. (2013) "Cost of Flood Insurance Rises, Along with Worries," *The New York Times*, Oct. 12. www.nytimes.com/2013/10/13/us/cost-of-flood-insurance-rises-along-with-worries.html?pagewanted=all&_r=0. Accessed December 12, 2014.

Association of State Floodplain Managers. (2000) "Mitigation Success Stories in the United States," Madison, WI. www.floods.org/PDF/Mitigation%20Success%20Stories%20III_print.pdf. Accessed December 14, 2014.

Berke, P. R. and French, S. P. (1994). "The influence of state planning mandates on local plan quality," *Journal of Planning Education and Research*, 13(4): 237–250.

Berke, P. R., W. Lyles, and G. Smith. (2014) "Impacts of Federal and State Hazard Mitigation Policies on Local Land Use Policy," *Journal of Planning Education and Research* 34(1): 60–76. doi: 10.1177/0739456X13517004.

Berke, P. R., D. Roegnik, E. Kaiser, and R. Burby. (1996) "Enhancing Plan Quality: Evaluating the Role of State Planning Mandates for Natural Hazard Mitigation," *Journal of Environmental Planning and Management* 39(1): 79–96.

Berke, P. R., G. Smith, and W. Lyles. (2012) "Planning for Resiliency: Evaluation of State Hazard Mitigation Plans under the Disaster Mitigation Act," *Natural Hazards Review* 13(2): 139–149. doi:10.1061/(ASCE)NH.1527-6996.0000063.

Birkland, T. A. (1997) *After Disaster: Agenda Setting, Public Policy, and Focusing Events*, Washington, DC: Georgetown University Press.

Birkland, T. A. (1998) "Focusing Events, Mobilization, and Agenda Setting," *Journal of Public Policy* 18(1): 53–74.

Birkland, T. A. and S. Waterman. (2008) "Is Federalism the Reason for Policy Failure in Hurricane Katrina?" *Publius: The Journal of Federalism* 38(4): 692–714.

Beer, B. P. (2014) "Underwater," *Earth Island Journal* 29(4), www.earthisland.org/journal/index.php/eij/article/underwater/. Accessed December 12, 2014.

Bogason, P. and J. A. Musso. (2006) "The Democratic Prospects of Network Governance." *The American Review of Public Administration* 36(1): 3–18. doi:10.1177/0275074005282581.

Brody, S. D., S. Zahran, P. Maghelal, H. Grover, and W. E. Highfield. (2007) "The Rising Costs of Floods: Examining the Impact of Planning and Development Decisions on Property Damage in Florida," *Journal of the American Planning Association* 73(3): 330–345.

Burby, R. J. (2006) "Hurricane Katrina and the Paradoxes of Government Disaster Policy: Bringing about Wise Governmental Decisions for Hazardous Areas," *The Annals of the American Academy of Political and Social Science* 604(1): 171–191.

Burby, R. J. and L. C. Dalton. (1994) "Plans Can Matter! The Role of Land Use Plans and State Planning Mandates in Limiting the Development of Hazardous Areas," *Public Administration Review* 54(3): 229–238.

Burby, R. J. and S. P. French. (1985) "Coping with Floods: The Land Use Management Paradox," *Journal of the American Planning Association* 47(3): 289–300.

Burby, R. J. and E. J. Kaiser. (1987) An Assessment of Urban Floodplain Management in the United States: The Case for Land Acquisition in Comprehensive Floodplain Management, Technical report prepared for Association of State Flood Plain Managers, Inc. Madison, Wisconsin. www.floods.org/ace-files/documentlibrary/Publications/ASFPMPubs-%20TechRep1-6%2787.pdf. Accessed December 14, 2014.

Burby, R. J. and P. J. May. (1989) "Intergovernmental Environmental Planning: Addressing the Commitment Conundrum," *Journal of Environmental Planning and Management* 41(1): 95–110.

Comfort, L. K. (2007) "Crisis Management in Hindsight: Cognition, Communication, Coordination, and Control," *Public Administration Review* 67(s1): 189–197.

Conlan, T. (2006) "From Cooperative to Opportunistic Federalism: Reflections on the Half- Century Anniversary of the Commission on Intergovernmental Relations," *Public Administration Review* 66(5): 663–676.

Cooper, C. and R. Block. (2006) *Disaster: Hurricane Katrina and the Failure of Homeland Security*, New York, NY: Times Books.

Cutter, S. L., L. Barnes, M. Berry, C. Burton, E. Evans, E. Tate, and J. Webb. (2008) "A Place-Based Model for Understanding Community Resilience to Natural Disasters," *Global Environmental Change* 18(4): 598–606.

Dauber, M. L. (2013) *The Sympathetic State: Disaster Relief and the Origins of the American Welfare State*, Chicago: University of Chicago Press.

Deyle, R. E., T. S. Chapin, and E. J. Baker. (2008) "The Proof of the Planning is in the Platting: An Evaluation of Florida's Hurricane Exposure Mitigation Planning Mandate," *Journal of the American Planning Association* 74(3): 349–370.

Federal Emergency Management Agency. (2011) "Mitigation Best Practices: Public and Private Sector Best Practice Stories for All Activity/Project Types in All States and Territories Relating to All Hazards," Washington DC: Federal Emergency Management Agency. www.hsdl.org/?view&did=683132. Accessed December 12, 2014.

Godschalk, D., T. Beatley, P. Berke, D. Brower, and E. J. Kaiser. (1998) *Natural Hazard Mitigation: Recasting Disaster Policy and Planning*, Washington DC: Island Press.

Goggin, M. L., A. Bowman, J. P. Lester, and L. J. O'Toole, Jr. (1990) *Implementation Theory and Practice: Toward a Third Generation*, Glenview, IL: Scott Foresman/Little Brown.

Green, R., L. K. Bates, and A. Smyth. (2007) "Impediments to Recovery in New Orleans' Upper and Lower Ninth Ward: One Year After Hurricane Katrina," *Disasters* 31(4): 311–335.

Hill, M. J. and P. L. Hupe. (2002) *Implementing Public Policy: Governance in Theory and Practice*, London: Sage.

Jones, C., W. S. Hesterly, and S. P. Borgatti. (1997) "A General Theory of Network Governance: Exchange Conditions and Social Mechanisms," *The Academy of Management Review* 22(4): 911–945. doi:10.2307/259249.

Kousky, C. and L. Shabman. (2013) "Does Disaster Aid Encourage People to Locate in Harm's Way? Thinking through the Moral Hazard Effects of Federal Disaster Aid," *Natural Hazards Observer* 37(5): 1, 12.

Lester, J. P. and M. L. Goggin. (1998) "Back to the Future: The Rediscovery of Implementation Studies," *Policy Currents* 8(3): 1–9.

Logan, J. R. and H. L. Molotch. (1988) *Urban Fortunes: The Political Economy of Space*, Berkeley CA: University of California Press.

Lyles, W., Berke, P., and Smith, G. (2014). "A comparison of local hazard mitigation plan quality in six states, USA," *Landscape and Urban Planning*, 122, 89–99.

May, P. J. (1985) *Recovering from Catastrophes: Federal Disaster Relief Policy and Politics*, Westport CT: Greenwood Press.

May, P. J. and W. Williams. (1986) *Disaster Policy Implementation: Managing Programs Under Shared Governance*, New York: Plenum Press.

Mileti, D. (1999) *Disasters by Design: A Reassessment of Natural Hazards in the United States*, Washington: National Academies Press.

Miskel, J. F. (2006) *Disaster Response and Homeland Security: What Works, What Doesn't*, Westport CT: Greenwood Publishing Group.

Multihazard Mitigation Council. (2005) *Natural Hazard Mitigation Saves: An Independent Study to Assess the Future Savings from Mitigation Activities*, Washington DC: National Institute of Building Sciences.

National Flood Insurance Program. (2014) "We Are the Mitigation Directorate," www.nfipiservice.com/watermark/weare_mitigation.html. Accessed December 12, 2014.

Olshansky, R. B. (2001) "Land Use Planning for Seismic Safety: The Los Angeles County Experience 1971–1994," *Journal of the American Planning Association* 67(2), 173–185.

Peterson, P. E. (1981) *City Limits*, Chicago IL: University of Chicago Press.

Platt, R. H. (1999) *Disasters and Democracy*, Washington DC: Island Press.

Posner, P. (2007) "The Politics of Coercive Federalism in the Bush Era," *Publius: The Journal of Federalism* 37(3): 390–412.

Roberts, P. S. (2013) *Disasters and the American State: How Politicians, Bureaucrats, and the Public Prepare for the Unexpected*, New York: Cambridge University Press.

Sylves, R. (2008) *Disaster Policy and Politics: Emergency Management and Homeland Security*, Washington DC: CQ Press.

Tierney, K. (2005) "The Red Pill. Understanding Katrina: Perspectives from the Social Sciences," *Social Science Research Center*, Available from http://understandingkatrina.ssrc.org/Tierney/. Accessed May 21, 2019.

Vale, L. J. and T. J. Campanella (eds.). (2005) *The Resilient City: How Modern Cities Recover from Disaster*, New York: Oxford University Press.

White, G. F. (1945) *Human Adjustment to Floods: A Geographical Approach to the Flood Problem in the United States*, Chicago IL: University of Chicago Department of Geography.

White, G. F. (1958) *Changes in the Urban Occupancy of Flood Plains in the United States*, Chicago IL: University of Chicago Department of Geography.

White, G. F. (1975) *Flood Hazard in the United States: A Research Assessment*, Boulder CO: University of Colorado.

White, G. F. (1977) *Environmental Effects of Complex River Development*, Boulder CO: Westview Press.

White, G. F. and R. W. Kates. (1968) *The Human Ecology of Extreme Geophysical Events*, Boulder CO: University of Colorado Natural Hazards Research and Applications Information Center.

6

A GENERAL FRAMEWORK FOR ANALYZING PLANNING FOR COMMUNITY RESILIENCY

Philip R. Berke and Ward Lyles

Introduction

Growth in disaster losses throughout the 20th century has been attributed to expansion of urbanization in hazardous locations, but the frequency and severity of climate-related natural hazard events remained steady during this period. For example, Pielke et al. (2008) concluded that growth in losses related to U.S. tropical cyclones since 1900 has been minimally influenced by changes in storm climatology and rather is primarily explained by the movement of people and accompanying wealth to areas that are at higher risk. This trend is changing as climate warming increases the intensity and frequency of hazards affected by the climate (e.g., inland flooding, tropical storms, sea level rise) (Melillo et al. 2014). The 2017 Hurricane Season, with Harvey, Irma, and Maria battering Texas, Louisiana, Florida, and Puerto Rico, elevated attention in the United States to the increasingly unavoidable connection between disaster losses and climate change.

The spatial planning approach has been characterized as the most promising long-term solution to reducing the destructive effects of disasters (Mileti 1999; NRC 2014: 117–158). Almost all communities engage disaster issues via a network of multiple planning activities that can influence the communities' vulnerability in multiple ways. These include avoiding or limiting new development in hazard locations, relocating existing development away from hazard areas (Burby et al. 1999; Burby 2006), supporting the health and safety of the resident population (Horney et al. 2012), and protecting ecosystems that reduce flooding (Brody et al. 2010). Communities that invest in planning are more resilient—a critical concept in hazards research—because they have "the ability to prepare and plan for, absorb, recover from, and more successfully adapt to adverse events" (NRC 2012: 1).

Planning for community resiliency is embedded in a broader complex social and ecological system (SES). SESs are composed of multiple interacting subsystems at multiple levels. In our view, the subsystems are hazard forces (floods, droughts, tsunamis), social and biophysical environments vulnerable to hazards (populations, economies, infrastructure, ecosystems), and governance systems that include government and other organizations that plan and manage land use. In this chapter, we identify practices aimed at planning for the resiliency of complex SESs that have been tried and empirically evaluated across jurisdictions vulnerable to hazards. Our focus is on formal government—whether a national, state, or local entity—that plays a critical role in managing these subsystems toward the public purpose of vulnerability reduction. These subsystems interact to ameliorate or amplify vulnerability outcomes (people and property exposed to hazards), which in turn feed back into and affect these subsystems at varying spatial scales (e.g., metro regions, watersheds, municipalities, neighborhoods).

Knowledge is needed to enhance adoption and implementation of effective proactive plans and actions that promote resiliency to sustain SESs, but the ecological and social sciences have developed independently and are not easily integrated (NRC 2006). For example, global climate change models do not readily fit the scale needed for assessing the social and physical vulnerability impacts of local land use decisions. Furthermore, scholars have tended to formulate underdeveloped theoretical arguments to analyze aspects of planning and to prescribe universal solutions. For example, state-mandated local planning is doomed to fail because "[it] undermine[s] ongoing planning efforts in communities that are serious about managing growth" (Susskind 1978: 17). The prediction of failure is based on lack of public support for the local plan, most localities' inability to provide funding to sustain credible planning, unworkable statewide standards, and disagreement in the planning profession on what makes a good plan. The gloomy predictions, however, are unwarranted when mandates are designed to build local commitment and capacity that enable conditions that facilitate initially reluctant community partners to self-organize through comprehensive planning (Burby and May 1997) and other forms of collaborative action such as economic development and urban service delivery (Feiock 2009).

A core challenge is to determine why some communities proactively enact and implement strong plans, whereas the vast majority of communities are reactive and do little to reduce vulnerability. To confront and overcome this challenge requires the identification and analysis of relationships among multiple systems of complex SESs (NRC 2012). Understanding a complex whole requires accumulation of knowledge about specific concepts focused on resiliency, as well as how data is collected, replicated and compared across multiple studies by multiple investigators. A common framework is needed to facilitate multidisciplinary efforts toward a better understanding of complex SESs (Ostrom 2009; Peacock et al. 2008).

This chapter introduces the largely untested promise of resiliency planning. It outlines the underlying logic and potential benefits of resiliency planning and then develops a common framework to guide future analyses of factors that define the causes and outcomes of planning within broader SESs. Without a framework to organize relevant factors identified in theories and empirical research, isolated knowledge acquired from studies of resiliency planning in diverse systems in different countries is not likely to cumulate (NRC 2006; Peacock et al. 2008). Next, the chapter identifies major challenges that constrain knowledge about the performance of plans in reducing risk. One challenge is to diagnose why some communities enact and implement a coordinated network of strong plans aimed at vulnerability reduction, whereas the vast majority do not. Another challenge is the need to derive indicators that can be used to monitor and measure change in vulnerability outcomes (e.g., built, natural, social, economic) and to gauge plan performance.

Why Resiliency Planning?

Proactive planning has been set forth as a critical way to broaden responsibility for managing and building the resiliency of SESs. Scholars and practitioners contend that rather than simply reacting to a disaster, local planning enables at-risk communities to anticipate, absorb, recover from, successfully adapt to future adverse events, and to build back to become safer, healthier, and more equitable (Godschalk 2003; Mileti 1999). Such planning considers a wide range of policy instruments that are derived from the ability to regulate, spend, tax, and acquire land, which are powerful tools available to state and local governments to guide development in the most appropriate locations (see Figure 6.1). Local mitigation planning processes also provide additional benefits, including public education, consensus building, and improved coordination (see Figure 6.2).

Land Use Approach	Description
Development Regulations	
Permitted Land Use	Provision regulating the types of land use (e.g. residential, commercial, industrial, open space, etc.) permitted in areas of community; may be tied to zoning code
Density of Land Use	Provision regulating the density of land use (e.g. units per acre); may be tied to zoning code
Subdivision Regulations	Provision controlling the subdivision of parcels into developable units and governing the design of new development (e.g. site storm water management)
Zoning Overlays	Provision related to using zoning overlays that restrict permitted land use or density of land use in hazardous areas; may be special hazard zones or sensitive open space protection zones
Setbacks or Buffer Zones	Provision requiring setbacks or buffers around hazardous areas (e.g. riparian buffers and ocean setbacks)
Cluster Development	Provision requiring clustering of development a way from hazardous areas, such as through conservation subdivisions
Density Transfer Provisions	
Density Transfer	Provision for transferring development rights to control density; may be transfer of development rights or purchase of development rights
Financial Incentives and Penalties	
Density Bonuses	Density bonuses such as ability to develop with greater density in return for dedication or donation of land in areas subject to hazards
Tax Abatement	Tax breaks offered to property owners and developers who use mitigation methods for new development
Special Study	Provision requiring impact fees or special study fees on development in hazardous areas; fees could cover costs of structural protection
Land Use Analysis and Permitting Process	
Land Suitability	Hazards are one of the criteria used in analyzing and determining the suitability of land for development
Site Review	Provision requiring addressing hazard mitigation in process of reviewing site proposals for development
Public Facility Locations	
Site Public Facilities	Provision related to siting public facilities out of hazardous areas in order to maintain critical services during and after hazard events
Post-Disaster Reconstruction Decisions	
Development Moratorium	Provision imposing a moratorium on development for a set period of time after a hazard event
Post-Disaster Land Use Change	Provision related to changing land use regulations following a hazard event; may include redefining allowable land uses after a hazard event
Post-Disaster Capital Improvements	Provision related to adjusting capital improvements to public facilities Following a hazard event

Figure 6.1 Land Use Approaches Useful for Mitigating Natural Hazard Risks

A vision of a disaster-resilient community and nation set forth in a recent report, *Disaster Resilience: A National Imperative*, by the National Research Council (2012) reflects these benefits by recognizing the importance of proactive planning. Core elements of the vision are:

- Every individual and community in the nation has access to [. . .] risk and vulnerability information they need to make their communities more resilient.

Godschalk et al. (1998) describe many benefits of local planning for hazard mitigation. Specifically, hazard mitigation planning:

1. Provides a systematic approach to gathering facts about hazards, the adequacy of existing hazard mitigation policy tools adopted by the community, and a variety of other tools;

2. Educates the community in the course of generating information necessary for decision making, and particularly those with a stake in the outcomes of plans;

3. Demonstrates the connection between the public interest and governmental policies that is critical for legal defensibility;

4. Fosters debate about the issues, and helps build consensus on a vision of resiliency, goals, and action;

5. Coordinates the actions of various federal, state, and local government agencies that affect vulnerability to foster synergy, and avoid duplication of effort and conflict;

6. Guides day-to-day decisions of public officials in the context of broader vision and goals;

7. Provides a means of implementing policy by serving as a reference for elected and appointed officials to use in reaching decisions about regulations, allocating funds for capital investments, and granting permits for development; and

8. Supports monitoring and evaluation of the performance of risk reduction practices based on measurable indicators to gauge goal achievement.

Figure 6.2   Benefits of Local Planning for Risk Reduction

- All levels of government, communities, and the private sector have designed resilience strategies and operation plans based on this information.
- Proactive investments and policy decisions have reduced loss of lives, costs, and socioeconomic impacts of future disasters.
- Community coalitions are widely organized, recognized, and supported to provide essential services before and after disaster occur (NRC 2012: 1).

In response to the underlying logic and promise of planning, over the past two decades there has been increasing societal weight placed on planning to reduce the rising levels of vulnerability. This rise in vulnerability has been accompanied by a steady expansion of planning requirements placed on communities in the United States. The federal government has steadily been shifting emphasis from reactive response after a disaster to more proactive policy. The Stafford Act of 1988 required states to plan following a disaster declaration, the Disaster Mitigation Act of 2000 requires state and local governments to plan prior to a disaster declaration, and the more recent Community Rating System requirements give added credit for planning that reduces property owner insurance rates under the National Flood Insurance Program. Twenty-five states now mandate local comprehensive planning and many states require a hazard mitigation element (Schwab 2009). Many local governments are acting on their own to respond to increased vulnerability.

Evidence shows that measurable indicators of the strength of local hazard mitigation plans have a positive effect on adoption of more robust regulatory and spending mitigation policies (Berke et al. 2006; Burby and May 1997). However, these and other studies found that state and local hazard mitigation plans are often poorly crafted (cf. Berke et al. 2012; Godschalk et al. 1999; Kang et al. 2010; Lyles et al. 2014a). The results show that few communities have prepared well for hazards. Most plans have a weak factual basis (i.e., inaccurate or inadequate vulnerability and risk assessments); unclear goals and objectives; ineffective policies; and few coordination, implementation, and monitoring mechanisms. Further, the average plan quality score was less than half of the maximum. While research on biophysical and socioeconomic outcomes of plans is one of the greatest gaps in planning research, most studies have found that local government plans with goals targeting vulnerability reduction have an important positive impact on local governments adopting land use regulations, tax incentives and capital improvements programs for infrastructure aimed at mitigating pre-disaster vulnerability (Berke et al., 2006; Burby 2006; Lyles et al. 2016), facilitating voluntary mitigation actions by households (Horney et al. 2015), and reducing property damage from disaster events (Burby 2006; Nelson and French, 2002). Brody and Highfield (2005), however, found that land use design in comprehensive plans has no effect on guiding development in open spaces that generate ecosystems services with vulnerability reduction values. Thus, the predominance of evidence supports the influence of plans.

A General Framework for Resiliency Planning

As noted earlier, resiliency planning is embedded in a complex SES composed of multiple subsystems, both external (global climate change and socioeconomic systems) and internal (systems associated with plans, local government capacity to plan, and how the plans are made and implemented). The process of examining complex SESs is complicated because entirely different theories and models are used by different disciplines to analyze the respective elements that comprise the SES. For example, lack of disciplinary interaction is evident in articles that focus on the common topic of plan evaluation. Baer's (1997) classic review of urban plan evaluation approaches in the *Journal of the American Planning Association* does not cite plan evaluation studies for bio-conservation. Meanwhile, Bottrill and Pressey's (2012) review of bio-conservation plan evaluations in a major journal of the Society of Conservation Biology, *Conservation Letters*, does not cite studies in urban plan evaluations. A common conceptual framework is needed to classify subsystem elements and facilitate multidisciplinary efforts toward a better understanding of resiliency planning in complex SESs.

We present an updated version of a multi-scale, nested framework for analyzing vulnerability outcomes achieved in SESs. The evolution of plan quality research over the past two decades has helped foster this common framework. Figure 6.3 provides an overview of the framework, showing the relationships among the four organizing subsystems of an SES that affect one another, as well as social, economic, and political settings. The organizing subsystems are (1) the planning context, (2) the planning process, (3) outputs of the planning process, and (4) vulnerability outcomes. Together, the planning context and the planning process subsystems comprise *adaptive capacity*. The outputs of the planning process (plans, land use regulations, infrastructure investment programs) are the *adaptations*, while social, physical, and environmental hazard vulnerability are the outcomes of interest. As Figure 6.4 indicates, each organizing subsystem is made up of multiple first-level concepts (e.g. state policy, public engagement, plan quality) and second-level concepts (e.g., staff capacity and commitment; plan goals, facts and policies; hurricane surge and sea level rise zones; number of residential units and elderly people in hazard zones), which are further composed of deeper third-level indicators.

This framework helps to identify relevant indicators for studying a given SES, such as the loss of wetlands and flood mitigation functions in Florida and the capacity of land-use planners to protect wetlands (Brody and Highfield 2005). It also provides a common set of variables for organizing studies of similar SESs, such as why the strength of hazard mitigation plans in some U.S. states is stronger than in others (Burby and May 1997), why some local district council plans do better than others in how well they aim to protect indigenous rights in New Zealand (Berke et al. 2002), or how the strength of land use plans affects the level of damage after the Northridge earthquake in California (Nelson and French 2002).

A framework is useful for organizing common measurable indicators, designing data collection instruments to evaluate plans, conducting fieldwork, collecting digital geospatial data, and analyzing findings about the resiliency of SESs. It guides the identification of important factors that are likely to influence plans for vulnerability reduction outcomes for one type of state policy regime and level of local government adaptive capacity, and not others. Without a framework to organize theories and empirical research, isolated knowledge derived from studies of different resiliency systems in various places by diverse researchers has limited potential to cumulate.

To illustrate one use of the SES framework, we focus on the following question: When will communities vulnerable to hazards and climate change adopt and implement a strong plan aimed at hazard vulnerability reduction? In a prominent national consensus study, Mileti (1999) argued that research is needed to understand why some communities adopt and implement strong plans while most do not. Since that time research on resiliency planning has accumulated and improved our ability to make predictions of action. Figure 6.4 lists the first-level concepts (e.g. state policy) and second-level concepts (e.g. mandates and oversight of local planning) identified in many empirical studies as influencing adaptations and hazard vulnerability. The choice of relevant third-level indicators to measure the concepts in analyses varies based on the particular questions under study, the location of the SES, and the spatial scales of analysis (regional or local). Of the third-level indicator, seven (indicated by asterisks in Figure 6.4) are frequently identified as positively or negatively affecting the likelihood of communities to act (cf. Berke et al. 1996; Brody 2003; Burby 2003; Burby and Dalton 1994, Dalton and Burby 1994). To explain why these indicators are potentially important for understanding resiliency and, in particular, for addressing the question of when proactive responses will occur, we briefly discuss the relevance for hazard vulnerability, adaptation, and adaptive capacity.

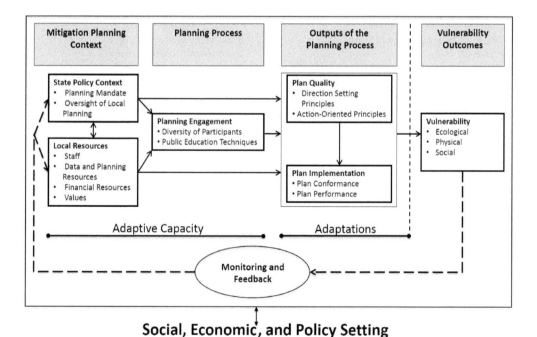

Figure 6.3 Conceptual Framework for Resiliency Planning

Planning Context

State Policy (SP)
 Mandates (SPM)
  SPM1 – Presence/Absence of Mandate *
  SPM2 – Mandate Design Features *
  SPM3 – Mandate Clarity
 Oversight of Local Planning (SPO)
  SPO1 – Mandate Enforcement Effort *
  SPO2 – Technical Assistance
  SPO3 – Guidance and Resources

Loca l Resources (LR)
 Staff (LRS)
  LRS1 – Number of Planning Staff
  LRS2 – Consultant Involvement
 Data and Planning Resources (LRDR)
  LRDR1 – Prior Plan Quality *
  LRDR2 – Geospatial Data and GIS Capabilities
 Financial Resources (LRB)
  LRB1 – Planning Budget (General)
  LRB2 – Planning Process Budget
 Values (LRV)
  LRV1 – Staff/Agency Commitment *
  LRV2 – Elected Official Commitment *

Planning Process

Planning Engagement (PE)
 Diversity of Participants (PED)
  PED1 – Number of Types of Stakeholder Groups *
  PED2 – Planner Involvement in Hazards Planning
 Public Participation Techniques (PEP)
  PEP1 – Information Sharing Techniques
  PEP2 – Public Education Techniques

Outputs of the Planning Process

Plan Quality (PQ)
 Direction Setting Principles (PQDS)
  PQDS1 – Vision
  PQDS2 – Goals
  PQDS3 – Fact Base
  PQDS4 – Policies
 Action -Oriented Principles (PQAO)
  PQAO1 – Participation
  PQAO2 – Inter-Organizational Coordination
  PQAO3 – Implementation
  PQAO4 – Monitoring and Evaluation

Plan Implementation (PI)
 Plan Conformance (PIC)
  PIC1 – Policy adoption
  PIC2 – Permit/Rule enforcement

 Plan Performance (PIP)
  PIP1 – Plan use in decision-making situations
  PIP2 – Affects-based measurements

Hazard Vulnerability

Ecological Vulnerability (EV)
 Ecosystem Vulnerability (EVE)
 Species Vulnerability (EVS)
 Individual Organism Vulnerability (EVIO)

Physical Vulnerability (PV)
 Residential Vulnerability (PVR)
 Commercial Vulnerability (PVC)
 Industrial Vulnerability (PVI)
 Public Facilities Vulnerability (PVPF)

Social Vulnerability (SV)
 Low -Income Populations Vulnerability (SVLI)
 Elderly Populations Vulnerability (SVE)
 Minority Populations Vulnerability (SVM)
 Children Vulnerability (SVC)
 People with Disabilities Vulnerability (SVPD)

Figure 6.4 Examples of Local Planning for Resiliency First-level Concepts (e.g. *State Policy—SP*), Second-level Concepts (e.g. Mandates—SPM), and Third-Level Indicators (e.g. SPM1—Presence/ Absence of Mandate)

Hazard Vulnerability

Hazard vulnerability is a function of hazard exposure and the characteristics of the natural and built environments and resident populations that may lead to particular ecological, physical, and socio-economic vulnerabilities. Hazard exposure is defined as a geographic area exposed to an event (e.g., flooding, wind, surge) at a given magnitude, sometimes including probabilities or return periods (NRC 2006: 72–73). First-level concepts are ecological vulnerability (EV), physical vulnerability (PV), and social vulnerability (SV). *Ecological* (or ecosystem) *vulnerability* is viewed as the potential (or lack thereof) of an ecosystem to be stressed over time (Williams and Kapustka 2000). *Physical vulnerability* is viewed as potential damage to the built environment (e.g., houses, apartments, schools, hospitals, other infrastructure) that will be sustained from a hazard event (NRC 2006: 72–73). Recent perspectives have expanded the vulnerability construct to include *social vulnerability*, which refers to the characteristics of people that can differentially make certain subpopulations vulnerable to disasters, such as poverty, race, disability, language barriers, and age (Cutter 1996; Cutter et al. 2003, 2008; Flanagan et al. 2011, Van Zandt et al. 2011). Examples of second-level concepts under physical vulnerability include residential vulnerability (PVR) and commercial vulnerability (PVC). (Third-level indicators for hazard vulnerability are not shown given the complexity and space limitations here.) As shown in Figure 6.3, hazard vulnerability is conceptualized as being directly influenced by the adaptations taken as a result of planning processes. These three forms of vulnerability interact. For example, degradation of wetlands can lessen floodwater storage capacity, which can increase downstream physical vulnerability in areas that may have high levels of social vulnerability, in part because of awareness that higher physical vulnerability results in lower land values. Planning for resiliency accounts for these interacting vulnerabilities and minimizes their overlaps.

Adaptations: Outputs of the Planning Process

Outputs of resiliency planning processes, which we also refer to as adaptations, are intermediate planning products and actions taken to reduce vulnerability—ecological, physical, and social—to natural hazards. First-level resiliency planning outputs include the quality of relevant plans, or *plan quality*, and the implementation of policies, programs and actions included in plans, or *plan implementation*.

Plan quality (PQ) refers to the characteristics that distinguish a good plan from a bad one (Berke et al. 2006: Chapter 3). While these characteristics may vary depending on the purposes for which and the context in which a plan is developed, in recent years consensus has emerged around a core set of principles that high-quality plans should address (Berke and Godschalk 2009; Lyles and Stevens 2014). These principles and the context-specific indicators used to measure them are the deeper-level variables in this framework.

The consensus about plan quality principles is grounded in empirical evidence from nearly 50 peer-reviewed publications, including more than a dozen publications dealing with vulnerability to natural hazards and climate change (Stevens et al. 2014). Core principles of plan quality have been categorized into two second-level concepts involving direction-setting principles (PQDS) that provide the foundation for future action and action-oriented principles (PQAO) that relate to the use and influence of plans (Berke et al. 2013). In the context of planning for long-term hazard mitigation, Berke et al. (2013: 140) have proposed that the third-level indicators for the direction-setting principles are goals (PQDS1), fact base (PQDS2), and policies (PQDS3) and, while the action-oriented principles are participation (PQAO1), inter-organizational coordination (PQAO2), implementation (PQAO3), and monitoring (PQAO4)—see Figure 6.5 for definitions of plan quality principles.

A high-quality plan that addresses all of these principles serves the function of guiding how a community anticipates and responds to disasters. Communities with high-quality plans are poised to proactively anticipate and reduce vulnerability rather than merely react following an event, as potential losses from hazards can be significantly affected by plan quality (Brody and Highfield 2013; Burby 2006; Nelson and French 2002). However, a wide body of evidence suggests that high-quality hazard mitigation and climate change adaptation plans are rare in the United States, at both the state and local levels (Berke et al. 1996; Berke et al. 2012; Burby and May 1997;

Direction-Setting Principles:
- "Goals are future desired conditions that reflect the breadth of values affected by the plan."
- "The fact base provides the empirical foundation to ensure that key hazard problems are identified and prioritized and mitigation policy making is well informed."
- "Policies (or actions) serve as a general guide to decisions about development and assure that plan goals are achieved."

Action-Oriented Principles:
- "Implementation and monitoring involves the assignment of organizational responsibilities, timelines, and funds to implement plan. It also involves tracking the extent to which policies are carried out."
- "Interorganizational coordination entails recognition of the interdependent actions of state and local organizations that need coordination for plan implementation."
- "Participation involves recognition offormal and informal actors engaged in preparing the plan, including other governmental bodies, private sector institutions, nonprofits, and individual citizens."

Figure 6.5   Hazard Mitigation Application of Plan Quality Principles

Godschalk et al. 1999; Horney et al. 2017; Kang et al. 2010; Lyles et al. 2017; Olonilua and Ibitayo 2011; Wheeler 2008; Woodruff and Stults 2016,).

Broadly speaking, *plan implementation* (PI) refers to whether a plan has an influence following its creation. In evaluating plan implementation, scholars such as Alexander and Faludi (1989), Talen (1997), and Oliveira and Pinho (2010) have conceived of conformance-based approaches (PIC) (second-level concept) focused on whether the policies (or actions) in a plan are carried out via policy adoption (PIC1) and permit rule enforcement (PIC2). Performance-based approaches (PIP) to implementation focused more on planning process, such as "whether the plan plays a role in those decision situations in which it was meant to be used" (PIP1) (Mastop and Faludi 1997: 820). The number of studies that have investigated hazard-related plan implementation is more limited than those of hazard-related plan quality, and most hazard-related plan implementation studies have taken a conformance-based approach (Lyles et al. 2016). Conformance-based studies have investigated the adoption of hazard risk reduction policies included in plans (Brody et al. 2006; Burby 2003; Burby and May 1997;Dalton and Burby 1994) and adherence of hazard-related permits to plan policies (Berke et al. 2006; Brody and Highfield 2005; Brody et al. 2006; Laurian et al. 2004).

Adaptive Capacity: Planning Processes and the Planning Setting

The adaptive planning and implementation activities that communities undertake are influenced by their *adaptive capacity*, i.e., their ability to anticipate and adjust to change, and to moderate vulnerability. Adaptive capacity is comprised of the planning processes, policy context, and local resources.

Planning processes have been a topic of central interest in planning scholarship in recent decades. Building on the work of Arnstein (1969) and others, planning scholars have argued for a less expert-driven, top-down, and technically-oriented approach to planning in favor of more bottom-up, participatory *planning engagement* (PE) processes (first-level concepts) with an emphasis on facilitating communication and collaboration (cf. Forester 1989; Forester 1993; Innes and Booher 2010). Similar calls for collaborative planning have been made in the hazards and disasters literature, including NRC (2011). A particular challenge in the realm of planning for resiliency is the lack of a public constituency for hazard risk reduction and the infrequent windows of opportunity to increase its saliency as a political priority (Birkland 1997; Birkland 2006; Burby et al. 1999; Godschalk et al. 2003; Prater and Lindell 2000). Research on hazard mitigation planning processes has focused on ways that local commitment can be increased in planning processes by diversifying participants (PED) and exploring use of public education techniques (PEP). Third-level indicators include the engagement of a broad, diverse array of local stakeholders (PED1) (Burby 2003), greater involvement of local planners (PED2) (Lyles et al. 2014b, Lyles 2015), targeted information sharing education programs (PEP1), and designing creative education techniques (PEP2) (Godschalk et al. 2003) (third-level indicators). An emerging approach to grappling with the unique challenges of gathering, organizing, and evaluating hazard-related information in the face of uncertainty, especially in the context of climate change adaptation, is the use of scenarios in adaptive and collaborative planning processes (Berke and Lyles 2013; Quay 2010).

The *planning context* subsystem consists of two first-level concepts: state policy and local resources. *State policy (SP) context* includes the second-level concepts of legal mandates for local hazard-related planning (SPM) and state agency oversight (SPO) of local planning efforts. State mandates for local hazards planning enhance the quality of local plans, while variations in the design of the mandates have differential impacts (Berke and French 1994; Berke et al. 1996; Berke et al. 2014a; Burby and May 1997). Examples of third-level indicators are variations in state agency enforcement (SPO1) of mandate features that influence local hazard-related planning (Deyle and Smith 1998), as are differences in state-level mandate design (SPM2) of local planning to meet federal hazard planning requirements (Smith et al. 2013).

The first-level concept of *local resources* (LR) consists of four second-level concepts (staff-LRS, data and planning resources-LRDR, financial resources-LRB, and values-LRV) that facilitate stronger planning processes, higher plan quality, and greater implementation. Third-level indicators that have been found to be influential include the commitment of planning staff and elected officials, as well as involvement of consultants (Burby and May 1997; Dalton and Burby 1994; Loh and Norton 2013, 2015; Norton 2005).

Monitoring and Feedback, and Broader Social, Economic, and Policy Settings

Two additional features of our conceptual framework merit attention. First, planning processes do not end with the adoption of a plan document or the implementation of a plan policy or action. Also influential is the opportunity for communities to engage in adaptive, social learning over time through repeated planning processes (Brody 2003) and to build overall organizational capacity (Brody et al. 2010). Successful planning must engage in regular monitoring of planning outputs and hazard vulnerability, as well as periodic feedback for evaluation and updating of plans. These processes are reflected in the feedback arrow in Figure 6.3 from hazard vulnerability to the mitigation planning section. Second, the subsystems of our framework are embedded in broader social, economic, and policy settings, as indicated at the bottom of Figure 6.3. Multiple studies have found a variety of contextual variables to influence hazards planning and outputs, including population density and growth, community wealth, development pressures in hazardous areas, and previous hazard losses (Brody 2003; Burby 2003; Burby and Dalton 1994; Norton 2005).

Local Case Studies

To further demonstrate the value of our resiliency planning framework, we use it to analyze the case of resiliency planning in New Hanover County, North Carolina, which has been a regional and, arguably, national leader in resiliency planning. We will also briefly compare New Hanover County with two similar coastal counties (Martin County, Florida, and Onslow County, North Carolina) in order to demonstrate how differences in certain variables can affect planning for resiliency. At the time these cases were conducted in 2011–2012, both North Carolina and Florida were states that mandated local planning for land use and hazards, albeit with different emphases in policy and state oversight (Berke et al. 2014a; Lyles 2015).

New Hanover County is a fast-growing, urbanized coastal county with a population of 160,307 (US Census 2000), one primary city (Wilmington), and three beachfront towns. In the wake of a series of major hurricanes in the 1990s (Bertha, Fran, Bonnie, and Floyd), New Hanover County and Wilmington made a major commitment to resiliency planning (LRV in Figure 6.4). Starting in 1997, they were pilot partners in the Federal Emergency Management Agency's (FEMA) Project Impact: Building Disaster Resilient Communities program, which provided funding for staffing and hazard mitigation projects (LRS1 and LRB1). The county developed its first hazard mitigation plan in 2002 in advance of federal Disaster Mitigation Act (DMA) requirements and updated the plan in 2005 and 2010 to meet DMA requirements (Lyles 2012). The 2010 plan update process involved 34 stakeholders (PED1), was co-led by the county planning and emergency management agencies, and employed the services of external planning consultants (LRS2) who had worked on the 2005 version as well (New Hanover County 2010).

The resulting plan, the New Hanover County Multi-Jurisdictional Hazard Mitigation Plan, which covers the county, Wilmington, and two of the three beachfront towns, is a highly readable and well-organized document. Its fact base (PQDS3—see Figure 6.4) includes detailed assessments of hazard vulnerability and existing planning capacities, as well as a diverse set of future-oriented actions and policies for which there is detailed information on capacities needed for implementation. The plan contains a broad range of innovative land use policies (PQDS4) aimed at reducing development in known hazardous areas. Reporting on implementation progress (PIC) provided to FEMA in 2011 indicated that the land use-oriented policies have been or are in the process of being adopted and enforced, although progress in completing actions dependent on funding is limited (Lyles 2012).

Returning to our resiliency planning framework, we can see the relevance of the first- and second-level concepts and the key third-level indicators. In terms of state policy, North Carolina is a planning mandate state (SPM1) that has provided a wide array of support for local land use and hazards planning (SPO1) (Lyles 2012; Smith et al. 2013). In terms of local resources, the 2002 and 2005 planning processes established prior plan quality (LRDR1) and both the local planning and emergency management agencies committed to working together on hazard mitigation planning (LRV1). In terms of planning engagement (PED1), a large and diverse array of stakeholders participated on the planning committee, and the planning and emergency management staff collaborated closely. The interplay of these factors played major roles in the high quality of the plan overall, as well as the implementation successes already observed. So, too, did the external support provided by participation in FEMA's Project Impact, which can be understood as a complement to state planning support.

Martin County, Florida provides an interesting contrast. Martin County has a long history of commitment by local staff (LRV1) and officials (LRV2) to aggressive land use planning that has been quite restrictive of development in hazardous areas (Lyles 2012). This history is reflected in its hazard mitigation plan's detailed review of existing hazard-related land use policies, (PQDS3) while the plan's future-oriented actions (PQDS4) essentially ignore land use approaches to steering developing out of hazardous areas. Moreover, although nearly three dozen stakeholders were invited to participate, only 11 chose to serve on the planning committee (PED1), none of whom were local planners (PED2).

Our framework highlights a major difference between the resiliency planning context in Martin County, Florida, and in New Hanover County, North Carolina. Florida's pioneering state laws requiring hazard mitigation planning (SPM1) heavily emphasized using hazard mitigation plans to prioritize a list of projects that could be approved quickly, efficiently, and equitably for federal post-disaster funds following a major disaster declaration (SPM2) (Berke et al. 2014a; Smith et al. 2013). In contrast, through state mandate features (SPM2) and oversight of local planning (SPO) North Carolina has emphasized the inclusion of a more comprehensive and integrated array of projects, programs, and policies. This prioritization of projects over policy by the State of Florida appears to have led stakeholders in Martin County to view mitigation as a dead-end project planning process that does not feed back into comprehensive land use planning and other local initiatives—a view expressed by a prominent county planning official (Lyles 2012). New Hanover County, by contrast, included a combination of policies and projects in its plan, which increased the relevance of its hazard mitigation plan for ongoing decision-making situations.

Onslow County, North Carolina, also provides an interesting contrast to New Hanover County. Similar in size to New Hanover County and separated by just one county along the coast, Onslow County is affected by the same state policies and oversight of local planning (SPM1, SPM2, and SPO1). Yet its hazard mitigation plan is of lower quality overall and includes just two very weak land use-related policies in the future-oriented policy section of the plan (PQDS4) (Lyles 2012).

Our conceptual framework highlights two key differences between the counties. First, whereas New Hanover County has had strong collaboration between its planners and emergency managers as well as diverse engagement (PED1 and PED2), collaboration and engagement have been much weaker in Onslow County. Onslow County had three different emergency management agency directors and two different planning agency directors during the late 2000s and the planning committee involved just 11 stakeholders, mostly representing local government (PED1). Emergency management officials expressed a very reluctant attitude towards engaging land use issues in hazard mitigation planning in order to avoid politicization and controversy in the planning process (LRV1) (Lyles 2012). Second, Onslow County has not received outside resources, such as federal Project Impact support, to build local capacity and commitment for local planning. While these comparisons are presented as being illustrative rather than definitive, they do highlight features of our conceptual framework that help explain why New Hanover County has succeeded in using its hazard mitigation planning efforts to build resiliency in ways that Martin County and Onslow County have not.

Limitations in Knowledge that Impede Resiliency Planning

Five decades ago, Alan Altshuler (1965) maintained that urban planning and the comprehensive plan are neither practically feasible nor politically viable. He argued that the comprehensive approach to planning has little effect on human settlement patterns because it is too general and requires more knowledge about interrelationships among the complex individual physical systems (land use, transportation, housing, and environment) of a city than any local government organization could grasp. Thus, it was not possible to produce workable strategies to guide future physical growth and development. The rationale for resiliency planning and the general framework for resiliency planning offered here suggest that it is now possible—as it was not in 1965—to undertake comprehensive approaches to planning. This suitability is especially the case for resiliency planning that requires a systems perspective. New knowledge and practices have since developed, and the literature has generated new knowledge about the plan quality and implementation and the factors that influence them.

However, several major gaps in knowledge require attention if resiliency planning is to move forward and further enable vulnerable communities to better anticipate, absorb, recover from, and successfully adapt to future adverse events. Discussed below are two of the most pressing limitations that must be addressed in order to enable scholars to build and test theoretical models of resiliency planning to reduce risk.

Isolated Plans or Networked Plans

Prior research has centered on a partial view of plans that does not account for the whole system of plans (Finn et al. 2007; Hopkins 2001). Studies only offer partial views of planning by focusing on the quality of comprehensive land use plans (Brody et al. 2010; Dalton and Burby 1994) or stand-alone plans for hazard mitigation (Tang et al. 2010) and disaster recovery (Berke et al. 2014a), but failing to capture the full complexity of networks of plans that are adopted by diverse organizations. As noted, over the past two decades there has been increasing societal weight placed on planning to reduce the rising levels of vulnerability, and this has been accompanied by a steady expansion of local planning requirements.

Some planning researchers have long called for better integration of multiple plans into a comprehensive plan (Burby and French 1984; Godschalk et al. 1998: 231–276). Alternatively, others contend that it is not practical to create a single comprehensive plan and that planners should learn to work effectively with the network of many plans (Finn et al. 2007). Yet we know little about the networks of plans and their ultimate influence on where and how local land use and development takes place in hazardous locations. Berke et al. (2015) developed a Plan Integration for Resilience Scorecard to evaluate the degree to which plans are coordinated in reducing hazard vulnerability in different geographic areas (downtowns, neighborhoods, waterfronts). The scorecard is not widely applicable as it has only been applied to a small coastal community. Furthermore, we know little about the type of plans that comprise the networks across communities, the strength of the plans based on plan quality principles, and the degree to which the plan policies are coordinated or in conflict.

Resiliency Planning: An Untested Assumption?

Although our conceptual framework is designed to contribute to the broader issues of plan implementation and indicator development, methods and indicators to track the effectiveness of plans are inadequate. The large body of research on the evaluation of policy implementation, as represented by Pressman and Wildavsky's classic book, *Implementation: How Great Expectations in Washington are Dashed in Oakland* (1984), is curiously not paralleled by deep inquiry into evaluation in the urban planning field. Planning researchers have not yet developed an equivalent ability to link plans to subsequent impacts, notably plans aimed at improving resiliency to disasters and adapting to the uncertainties of climate change. Given the lack of methods to empirically evaluate plan implementation, many plans are impressionistically rather than empirically assessed (Laurian et al. 2004; Talen 1997). As a consequence, planners know very little about the effects of a specific plan on the community development process. Although measuring the effect of plans on urban development is a formidable empirical challenge, and differences between local institutions and across metropolitan areas make it difficult to compare the planning implementation outcomes, a fuller understanding of the relationship between plans and their outcomes should help policy makers both better understand the likely impacts of plans and tailor them to achieve desired outcomes..

A recent NRC (2012) committee concluded that a national priority should be to develop indicators to measure plan performance and progress toward increasing resilience in communities. It noted that development and application of indicators should be initiated at a national level through a mechanism called a "national resilience scorecard." The NRC committee envisioned that such a scorecard would identify areas that merit priority, provide a baseline from which to measure change, and offer a systematic approach for measuring progress in building resilient communities (NRC 2012). The report recommendation identified the process of developing a scorecard as one that would involve engagement by all levels of government (federal, state, local), the private sector, and community groups and individuals.

Conclusions

Building on planning scholarship in recent years, we have developed a framework for planning for community resilience. To revise and further develop the resiliency planning framework presented here, researchers need to establish comparable databases to enhance the gathering of research findings about national (and state) policy, local resources, and internal planning processes affecting community vulnerability around the world. Such work is underway and expanding, and beginning to address the major challenges discussed here that constrain knowledge about the performance of planning to build resiliency. We anticipate that research across disciplines will cumulate more rapidly and increase the knowledge needed to enhance the resiliency of communities in the context of complex SESs. Ongoing refinement of this framework can inform collection of quantitative and qualitative data about the core resiliency planning indicators to enable researchers to build and test models of planning aimed at building the resiliency of communities to future disasters and climate change.

References

Adger, W. N. (2006) "Vulnerability," *Global Environmental Change* 16: 268–281.

Alexander, E.R. and A. Faludi. (1989) "Planning and Planning Implementation: Notes on Evaluation Criteria," *Environment and Planning B: Planning and Design* 16(2): 127–140.

Altshuler, A. (1965) "The Goals of Comprehensive Planning," *Journal of the American Institute of Planners* 31(3): 186–197.

Arnstein, S. (1969) "A Ladder of Citizen Participation," *Journal of the American Planning Association* 35(4): 216–224.

Baer, W. C. (1997) "General Plan Evaluation Criteria: An Approach to Making Better Plans," *Journal of the American Planning Association* 63(3): 329–345.

Berke, P. R., M. Backhurst, M. Day, N. Ericksen, L. Laurian, J. Crawford, and J. Dixon. (2006) "What Makes Plan Implementation Successful? An Evaluation of Local Plans and Implementation Practices in New Zealand," *Environment and Planning B: Planning and Design* 33(4): 581–600.

Berke, P. R., N. Ericksen, J. Crawford, and J. Dixon. (2002) "Planning and Indigenous People: Human Rights and Environmental Protection in New Zealand," *Journal of Planning Education and Research* 22(2): 115–134.

Berke, P. R. and S. P. French. (1994) "The Influence of State Planning Mandates on Local Plan Quality," *Journal of Planning Education and Research* 13(4): 237–250.

Berke, P. R. and D. Godschalk. (2009) "Searching for the Good Plan: A Meta-Analysis of Plan Quality Studies," *Journal of Planning Literature* 23(3): 227–240.

Berke, P. R., D. R. Godschalk, E. J. Kaiser, and D. A. Rodriguez. (2006) *Urban Land Use Planning*, Urbana IL: University of Illinois Press.

Berke, P. R., J. Lee, G. Newman, T. Combs, C. Kolosna, and D. Salvesen. (2015) "Evaluation of Networks of Plans and Vulnerability to Hazards and Climate Change: A Resilience Scorecard," *Journal of the American Planning Association* 81(4): 287–302.

Berke, P. R. and W. Lyles. (2013) "Public Risks and the Challenges to Climate-Change Adaptation: A Proposed Framework for Planning in the Age of Uncertainty," *Cityscape* 15(1): 181–208.

Berke, P. R., W. Lyles, and G. Smith. (2014a) "Impacts of Federal and State Hazard Mitigation Policies on Local Land Use Policy," *Journal of Planning Education and Research* 34(1): 60–76.

Berke, P. R., D. Roenigk, E. Kaiser, and R. Burby. (1996) "Enhancing Plan Quality: Evaluating the Role of State Planning Mandates for Natural Hazard Mitigation," *Journal of Environmental Planning and Management* 39(1): 79–96.

Berke, P. R., G. Smith, and W. Lyles. (2012) "Planning for Resiliency: Evaluation of State Hazard Mitigation Plans Under the Disaster Mitigation Act," *Natural Hazards Review* 13(2): 139–150.

Berke, P. R., D. Spurlock, G. Hess, and L. Band. (2013) "Local Comprehensive Plan Quality and Regional Ecosystem Protection: The Case of the Jordan Lake Watershed, North Carolina, USA," *Land Use Policy* 31: 450–459.

Birkland, T. A. (1997) *After Disaster: Agenda Setting, Public Policy, and Focusing Events.* Washington DC: Georgetown University Press.

Birkland, T. A. (2006) *Lessons of Disaster: Policy Change after Catastrophic Events.* Washington DC: Georgetown University Press.

Bottrill, M. and R. Pressey. (2012) "The Effectiveness and Evaluation of Conservation Planning," *Conservation Letters* 5(6): 407–420.

Brody, S. D. (2003) "Are We Learning to Make Better Plans? A Longitudinal Analysis of Plan Quality Associated with Natural Hazards," *Journal of Planning Education and Research* 23(2): 191.

Brody, S. D., V. Carrasco, and W. E. Highfield. (2006) "Measuring the Adoption of Local Sprawl Reduction Planning Policies in Florida," *Journal of Planning Education and Research* 25(3): 294–310.

Brody, S. D. and W. E. Highfield. (2005) "Does Planning Work? Testing the Implementation of Local Environmental Planning in Florida," *Journal of the American Planning Association* 71(2): 159–175.

Brody, S. D. and W. E. Highfield. (2013) "Open Space Protection and Flood Mitigation: A National Study," *Land Use Policy* 32: 89–95.

Brody, S. D., W. E. Highfield, and S. Thornton. (2006) "Planning at the Urban Fringe: An Examination of the Factors Influencing Nonconforming Development Patterns in Southern Florida," *Environment and Planning B: Planning and Design* 33(1): 75–96.

Brody, S. D., J. E. Kang, and S. Bernhardt. (2010) "Identifying Factors Influencing Flood Mitigation at the Local Level in Texas and Florida: The Role of Organizational Capacity," *Natural Hazards* 52(1): 167–184.

Burby, R. J. (ed.). (1998) *Cooperating with Nature: Confronting Natural Hazards with Land-Use Planning for Sustainable Communities.* Washington DC: Joseph Henry/National Academy Press.

Burby, R. J. (2003) "Making Plans That Matter: Citizen Involvement and Government Action," *Journal of the American Planning Association* 69(1): 33–49.

Burby, R. J. (2006) "Hurricane Katrina and the Paradoxes of Government Disaster Policy: Bringing About Wise Governmental Decisions for Hazardous Areas," *The ANNALS of the American Academy of Political and Social Science* 604(1): 171–191.

Burby, R. J., T. Beatley, P. R. Berke, R. E. Deyle, S. P. French, D. R. Godschalk, E. J. Kaiser, J. D. Kartez, P. J. May, R. Olshansky, R. G. Paterson, and R. H. Platt. (1999) "Unleashing the Power of Planning to Create Disaster-Resistant Communities," *Journal of the American Planning Association* 65(3): 247–258.

Burby, R. J. and L. C. Dalton. (1994) "Plans Can Matter: The Role of Land Use Plans and State Planning Mandates in Limiting the Development of Hazardous Areas," *Public Administration Review* 54(3): 229–238.

Burby, R. J. and P. J. May. (1997) *Making Governments Plan.* Baltimore MD: Johns Hopkins University Press.

Cutter, S. L. (1996) "Vulnerability to Environmental Hazards," *Progress in Human Geography* 20(4): 529–539.

Cutter, S.L., L. Barnes, M. Berry, C. Burton, E. Evans, E. Tate, and J. Webb. (2008) "A Place-based Model for Understanding Community Resilience to Natural Disasters," *Global Environmental Change* 18(4): 598–606.

Cutter, S. L., B. J. Boruff, W. L. Shirley. (2003) "Social Vulnerability to Environmental Hazards," *Social Science Quarterly* 84(2): 242–261.

Dalton, L. and R. J. Burby. (1994) "Mandates, Plans and Planners: Building Local Commitment to Development Management," *Journal of the American Planning Association* 60(4): 444–461.

Deyle, R. E. and R. A. Smith. (1998) "Local Government Compliance with State Planning Mandates: The Effects of State Implementation in Florida," *Journal of the American Planning Association* 64(4): 457–469.

Feiock, R. C. (2009) "Metropolitan Governance and Institutional Collective Action," *Urban Affairs Review* 44(3): 356–377.

Finn, D., L. Hopkins, and M. Wempe. (2007) "The Information System of Plans Approach: Using and Making Plans for Landscape Protection," *Landscape and Urban Planning* 81(1–2): 132–145.

Flanagan, B. E., E. W. Gregory, E. J. Hallisey, J. L. Heitgerd, and B. Lewis. (2011) "A Social Vulnerability Index for Disaster Management," *Journal of Homeland Security and Emergency Management* 8(1): 1–22.

Forester, J. (1989) *Planning in the Face of Power*, Berkeley CA: University of California Press.

Forester, J. (1993) *Critical Theory, Public Policy, and Planning Practice: Toward a Critical Pragmatism*, Albany NY: State University of New York Press.

Godschalk, D. R. (2003) "Urban Hazard Mitigation: Creating Resilient Cities," *Natural Hazards Review* 4(3): 136–43.

Godschalk, D. R., T. Beatley, P. R. Berke, D. Brower, and E. Kaiser. (1999) *Natural Hazard Mitigation: Recasting Disaster Policy and Planning*, Washington DC: Island Press.

Godschalk, D. R., S. D. Brody, and R. J. Burby. (2003) "Public Participation in Natural Hazard Mitigation Policy Formation: Challenges for Comprehensive Planning," *Journal of Environmental Planning and Management* 46(5): 733–754.

Godschalk, D. R., E. J. Kaiser, and P. R. Berke. (1998) "Integrating Hazard Mitigation and Local Land Use Planning," 85–118, in R. J. Burby (ed.), *Cooperating with Nature: Confronting Natural Hazards with Land-Use Planning for Sustainable Communities*, Washington DC: Joseph Henry Press.

Hopkins, L. (2001) *Urban Development: The Logic of Making Plans*, Washington, DC: Island Press.

Horney, J. A., A. I. Naimi, W. E. Lyles, M. Simon, D. Salvesen, P. R. Berke. (2012) "Assessing the Relationship Between Hazard Mitigation Plan Quality and Rural Status in a Cohort of 59 Counties from 3 States in the Southern United States," *Challenges* 3(2): 183–193.

Horney, J., M. Nguyen, D. Salvesen, C. Dwyer, J. Cooper, and P. Berke. (2017) "Assessing the Quality of Rural Hazard Mitigation Plans in the Southeastern United States," *Journal of Planning Education and Research* 37(1), 56–65.

Horney, J., M. Simon, S. Grabich, P. R. Berke. (2015) Measuring Participation by Socially Vulnerable Groups in Hazard Mitigation Planning, Bertie County, NC. *Journal of Environmental Planning and Management* 58(5): 802–818.

Innes, J. E. and D.E. Booher. (2010) *Planning with Complexity: An Introduction to Collaborative Rationality for Public Policy*. Oxford: Routledge.

Kang, J. E., W. G. Peacock, and R. Husein. (2010) "An Assessment of Coastal Zone Hazard Mitigation Plans in Texas," *Journal of Disaster Research* 5(5): 520–528.

Laurian, L., M. Day, M. Backhurst, P. R. Berke, J. Crawford J. Dixon, S. Chapman. (2004) "What Drives Plan Implementation? Plans, Planning Agencies and Developers," *Journal of Environmental Planning and Management* 47(4): 555–577.

Loh, C. G., and Norton, R. K. (2013) "Planning Consultants and Local Planning: Roles and Values," *Journal of the American Planning Association*, 79(2), 138–147.

Loh, C. G., and Norton, R. K. (2015) "Planning Consultants' Influence on Local Comprehensive Plans," *Journal of Planning Education and Research* 35(2), 199–208.

Lyles, L. W. (2012) *Stakeholder Network Influences on Local-Level Hazard Mitigation Planning Outputs*, PhD Dissertation, The University of North Carolina, Chapel Hill.

Lyles, W. (2015) "Using Social Network Analysis to Examine Planner Involvement in Environmentally Oriented Planning Processes Led by Non-Planning Professions," *Journal of Environmental Planning and Management* 58(11), 1961–1987.

Lyles, W., P. Berke, and G. Smith. (2014a) "A Comparison of Local Hazard Mitigation Plan Quality in Six States, USA," *Landscape and Urban Planning* 122, 89–99.

Lyles, L. W., P. R. Berke, and G. Smith. (2014b) "Do Planners Matter? Examining Factors Driving Incorporation of Land Use Approaches into Hazard Mitigation Plans," *Journal of Environmental Planning and Management* 57(5): 792–811.

Lyles, W., P. Berke and G. Smith. (2016) "Local Plan Implementation: Assessing Conformance and Influence of Local Hazard Plans in the United States," *Environment and Planning B* 43(2): 381–400.

Lyles, W., and M. Stevens. (2014) "Plan Quality Evaluation 1994–2012: Growth and Contributions, Limitations, and New Directions," *Journal of Planning Education and Research* 34(4), 433–450.

Lyles, W., P. Berke, and K. H. Overstreet. (2017) "Where to Begin Municipal Climate Adaptation Planning? Evaluating Two Local Choices," *Journal of Environmental Planning and Management*, 1–21.

Mastop, H. and A. Faludi. (1997) "Evaluation of Strategic Plans: The Performance Principle," *Environment and Planning B: Planning and Design* 24(6): 815–832.

Melillo, J. M., T. T. Richmond, and G. Yohe. (2014) Climate change impacts in the United States. *Third National Climate Assessment*. Accessed May 20, 2019 at www.globalchange.gov/nca3-downloads-materials

Mileti, D. (1999) *Disasters by Design: A Reassessment of Natural Hazards in the United States*, Washington DC: National Academies Press.

NRC—National Research Council. (2006) *Facing Hazards and Disasters: Understanding Human Dimensions*, Washington DC: National Academies Press.

NRC—National Research Council. (2011). *Building Community Disaster Resilience Through Private–Public Collaboration*. Washington DC: National Academy Press.

NRC—National Research Council. (2012) *Disaster Resilience: A National Imperative.*, Washington DC: National Academies Press.

NRC—National Research Council. (2014) *Reducing Coastal Risks on the East and Gulf Coasts*. Washington DC: National Academies Press.

Nelson, A. and S. P. French. (2002) "Plan Quality and Mitigating Damage from Natural Disasters: A Case Study of the Northridge Earthquake with Planning Policy Considerations," *Journal of the American Planning Association* 68(2): 194–207

New Hanover County. (2010) *New Hanover County Multi-Jurisdictional Hazard Mitigation Plan*. New Hanover County, NC: New Hanover County Department of Emergency Management.

Norton, R. K. (2005) "Local Commitment to State-Mandated Planning in Coastal North Carolina," *Journal of Planning Education and Research*. 25(2): 149–171.

Oliveira V. and P. Pinho. (2010) "Evaluating in Urban Planning: Advances and Prospects," *Journal of Planning Literature* 24(4): 343–361.

Olonilua, O. O. and O. Ibitayo. (2011) "Toward Multihazard Mitigation: An Evaluation of FEMA-Approved Hazard Mitigation Plans under the Disaster Mitigation Act of 2000," *Journal of Emergency Management* 9(1): 37–49.

Ostrom, E. (2009) "A General Framework for Analyzing Sustainability of Social-Ecological Systems," *Science*, 325(5939), 419–422.

Peacock, W. G., H. Kunreuther, W. H. Hooke, S. L. Cutter, S. E. Chang, and P. R. Berke. (2008) *Toward a Resiliency and Vulnerability Observatory Network: RAVON*, Report #08-02R, College Station, TX: Hazard Reduction and Recovery Center.

Pielke Jr, R. A., J. Gratz, C. W. Landsea, D. Collins, M. A. Saunders, and R. Musulin. (2008) "Normalized Hurricane Damage in the United States: 1900–2005," Natural Hazards Review 9(1): 29–42.

Prater, C. S. and M. K. Lindell. (2000) "Politics of Hazard Mitigation," *Natural Hazards Review*, 1(2): 73–82.

Pressman, J. and A. Wildavsky. (1984) *Implementation: How Great Expectations in Washington are Dashed in Oakland*, Oakland CA: University of California Press.

Quay, R. (2010) "Anticipatory Governance: A Tool for Climate Change Adaptation," *Journal of the American Planning Association* 76(4): 496–511.

Schwab, J. C. (2009) Survey of State Land Use Planning and Natural Hazards Laws. Chicago: American Planning Association.

Smith, G., W. Lyles, P. R. Berke. (2013) "The Role of Hazard Mitigation Planning in Building Local Capacity and Commitment: A Tale of Six States," *International Journal of Mass Emergencies and Disasters* 31(2): 178–203.

Stevens, M. R., W. Lyles, and P. R. Berke. (2014) "Measuring and Reporting Intercoder Reliability in Plan Quality Evaluation Research," *Journal of Planning Education and Research* 34(1): 77–93.

Susskind, L. (1978) "Should State Government Mandate Local Planning? . . . No," *Planning* 44(6): 17–20.

Talen, E. (1997) "Success, Failure and Conformance: An Alternative Approach to Planning Evaluation," *Environment and Planning B: Planning and Design* 24: 573–587.

Tang, Z., S. D. Brody, C. Quinn, L. Chang, and T. Wei. (2010) "Moving from Agenda to Action: Evaluating Local Climate Change Action Plans," *Journal of Environmental Planning and Management* 53(1): 41–62.

United States Census. (2000) American Fact Finder. URL: http://factfinder2.census.gov. Accessed May 9, 2012.

Van Zandt, S., W. G. Peacock, D. W. Henry, H. Grover, W. E. Highfield, and S. D. Brody. (2011) "Mapping Social Vulnerability to Enhance Housing and Neighborhood Resilience," *Housing Policy Debate* 22(1): 29–55.

Wheeler, S. M. (2008) "State and Municipal Climate Change Plans: The First Generation," *Journal of the American Planning Association* 74(4), 481–496.

Williams, L. R. R. and L. A. Kapustka. (2000) "Ecosystem Vulnerability: A Complex Interface with Technical Components," *Environmental Toxicology and Chemistry* 19(4): 1055–1058.

Woodruff, S. C. and M. Stults. (2016) "Numerous Strategies but Limited Implementation Guidance in US Local Adaptation Plans," *Nature Climate Change*, 6(8), 796–802.

7

THE ADOPTION OF HAZARD MITIGATION AND CLIMATE CHANGE ADAPTATION POLICIES, PROGRAMS, AND ACTIONS BY LOCAL JURISDICTIONS ALONG THE GULF AND ATLANTIC COASTS

Walter Gillis Peacock, Michelle Annette Meyer, Shannon Van Zandt, Himanshu Grover, and Fayola Jacobs

Introduction

Trends in disaster losses—particularly those related to metrological, hydrological, and climatological events—continue to rise in the United States as well as throughout the rest of the world (Hoeppe 2016). The last ten years, since 2008, have seen communities throughout the Gulf and Atlantic coasts on the mainland and Puerto Rico devastated by five of the ten costliest hurricanes to impact the United States—Harvey (2017), Maria (2017), Irma (2017), Sandy (2012), and Ike (2008) (NHC 2017). Together, these five storms are estimated to have caused $370 billion in damage. It is important to note that two of these were not major, Category 3 or higher, hurricanes; nevertheless, they caused coastal and inland flooding that resulted in major damage. It is also sobering to consider that this list does not include storms, such as Matthew (2016), Irene (2011), Gustav (2008) or Dolly (2008), that had major impacts on communities from south Texas to the northeastern United States, but simply fell below the substantial and continually increasing damage threshold, currently at $23.7 billion (in 2017 dollars), to be classified as a top ten costliest hurricane.

The risks that coastal areas from Texas to Maine are likely to encounter due to extremes in weather and climate change are not anticipated to diminish in the future. All indications are that these risks will continue to increase, particularly with respect to hurricanes. As is often noted, it is the general consensus of the scientific community that climate change and its associated impacts are projected to become more severe in the future (IPCC 2012). Indeed, most climate models focusing on hurricanes suggest that, although there may be a decrease in the overall numbers of hurricanes each year, the frequency of particularly intense storms and their duration is likely to increase (Bacmeister et al. 2018; Knutson et al. 2010, 2015). Another key factor, related to increased risk of severe storms, is the concomitant issue of sea-level rise. Communities along the Gulf and Atlantic coasts such as Miami, Norfolk and the Hampton Roads region, and the entire Louisiana coast, are already having to address sea-level rise. In addition, Louisiana faces coastal subsidence as well. Recent research has also suggested that sea-level rise will increase and exacerbate flooding and storm surge due to tropical cyclones for large areas along the Gulf and Atlantic coasts (Neumann et al. 2015), particularly those having relatively shallow bathymetry (Woodruff et al. 2013).

The other major factor contributing to the growth in coastal disaster losses and increasing risk is, of course, the increasing concentrations of population, housing, infrastructure, and economic activities along the Gulf and Atlantic coasts (Crossett et al. 2013; Pielke et al. 2008). Coastal states from Texas to Maine have a population of 144.6 million people (44.8 percent of the nation's total), which grew by 10.1 percent since 2006 (US Census 2017). Of that population, 59.6 million (52 percent of the coastal states' total) are located in coastal counties within those states and that population grew by 14.5 percent since 2006. The Census (2017) also estimates that 61 million or 45 percent of our nation's housing units are located in Gulf and Atlantic coastal states, along with 3.4 million business establishments with nearly 54.9 million paid workers. Interestingly, recent conservative estimates suggest that, based on 2010 population levels, the estimated population located in areas likely to be inundated by a 0.9 and 1.8-meter sea-level rise by 2100 along the Atlantic and Gulf coasts would be 2.0 and 6.0 million, respectively. Worse yet, if current growth trends continue, the estimates are 3.8 and 11.9 million, respectively (Hauer et al. 2016). Of course, population forecasts for that amount of time are highly uncertain, but even shorter-term projections, using higher potential sea-level rise estimates, suggest that these numbers could reach 20 million as early as mid-century (Curtis and Schneider 2011). Furthermore, when coupled with the internal migrations suggested by both projections, the consequences for local jurisdictions in coastal counties, coastal states, and our nation as a whole are profound (Curtis and Schneider 2011; Hauer 2017).

The chronic and growing issue of sea-level rise—aggravated by the acute and potentially devastating impacts of surge, inland flooding, and wind brought about by tropical storms and hurricanes—indicates that coastal jurisdictions should be placing greater emphasis on hazard mitigation and climate change adaptation issues to reduce vulnerabilities and potentially future deaths and losses. Indeed, as a number of researchers have noted, although federal and state level governments may adopt various policies to promote hazard mitigation and climate change adaptation, local jurisdictions will be the critical focus for addressing these issues (Bedsworth and Hanak 2010; Farber 2009; Nordgren et al. 2016; Woodruff and Stults 2016).

Since the 1990s, hazard mitigation studies have primarily focused on the existence and quality of hazard mitigation plans (Berke et al. 2006) and more recently climate change adaptation plans (Hu et al. 2018; Woodruff and Stults 2016). It has been rare to find research that actually focuses on the adoption and implementation of programs, policies, and actions at the local jurisdiction level, whether considering hazard mitigation or climate change adaptation (Berke et al. 2006; Berke and Beatley 1992a, 1992b; Brody et al. 2010; Burby and May 1997; Godschalket al. 1989; Henstra and McBean 2004; Nordgren et al. 2016; Norton 2005a; Wilson 2009). The increasing number of jurisdictions participating in hazard mitigation and climate change adaptation planning activities has not guaranteed the implementation of hazard mitigation and climate change adaptation strategies and practices at the local level (Nordgren et al. 2016; Wise et al. 2014; Woodruff and Stults 2016).

To the extent that actions are actually studied, it is often in the context of actions *proposed* as part of a hazard mitigation or climate change adaptation plan, but not actual policies *adopted and implemented* by local jurisdictions. However, looking at only plans and their proposed actions can be problematic. For example, Rovins' (2009) study of hazard mitigation planning in Florida suggested that hazard mitigation plans, in particular, are undertaken as a bureaucratic step in mitigation grant funding process rather than a step toward adopting mitigation practices. Kang et al. (2010), in their study of coastal hazard mitigation plans in Texas, found that many proposed actions targeted emergency management and not hazard mitigation issues, particularly neglecting issues such as natural resource protection and land use planning. Woodruff and Stults (2016), in their studies of climate change adaptation plans, found that they rarely identified or assessed impacts, let alone identified potential strategies to address impacts or provided details about implementation. Although these studies suggest a disconnect between mitigation planning and practice, they actually depend upon an assessment of *planned actions*, not actual *mitigation practice*. As a consequence, there is a fundamental blind spot in the literature—little is known about the actual adoption and implementation of mitigation measures by local jurisdictions. This chapter is an exception in that it will present findings from a survey of Atlantic and Gulf coastal jurisdictions (counties and municipalities) on the adoption and the implementation of broad-based hazard mitigation and climate change adaptation policies, paying primary attention to land use and development practices that can enhance hazard mitigation and climate change adaptation actions.

Hazard Mitigation and Climate Change Adaptation

In the climate change literature, the term mitigation takes on quite a particular meaning that diverges from its usage in the hazard/disaster literature. In the climate change literature, mitigation generally is concerned with anthropogenic elements of climate change; hence, mitigation refers to "the reduction of the rate of climate change via the management of its causal factors (the emission of greenhouse gases from fossil fuel combustion, agriculture, land use changes, cement production etc.)" (IPCC 2012: 48). By contrast, the term in the disaster/hazards literature generally reflects protecting, reducing, or eliminating hazard risks and their associated impacts by taking action before a hazard event (FEMA 2007; Godschalk et al. 1998; Lindell et al. 2006; Schwab et al. 2007; UNISDR 2002; Williams and Micallef 2009). This latter definition is more closely aligned with definitions of *adaptation* in the climate change literature, which refers to "the process of adjustment to actual or expected climate and its effects, in order to moderate harm or exploit beneficial opportunities" (IPCC 2012: 48). In a very real sense, this idea of adaptation harkens back to the early hazards literature in the 1970s and 1980s, where the focus was on "adjustments" by human populations to address or lessen the risks of potential hazard impacts (Bassett and Fogelman 2013; Burton et al. 1978). Indeed, much of the hazard mitigation literature when addressing household and individual response to hazards, characterizes these responses as "adjustments" (e.g., Lindell and Perry 2000, 2012).

The Intergovernmental Panel on Climate Change's (IPCC 2012) special report on *Managing the Risk of Extreme Events and Disasters to Advance Climate Change Adaptation* has sought to link these different perspectives of hazard mitigation and climate change adaptation within a common framework by addressing *disaster risk reduction* through *disaster risk management*. Disaster risk reduction, the report states, "denotes both a policy goal or objective and the strategic and instrumental measures employed for anticipating future disaster risk, reducing existing exposure, hazard, or vulnerability and improving resilience" (IPCC 2012: 34). Indeed, from the perspective of this report, "adaptation is the goal to be advanced and extreme events and disaster risk management are methods for supporting and advancing that goal" (IPCC 2012: 48). What are these disaster risk reduction management measures? They are, in fact, many of the same broad-based strategies for addressing disaster risk reduction and environmental planning to promote sustainable and resilient communities (Bedsworth and Hanak 2010; Daniels 2014; Masterson et al. 2014; Wisner et al. 2004).

Hazard mitigation strategies have commonly been classified into structural and non-structural mitigation (Godschalk et al. 1998) and one finds similar patterns in the climate change adaptation literature (IPCC 2012: 304–306). Structural mitigation generally is associated with large-scale engineered infrastructure features to provide protection for communities or even regions from disaster impacts such as levees, dams, seawalls, dykes, and riprap (Godschalk et al. 1998; IPCC 2012; Klee 1999). Codes specifying "building designs and construction materials to increase the ability of an individual structures' foundation and load bearing framework to resist environmental extremes" (Lindell et al. 2006: 194) might also be included. Non-structural hazard mitigation and climate change adaptation includes a rather broad set of regulations and policies such as various forms of development regulations in environmentally sensitive and hazardous areas, educating the public to reduce any impact of hazards and even various forms of building codes and standards such as installing window shutters for buildings located on hurricane-prone coastlines (Burby 1998; Burby et al. 2000, 2007; Godschalk et al. 1998; IPCC 2012; Lindell et al. 2006; Peacock 2003). The degrees of overlap in these categories and its failure to capture the diversity of practices have led some researchers to urge caution in the use of this simple classification (Lindell et al. 2006).

Many have noted that non-structural approaches are more diverse and can offer a comprehensive approach with fewer negative effects when compared to large-scale community protection works, at least with respect to promoting appropriate development in risky areas and with respect to the minimizing impacts, often negative, on the natural environment (Burby 1998; Dalton and Burby 1994; Klee 1999; White et al. 2001). Nonstructural approaches can also be relatively less costly and provide more sustainable and resilient hazard mitigation and climate change adaptation at the local level (Brody et al. 2011). They offer a more obvious and direct solutions to avoid many disasters and lessen losses, by actually keeping people and the built environment out of harm's way (Burby 1998; Hyndman and Hyndman 2006). Conceptually, these strategies offer a portfolio for adjusting and adapting human activities, particularly development and building activities, by promoting development out of known hazardous areas, or by utilizing appropriate building standards and techniques that explicitly address natural hazard risks, while still preserving environmental resources and ecosystem services that can lessen potential impacts.

Although the distinction between structural versus non-structural hazard mitigation and climate change adaptation strategies provides a simple, straightforward classification scheme, it fails to capture the full spectrum of policies, strategies, and tools that local jurisdictions might undertake to lessen disaster risks. The hazard mitigation and climate change adaptation literatures offer a great variety of ways to classify these strategies at international, national, state, and local levels (Beatley 2003; Bedsworth and Hanak 2010; Berke and Beatley 1992a, 1992b; Daniels 2014; Daniels and Daniels 2002; Godschalk et al. 1989; IPCC 2012; Lindell et al. 2007; Masterson et al. 2014; Nordgren et al. 2016; Peacock and Husein 2011). Drawing on this literature and focusing on policies, strategies, and tools that might be adopted at the local jurisdictional level, we developed a

Table 7.1 Climate Change Adaptation and Hazard Mitigation Tools

Strategy	Goals	Tools
1 Development regulation and land use management	– Restrict occupancy in hazardous areas – Enhance density in safer areas – Discourage development in environmentally sensitive and hazardous areas	1 Residential subdivision ordinance 2 Zoning 3 Performance-based zoning 4 Planned unit development 5 Special overlay districts 6 Agricultural/open space zoning 7 Hazard setback ordinances 8 Storm water retention requirements
2 Limiting shoreline development	– Limit shoreline use – Preserve ecologically sensitive coastal areas and ecosystem services – Restrict activities in hazardous areas	9 Limit shoreline development to water-dependent uses 10 Restrictions on shoreline armoring 11 Restrictions on dredging/filling

3 Building standards	– Design regulations that reduce loss and damage and reflect hazard risks	12 Building codes 13 Wind hazard resistance standards 14 Flood hazard resistance for new homes 15 Retrofit for existing buildings 16 Special utility codes
4 Natural resource protection	– Preserve ecologically sensitive coastal areas and ecosystem services	17 Dune protection 18 Wetland protection 19 Coastal vegetation protection 20 Habitat protection/restoration 21 Protected areas 22 Environmental impact asses
5 Public information and awareness	– Disseminate information, advice and training about hazards, hazardous areas, climate, and mitigation/ adaptation techniques and goals	23 Public education for hazard mitigation 24 Citizen involvement in hazard mitigation planning 25 Seminars or workshops on hazard mitigation practices for developers and builders 26 Hazard disclosure 27 Hazard zone signs
6 Local Incentive tools	– Encourage land owners and developers to avoid environmentally sensitive and hazardous areas	28 Transfer of development rights 29 Density bonuses 30 Clustered development
7 Transferring risks/federal incentives	– Encourage land owners to avoid development in hazardous areas – Risk dispersion and risk reduction	31 Participation in the National Flood Insurance Program 32 Participation in the FEMA Community Rating System
8 Property acquisition programs	– Acquire and hold property for public benefit and use – Remove at-risk property from the private market	33 Fee simple purchases of undeveloped lands 34 Acquisition of developments and easements rights 35 Relocation of existing structures out of hazardous areas
9 Financial tools	– Distribute more fairly the public costs of private development	36 Lower tax rates 37 Special tax assessments 38 Impact fees or special assessments
10 Critical public and private facilities policies and requirements	– Direct the location of infrastructure and critical facilities away from hazardous areas	39 Requirements for locating public facilities and infrastructure 40 Requirements for locating critical private facilities 41 Using municipal service areas to limit development
11 Private-public sector initiatives	– Work with private for- and non-profit entities to mitigate hazard and climate impacts	42 Land trusts 43 Public-private partnerships
12 Use of Professionals	– Working with professionals to provide skill sets and technical expertise	44 Identify suitable building sites 45 Develop special building techniques 46 Conduct windstorm/roof inspection 47 Writing plans

classification of various types of strategies, along with their goals and detailed examples of the tools for implementing these strategies that have been identified in the literature (Table 7.1). In total 12 sets of strategies are identified including: (1) development regulations and land use management tools, (2) limiting shoreline development, (3) building standards, (4) natural resource protection,

(5) public information and awareness tools, (6) local incentive tools, (7) risk transfer/federal incentives, (8) property acquisition programs, (9) financial tools, (10) critical public and private facilities policies and requirements, (11) private-public sector initiatives, and (12) use of professionals. It should be noted that more traditional structural approaches associated with public works or engineered structural approaches such as major dams or levee systems were not included in this listing in part because these usually require major capital investments far beyond those of most local jurisdictions. The focus here is on strategies that can be adopted by local jurisdictions to address their hazard risks.

In total there are 47 different tools included within these 12 strategy sets or categories. The exact placement of a particular tool, such as hazard setbacks, is somewhat arbitrary in that strategies are often related, and tools can be employed in a variety of ways and for a variety of purposes. Hence, the focus of this table is not to definitively categorize tools under particular strategies, but rather to provide a convenient method for identifying different sets of strategies and the tools that are generally associated with those strategies. Each of these strategies and their respective tools will be discussed along with the findings of the survey on their adoption and implementation across coastal jurisdictions in following sections. However, before those discussions can take place, it is important to discuss more completely the survey methodology and sample to provide the context in which to consider the findings.

The Survey Sample and Methodology

The target population for our sample was all jurisdictions with a governmental entity located in NOAA defined coastal watershed counties along the Gulf and Atlantic Coasts. Figure 7.1 identifies these counties in darker gray, along with those areas within the coastal management zones for each of the 19 states boarding the Gulf and Atlantic coasts. Although there were a few counties in Florida that are not strictly identified as coastal watershed counties, all were included because the state includes them within its coastal management zone. In total, the project identified 4,364 potential jurisdictions ranging from incorporated places, census designated places (CDPs), county subdivisions, and counties within this area.[1] Unlike previous studies, this one did not employ an *a priori* arbitrary limit on the jurisdiction's population size; rather, the focus was on whether the jurisdiction had some form of governing authority over the area and a population within its boundary. Extensive investigations—undertaken using methods such as web-based searches, city/county data books, state websites, and the US Census—determined that 2,977 jurisdictions had some form of governing authority/structure. These jurisdictions included 1,382 incorporated places (cities and towns), 99 census designation places, 1,113 county subdivisions, and 383 counties.

Our survey methodology was an Internet survey that was tested as part of a pilot study conducted in Texas (Peacock and Husein 2011). An Internet survey was feasible because most local government officials have Internet access to email and web browsers on the computers they utilize in carrying out their official duties. The survey instrument was initially developed based on previous instruments employed by other researchers and modified by findings from the pilot Texas study. Questions were discussed with mitigation experts; the instrument was sent to some planning practitioners to get their feedback, and it was pretested with a selected set of local officials. After gaining feedback from these groups, we revised the instrument and then pretested it as a web-survey using Qualtrics online survey management software.

The critical task in undertaking the survey was identifying key informants in each jurisdiction who were knowledgeable about hazard mitigation policies related to land use, community development, environmental controls, building code regulations, and the characteristics of local government and communities in general. Given the types of data being collected, we initially targeted professionals in city and county planning departments or their equivalent. The great heterogeneity of jurisdictional governing structures found among our target jurisdictions required a broader consideration of other

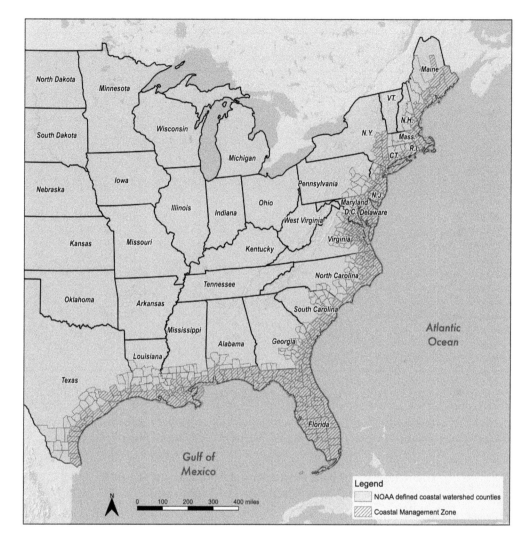

Figure 7.1 Coastal Jurisdictions Targeted for Study

officials as key informants. The development of the sample therefore required extensive investigative work using web-based sources and telephone conversations to properly identify key informants. We called each jurisdiction to determine key informants and obtain consent to email them a weblink to the survey. Contact was made with 2,139 (71.9 percent) of the 2,977 jurisdictions identified and contact information for key informants was obtained for 1,923 jurisdictions (the remaining jurisdictions either refused to participate or used external consultants).

The survey was conducted following Dillman's three-tiered approach for Internet surveys from mid-2013 through the fall of 2014, with a short period in early 2015 to finalize any outstanding survey responses (Dillman et al. 2008). Ultimately, we were able to obtain survey data from 1,172 jurisdictions yielding an overall response rate of 61.0 percent based on the 1,923 jurisdictions with identified informants, 54.8 percent if based on the 2,139 jurisdictions for which some contact was made. Table 7.2 presents data on the responding jurisdictions, broken down by state and jurisdictional type. Focusing on the most comparable response rate of 61.0 percent, this was better than the

Table 7.2 Responding Jurisdiction Types by State

State	Incorporated Places	Census Designated Places	County Subdivisions	Counties	Total
Alabama	12	0	0	5	17
Connecticut	10	7	54	0	71
Delaware	9	0	0	3	12
Florida	154	0	0	41	195
Georgia	17	0	0	10	27
Louisiana	34	0	0	21	55
Maine	11	3	31	0	45
Maryland	15	0	0	8	23
Massachusetts	19	28	66	2	115
Mississippi	6	0	0	2	8
New Hampshire	3	2	35	1	41
New Jersey	50	3	52	8	113
New York	42	2	34	7	85
North Carolina	30	0	0	23	53
Pennsylvania	33	0	98	6	137
Rhode Island	1	1	7	0	9
South Carolina	24	0	0	14	38
Texas	72	0	0	20	92
Virginia	13	0	0	23	36
Total	555	46	377	194	1,172

46 percent response rate obtained in the pilot Texas study and comparable to the 66.9 percent rate reported by Godschalk et al. (1989). On the whole, the response rates are quite strong given today's standards and particularly considering that the full spectrum of jurisdictions targeted.

The Adoption of Risk Management Methods and Tools by Coastal Jurisdictions: The Findings

The following sections provide a brief discussion of the 12 strategy sets that were considered in the survey of local jurisdictions, along with the tools or measures specifically examined with respect to each set. Following the discussion, the frequencies with which each tool or measure has been adopted by local jurisdictions will be discussed. Our discussion begins with what many consider the most traditional set of planning tools, land use and development regulations.

1) Land Use and Development Regulations

Land use and development regulations are by far the most traditional planning tools utilized by local jurisdictions to shape their development patterns (Burby 1998; Olshansky and Kartez 1998). It is ironic the same tools that often shape "normal" development patterns that have resulted in so many of our communities being highly vulnerable to hazards have also been touted by so many scholars in the hazards, disaster, and climate communities as significant tools for reducing risk and ultimately the loss of lives and property (Burby 1998; Berke and Beatley 1992b; Deyle et al. 2008; Godschalk et al. 1989; White 1936). The problem, of course, is that these tools are not being properly employed to address hazard risks and reduce vulnerabilities. If effectively employed, research has found that these regulatory tools keep population and development away from high-risk locations, preserve environmental resources and ecosystem services, and impose performance standards to reduce vulnerability in exposed areas. For instance, requiring new development to be set back a minimum distance from high erosion shorelines, not only keeps structures out of harm's way, but better preserves natural environmental features such as dunes and mangroves (Beatley 2009: 30).

The land use and development regulations considered in this study include eight different policy tools: (1) residential subdivision ordinances, (2) zoning, (3) performance-based zoning, (4) planned unit development, (5) special overlay districts, (6) agricultural or open space zoning, (7) hazard setback ordinances, and (8) storm water retention requirements. Jurisdictions can employ subdivision regulations to shape how property is subdivided into smaller units for development and ultimate resale. This tool is highly flexible—capable of specifying infrastructure requirements, amenities such as public spaces and parks and other green spaces, along with densities, and street layouts and designs. Planned unit development could be thought of as a form of zoning, but it is highly flexible, in that the developer presents a plan for the development of the area, whether it is for commercial, residential, or mixed-use development. Flexibility is achieved by allowing greater impacts or densities than might be normally allowed, as compensation for designating other areas within the development as open spaces, for other public amenities, or some other public or environmental amenity. Zoning is, of course, the division of a jurisdiction's areas into zones specifying land uses for each area, which can be flexible, as well as constraining. Performance-based zoning is a particularly flexible form of zoning that specifies a set of performance standards to limit, for example, the impacts of development on ecosystem services or surrounding parcels. Developers can then apply various approaches—such as pervious land cover options, open spaces, and protection of wetlands—to meet performance standards. Overlay zoning is also a highly flexible technique in which new district or zonal boundaries are superimposed on existing zoning to establish additional land use requirements. These new requirements can be as diverse as additional flood standards, green roof standards, requirements to preserve historical or natural resources, or special incentives to promote redevelopment and revitalization.

From the above discussion, subdivision ordinances, planned unit development, zoning, overlay zoning, and performance-based zoning all have the potential to address hazard risk reduction possibilities for hazard mitigation or climate change adaptation by shifting development out of risk areas or minimizing developmental impacts that could increase vulnerabilities to people or the built environment in specified areas, surrounding areas, or for the entire jurisdiction. The final land use tools, however, more directly and simply address risk reduction strategies. Agricultural or open space zoning ordinances can be utilized to preserve a host of environments including forest habitats, watersheds, hilltops, and slopes, thereby limiting development in areas that are likely to experience various forms of hazards from floods to landslides, as well as prevent erosion and avoid compromising wetlands. Hazard setbacks simply require that development be prohibited in high hazard areas or close to natural resources such as dunes, vegetation, shorelines, and wetlands. Last, storm water retention requirements ensure that runoff from a development is limited and contained within the development itself, thereby reducing flooding potentials downstream.

Figure 7.2 displays the data on the percent of jurisdictions along the Gulf and Atlantic coasts that have adopted each of the eight land use management and development regulations. These tools have been ordered from highest to lowest based on the percent of jurisdictions that have adopted each type of tool. The number of jurisdictions for which data are being reported appears in parentheses after each tool's label, as some respondents skipped questions. Not surprisingly, the most prevalently adopted approach for guiding development is residential subdivision regulations; 96.3 percent of jurisdictions utilize this planning tool. This is followed closely by general zoning at 92.8 percent, and storm water retention which is at 92.7 pecent. Some form of hazard setback is employed by 85 percent of jurisdictions, with 82.8 percent employing special overlay districts, and 80.7 percent using planned unit developments. Agricultural/open space zoning is employed by just over 75 percent of these jurisdictions, but performance-based zoning, which can be quite demanding on the planning staff, is employed in less than 40 percent of jurisdictions.

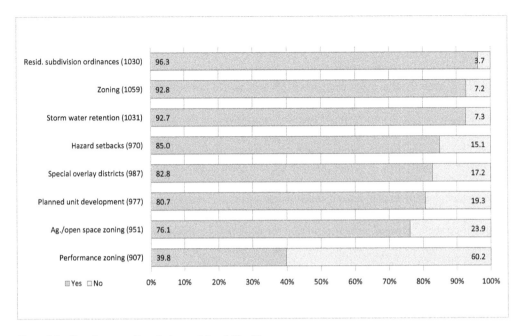

Figure 7.2 Development Regulation and Land Use Management

Given the flexibility of several of these general land use and development regulations, particularly subdivision ordinances, planned unit development, zoning, and overlay zoning, the high percentage of jurisdictions that have adopted and are using these tools bodes well for their potential application in addressing hazard risk reduction across many jurisdictions. Although widespread, it is probably safe to say that these tools are not currently being employed to directly address hazard mitigation and risk reduction issues. Nevertheless, even if not currently employed in this fashion, the potential is there for them to be adapted for this purpose. Additionally, there also appears to be relatively high adoption of development regulations that directly and specifically address hazards such as storm water retention, hazards setbacks, even open space zoning.

2) Shoreline Regulations

Shoreline regulations include restrictions on shoreline armoring, restrictions on dredging and filling, and limiting shoreline development to water-related uses only. Shoreline armoring uses physical structures to protect shorelines from coastal erosion. Although armoring may well slow coastline erosion, it also can destroy natural habitats and restrict natural sedimentation processes. Armoring can also promote development behind armored shorelines by fostering a false sense of security that armoring has eliminated erosion—the coastal equivalent of the "levee effect" on inland rivers (Tobin 1995). Dredging removes sediment from the bottom of bodies of water and filling uses material, often dredged material, to fill in the coastline. Although this process has benefits, like armoring it disrupts the natural ecosystem. Restricting shoreline development to water-related uses such as boat docks can limit the types of property development along shorelines, keeping dense housing and other development from these areas.

Figure 7.3 displays the relative frequencies for shoreline-related land use tools. It should be noted that jurisdictional informants could choose not to respond to specific tools or simply indicate that mitigation/adaptation measures were not relevant to their jurisdictions. In the case of these shoreline

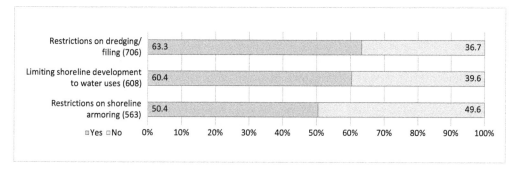

Figure 7.3   Limiting Shoreline Development

measures, the number of jurisdictions responding is much lower, because respondents indicated that these measures were not relevant to their jurisdictions. So, when considering jurisdictions among which respondents presumably thought such measures were relevant for their jurisdictions, just over 63 percent indicated they had restrictions on dredging and filling in waterways, just over 60 percent indicated they have adopted policies restricting shoreline development to water uses, and just over 50 percent restricted shoreline armoring. On the whole, although a far from insignificant percentage of jurisdictions have adopted regulations that limit development along their waterways and disruptions to those waterways themselves, there are nevertheless substantial percentages of jurisdictions that have little or no regulation within these often highly sensitive shoreline areas.

3) Building Standards

Implementing building standards and codes can reduce hazard vulnerabilities and address climate change adaptation. For example, the Insurance Institute for Busines and Home Safety (IBHS), in its report rating state building codes, notes that the 2009 International Residential Code (IRC) offers effective standards for addressing potential structural vulnerabilities to wind hazards and the 2009 International Energy Conservation Code (IECC) enhances energy efficiency, which can achieve climate change mitigation by reducing dependence on fossil fuel-based electric production (IBHS 2011). With respect to hazard risk reduction, as Beatley (2009) noted, it is rarely possible to completely avoid development in hazardous areas and, even when possible, it can be politically difficult to impose other land use and development regulations that create restrictions on this development. For example, land development rights are strong in the US in general and are particularly strong in some states such as Texas. Hence, local jurisdictions often display little or no willingness to limit development in high hazard areas and some states restrict local juridictions' ability to do so. In addition, limited choices have historically resulted in development for low income and minority households being located in environmentally sensitive or hazard-prone areas simply because the land is more affordable (Jacobs 2018). In other cases, pressure from development interests can produce development in high hazard areas along the coast, on hillsides, and along rivers because of their attractiveness and recreational and economic opportunities. Consequently, high income households' homes are sometimes located in areas subject to high winds, floodwaters, surge, landslides, slope failures, and earthquakes.

Building standards and code requirements can be critical for reducing the vulnerability of structures that are built in hazardous areas. Furthermore, although building codes can address many issues, they are generally directed at new construction and frequently ignore hazard-related issues. Hence, a broad-based package of building codes and standards might be necessary. For example,

a full complement might include a building code for new construction, floodproofing requirements in particular neighborhoods, and retrofitting requirements for existing buildings (Olshansky
and Kartez 1998). Such a package might also include wind hazard standards for new housing and
special codes for adding wind resistance technology, such as shutter or glazing systems for existing
homes (Beatley 2009), as well as energy efficiency and other utility-related standards. The building
regulation standards and codes data collected by our survey focused on five policy areas: (1) general
building code adoption by local jurisdictions, (2) flood hazard standards for new homes, (3) wind
hazard standards for new homes, (4) retrofitting for existing building, and (5) special utility codes.

The findings for building codes and standards are presented in Figure 7.4. It is noteworthy to
see that 96.0 percent of jurisdictions report having adopted a building code. However, if one looks
a little deeper, there is a good deal of room for improvement. Building codes have increasingly
become based on the ICC's model codes, so all codes that are currently adopted along the Gulf and
Atlantic coasts are based on ICC model codes that have been released every three years since 2000.
However, there is considerable variability in timing and nature of code implementation, the extent
to which states and/or local jurisdictions amend or exclude sections of each code, and the extent to
which the codes are enforced (IBHS 2011, 2015, 2018). In our sample, some jurisdictions were still
employing the IRC, which applies to one- and two-unit dwellings, for 2000 or 2003. Furthermore,
although just over 70 percent had adopted some version of the IRC 2009, which the IBHS suggests
has strong wind standards, those standards were explicitly excluded by some states and local jurisdictions (IBHS 2011, 2015).

In addition to general building codes, flood standards for new home construction have been
adopted by 66.4 percent of the jurisdictions, with an additional 52.6 percent adopting special wind
standards for new construction. Finally, 34.6 percent of jurisdictions have adopted retrofitting standards. The latter can be important because, as essential as building codes are, they only apply to new
construction. Substantial proportions of a community's residential housing stock may be quite old,
and thus built under standards that are much weaker. For example, Highfield et al. (2014), in their
analysis of residential housing following Hurricane Ike, found that homes built during the 1940s
and 1950s suffered significantly greater damage than new homes and even homes built prior to that
period, in part because of very poor building codes and standards that guided construction during

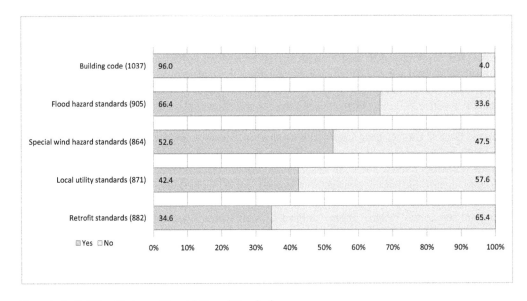

Figure 7.4   Building Codes and Special Hazard Standards

and immediately after World War II. This situation makes establishing standards for retrofitting existing structures just as important as adopting a new code.

4) Natural Resource Protection

Many scholars have discussed protecting natural resources for coastal hazard mitigation and climate change adaptation (Beatley 2009; Beatley et al. 2002; Brody et al. 2011; Daniels 2014; IPCC 2012; Mileti 1999). There are a number of advantages in protecting natural resources such as wetlands, barrier islands, estuaries, water supply reservoirs, dunes, and forests. First, to the extent that these natural resources areas are themselves high hazard areas, preservation keeps development and concomitant human occupation low, reducing the potential loss of life and property. Additionally, protecting these areas avoids deleterious secondary consequences by preserving their ecosystem services for reducing the impacts of flooding, hurricanes, storm surges, and coastal erosion (Beatley 2009; Bernd-Cohen and Gordon 1999; Klee 1999; Mileti 1999; Williams and Micallef 2009).

Williams and Micallef (2009), for example, noted the protective features of natural environments, such as vegetation's ability to reduce wave action, current energy, and erosion, as well as trap sediment helping regenerate shorelines. A host of researchers have shown the ability of estuarine and coastal ecosystems such as marshes, coastal wetlands, barrier islands, and beaches, to reduce damage and life loss in a variety of coastal environments along the Gulf and Atlantic coasts (Barbier 2015; Barbier and Enchelmeyer 2014; Barbier et al. 2011; Costanza et al. 2008; Narayan et al. 2017; Petrolia et al. 2014; Petrolia and Kim 2009, 2011; Pompe and Rinehart 1994; Shepard et al. 2011). For example, Highfield et al. (2018) have shown a strong association between estuarine wetlands and reduced flood losses due to storm surge. The long-term work by Sam Brody and his colleagues has shown a strong, consistent, and direct relationship between wetland loss and flood-related damage and loss of life. Their studies have shown that altering wetlands by permitting development has resulted in higher flood damage loss and loss of life and these effects are cumulative over the long run (Brody et al. 2007, 2008, 2010, 2015; Brody et al. 2011; Zahran et al. 2008). This research clearly suggests that preservation of wetlands can reduce risks to loss of life and property. There are also benefits for carbon sequestration by preserving natural resources such as forests; hence wetland preservation can produce a double payoff for hazard and climate change mitigation.

Figure 7.5 displays the findings with respect to natural resource protection policies and programs. Over 86 percent of the jurisdictions have implemented some form of wetland protection in their areas, with nearly 83 percent employing environmental impact assessments for development, 79.2 percent employing protected areas policies, and just over 71 percent implementing habitat protection and restoration programs. For those with coastal or shoreline areas, 59 percent are engaging in vegetation protection programs within these areas and 36.1 percent have dune protection policies. On the whole, there appears to be widespread recognition of the need to address wetland policies and other programs to protect environmentally sensitive areas and perhaps gain the benefits from the ecosystem services they can provide. However, it must be remembered that the extent to which these policies have been developed, utilized, and enforced is not being captured in a simple measure based on adoption. Indeed, IBHS (2011, 2015, 2018), although not ranking jurisdictions, has offered a methodology for ranking states, not only on the adoption and mandating of residential building codes, but also on enforcement and training of enforcement officials. Training is a topic addressed in our assessment of public information and awareness programs.

5) Public Information and Awareness Programs

There are many reasons why public information and awareness programs can be important components of a local jurisdiction's broad-based hazard mitigation and climate change adaptation strategy.

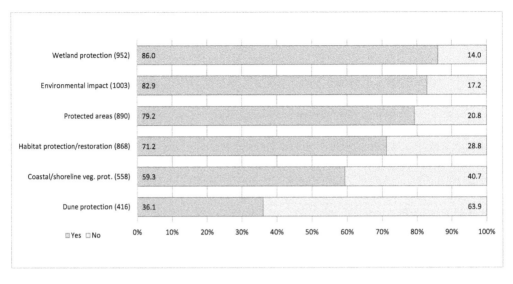

Figure 7.5 Natural Resource Protection

One important reason is simply to provide developers, builders, homeowners, and renters through-out a community with a better understanding of their hazard risks and potential vulnerabilities so they can make better informed decisions. These programs can potentially be an important step toward promoting voluntary adoption of hazard adjustments (Ge et al. 2011; Hyndman and Hyndman 2006; Lindell and Perry 2004, 2012; Lindell et al. 2006; Peacock 2003). These programs can be important not only to educate longtime residents about the evolving nature of risks and vulnerabilities due to climate change and sea-level rise, and hence the possible need to rethink their risks and vulnerabilities, but also for the many new residents that are moving to coastal communities (Godschalk et al. 2003; IPCC 2012; Lindell and Perry 2004; Olshansky and Kartez 1998). Even in situations where these hazard awareness programs do not produce a voluntary cesssation of development in hazardous areas, hazard disclosure policies and hazard zone signage can increase awareness and perhaps spur action to reduce vulnerabilities (Beatley 2009). Importantly, enhancing overall community aware-ness of hazard risks can also promote stakeholders' commitment toward governmental actions with respect to mitigation policies—ranging from improvements in building codes to various land use and development regulations, as well as involvement in hazard mitigation and climate change adaptation planning (Norton 2005b; Robins 2008).

Public information and awareness programs for hazard mitigation and climate change adaptation can be quite diverse. Examples might include straightforward public education programs, hazard disclosures in property and real-estate transactions in addition to the mandatory flood disclosures statements for homes purchases; hazard zone signage; programs encouraging insurance for environ-mental hazards (wind, flood, earthquake); technical assistance and training to builders, developers, and property owners for hazard mitigation; hazard information centers; and training materials pro-vided in multiple languages (Beatley 2009; Berke 1996; Brody and Highfield 2005; Burby 1998; Godschalk et al. 1998; Olshansky and Kartez 1998; Srivastava and Laurian 2006). For the purposes of this research, jurisdictional informants were asked about public education programs, citizen involve-ment in hazard mitigation planning, hazard mitigation training for builders and developers, hazard disclosure statements, and hazard zone signage.

Our survey results for public information and awareness programs are presented in Figure 7.6. Public education programs, which range from simple brochures and posters to public service

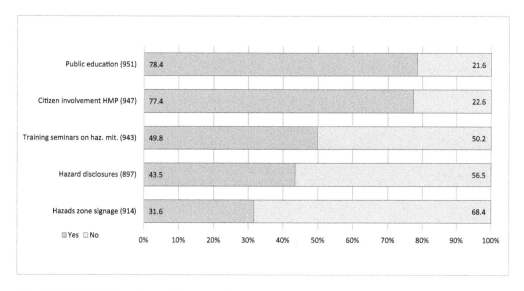

Figure 7.6   Public Information and Awareness Programs

announcements and special events, have been implemented by 78.4 percent of the jurisdictions. Just over 77 percent of them also claim to have citizen involvement in hazard mitigation planning activities, but it should be noted that involvement can take on a variety of meanings in this context—ranging from informing to citizen control (Arnstein 1969). Hazard disclosures for all real estate transitions are required by only 43.5 percent of responding jurisdictions. And finally, and perhaps not surprisingly, just over 30 percent of the sampled jurisdictions indicated that they provide signage indicating hazard zones within their communities. This is a comparatively simple form of public information, but also one that can be quite difficult to enact at the local level, particularly given local concerns about its potentially negative impacts on development.

6) Local Incentive Tools

Incentive tools are non-prescriptive approaches that seek to limit or eliminate development in hazardous areas, as well as preserve natural resources and the ecosystem services they provide by motivating land owners, builders, developers, and others to shape their developmental activities in a more sustainable and resilient manner (Daniels 2014; Daniels and Daniels 2003; Tang et al. 2011). Strategies such as these provide flexibility by allowing existing land use regulations or limitations to be exceeded or partially waived if developers make concessions. These programs might take the form of density bonuses, which allow for greater residential densities than would otherwise be possible, if additional mitigation features are included in the development. Similarly, cluster development provides the opportunities to exceed land use regulations if development is clustered within less hazardous areas or out of environmentally sensitive areas of a subdivision. Another incentive-based program is based on the transferring of development rights (TDR). In these programs, a community generally designates a particular area such as a wetland or green space area as a conservation zone (sending area) while another area is designated as a development (receiving) area (Beatley 2009). Land owners in sending areas can "sell" their development rights and in turn receive additional concessions for developing in the receiving area.

Figure 7.7 displays the adoption rates for the three local incentive programs—cluster development, density bonuses, and TDR programs—that were addressed in our survey. Of these three

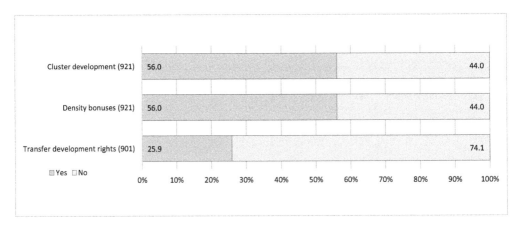

Figure 7.7 Local Incentive Tools

programs, cluster development and density bonuses have been adopted at much higher rates than TDR programs. Although 56 percent of jurisdictions utilized cluster development and density bonuses, only 25.9 percent of jurisdictions report using TDRs. This is unsurprising because Beatley (2009) has reported that TDR programs can be quite difficult to implement, particularly in relatively small communities.

7) Federal Incentive Programs

In addition to local incentive programs, there are federal programs that promote hazard mitigation, particularly at the community level. However, these can also have indirect consequences for household level activities. The National Flood Insurance Program (NFIP) was established in 1968 in response to the private sector pulling out of underwriting flood risk. The NFIP allows development in hazardous areas (i.e., flood plains) by spreading the risks associated with such development. However, it also promotes mitigation activities by requiring jurisdictions to adopt and enforce minimum land use and building construction standards (Holway and Burby 1990; Lindell et al. 2016; Schwab 2010; Schwab et al. 2007). This can be an important issue after a flood, whereby properties that have suffered 50 percent or more damage must be brought up to the current flood standards. In many cases, this means elevating a home above the 100-year flood standard, which can put rebuilding costs beyond the reach of many households. This is particularly problematic given that there is also a ceiling or maximum payout with flood events.

A related federal incentive program is FEMA's Community Rating System (CRS), which is part of the NFIP. This program targets jurisdictions that are participating in the NFIP by promoting the adoption of a host of policies that exceed minimum NFIP requirements and, ultimately lessen flood risk in the community (Schwab et al. 2007). The payoff is that the jurisdiction can earn community-wide discounts as high as 25 percent on flood insurance premiums by engaging in approved mitigation activities. These activities include flood mapping and regulations, flood damage reduction activities such as building codes, flood response planning, and mitigation works such as dam and levee maintenance (FEMA 2007). Research has shown that involvement in the CRS can substantially reduce local flood losses (Brody et al. 2011).

As can be seen in Figure 7.8, participation in the NFIP by jurisdictions in our survey is extremely high, with 94 percent of the sampled jurisdictions reporting participating in the program. Given early concerns about how slowly growth in the adoption of this program was, this 94 percent figure

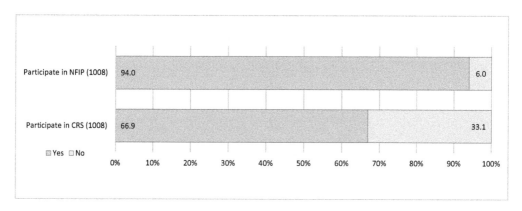

Figure 7.8   Federal Incentive Programs

is remarkable, but also suggests that many communities and their residents are highly dependent on this government insurance program. The percentage of jurisdictions participating in the CRS is relatively high compared to other mitigation tools examined in this research, with nearly 67 percent of jurisdictions participating, but there is clearly room for more participation. Given the research showing the effectiveness and cost benefits of this program, it would make sense to try to increase participation in the future.

8) Property Acquisition Programs

Government acquisition of private property is not without its challenges; however, local governments can acquire and hold property for public benefit such as attempts to limit or preclude development in hazardous areas (Beatley 2009). Similarly, local governments can acquire property to protect critical ecosystems or natural features such as wetlands, maritime forests, and estuaries, or more simply to provide open space for community recreational benefits (Beatley 2009; Schwab et al. 2007). The specific tools included in our survey are fee simple purchases, acquisition of development and easement rights, and relocation of existing structures out of hazardous areas. The fee simple purchase of property simply transfers full ownership of a property to the local jurisdiction (Beatley 2009; Schwab et al. 2007). Acquiring development rights or easements is a legal agreement between a landowner and an eligible easement holder that restricts future activities on the land to protect its value for natural protection or conservation (Beatley 2009; Daniels 2014; Schwab et al. 2007). The goal of an easement or purchase is simply to remove a property that is at risk or critical for mitigation services from the private market and thereby reduce the possibility of inappropriate development that will heighten vulnerabilities and risks of future disasters (Beatley 2009; Schwab et al. 2007). Although Schwab et al. (2007: 263) note that "in the long run it is often less expensive to acquire and demolish a building than to repeatedly provide for its construction," these programs are nevertheless expensive for local governments that rarely have the capital reserves for such purchases.

Figure 7.9 indicates that, of these three tools, the fee simple purchase of property is the most frequently adopted, with 57.4 percent of jurisdictional informants indicating that their jurisdictions purchased properties to remove them from market development. Nearly 50 percent of responding jurisdictions reported acquiring development rights/easements but only 28.7 percent reported acquiring and relocating structures. In many respects these percentages are perhaps comparable to those seen for the adoption and utilization of local incentive tools.

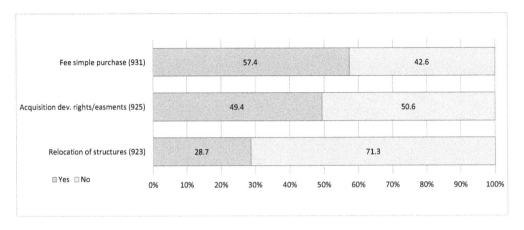

Figure 7.9 Property Acquisition Programs

9) *Financial Tools*

Financial tools offer a mixture of both incentives for improving, and disincentives for discouraging, particular forms of development in environmentally sensitive or hazardous areas. These approaches also provide an opportunity to address the differential impacts that development in high hazards areas can have on communities. They also can offer a mechanism for partially addressing externalities associated with development in environmentally sensitive and hazardous areas. The survey addressed lower tax rates, special tax districts, and impact fees. Lower tax rates may be awarded to developers and builders that implement hazard mitigation measures or preserve natural resources. Special tax districts may be imposed on development in environmentally sensitive or hazardous areas to generate additional revenue that can address the extra expenses associated with these developments. Similarly, impact fees associated with new construction can be assessed to help defray additional public expenses associated with the development of properties on or near environmentally sensitive or hazardous areas. In Florida, for example, impact fees associated with new development help fund emergency preparedness activities that will ultimately support emergency response to hazard events.

As the results in Figure 7.10 indicate, financial tools are clearly not often employed by local jurisdictions. Only 19.9 percent of jurisdictions seek to incentivize developers and builders by providing reduced tax breaks for development that incorporates higher building standards or other mitigation design features. Impact fees are even more rarely employed, with only 15.6 percent of responding jurisdictions indicating that they have adopted such a fee system. Equally unlikely are special tax assessments to discourage development; only 13.4 percent of jurisdictions report adopting these forms of disincentives for discouraging development. Of all the hazard mitigation approaches investigated thus far, financial tools are clearly the least frequently adopted.

10) *Critical Public/Private Facilities and Infrastructure Requirements*

Critical facilities include police and fire stations, hospitals and emergency care and medical centers, as well as schools, governmental offices, and evacuation centers (Schwab et al. 2007). Keeping critical facilities out of hazardous areas is appropriate for a variety of reasons. First and foremost, the nature of critical facilities is such that they provide essential resources during emergency response and

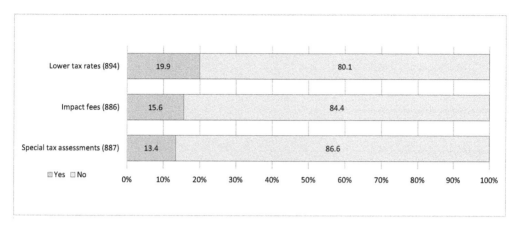

Figure 7.10   Financial Tools

disaster recovery. If these facilities are damaged by disaster impact, lose essential infrastructure such as electric power and telecommunications, or become inaccessible because of blocked roads, then they are useless in helping communities address their many needs. Oftentimes these facilities—and the equipment, technology, and records they contain—can also be quite expensive to build and maintain. Hence, keeping them out of harm's way can reduce the cost of repairs and replacement for communities that are often hard pressed financially following a disaster. As a consequence, Beatley (2009: 74) suggests that these facilities should be sited outside of hazardous areas so they will remain functional during major community disruptions. Establishing these policies, implementing them, and publicizing their existence can also set an important example about taking hazard reduction seriously as an integral part of community development.

The issues of critical infrastructure, such as electrical power, fuel (e.g., natural gas pipelines), water, wastewater, telecommunications, and transportation are also important to consider. These elements should be part of a community's planning efforts and taken into consideration so that elements of these systems, such as lift stations or generator substations, are kept out of hazardous areas. Similarly, decisions about distribution lines, as to whether or not they are hanging or buried, should be considered in wind hazard areas. However, it is in the more general decisions regarding municipal service areas that communities can seek to limit or keep development from expanding into hazardous or environmentally sensitive areas (Beatley 2009). In other words, jurisdictions, through their land use planning efforts can decide to limit or simply not provide infrastructure or other services such as schools in areas that are designated as protected areas or high hazard areas and thereby limit development in these areas as well.

The percentages of jurisdictions utilizing these policies and programs is quite large, particularly when considering the requirements for critical public and private facilities. Figure 7.11 indicates that just over 85 percent of jurisdictions surveyed have requirements with respect to location of their public critical facilities in hazardous areas. Similarly, 81.5 percent place such requirements on critical private facilities as well. Additionally, 66.4 percent are utilizing municipal service area decisions to limit development within their communities, which would probably extend to extra-territorial jurisdictional areas as well. These findings are quite different and much more robust than those found by Masterson et al. (2014) when only considering coastal jurisdictions in Texas and much more consistent with research reported by other researchers (Olashansky and Kartez 1998; Schwab et al. 2007).

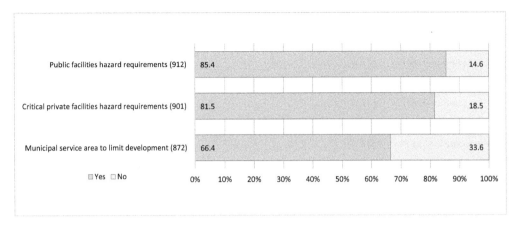

Figure 7.11 Critical Facilities and Infrastructure Requirements

11) Public–Private Sector Initiatives

Local jurisdictions often find that they have limited budgets for addressing many of the activities that need to be undertaken, but simply cannot because of financial constraints. Additionally, even when funds are available, partnering with the private sector can greatly leverage the resources available to address hazard mitigation and climate change adaptation. Moreover, many private entities can gain from these activities, in terms of advertising and also showing a commitment to their communities' concerns and well-being. In the case of hazard mitigation and climate change adaptation, local hardware and building supply stores, as well as local contractors, will often partner with their communities on public education programs related to retrofitting for wind hazards, such as adding shutters and proper roof installation. Similarly, private electric companies can partner with their communities to offer energy efficiency audits, promote wind and solar power systems, and provide credits for installation of more efficient cooling or water heating systems. Equally important can be public-private partnerships for developing land trusts and obtaining development rights, which can be quite difficult for local jurisdictions to implement but can become viable options with private partnerships and funding.

Figure 7.12 displays the results for the two public-private sector partnership questions asked of jurisdictional informants related to land trusts explicitly, and then more generally, any form of partnership. Perhaps somewhat surprisingly, only about half of jurisdictions made use of these possibilities. Specifically, 49.7 percent indicate that these partnerships are being employed to set up land trusts. Additionally, nearly 45 percent suggest that they are employing public-private partnerships in general. The latter is perhaps the more surprising of these findings, in that this includes the full range of partnerships that can be mutually beneficial for both local jurisdictions and the private sector.

12) Use of Professionals

The final strategy addressed in the survey was local jurisdictions' use of professionals to enhance the effectiveness of their activities related to hazard mitigation and climate change adaptation. Contracting with professionals such as engineering, planning, and consulting firms can be important for jurisdictions, since many lack the financial ability to retain such professionals as full-time staff. Schwab (2010) notes that many communities without a permanent planning staff hire planning

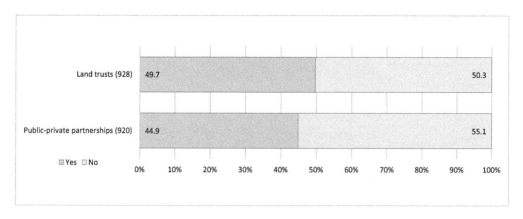

Figure 7.12   Public–Private Sector Initiatives

consultants or companies to draft comprehensive plans, zoning regulations, and building codes and to assist in implementing them. This is especially likely when addressing hazard mitigation and climate change adaptation issues because these can demand more technical and scientific information and data sources (Schwab 2010). For example, geological consultants may be necessary to help identify suitable building sites and engineering consultants are needed to address or develop special building techniques in hazard-prone areas (Tang et al. 2011). The Louisiana Coastal Protection and Restoration (LACPR) program used outside technical expertise in guiding initiatives for post-Katrina and post-Rita hazard mitigation (Cigler 2009). Research evaluating local hazard mitigation plans, for example, has found that a single firm has undertaken the majority of these planning efforts for county and regional hazard mitigation plans in particular states (Kang et al. 2010; Peacock et al. 2009). Indeed, when undertaking our survey work for this research we found that a number of communities outsourced much of their planning and permitting processes to private consultants rather than maintaining their own full-time planning staffs.

Our survey assessed four ways professionals might be employed by local jurisdictions. These include: (1) identifying suitable building sites for public infrastructure and facilities, (2) providing technical expertise from geographic information systems (GIS) to engineering-related services and building inspections, (3) conducting windstorm/roof inspections, and (4) developing and writing plans or other planning documents and policies. The survey results are displayed in Figure 7.13. Clearly, an overwhelming percentage of jurisdictions are making use of professionals for writing plans and planning documents (87.2 percent) and for technical expertise (87 percent). Although significantly smaller percentages use professionals for siting buildings (49.2 percent) and conducting windstorm or roof inspections (47.3 percent), these percentages are nevertheless significant. It is interesting to note that so many jurisdictions depend upon outside professionals for their plan writing activities. This, of course, can be critically necessary as local staffs are often so involved in their day-to-day duties that taking on a major additional task, such as developing a hazard mitigation plan, can be quite difficult. However, the other side of the coin is that these activities can become rather formulaic efforts in which the external consultants produce "fill-in-the-blanks" plans from templates. Such plans meet legal requirements and mandates but fail to actually engage the community in a broad-based planning process or fully integrate the plan with other planning activities. Recent work by Berke et al. (2015, 2018) has clearly shown that there can be major disconnects between plans and planning activities, resulting in major inconsistencies between plans that weaken overall attempts at reaching sustainability and resiliency goals.

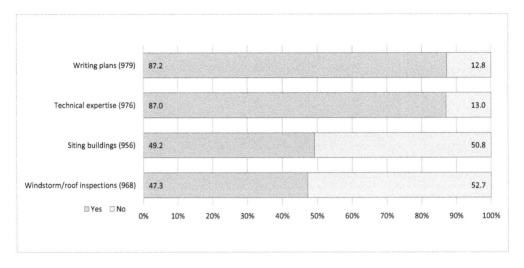

Figure 7.13 Use of Professionals

Putting the Pieces Back Together

These results show the adoption of 47 different tools for hazard mitigation and climate change adaptation by local jurisdiction along the Gulf and Atlantic coasts. Figure 7.14 combines the survey results on the adoption of these 47 planning activities/tools and orders them from the most frequently adopted (residential subdivision ordinances at 96.3 percent) to the least frequently adopted (special tax assessments at 13.4 percent). The top ten, which range in adoption rates from 96.3 percent to 85 percent are residential subdivision ordinances, building codes, participation in the NFIP, zoning, storm water retention, utilizing professionals for plan writing and technical expertise, wetland protection, hazard requirements for public critical facilities, and hazard setbacks. Clearly four of these top ten are directly targeting hazard mitigation issues. The fact that the other six are more general planning tools that can potentially address hazard mitigation and climate change adaptation issues, begs the question about the extent to which they are actually directly addressing these issues as well. Indeed, the fact that we can clearly see that residential subdivision ordnances are the chief planning tool being employed by jurisdictions along the Gulf and Atlantic coasts suggests these tools could have major impacts on hazard mitigation and climate change adaptation if they would incorporate some elements that are relevant to these issues. Similarly, it is significant to see that building codes in general are being adopted by local jurisdictions; however, as noted above, there can be a good deal of variation in the nature of these codes and whether or not they incorporate critical wind mitigation and energy conservation features. Indeed, there are also significant variations from state to state in the degree to which adoption is mandated and enforced (IBHS 2011, 2015, 2018; Masterson et al. 2014).

The bottom ten in adoption rates are equally interesting. They are local utility standards, performance-based zoning, dune protection, retrofitting standards, hazard zone signage, relocation of structures out of hazard areas, transfer of development rights, lower tax rates, impact fees, and special tax assessments. It is perhaps not surprising that in a nation and at a time when discussion of taxes and fees must also be accompanied by the concept of "cutting spending elsewhere," reducing debt, and balancing budgets, that these have the lowest adoption rates of below 20 percent. Lower tax rates are not effective tools when rates are already low, for example, and, conversely, the possibility of adding fees and increasing tax rates, becomes an impossibility because "increasing" any development costs becomes a political liability for those promoting them. Although quite low in adoption, these tools still can be effective in particular locations and under unique local conditions.

As noted above, the last major effort to understand the degree to which specific planning tools and approaches were being adopted by local jurisdictions to address hazard mitigation was undertaken by Godschalk et al. (1989) and published in their book entitled, *Catastrophic Coastal Storms: Hazard Mitigation and Development Management*. Their work focused only on jurisdictions directly on the coast—those in V-zones, which are areas that experience coastal wave action associated with storm surge. Hence, our survey was more expansive, including jurisdictions well inland from the coast, where flooding and consequential damage and deaths frequently occur. Additionally, our work considered many more planning tools that might be employed for hazard mitigation and climate change adaptation; thus, a direct comparison of the two studies' results would be misleading. Nevertheless, despite these differences, it is perhaps worth comparing our results with theirs for the 12 planning tools that were included in both surveys. It should be noted that these include whether or not a jurisdiction has a comprehensive or general plan which was not included in our discussion above.

Figure 7.15 displays the percentage of jurisdictions reporting that they have adopted or utilized the 12 tools in common to the two surveys. Ranking the planning tools from lowest to highest adoption rates based on the Godschalk et al. (1989) findings reveals two patterns. First, there is a general tendency to find that the relative popularity of these tools in the late 1980s seems to hold in the more recent survey. Those activities that were not extensively adopted in the early survey, such as preferential property tax assessments and the transfer of development rights, remain at lower adoption rates in the later survey. In addition, those actions that were adopted at higher rates such as zoning and residential ordinances, remain at the highest adoption rates and many of those falling in between these extremes display the same patterns of relative adoption. The second pattern is that, with the exception of dune protection, adoption rates are generally higher in the most recent survey than in the late 1980s. Even the activities that were highly adopted in the Godschalk et al. survey (e.g., having a comprehensive plan, subdivision ordinances, and zoning) have adoption rates that are slightly higher today. The greatest differences between the two surveys are with respect to setbacks, locating public facilities, using public services to limit development, fee simple purchases, hazard disclosures, and construction training seminars. With respect to each of these tools we find much higher adoption rates today. These changes suggest that jurisdictions are addressing, at least partially, the increasing risks they face with respect to natural hazards. Nevertheless, given the continuing increases in losses due to coastal hazards witnessed in just the past ten years, it is questionable whether these changes are coming quickly enough throughout the region to keep pace with disaster impacts and coastal population growth.

Variations and Consistencies Among Coastal States

With respect to similarities and differences in the patterns of adoption among the 19 coastal states, Tables 7.3 and 7.4 present the average percent of jurisdictions within each state that have adopted the tools associated with each of the 12 sets of strategies. It should be recalled (see Table 7.1), that there is a great deal of variation across states in terms of the numbers of jurisdictions (counties or municipalities) that were sampled, depending on the number of NOAA-defined coastal watershed counties and their associated jurisdictions in the state. In Mississippi, for example, there are only eight jurisdictions whereas in Florida there are 195. Hence, the average adoption rates for some states are sometimes based on very few observations, whereas for other states they are based on many jurisdictions. Nevertheless, these average rates do provide some insights into the similarities and differences in the adoption patterns for the 12 strategy sets among Gulf and Atlantic States.

These strategy sets, along with a total adoption rates for all 47 tools, are split across Tables 7.3 and 7.4. The strategy sets are ordered from left to right based on that set's average adoption rate

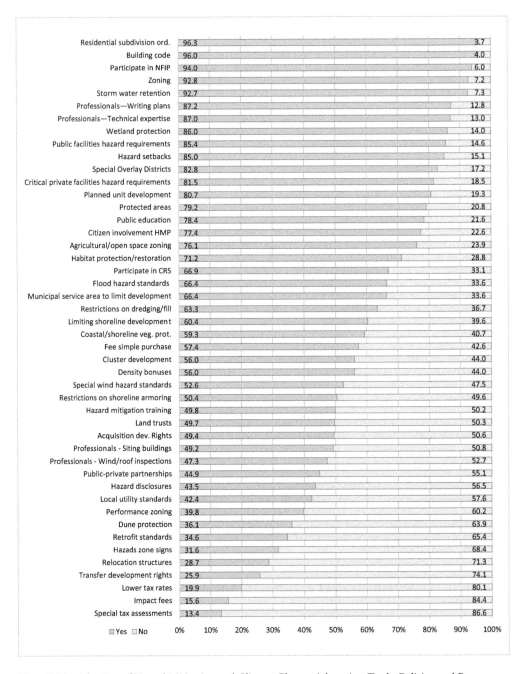

Figure 7.14 Adoption of Hazard Mitigation and Climate Change Adaptation Tools, Policies, and Programs

(see the *total* row at the bottom of each table). Hence, the first column in Table 7.3 is for development regulations and land use management tools since they had the highest average adoption rate of 81.8 percent (i.e., on average 81.8 percent of the eight tools were adopted across all jurisdictions) and the second to last column in Table 7.4 is for financial tools that were only adopted at a rate of 17.2 percent. The states themselves are ordered from top to bottom, based on their average jurisdictional adoption rate

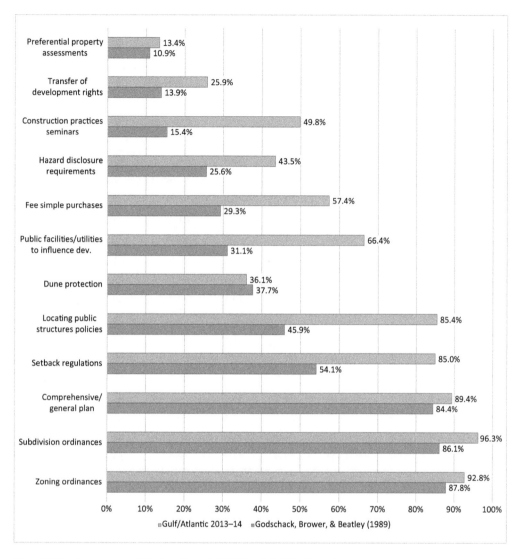

Figure 7.15 Comparison of Godschalk et al. (1989) and Gulf/Atlantic Surveys

across all 47 tools and this adoption rate is presented in the last column on Table 7.4. Rhode Island had an adoption rate of 74.6 percent so its rates are presented in the first row in each table, whereas Georgia's adoption rate was only 56.1 percent so it appears in the last row in each table. Within each strategy set, cells with darker gray shading indicate the top four adoption rates, with the highest having a bold font, and cells with lighter gray shading indicate the bottom four adoption rates, with the lowest also being in bold font.

As mentioned above, the development regulations and land use management policies strategy set has the highest relative adoption rates among jurisdictions at 81.8 percent. Within this strategy set, Rhode Island's jurisdictions have the highest adoption rate of 94.2 percent followed closely by Delaware, Virginia, and Mississippi. The lowest rates are found in Texas at 66.9 percent, followed by Louisiana. Average adoption rates are quite high in general for federal incentive

Table 7.3 Average Adoption Rates for Each Strategy Set Across States (Part 1)

State	Dev. Reg. Land Use	Federal Incentives	Critical Facilities	Natural Resources	Use professionals	Building Standards	Limiting Shoreline dev
Rhode Island	**94.2**	83.3	**91.7**	80.7	**72.2**	66.7	46.7
Connecticut	83.5	77.4	83.9	82.0	63.5	65.6	71.0
Virginia	90.2	78.8	79.2	83.1	66.9	67.5	**84.6**
Florida	85.4	89.9	80.9	84.6	70.5	71.7	75.9
Mississippi	87.5	**92.9**	76.2	62.5	64.3	**85.7**	50.0
Alabama	79.8	78.6	83.3	82.9	67.9	77.1	51.7
Massachusetts	83.8	72.1	79.9	**89.0**	65.8	47.7	73.6
Maine	84.7	80.5	78.8	83.3	61.4	49.5	77.4
North Carolina	83.0	80.6	79.4	66.8	68.9	67.9	52.7
South Carolina	83.2	85.5	82.1	69.3	68.5	81.0	44.0
New Hampshire	85.8	72.2	78.7	79.1	64.6	**47.0**	74.8
New York	76.1	80.8	75.4	70.4	72.0	60.6	52.7
Pennsylvania	86.2	76.2	76.0	75.5	70.8	58.9	**31.7**
New Jersey	80.6	80.1	76.0	67.1	70.3	61.2	37.6
Maryland	76.4	**67.5**	86.0	70.5	63.8	50.8	65.7
Delaware	90.4	72.7	74.1	81.2	**60.0**	57.7	38.9
Louisiana	70.9	88.3	**68.3**	59.0	64.7	81.0	44.1
Texas	**66.9**	82.4	69.7	**57.5**	70.7	65.9	55.3
Georgia	82.2	78.3	75.4	71.6	65.2	68.3	40.0
Total	81.8	80.5	78.0	75.7	68.2	63.3	61.4

programs (NFIP and CRS), followed by critical facility regulations, and natural resource protection, with average rates at 75.7 percent. Adoption rates fall below 50 percent with respect to public-private initiatives, local incentive programs, and property acquisition programs. But, the lowest adoption rates are found with financial tools, where the average rate is only 17.2 percent.

A cursory view of these tables fails to reveal any strong systematic pattern of differences across the states. This is not particularly surprising because there is no necessary connection among these strategy sets, such that if a state's jurisdictions are high adopters for one set, they would be expected to also be high adopters for all others. Indeed, when examining adoption patterns among jurisdictions in Texas, Masterson et al. (2014) noted that jurisdictions may have legal, capacity, or commitment constraints that can result in very different patterns of strategy adoption to address hazard risks. Nevertheless, some patterns do emerge. First, as noted above, Rhode Island's jurisdictions are consistently high adopters. Indeed, Rhode Island is not only in the top four on seven of the 12 sets but, it has the highest adoption rate in each of these seven sets—development regulations and land use, critical facilities, use of professionals, public-private initiatives, local incentives, property acquisition, and financial tools. On the other hand, Georgia, Texas, and Louisiana consistently rank as the lowest adopters, with overall adoption rates of 56.5 percent or less for all 47 tools. Louisiana ranks in the bottom four on seven of 12 sets and Texas and Georgia rank in the bottom four on six of the

Table 7.4 Average Adoption Rates for Each Strategy Set Across States (Part 2)

State	Information/ Awareness	Public- Private	Local Incentives	Property Acquisition	Financial Tools	All 47 tools
Rhode Island	60.2	100.0	74.1	81.5	25.9	74.6
Connecticut	57.2	76.6	68.9	63.8	22.2	69.4
Virginia	66.9	50.0	52.5	36.4	22.9	68.3
Florida	55.9	33.6	54.8	49.5	14.7	68.2
Mississippi	90.7	64.	38.9	54.2	14.3	67.8
Alabama	71.4	55.0	30.6	59.0	8.3	66.5
Massachusetts	57.1	76.5	55.8	61.6	17.0	66.0
Maine	50.2	70.5	49.5	38.9	20.7	65.6
North Carolina	66.6	45.7	37.0	51.4	21.7	64.2
South Carolina	61.5	41.1	42.9	32.7	10.1	63.7
New Hampshire	52.4	66.7	56.5	48.6	18.1	63.2
New York	67.1	51.7	49.2	40.2	23.0	62.6
Pennsylvania	45.7	42.3	44.3	38.2	16.7	61.9
New Jersey	48.3	39.1	43.0	42.6	22.2	61.1
Maryland	65.0	41.7	54.4	59.6	7.4	60.6
Delaware	62.2	18.8	41.7	40.7	12.5	60.3
Louisiana	59.9	17.1	11.7	23.8	7.1	56.5
Texas	60.8	23.1	18.8	30.4	16.9	56.2
Georgia	60.0	30.4	25.8	21.7	7.6	56.1
Total	57.1	47.5	46.5	45.4	17.2	63.7

12 sets. These low adoption rate patterns are most consistent with respect to critical facilities regulations, public-private initiatives, local incentives, and property acquisition programs. The ultimate question, of course, is what factors shape these adoption patterns.

Assessing Factors Shaping Hazard Mitigation and Climate Change Adaptation Policy Adoption at the Local Level

In addition to the few studies that specifically address the adoption and use of hazard mitigation practices, the plan quality literature also identifies a host of factors influencing planning efforts at the local level. In future years, our research team will be using the data collected in this survey to explore factors shaping the adoption of hazard mitigation and climate change adaptation tools. Figure 7.16 provides a conceptual model seeking to broadly capture six groups of factors suggested by the literature as influencing mitigation practices at the local level. We begin with the state planning environment. A number of researchers have noted the importance of state planning mandates (Berke and French 1994; Berke et al. 1999, 2006; Berke and Roenigk 1996; Burby et al. 1993; Burby and Dalton 1994; Dalton and Burby 1994), the state's political culture (Berke and Beatley 1992b; Mileti 1999) and its activities promoting mitigation (Burby and May 1997; Godschalk et al. 1989). A major area of neglect that begins at the state level but ultimately relates to jurisdictional characteristics is the issue of enabling legislation that provides the legal authority for local jurisdictions to adopt and implement many mitigation practices, particularly those normally associated with land use planning and a variety of development regulations such as zoning, subdivision regulations, and investment policies.

Researchers and practitioners often assume that local governments such as counties have "considerable autonomy in their choice of approaches and policy instruments for land use" (Feiock et al. 2008: 462). Yet there can be much variation across jurisdictional types within and across states in terms of the legal latitude that local government has to undertake land use management, taxation, and other policies. Some have sought to capture these variations in the distinction between "home rule" and "Dillon rule" states (Krane et al. 2001; Salvino 2007a, 2007b; Turnbull and Geon 2006; Wood 2010). In short, the "Dillon rule," stems from a court ruling in 1868 by John Forrest Dillon, which found that local governments are a creation of states and hence their powers are granted by those states. "Dillon rule" states are characterized as states that have retained centralized power, whereas "home rule" states have devolved power by granting some degree of local discretion or autonomy to various forms of local government. This simple dichotomy and the utility of employing "Dillon" and "home" rule distinctions as dichotomous is challenged in the literature both conceptually and empirically (Bluestein 2006; Richardson 2011; Richardson et al. 2003; Wolman et al. 2010). As noted by Richardson (2011), this distinction—Dillion vs. home rule—may be a starting point but there are far too many variations found within and across states in terms of local jurisdictional authority, structure, and capacity to such coarsely categorize local jurisdictions' capacity to adopt land use strategies and policies.

This does not mean that local jurisdictional discretion is not important for understanding planning practice that affects hazard and climate change risk. Indeed, Zimmerman (1981, 1995) developed a comprehensive approach to capture the degree to which states have granted local jurisdictions (counties, cities, town, villages, townships, and boroughs) discretionary authority based on their abilities to determine their structure, functions, finances, and personnel. The results revealed considerable variation among states. For example, when considering how Gulf and Atlantic states ranked among all states, Maine and North Carolina are ranked second and third, respectively, in granting the most discretion across all local jurisdictions, whereas Massachusetts and Mississippi are 43rd and 45th out of the 50 states. Texas is one of the most internally diverse. Although it is ranked 11th on the combined jurisdictional ranking for all 50 states, it is number one for cities and yet number 43 for counties. To elaborate, although municipalities have broad discretion for land use planning, counties are severely limited as our own research on mitigation practices confirms (Masterson et al. 2014; Peacock and Husein 2011). Not surprisingly then, Texas municipalities displayed significantly higher rates of adoption and implementation with respect to subdivision ordinances, plan unit development, building codes, wind hazard standards, and special local utility standards than Texas counties (Peacock and Husein 2011). Similarly, Husein (2012) found that the county/city distinction, because of varying degrees of discretion, was a significant predictor of local jurisdictions' adoption of development and land use regulations, building standards, and overall mitigation practices in a multivariate analysis that controlled for commitment, capacity, vulnerability, and other factors. The multi-state nature of the present study provides a more robust test of the significance of jurisdictions' legal authority as a determinant of hazard mitigation practice.

Additional jurisdictional characteristics that can be addressed include location in the coastal management zone (CMZ) and urban/rural status. CMZ designation is important not because of higher risk levels, as well as factors to be controlled and discussed elsewhere, but because the Coastal Zone Management Act provides for much greater resources and emphasis to plan and implement mitigation actions by jurisdictions located in these areas (Beatley et al. 2002). Indeed, Peacock and Husein (2011) found higher adoption and implementation of natural resource protection and building standards and regulations in CMZ jurisdictions than in non-CMZ jurisdictions. Research has also shown

- Number of personnel
- Training
- Support within jurisdiction
- Stakeholder support for general planning
- Budget allocation
- Financial sources
- Data sources

- Coordination & work with other jurisdiction/s
- Intra- and Inter-jurisdictional Agency Agreements
- Administrative staff-time allocation
- Stakeholder/positional leader commitment

- Planning authority/discretion
- Jurisdiction type (county and municipality)
- CMZ/non-CMZ
- Rural/Urban

Capacity

Commitment

Jurisdiction Characteristics

Mitigation Policies and Strategies
- Land use/ development regulations
- Shoreline regulations
- Natural resource protection
- Building standards
- Information dissemination/ awareness
- Property acquisition
- Financial tools
- Local incentives
- Federal incentive programs
- Critical public and private facilities
- Private-public initiatives

State Planning Environment

Hazard Exposure

Socio-Demographic Profile

- Planning mandate
- Enabling legislation
- Policy environment

- Hazard experience (10 coastal hazards)
- Hazard vulnerability/risk profile (flooding and surge)

- Population size
- Social vulnerability
- Population change
- Median home value

Figure 7.16 Conceptual Model of Factors Influencing Mitigation Actions

that rural jurisdictions display higher levels of economically- and socially-generated vulnerability (Saenz and Peacock 2006) and hence should be addressed in relation to these findings.

Capacity and commitment have emerged as key factors particularly in plan quality analysis (Berke et al. 1996; Burby and May 1998; May 1993; Norton 2005a; Tang 2008, Tang et al. 2008; Tang et al. 2011). Burby and Dalton (1994) addressed staff capacity and commitment for the adoption of measures to limit development in hazardous areas. Measures of commitment such as local political support and opposition were included by Godschalk and his colleagues (1989) and Burby and May (1998, 1999) and the latter also included capacity measures such as personnel training. More recently, researchers have used measures of capacity and commitment to explain the adoption of these planning tools (Berke et al. 2006; Norton 2005b). Brody and his colleagues (2010, 2011) collapsed the capacity and commitment issues into a single index seeking to capture what they termed organizational capacity. In our future research, we will seek to address and measure each concept separately.

A community's socio-demographic profile, as captured by population size and growth, has been included by researchers not only to capture footprint issues, but also development pressures that can negatively impact the adoption of mitigation practices (Berke et al. 2006; Berke and Roenigk 1996; Burby and May 1998, 1999; Godschalk et al. 1989; Norton 2005b). Additionally, economic status indicators such as education and median home values have been introduced, often with the expectation that higher status/class communities are more willing to introduce land use constraints to preserve environmental amenities and ecosystem services, as well as ensure safety (Burby and May 1999; Brody et al. 2010; Feiock et al. 2008; Godschalk et al. 1989). Interestingly, Feiock and his colleagues (2008) also included race/ethnic compositional factors based on logic consistent with social vulnerability arguments (Boruff et al. 2005; Cutter et al. 2003; Heinz Center 1999; Wisner et al. 2004). Social vulnerability issues are a neglected element in the mitigation practice literature that will be addressed in our future research.

The final set of factors that must be considered relate to the community's hazard profile. A community's hazard vulnerability and risk profile—which are captured by measures such as the percent of a jurisdiction in the floodplain, surge zone, and hurricane or seismic hazard zone—have generally been found to have positive effects on mitigation practice adoption (Brody et al. 2011; Brody et al. 2010; Burby and May 1999; Godschalk et al. 1989). In addition, hazard experiences or what Godschalk and his colleagues (1989) term *policy catalyst factors* have also been found to have positive effects on mitigation practice adoption (cf. Brody et al. 2011; Brody et al. 2010; Burby and May 1999). As mentioned above, our future analyses will explore how these factors shape the adoption of the 12 strategy sets among our sample of states and jurisdictions. We hope that this work will help shape polices that can enhance the capacity of all local jurisdictions to be better prepared to face a future in which coastal hazards, and ultimately disaster impacts, are likely to be increasing.

Acknowledgment

This chapter is based, in part, on work supported by the National Science Foundation (CMMI-1235374) and the National Institute of Standards and Technology (70NANB15H044). Any opinions, findings, and conclusions or recommendations expressed in this chapter are those of the authors and do not necessarily reflect the views of the National Science Foundation or the National Institute of Standards and Technology.

Note

1 Incorporated places are legally designated jurisictions established by state governments to provide various governmental functions to the people residing in a designated area. Census designated places are concentrations of population that are not legal jurisdictions but identified and recognized by local populations.

They may have some form of governing status, but generally are not fully incorporated places. County subdivisions are generally statistical subdivisions of counties, but in some areas of the country, particularly in New England, these subdivisions are often recognized municipalities. For more detailed discussions of these designations see the US Census website (www.census.gov).

References

Arnstein, S. R. (1969) "A Ladder of Citizen Participation," *Journal of the American Institute of Planners* 35(4): 216–224.

Bacmeister, J.T., K.A. Reed, C. Hannay, P. Lawrence, S. Bates, J.E. Truesdale, N. Rosenbloom, and M. Levy. (2018) "Projected Changes in Tropical Cyclone Activity Under Future Warming Scenarios Using a High-Resolution Climate Model," *Climatic Change* 146: 547–560.

Barbier, E.B. (2015) "Valuing the Storm Protection Service of Estuarine and Coastal Ecosystems," *Ecosystems Services* 11: 32–38.

Barbier, E.B., and B. Enchelmeyer (2014) "Valuing the Storm Surge Protection Service of US Gulf Coast Wetlands," *Journal of Environmental Economics and Policy* 3(2): 167–185.

Barbier, E.B., S.D. Hacker, C. Kennedy, E.M. Koch, A.C. Stier, and B.R. Silliman. (2011) "The Value of Estuarine and Coastal Ecosystem Services," *Ecological Monographs* 81(2): 169–183.

Bassett, T.J., and C, Fogelman. (2013) "Deja Vu or Something New? The Adaptation Concept in the Climate Change Literature," *Geoforum* 48: 42–58.

Beatley, T. (2009) *Planning for Coastal Resilience: Best Practices for Calamitous Times*. Washington DC: Island Press.

Beatley, T., D.J. Brower, and A.K. Schwab. (2002) *An Introduction to Coastal Zone Management*, 2nd ed. Washington DC: Island Press.

Bedsworth, L.W. and E. Hanak. (2010) "Adaptation to Climate Change: A Review of Challenges and Tradeoffs in Six Areas," *Journal of the American Planning Association* 76(4): 477–495.

Berke, P.R., M. Backhurst, L. Laurian, J. Crawford, and J. Dixon. (2006) "What Makes Plan Implementation Successful? An Evaluation of Local Plans and Implementation Practices in New Zealand," *Environment and Planning B: Planning and Design* 33(4): 581–600.

Berke, P.R. and T. Beatley. (1992a) "A National Assessment of Earthquake Mitigation: Implications for Land Use Planning and Public Policy," *Earthquake Spectra* 8(1): 1–17.

Berke, P.R. and T. Beatley. (1992b) *Planning for Earthquakes: Risk, Politics and Policy*. Baltimore MD: Johns Hopkins University Press.

Berke, P.R., J. Crawford, J. Dixon, and N. Ericksen. (1999) "Do Cooperative Environmental Planning Mandates Produce Good Plans? Empirical Results from the New Zealand Experience," *Environment and Planning B* 26(5): 643–664.

Berke, P.R., and S. French. (1994) "The Influence of State Planning Mandates on Local Plan Quality," *Journal of Planning Education and Research* 13(4): 237–250.

Berke, P.R., M.L. Malecha, S. Yu, J. Lee, and J.H. Masterson. (2018) "Plan Integration for Resilience Scorecard: Evaluating Networks of Plans in Six US Coastal Cities," *Journal of Environmental Planning and Management* 1–20.

Berke, P.R., G. Newman, J. Lee, T. Combs, C. Kolosna, and D. Salvesen. (2015) "Evaluation of Networks of Plans and Vulnerability to Hazards and Climate Change: A Resilience Scorecard," *Journal of the American Planning Association* 81(4): 287–302.

Berke, P.R., & Roenigk, D. J. (1996) "Enhancing Plan Quality: Evaluating the Role of State Planning Mandates for Natural Hazard Mitigation," *Journal of Environmental Planning & Management* 39(1), 79–96.

Bernd-Cohen, T., and M. Gordon. (1999) "State Coastal Program Effectiveness in Protecting Natural Beaches, Dunes, Bluffs, And Rocky Shores," *Coastal Management* 27(2–3): 187–217.

Bluestein, F. S. (2006) "Do North Carolina Local Governments Need Home Rule?" *North Carolina Law Review* 84: 1983–2029.

Boruff, B.J., C. Emrich, and S.L. Cutter. (2005) "Erosion Hazard Vulnerability of US Coastal Counties," *Journal of Coastal Research* 21(5): 932–842.

Brody, S., J. Gunn, W.G. Peacock, and W. Highfield. (2011) "Examining the Influence of Development Patterns on Flood Damages Along the Gulf of Mexico," *Journal of Planning Education and Research* 31(4): 438–348.

Brody, S.D., and W.E. Highfield. (2005) "Does Planning Work? Testing the Implementation of Local Environmental Planning in Florida," *Journal of the American Planning Association* 71(2): 159–175.

Brody, S.D., W. Highfield, and R. Blessing. (2015) "An Empirical Analysis of the Effects of Land Use/Land Cover on Flood Losses Along the Gulf of Mexico Coast from 1999 to 2009," *Journal of the American Water Resources Association* 51(6): 1556–1567.

Brody, S.D., W.E. Highfield, and J.E. Kang. (2011). *Rising Waters: The Causes and Consequences of Flooding in The United States.* Cambridge: Cambridge University Press.

Brody, S.D., J.E. Kang, and S. Bernhardt. (2010) "Identifying Factors Influencing Flood Mitigation at the Local Level in Texas and Florida: The Role of Organizational Capacity," *Natural Hazards* 52(1): 167–184.

Brody, S., S. Zahran, W. Highfield, H. Grover, and A. Vedlitz. (2008) "Identifying the Impact of the Built Environment on Flood Damage in Texas," *Disasters* 32(1): 1–18.

Brody, S., S. Zahran, P. Maghelal, H. Grover, and W.E. Highfield. (2007) "The Rising Costs of Floods: Examining the Impact of Planning and Development Decisions on Property Damage in Florida," *Journal of the American Planning Association* 73(3): 330–345.

Burby, R. J. (1998) *Cooperating with Nature: Confronting Natural Hazards with Land-Use Planning for Sustainable Communities.* Washington DC: National Academies Press.

Burby, R. J., T. Beatley, P.R. Berke, R.E. Deyle, S.P. French, D.R. Godschalk, E.J. Kaiser, J.D. Kartez, P.J. May, R. Olshansky, and R.G. Paterson (2007). "Unleashing the Power of Planning to Create Disaster-Resistant Communities," *Journal of the American Planning Association* 65(3): 247–258.

Burby, R.J., P. Berke, L.C. Dalton, J.M. DeGrove, S.P. French, E.J. Kaiser, P.J. May, and D. Roenigk. (1993) "Is State-Mandated Planning Effective?" *Land Use Law and Zoning Digest* 45(10): 3–9.

Burby, R. J., and L. C. Dalton. (1994) "Plans Can Matter! The Role of Land Use Plans and State Planning Mandates in Limiting the Development of Hazardous Areas," *Public Administration Review* 54(3): 229–238.

Burby, R.J., R.E. Deyle, D.R. Godschalk, and R.B Olshansky. (2000) "Creating Hazard Resilient Communities through Land-Use Planning," *Natural Hazards Review* 1(2): 99–106.

Burton, I., R.W. Kates, and G.F. White. (1978) *The Environment as Hazard.* New York: Oxford University Press.

Burby, R.J., and P.J. May. (1997) *Making Governments Plan: State Experiments in Managing Land Use.* Baltimore MD: Johns Hopkins University Press.

Burby, R., and P.J. May. (1998) "Intergovernmental Environmental Planning: Addressing the Commitment Conundrum," *Journal of Environmental Planning and Management* 41(1): 95–110.

Burby, R., and P.J. May. (1999) "Making Building Codes and Effective Tool for Earthquake Hazard Mitigation," *Environmental Hazards* 1(1): 27–37.

Cigler, B.A. (2009). "Post-Katrina Hazard Mitigation on the Gulf Coast," *Public Organization Review* 9(4): 325.

Costanza, R., O. Pérez-Maqueo, M.L. Martinez, P. Sutton, S.J. Anderson, and K. Mulder. (2008) "The Value of Coastal Wetlands for Hurricane Protection," *Ambio* 37(4): 241–248.

Crossett, K., B. Ache, P. Pacheco, and K. Haber. (2013) *National Coastal Population Report: Population Trends from 1970 to 2020. National Coastal Population Report.* NOAA, National Ocean Service. National Oceanic and Atmospheric Administration, Department of Commerce. Retrieved, May 6, 2018, from https://coast.noaa.gov/digitalcoast/training/population-report.htmlhttp://oceanservice.noaa.gov/programs/mb/pdfs/coastal_pop_trends_complete.pdf.

Curtis, K.J., and A. Schneider. (2011) "Understanding the Demographic Implications of Climate Change: Estimates of Localized Population Predictions Under Future Scenarios of Sea-Level Rise," *Population and Environment* 33(1): 28–54.

Cutter, S.L., B.J. Boruff, and W.L. Shirley. (2003) "Social Vulnerability to Environmental Hazards," *Social Science Quarterly* 84(2): 242–261.

Dalton, L.C., and R.J. Burby. (1994) "Mandates, Plans, and Planners: Building Local Commitment to Development Management," *Journal of the American Planning Association* 60(4): 444–461.

Daniels, T. (2014) *The Environmental Planning Handbook for Sustainable Communities and Regions,* 2nd ed. Chicago IL: American Planning Association.

Daniels, T. and K. Daniels. (2003) *The Environmental Planning Handbook for Sustainable Communities and Regions.* Chicago IL: American Planning Association.

Deyle, R. E., T. S. Chapin, and E. J. Baker. (2008) "The Proof of the Planning Is in the Platting: An Evaluation of Florida's Hurricane Exposure Mitigation Planning Mandate," *Journal of the American Planning Association* 74(3): 349–370.

Dillman, D.A., J.D. Smyth, and L.M. Christian. (2008) *Internet, Mail, and Mixed-Mode Survey.* New York: John Wiley.

Farber, D.F. (2009) "Climate Adaptation and Federalism: Mapping the Issues," *San Diego Journal of Climate and Energy Law* 1(1): 259–286.

Feiock, R.C., A.F. Tavares, and M. Lubell. (2008). "Policy Instrument Choices for Growth Management and Land Use Regulation," *Policy Studies Journal* 36(3): 461–480.

FEMA—Federal Emergency Management Agency. (2007) *Local Multi-Hazard Mitigation Planning Guidance.* Washington DC: Author. Retrieved April 25, 2011, from http://training.fema.gov/EMIWeb/CRS/2007%20CRS%20Coord%20Manual%20Entire.pdf.

Ge, Y., W.G. Peacock, and M.K. Lindell. (2011) "Florida Households' Expected Responses to Hurricane Hazard Mitigation Incentives," *Risk Analysis* 31(10): 1676–1691.

Godschalk, D., T. Beatley, and P. Berke. (1998) *Natural Hazard Mitigation: Recasting Disaster Policy and Planning.* Washington DC: Island Press.

Godschalk, D.R., S. Brody and R. Burby. (2003) "Public Participation in Natural Hazard Mitigation Policy Formation: Challenges for Comprehensive Planning," *Journal of Environmental Planning and Management* 46(5): 733–754.

Godschalk, D.R., D.J. Brower, and T. Beatley. (1989) *Catastrophic Coastal Storms: Hazard Mitigation and Development Management.* Durham NC: Duke University Press

Hauer, M.E. (2017) "Migration Induced Sea-Level Rise Could Reshape the US Population Landscape," *Nature Climate Change* 7(5): 321–327.

Hauer, M.E., J.M. Evans, and D.R Mishra. (2016) "Millions Projected to Be at Risk from Sea-Level Rise in the Continental United States," *Nature Climate Change* 6(7): 691–695.

Heinz Center. (1999) *The Hidden Costs of Coastal Hazards: Implications for Risk Assessment and Mitigation.* Washington DC: Island Press.

Henstra, D., and G. McBean. (2004) *The Role of Government in Services for Natural Disaster Mitigation.* Toronto: Institute for Catastrophic Loss Reduction.

Highfield, W., S. Brody, and C. Shepard. (2018) "The Effects of Estuarine Wetlands on Flood Losses Associated with Storm Surge," *Oceans and Coastal Management* 157: 50–55.

Highfield, W., W.G. Peacock, and S. Van Zandt. (2014) "Mitigation Planning: Why Hazard Exposure, Structural Vulnerability, and Social Vulnerability Matter," *Journal of Planning Education and Research* 34(3): 287–300.

Hoeppe, P. (2016). "Trends in Weather Related Disasters: Consequences for Insurers and Society," *Weather and Climate Extremes* 11: 70–79.

Holway, J. M., and R. J. Burby. (1990) "The Effects of Floodplain Development Controls on Residential Land Values," *Land Economics* 66(3): 259–271.

Hu, Q., Z. Tang, L. Zhang, Y. Xu, X. Wu, and L. Zhang. (2018) "Evaluating Climate Change Adaptation Efforts on the US 50 State's Hazard Mitigation Plans," *Natural Hazards* 92(2), 783–804.

Husein, R. 2012. *Examining Local Jurisdictions' Capacity and Commitment for Hazard Mitigation Policies and Strategies Along the Texas Coast.* Dissertation. Texas A&M University.

Hyndman, D., and D. Hyndman. (2006) *Natural Hazards and Disasters.* Independence KY: Cengage Learning.

IBHS—Insurance Institute for Building and Home Safety. (2011) *Rating the States: An Assessment of Residential Building Code and Enforcement Systems for Life Safety and Property Protection in Hurricane-Prone Regions.* Retrieved June 17, 2018 from https://disastersafety.org/wp-content/uploads/ibhs-rating-the-states.pdf.

IBHS—Insurance Institute for Building and Home Safety. (2015) *Rating the States:2015 – An Assessment of Residential Building Code and Enforcement Systems for Life Safety and Property Protection in Hurricane-Prone Regions.* Retrieved June 17, 2018 from http://disastersafety.org/wp-content/uploads/2015/07/rating-the-states-2015-public.pdf.

IBHS—Insurance Institute for Building and Home Safety. (2018) *Rating the States:2018 – An Assessment of Residential Building Code and Enforcement Systems for Life Safety and Property Protection in Hurricane-Prone Regions.* Retrieved June 17, 2018 from http://disastersafety.org/wp-content/uploads/2018/03/ibhs-rating-the-states-2018.pdf.

IPCC—Intergovernmental Panel on Climate Change. (2012) *Managing the Risks of Extreme Events and Disasters to Advance Climate Change Adaptation. A Special Report of Working Groups I and II of the Intergovernmental Panel on Climate Change.* Cambridge: Cambridge University Press.

Jacobs, F. (2018) "Black Feminism and Radical Planning: New Directions for Disaster Planning Research," *Planning Theory* 18(1): 24–39.

Kang, J.E., W.G. Peacock, and R. Husein. (2010) "An Assessment of Coastal Zone Hazard Mitigation Plans in Texas," *Journal of Disaster Research* 5(5): 526–534.

Klee, G. (1999). *The Coastal Environment: Toward Integrated Coastal and Marine Sanctuary Management.* Upper Saddle River NJ: Prentice-Hall, Inc.

Knutson, T.R., J. McBride, J. Chan, , K. Emanuel, G. Holland, C. Landsea, I. Held, J.P. Kossin, K. Srivastava, and M. Sugi. (2010) "Tropical Cyclones and Climate Change," *Nature Geoscience* 3(3):157–163.

Knutson, T.S., J.J. Sirutis, M. Zhao, R.E. Tuleya, M. Bender, G.A. Vecchi, G. Villarini, and D. Chavas. (2015) "Global Projections of Intense Tropical Cyclone Activity for the Late Twenty-First Century from Dynamical Downscaling of CMIP5/RCP4.5 Scenarios," *Journal of Climate* 28(18): 7204–7224.

Krane, D., P.N. Rigos, and M. Hill. (2001). *Home Rule in America: A Fifty-State Handbook.* Washington DC: CQ Press.

Lindell, M.K., S.D. Brody, and W.E. Highfield. (2016) "Financing Housing Recovery Through Hazard Insurance: The Case of the National Flood Insurance Program," 50–65 in A. Sapat and A-M. Esnard (eds.) *Coming Home After Disaster: Multiple Dimensions of Housing Recovery*. Boca Raton FL: CRC Press.

Lindell, M. K., and R. W. Perry. (2000) "Household Adjustment to Earthquake Hazard," *Environment and Behavior* 32(4): 461.

Lindell, M. K., and R. W. Perry. (2004) *Communicating Environmental Risk in Multiethnic Communities*. Thousand Oaks CA: Sage Publications.

Lindell, M. K., and R. W. Perry. (2012) "The Protective Action Decision Model: Theoretical Modifications and Additional Evidence," *Risk Analysis* 32(4): 616–632.

Lindell, M., C.S. Prater, and R. W. Perry. (2006) Fundamentals of Emergency Management. Emmitsburg MD: Federal Emergency Management Agency Emergency Management Institute. [Available at training.fema.gov/EMIWeb/edu/fem.asp.]

Masterson, J.H., Peacock, W.G., Van Zandt, S.S., Grover, H., Schwarz, L.F., & Cooper, J.T. (2014) *Planning for Community Resilience*. Washington DC: Island Press.

May, P.J. (1993) "Mandate Design and Implementation: Enhancing Implementation Efforts and Shaping Regulatory Styles," *Journal of Policy Analysis and Management* 12(4): 634–663.

Mileti, D. S. (1999) *Disasters by Design*. Washington DC: Joseph Henry Press.

Narayan, S., M.W. Beck, P. Wilson, C.J. Thomas, A. Guerrero, C.C. Shepard, and D. Trespalacios. (2017) "The Value of Coastal Wetlands for Flood Damage Reduction in the Northeastern USA," *Scientific Reports* 7(1): Article 9463.

Neumann, J.E., K. Emanuel, S. Ravela, L. Ludwig, P. Kirshen, K. Bosma, and J. Martinich. (2015) "Joint Effects of Storm Surge and Sea-Level Rise on US Coasts: New Economic Estimates of Impacts, Adaptation, and Benefits of Mitigation Policy," *Climage Change* 129(1–2): 337–349.

NHC—National Hurricane Center. (2017) *Costliest U.S. Tropical Cyclones Tables Updated*. Retrieved June 17, 2018 from www.nhc.noaa.gov/news/UpdatedCostliest.pdf.

Nordgren, J., M. Stults, and S. Meerow. (2016) "Supporting Local Climate Change Adaptation: Where We Are and Where We Need to Go," *Environmental Science and Policy* 66: 344–352.

Norton, R.K. (2005a) "More and Better Local Planning," *Journal of the American Planning Association* 71(1): 55–71.

Norton, R.K. (2005b) "Local Commitment to State-Mandated Planning in Coastal North Carolina," *Journal of Planning Education and Research* 25(2): 149–171.

Olshansky, R.B., and J.D. Kartez. (1998) "Managing Land Use to Build Resilience," 167–202 in R.J. Burby (ed.) *Cooperating with Nature: Confronting Natural Hazards with Land-Use Planning for Sustainable Communities*. Washington DC: Joseph Henry Press.

Peacock, W.G. (2003) "Hurricane Mitigation Status and Factors Influencing Mitigation Status Among Florida's Single-Family Homeowners," *Natural Hazards Review* 4(3): 149–158.

Peacock, W.G., and R. Husein. (2011) *The Adoption and Implementation of Hazard Mitigation Policies and Strategies by Coastal Jurisdictions in Texas: The Planning Survey Results*. College Station, Texas: Hazard Reduction and Recovery Center.

Peacock, W.G., J.-E. Kang, R. Husein, G.R. Burns, C.S. Prater, S. Brody, and T. Kennedy. (2009) *An Assessment of Coastal Zone Hazard Mitigation Plans in Texas*. College Station: Texas A&M University Hazard Reduction and Recovery Center.

Petrolia, D.R., M.G. Interis, and J. Hwang (2014) "America's Wetland? A National Survey of Willingness to Pay for Restoration of Louisiana's Coastal Wetlands," *Marine Resource Economics* 29(1): 17–37.

Petrolia, D.R., and T.-G. Kim. (2009) "What Are Barrier Islands Worth? Estimates of Willingness to Pay for Restoration," *Marine Resource Economics* 24(2): 131–146.

Petrolia, D.R., and T.-G. Kim. (2011) "Preventing Land Loss in Coastal Louisiana: Estimates of WTP and WTA," *Journal of Environmental Management* 92(3): 859–865.

Pielke, Jr., R.A., J. Gratz, C.W. Landsea, D. Collins, M.A. Saunders, and R. Musulin. (2008) "Normalized Hurricane Damage in the United States: 1900–2005," *Natural Hazards Review* 9(1): 29–42.

Pompe, J.J., and J.R. Rinehart. (1994) "Estimating the Effect of Wider Beaches on Coastal Housing Prices," *Ocean and Coastal Management* 22(2): 141–152.

Richardson, J.J. (2011) "Dillon's Rule is from Mars, Home Rule is from Venus: Local Government Autonomy and the Rules of Statutory Construction," *Publius* 41(4): 662–685.

Richardson, J.J., M.Z. Gough, and R. Puentes. (2003) *Is Home Rule the Answer? Clarifying the Influence of Dillion Rule on Growth Management*. Washington, DC: The Brookings Institution.

Robins, L. (2008) "Capacity Building for Natural Resource Management: Lessons from Risk and Emergency Management," *Australasian Journal of Environmental Management* 15(1): 6–20.

Rovins, J. E. (2009) *Effective Hazard Mitigation: Are Local Mitigation Strategies Getting the Job Done?* Emmitsburg MD: Federal Emergency Management Agency. Retrieved from http://training.fema.gov.

Saenz, R. and W.G. Peacock. (2006) "Rural People, Rural Places: The Hidden Costs of Hurricane Katrina," *Rural Realities* 1(2):1–11.

Salvino, R.F. (2007a) *Home Rule, Selectivity and Overlapping Jurisdictions: Effects on State and Local Government Size.* Andrew Young School of Policy Studies, Georgia State University. Dissertation. http://digitalarchive.gsu.edu/econ_diss/46.

Salvino, R.F. (2007b) *Home Rule Effects on State and Local Government Size.* Working Paper 701. Atlanta GA: Georgia State University Andrew Young School of Policy Studies.

Schwab, J. (2010) *Hazard Mitigation: Integrating Best Practices into Planning.* Planning Advisory Service Report 560. Chicago IL: American Planning Association.

Schwab, A., K. Eschelbach, and D. Brower. (2007) *Hazard Mitigation and Preparedness.* Hoboken NJ: John Wiley.

Shepard, C.C., C.M. Crain, and M.W. Beck. (2011) "The Protective Role of Coastal Marshes: A Systematic Review and Meta-Analysis," *PLoS One* 6(11): e27374.

Srivastava, R., and L. Laurian. (2006) "Natural Hazard Mitigation in Local Comprehensive Plans: The Case of Flood, Wildfire and Drought Planning in Arizona," *Disaster Prevention and Management* 15(3): 461–483.

Tang, Z. (2008) "Evaluating Local Coastal Zone Land Use Planning Capacities in California," *Ocean and Coastal Management* 51(7): 544–555.

Tang, Z., M.K. Lindell, C.S. Prater, and S.D. Brody. (2008) "Measuring Tsunami Planning Capacity on US Pacific Coast," *Natural Hazards Review* 9(2): 91.

Tang, Z., M.K. Lindell, C.S. Prater, T. Wei, and C.M. Hussey. (2011) "Examining Local Coastal Zone Management Capacity in US Pacific Coastal Counties," *Coastal Management* 39(2): 105–132.

Tobin, G. (1995) "The Levee Love Affair: A Stormy Relationship?" *Water Resources Bulletin* 31(3): 359–367.

Turnbull, G.K., and G. Geon. (2006). "Local Government Internal Structure, External Constraints and the Median Voter," *Public Choice* 129(3–4): 487–506.

US Census Bureau. (2017) *2017 Hurricane Season Begins. Profile America Fact for Features: CB17-FF13.* Retrieved June 17, 2018 from https://census.gov/newsroom/facts-for-features/2017/cb17-ff13-hurricane.html?intcmp=s2hurricane.

UNISDR—UN Inter-Agency Secretariat for the International Strategy for Disaster Reduction. (2002) *Living with Risk. A Global Review of Disaster Reduction Initiatives.* Geneva, Switzerland: UNISDR. Retrieved June 17, 2018 from www.unisdr.org/files/657_lwr1.pdf.

White, G.F. (1936) "The Limit of Economic Justification for Flood Protection," *The Journal of Land & Public Utility Economics* 12(2): 133–148.

White, G.F., Kates, R.W., and Burton, I. (2001) "Knowing Better and Losing Even More: The Use of Knowledge in Hazards Management," *Global Environmental Change Part B: Environmental Hazards* 3: 81–92.

Williams, A., and A. Micallef. (2009) *Beach Management: Principles and Practice.* Sterling VA: Earthscan.

Wilson, J. P. (2009) *Policy Actions of Texas Gulf Coast Cities to Mitigate Hurricane Damage: Perspectives of City Officials.* San Marcos TX: Texas State University. Accessed 12 June, 2018 at http://ecommons.txstate.edu/arp/312.

Wise, R.M., I. Fazey, M. Stafford Smith, S.E. Park, H.C. Eakin, E.R.M. Archer Van Garderen, and B. Campbell. (2014) "Reconceptualizing Adaptation to Climate Change as Part of Pathways to Change and Response," *Global Environmental Change* 28: 325–336.

Wisner, B., P. Blaikie, T. Cannon, and I. Davis. (2004). *At Risk. Natural Hazards, People's Vulnerability and Disasters.* New York: Routledge.

Wolman, H., R. McManmon, M. Bell, and D. Brunori. (2010) "Comparing Local Government Autonomy Across States," 69–114, in M. E. Bell, D. Brunori, and J. Youngman (eds.) *The Property Tax and Local Autonomy.* Cambridge MA: Lincoln Institute of Land Policy.

Wood, C. (2010) "Understanding the Consequences of Municipal Discretion," *The American Review of Public Administration* 41(4):411–427.

Woodruff, J.D., J.L. Irish, and S.J. Camargo. (2013) "Coastal Flooding by Tropical Cyclones and Sea-Level Rise," *Nature* 504(7478): 44–52.

Woodruff, S. C., & Stults, M. (2016). "Numerous Strategies but Limited Implementation Guidance in US Local Adaptation Plans," *Nature Climate Change* 6(8): 796–802.

Zahran, S., S.D. Brody, W.G. Peacock, H. Grover, and A. Vedlitz. (2008) "Social Vulnerability and the Natural and Built Environment: A Model of Flood Casualties in Texas," *Disasters* 32(4): 537–560.

Zimmerman, J.F. (1981) *Measuring Local Discretionary Authority.* Washington DC: U.S. Advisory Commission on Intergovernmental Relations.

Zimmerman, J.F. (1995) *State-Local Relations: A Partnership Approach*, 2nd ed. New York: Praeger.

8

RECOVERY VERSUS PROTECTION-BASED APPROACHES TO FLOOD RISK REDUCTION

Working Towards a Framework for More Effective Mitigation in the United States

Samuel D. Brody, Wesley E. Highfield, William Merrell, and Yoonjeong Lee

Introduction

Despite efforts to mitigate the adverse impacts of flooding across the United States, losses from both acute and chronic events continue to rise, particularly in low-lying coastal areas. From 1999 to 2009, the U.S. suffered approximately $33.5 billion in insured flood losses alone. Counties and parishes along the Gulf of Mexico coastline reported almost $21.5 billion of this total. These property damage estimates help solidify what has been generally understood for years: that floods pose a major risk to communities and with increasing development in coastal areas the problem is growing worse.

Current U.S. policy is rooted in a recovery approach to flood risk reduction. The centerpiece of this strategy is the National Flood Insurance Program (NFIP), which offers federally subsidized insurance to residents living within 24,700 participating communities. At the end of 2013, the NFIP had approximately 5.48 million flood insurance policies in force covering over $1.28 trillion in assets. There are two major assumptions underlying the NFIP. First, residents are at risk of flooding and must have a fiscal mechanism to react and recover from an inundation event. Second, the primary standard for mitigating this risk is the 100-year floodplain. For example, required floodplain management activities and flood insurance purchase requirements are all centered on the 100-year floodplain boundary as identified by the Federal Emergency Management Agency (FEMA).

In contrast, a protection-based strategy focuses on mitigating flood risk before an event takes place, or eliminating it altogether. This more proactive approach to risk reduction assumes that residents should never bear the burden of inundation and associated loss, regardless of where they are located within coastal landscapes. Such an approach favors both systems-based structural interventions and land use planning techniques that seek to either remove structures from areas most at risk or avoid constructing them there in the first place.

Protective flood mitigation strategies have been widely pursued by European countries but remain an afterthought in the U.S. Structural and non-structural flood protection techniques are often conceived and implemented in a haphazard manner across various scales and governmental jurisdictions. *Ad hoc* strategies implemented as a reaction to specific storms have resulted in a patchwork of flood defenses and development policies that ignore the broader systemic nature of flood risk nationwide. In recent years, more comprehensive programs (such as FEMA's Community Rating System—CRS) have emerged, but they represent only a fraction of eligible communities in the U.S.

This chapter critically examines recovery versus protection-based approaches to flood risk reduction within the context of setting a strategy to more effectively decrease human exposure and the resulting adverse economic impacts in the U.S. First, we assess the current approach to mitigation in the U.S. with respect to both benefits and costs. Next, we propose an alternative model for flood risk reduction, building on previous research findings and case studies from around the world. Finally, we outline key principles for an integrated flood program in the U.S. that may better reduce vulnerability and associated flood losses in the future.

Increasing Flood Impacts in the United States

Floods continue to be the costliest and most disruptive natural hazard in the U.S. Increasing physical risk due to climate change, combined with rapid land use change and development in flood-prone areas, have amplified the adverse economic and human impacts in recent years. Never before have the repercussions from storms driven by both surge and rainfall been so damaging to local communities and the built environment. Losses from both acute and chronic flood events are especially problematic in the U.S., where development in low-lying coastal areas has accelerated in recent decades (Brody et al. 2011b). These damage estimates help solidify what has been generally understood for years: that floods pose a major risk to property and safety of local communities, and despite isolated attempts to mitigate the impacts, the problem appears to be getting worse.

Given the increasing impacts from both acute and chronic storm events, it has become clear that the rising cost of floods is not solely a function of changing weather patterns or inflationary monetary systems. Rather, flood risk and associated losses can only be addressed by understanding built-environment patterns across flood-prone landscapes. Financial incentives, along with population growth and associated sprawling development within and adjacent to the 100-year floodplain, are placing more structures at risk of flooding (Brody et al. 2011b). All mitigation strategies, both structural and non-structural, must consider the cumulative impacts of rapid development and land use change in coastal margins well into the future. More than ever, there is a critical need to understand how to reduce flood risk at the local level and facilitate the establishment of more resilient communities in the U.S. and around the world.

A Recovery-Based Approach to Mitigation

With the adoption of the NFIP in 1968, the U.S. formally embraced a recovery-based approach to flood mitigation for both fresh and saltwater inundation events. The prevailing logic at the time was that a government-based program could successfully offer insurance to homeowners that the private industry could not because it could: a) pool risks more broadly, b) spread losses over time, c) potentially borrow money from the federal government if there was a deficit in a given year, and most importantly, d) subsidize the true costs of the policies (Michel-Kerjan and Kousky 2010). This program was thought to be the most effective way to financially protect residents living in flood-prone areas without creating an undue financial burden. The NFIP soon became the primary vehicle for providing flood insurance to residents and businesses. Before this time, the only way to assist flood victims was through federal relief in the form of disaster loans and grants. However, the increased burden on the federal treasury caused policymakers to embrace the concept of insurance policies against flood losses as an alternative to federal aid (Pasterick 1998).

Currently, FEMA writes or underwrites flood insurance for participating NFIP communities. Individuals can purchase flood insurance directly through authorized FEMA representatives or through a traditional private insurer in what is known as the "Write Your Own" (WYO) program. Residents in non-participating NFIP communities do not have the opportunity to purchase insurance through the NFIP. Several characteristics distinguish the NFIP from the private insurance industry. First, flood insurance purchasers are held to a 30-day waiting period before the flood insurance coverage goes into effect. This policy effectively eliminates the possibility that a party can purchase flood insurance when there is an imminent risk of flooding. Second, the residential coverage amount is capped at $250,000 for buildings and $100,000 for contents. In many cases, this coverage ceiling is less than the total value of many structures, especially along the coast where storm surge-based flooding can result in catastrophic impacts. Finally, there is a mandated requirement to purchase flood insurance for structures located within the 100-year floodplain that are being secured via bank loan. As a result of habitual noncompliance, this requirement has been more forcefully implemented through lenders and loan servicers, requiring them to determine and document whether a structure is in the 100-year floodplain and ensure that the mortgager maintains flood insurance throughout the life of the loan.

The NFIP has produced a number of significant achievements in floodplain management, including more widespread public identification of flood hazards and increased development standards in floodplain areas (Holway and Burby 1990; U.S. Interagency Floodplain Management Review Committee 1994). However, it continues to suffer from a fatal flaw in that it focuses on economic recovery post-flood event, rather than proactive mitigation to reduce the risk in the first place. In fact, the vast majority of funding for coastal flood-related issues is provided by the federal government only after a disaster occurs, through emergency supplemental appropriations (NRC 2014). A recovery-based approach predicated upon insurance payments is, at its essence, an acceptance of failure when it comes to avoiding adverse impacts from floods. As the cornerstone of flood mitigation in the U.S., the NFIP creates an expectation from the federal down to the household level that residents will flood, incur damage, and need constant financial assistance to recover from their losses. This is a self-defeating strategy that has ultimately led to five unintended and undesirable consequences.

1   Subsidized insurance has made it more affordable to purchase a home in a flood zone and has increased overall household exposure to flood risk over the long term. Artificially low insurance rates create a "perverse incentive" (Beatley 2009) to locate in risky areas because the resident will receive financial recovery assistance even if a home is flooded. This incentive often overpowers local planning policies to avoid development in the floodplain and other risk zones. Thus, in an effort to protect residents against expected losses, the NFIP has actually made the flood problem worse.

2   The NFIP has encouraged sprawling development patterns and associated adverse environmental impacts in sensitive coastal areas. Subsidized insurance has enabled builders and homeowners to more affordably locate in floodplain areas outside of traditional urban cores that have historically been left undeveloped. These residential subdivisions often have *ad hoc* or substandard local drainage and storm-water management systems that further exacerbate the flooding problem (Brody et al. 2011a). Federally backed insurance has been a major driver in the proliferation of second homes situated on the beach front or barrier islands. In general, sprawling development patterns across flood-prone regions places more people and structures at risk when a major event

takes place. Also, low-density development spiraling outward from urban areas requires larger amounts of impervious surfaces, increase surface runoff, loss of critical natural habitats, and reductions in water quality, among other negative environmental impacts (Brody 2013).

3 Sprawling development in low-lying areas can change the spatial extent of floodplain boundaries faster than they can be officially mapped and put downstream communities at greater risk. Elevating neighborhoods using fill and the conversion of land to impervious surfaces often alters the natural drainage pattern of a watershed and puts adjacent communities at risk (Brody et al. 2013). Older structures that have never flooded before or are outside of the 100-year floodplain boundary are increasingly reporting inundation and associated property damage (Schuster et al. 2005).

4 The NFIP forces homeowners and communities into a constant repetitive losses and disaster-recovery cycle. Once a structure is flooded, insurance payouts require the owner to repair or rebuild in the same way (unless there is a local regulation that mandates a structural change). This process can occur up to three times before a property buy-out is initiated. Because flooding tends to be chronic and spatially repetitive, homeowners in vulnerable areas are often trapped in a repetitive damage-rebuild process. These so-called "repetitive loss" (RL) properties are defined by FEMA as any insurable building for which two or more claims of more than $1,000 were paid by the NFIP within any rolling ten-year period since 1978. Currently, there are over 122,000 RL properties nationwide which over time creates cumulative financial burdens (see www.fema.gov/txt/rebuild/repetitive_loss_faqs.txt).

In Harris County, TX (Houston) alone, we catalogued 9,521 insured claims over 2,896 properties from 1978 to 2008 totaling approximately $351 million in paid repetitive losses (Highfield et al. 2012). On average, each property in the county claimed 3.23 losses and received $36,881 per claim. Almost 35 percent of losses during this period were caused by a hurricane or tropical storm, most recently Hurricane Ike in 2008, which generated $20.4 million in RL damage. Of all losses paid, over 70 percent went to repair building-related damage (vs. contents), of which 94 percent was from single-family structures.

5 Repetitive and one-time insurance payments from the NFIP have consistently exceeded the income generated from premiums. This fiscal problem was initially intended to be remedied by raising premiums and/or dropping coverage. However, a combination of payouts from large events and the lack of actuarially sound rates have forced FEMA to regularly borrow money from the federal treasury to cover its deficit. As of August 2013, the NFIP had borrowed approximately $24 billion (Kousky and Kunreuther 2013); $16.6 billion of the cumulative amount followed Hurricanes Katrina and Rita in 2005 (King 2011). Additional expense is also incurred by the federal government through disaster aid granted to communities following a major disaster event.

To address these fiscal dilemmas, Congress passed the Flood Insurance Reform Act in 2012, which focused on raising premiums based on actual risk in an effort to help the program become more actuarially sound. Since that time, many of the requirements have been relaxed or postponed, particularly for cases where increased premiums would place a substantial financial burden on a property owner. However, while these fixes may help the NFIP reach more sound financial footing, the system itself is still broken because it unilaterally increases flood vulnerability by encouraging property owners to locate in flood-prone areas.

A Protection-Based Approach to Flood Mitigation

By contrast, a protection-based approach to flood mitigation focuses on avoiding losses ahead of a disaster. Damage to property or other adverse impacts are considered failures in the system rather than expected consequences. Structural and non-structural flood mitigation techniques are imple-mented to eliminate vulnerability to flood impacts as much as possible, as well as to incorporate contingencies if a disaster were to occur. This approach lends itself more towards the implementation of both avoidance and systematic resistance strategies that address flood impacts over the long term.

Non-Structural Mitigation Strategies

Although a protection-based approach to flood risk reduction is not the mainstay of federal policy in the U.S., there are multiple local-level initiatives that demonstrate its effectiveness. The most notable program focusing on flood protection rather than recovery is the CRS, which was estab-lished in 1990 as a way to encourage communities to exceed the NFIP's minimum standards for floodplain management. Communities participating in the CRS adopt various flood mitigation measures in exchange for federal flood insurance premium discounts ranging from 5 to 45 percent. The non-structural orientation of the CRS program categorizes planning and management activi-ties into the following four "series" containing 18 mitigation "activities":

1   Public information (Series 300) activities informing residents about flood hazards, the availabil-ity of insurance, and household protection measures,
2   Mapping and regulation activities (Series 400) containing both critical data needs and regulations that exceed NFIP minimum standards,
3   Damage reduction (Series 500) activities focusing on reducing flood damage to existing build-ings, which may entail acquiring, relocating, or retrofitting existing structures, and
4   Flood Preparedness (Series 600) activities implementing strategies associated with warning and response to minimize the adverse effects of floods (for more information on the CRS, see: http://training.fema.gov/EMIWeb/CRS/).

Credit points are assigned to participating CRS communities based on the degree to which different flood mitigation activities are implemented, but not all activities carry the same number of credit points (FEMA 2013). The total number of credit points obtained by a participating locality deter-mines the size of insurance premium discounts. Credit points are aggregated into "classes," from 9 (lowest) to 1 (highest), where communities must implement a larger number and scope of flood mitigation measures to be awarded a higher CRS class. Insurance premium discounts range from 5 (class 9) to 45 percent (class 1), depending on the degree to which a community plans for the adverse impacts of floods (FEMA 2007). Although the local jurisdiction takes responsibility for adopting and implementing each mitigation activity, the individual homeowner receives the discount on their flood insurance premium.

As described above, the CRS program incentivizes local communities to focus on avoiding flood risk at the outset, to increase public awareness of this risk, to encourage protective household behav-iors, and to prepare for the impact before a flood occurs. In doing this, communities can become more proactive in addressing flood problems over time.

Recent empirical research demonstrates the effectiveness of the CRS program in reducing prop-erty damage from floods. For example, two related cross-sectional studies examined all 67 counties in Florida and 37 coastal counties in Texas over a five-year period (1997–2001) to test the effect of the CRS program on observed flood loss. These studies found that communities in both states implementing mitigation activities under the CRS experienced significant reductions in flood damage, even when controlling for multiple natural environment, built environment, and socio-economic contextual characteristics. Specifically, in Florida, a real-unit change in CRS class (e.g., moving from Class 8 to Class 7) equaled, on average, a $303,525 decrease in the amount of damage per flood (see Brody et al. 2007 for more details). Results from the coastal Texas study showed that from 1997 to 2001, a real-unit increase in CRS class translated into a $38,989 reduction in the average property damage per flood event (Brody et al. 2008).

Building on these regional studies examining CRS classes, Highfield and Brody (2013) conducted a national study on the flood loss-reducing effect of specific CRS activities by tracking point totals on a yearly basis over an 11-year period from 1999 to 2009. Two avoidance-based mitigation activities were found to be most effective in reducing observed flood losses: freeboard (vertical avoidance by elevating structures in the floodplain above the base flood elevation) and open space protection (horizontal avoidance by protecting open space). The total dollar savings of a one-point increase in the freeboard element for total losses was equivalent to, on average, nearly $8,300 per community per year. Taking into account the average amount of credit points communities in the sample received for each activity in 2009 (the final year of the study period), freeboard requirements led to the highest overall reduction in flood damages with an estimated average of $800,000 per year. Concurrently, the dollar savings of a one-point increase in open space protection was equal to, on average, $3,532 per community, per year. Considering the average number of points accrued for open space protection among communities in the study sample, the total savings per year for this activity was equivalent to approximately $591,436. Empirical evidence also supports the notion that CRS-based mitigation activities at the community level significantly reduce losses incurred at the household level (Highfield et al. 2014).

Although initial studies demonstrate that a flood mitigation program focused on protection can result in significant reductions in losses, it is important to note that community participation in the CRS is minimal. In 2013, for example, there were 1,211 participating CRS communities, less than 5 percent of all NFIP-eligible communities. However, these jurisdictions contained approximately 3.8 million NFIP policyholders, more than 67 percent of all flood insurance policies in the U.S. (FEMA 2012).

Avoidance planning strategies outside of the CRS program can also reduce risk of flood impacts, particularly when they are implemented through local development regulatory frameworks. For example, policies that direct new development away from flood-prone areas, such as clustering, density bonuses, transfer of development rights, and strategic placement of public infrastructure, can help protect future structures from damaging floods. Spatially targeted development strategies that set back from or create a buffer around areas most at risk to flooding tend to be most effective (Beatley 2009; Brody et al. 2011a).

By avoiding critical flood-prone areas, development and the associated placement of impervious surfaces can proceed without unduly compromising hydrologic functions. In particular, protecting naturally occurring wetlands can lead to significant reductions in flood impacts, especially for precipitation-based events. Even when controlling for socioeconomic and geophysical contextual characteristics, Brody et al. (2008) showed that the loss of naturally occurring wetlands across 37 coastal counties in Texas from 1997 to 2001 significantly increased the amount of observed property damage from floods. Based on the number of wetland permits granted over the study period, the authors found that, on average, wetland alteration added over $38,000 in property damage per flood. A parallel analysis for all counties in Florida showed even greater economic value of wetlands (Brody et al. 2007). In this case, the alteration of wetlands increased the average property damage per flood at the county level by over $400,000. Based on this rate of change, wetland development cost the state over $30 million per year in flood losses.

At a regional scale, naturally occurring wetlands also appear to act as a key ecological indicator of resiliency when dealing with floods. Along the Gulf of Mexico, an acre loss of naturally occurring wetlands from 2001 to 2005 increased property damage caused by flooding by an average of $7,457,549—which amounts to approximately $1.5 million per year across the study area (Brody et al. 2012). The role of wetlands in attenuating the adverse impacts of floods at the landscape scale should not be overlooked by planners and policy makers as they clearly provide an important protective and economic benefit to coastal communities. These natural features provide not only wildlife habitat and increased biological diversity in coastal areas, but also a critical buffer that can reduce flooding of property, which, as our models demonstrate, translates into significant dollar savings.

Multi-Functional Structural Mitigation Strategies

Avoidance strategies are most effective in protecting communities from the adverse impacts of floods if there is no existing development in place, if structures can be elevated above the base flood elevation, or if they can be relocated to a less hazard-prone area. When there is already extensive development, critical infrastructure, and commerce, however, comprehensive resistance strategies become the preferred approach for flood protection. For these reasons, hard structures are likely to become increasingly important options for reducing coastal flood risk, especially in densely populated urban areas (NRC 2014).

Early efforts at flood mitigation in the U.S. focused mostly on structural techniques, such as those implemented after the Mississippi River flood in 1927 (Birkland et al. 2003). The Flood Control Act of 1930 subsequently dedicated funds to buildimg structural flood control works including levees, floodwalls, and fills, many of which are still standing today. Other structural mitigation approaches that actively alter the physical landscape include the use of channel- and land-phase structures to control floods. Channel-phase structures include dikes, dams, reservoirs, reducing bed roughness, and altering stream channels (Alexander 1993). Since the 1940s, the U.S. Army Corps of Engineers (USACE) has spent over $100 billion (in 1999 dollars) on structural flood control projects (Stein et al. 2000). Structural approaches to flood mitigation have been shown to reduce the adverse impacts of floods. For example, according to USACE, although flood damages in the U.S. totaled approximately $45 billion from 1991 to 2000, structural flood control measures averted an additional $208 billion of damage (USACE 2002).

A major problem with the historical implementation of structural flood mitigation is that it was done in a reactive and *ad hoc* manner. That is, structures were usually constructed at the site level in response to a specific storm without taking into consideration the dynamics of the larger physical and social system. Thus, this approach has several limitations.

1   When floods exceed the capacity of a flood control structure, the resulting flood damages are significantly higher than if the area had been unprotected and thus less populated (Burby et al. 1985; Larson and Pasencia 2001; Stein et al. 2000; White et al. 1975;).

2   Structures such as levees can increase the volume and velocity of water pulsing downstream. By constricting a waterway and hardening its banks, these structures increase the probability of downstream flooding (Birkland et al. 2003).

3   Structural approaches to flood mitigation, such as dams or dikes, can bring a false sense of security to residents living downstream (Burby and Dalton 1994). The perception that these protected areas are completely safe can encourage new development, increasing the risk of human casualties or property damage if the structure either under-performs or is breached during a storm event (Burby et al. 1985).

4   If placed improperly, structural mitigation measures can cause negative environmental impacts, such as the alteration of naturally occurring wetlands, degradation of fish and wildlife habitats, reduced water quality, and compromised function of hydrological systems (Abell 1999).

The potential adverse effects of *ad hoc* site-specific structural flood mitigation measures were partly responsible for the establishment of a federal insurance-based mitigation program in 1968. However, with changing risk profiles (especially in coastal areas) and increasing human development in flood-prone areas, a new breed of large-scale structural protection devices has emerged, particularly to combat storm surge directly on the coast.

The Netherlands, the most flood-vulnerable developed country in the world, has been a leader in constructing multi-functional, comprehensive, environmentally conscious flood barriers. Most notably, in 1997, the Delta Works organization completed the Maeslant Barrier to protect the city of Rotterdam from storm surges off the North Sea. The structure consists of two immense movable gates that automatically close during a storm threat. At all other times, the gates are open to permit shipping activity, recreation, and natural water flow. Similarly, in 1986, the Eastern Scheldt storm surge barrier (Oosterscheldekering) was completed as a linear series of large sluice gates. This structure was placed across the mouth of Oosterschelde Estuary to block damaging storm surges, but still permit tidal exchange, maintain fishery habitats, and keep the bay interior free from a disconnected scattering of mitigation structures. Most recently, the Netherlands has begun to implement its "Room for the River" program, which seeks to widen inland rivers to allow for natural water storage and reduce the probability of downstream flooding. For example, additional space is being created for the IJssel River by constructing two dikes through existing agricultural land so that flood water can be discharged during high water periods. An area of 713 hectares will be flooded during high water periods, dropping the river level by 71 cm. Also, a dike in Lent is being relocated approximately 350 meters inland to create space for a new channel in the floodplain of the Waal River. This new channel will provide additional capacity for water discharge during floods, which will reduce the water levels and potential overtopping. Excavation of the channel and the relocation of the dike will also create an island, which will be used for recreational and environmental restoration purposes (Netherlands Ministry of Infrastructure and Environment 2014). These and other flood protection structures across the Netherlands take into consideration the natural and social systems, and seek to protect communities from different types of floods over the long term. Barriers across the country are also being explicitly integrated into the landscape as multifunctional devices supporting tourism, recreation, and economic development.

Inspired by the Dutch approach, Italy initiated the MOSE (Modulo Sperimentale Elettromeccanico) project, a system of mobile barriers, to protect Venice from damaging high tides in 2001 (Fontini et al. 2010). This storm surge reduction project follows the Dutch approach by designing the barrier to shorten the coast. The plan includes constructing 78 high walled flap gates across the three inlets separating the lagoon and the city from the water which can cause flooding from the Adriatic Sea (Keane 2013). The barrier is designed to be flexible, depending on the severity of the tidal event. When a tide is expected to be above to a certain level, these floodgates lying on the bottom of the sea can be raised above the water level to block the lagoon. The gates are kept in specially devised containers when they are not needed. When operating, the water inside the flaps is replaced with air so they can rise above the surface (Fontini et al. 2010; Shilling 2012). Except in times of emergency, large vessels are allowed to pass the barrier and natural water flow is maintained (Squires 2008). The barrier is now expected to be completed sometime between 2018–2020.

The first large-scale surge barrier system in the U.S. was recently completed around New Orleans. Following Hurricane Katrina, the U.S. Congress appropriated $14.8 billion to build the Greater New Orleans Hurricane and Storm Damage Risk Reduction System (HSDRRS). This System was started in 2008 and achieved 100-year surge event protection in June 2011. Greater New Orleans is now protected by a massive ring dike, a 133-mile perimeter of levees, flood walls, and gated barriers. The strategy is to keep massive surges from entering the system by using four gated passages. Features of the HSDRRS include 350 miles of levees and floodwalls (including interior levees and floodwalls), hundreds of gates and structures for sealing the system, 78 pumping stations (federal and non-federal), and the four major surge gate systems. As of 2013, the U.S. Army Corps of Engineers (USACE) had spent $11.8 out of a projected $14.45 billion on building the system

(Poirier 2014). Although the HSDRRS achieves adequate protection from storm surge threatening the New Orleans area, it does not attempt to integrate into the environmental landscape or provide multi-functional benefits associated with tourism, recreation, or ecosystem services.

Storm surge protection has also been proposed for the Houston-Galveston Region in response to Hurricane Ike, which devastated the region in 2008. To prevent major surges in the future, the "Ike Dike," a surge barrier along the coast that protects the entire area, has been proposed for the Galveston Bay region. It consists of the present Galveston seawall, sand-covered revetments and extensions along the beaches of west Galveston Islands and the Bolivar Peninsula, a small surge gate at San Luis Pass, and a major gate system at Bolivar Roads. This coastal spine is designed to keep the hurricane-induced surge out of Galveston Bay. The Ike Dike proposal seeks to protect the Bay itself, all the area's ports, and all coastal communities from the adverse effects of storm surges. Preliminary cost estimates for this coastal spine are $4 to $6 billion, with additional annual maintenance costs of $25 million. Proponents argue that it is in the federal interest to protect the Houston/Galveston region, especially when compared to the cost of continued storm recovery and risks to the national economy and homeland security. In particular, they point to the extraordinary concentration of industrial facilities that have been drawn to the water by transportation needs. For example, nearly half of U.S. refining capacity is located on the Texas/Louisiana Gulf Coast. Their locations near water leave these plants vulnerable to surge flooding, posing a threat via disruption of national and global petrochemical markets.

Setting a Framework for Flood Risk Reduction in the U.S.

As described above, there are multiple local level examples of how a protection-based approach to flood mitigation can effectively reduce losses over the long term. However, no concerted national policy exists with this focus, and localities are constantly straining against the insurance-based model FEMA has embraced since 1968. For example, local avoidance strategies and up-front investments to significantly reduce risk are undercut by the availability of subsidized insurance and the assurance of rebuilding after a storm. Insurance premiums are increasing (despite delays in Congress in implementing the Biggert-Waters Act) and will eventually reach a more nearly actuarial rate. Nevertheless, we assert that the system itself needs to be fundamentally altered. As a first step in this process, we propose that a national protection-based approach to flood mitigation in the U.S meet six criteria.

1   *Regional and Systematic*: major flooding events are defined by hydrological systems (i.e., watersheds) and involve regional impacts, particularly surge-based inundation. Both avoidance- and resistance-based strategies cannot be aimed solely at the neighborhood or community scale. Policies must instead consider what may be transboundary ecosystems, hydrologic processes, and regional economies. Fragmented and *ad hoc* interventions can impose adverse impacts on neighboring communities, weaken the overall functionality of these systems over the long term, and lead to more damage overall.

2   *Cumulative and Long Term*: flood risk reduction is a long-term proposition. Decision makers must think beyond normal political cycles and consider changing physical, socioeconomic and built environment conditions over time. This involves small impacts that accumulate over years as well as the big events that provoke sudden shifts in the system. In the same way, the effectiveness of multiple mitigation strategies should be measured over large spatial and temporal scales. The value of these interventions must be measured cumulatively and for long-range flood probabilities.

3   *Integrated and Synergistic*: a protection-based approach to flood mitigation seeks to first avoid the risk, and then resist the threat of inundation using structural measures. However, given the fact

that there are few, if any, completely undeveloped coastlines in the U.S., an integrated approach involving a synergistic mix of structural and non-structural strategies is critical. For example, avoiding vulnerable or ecologically important areas for future development may be combined with the elevation of existing structures. Or, setbacks and buffers can be established even when an area is protected by a seawall. In general, a hybrid strategic approach to mitigating risk will be the most effective over the long term.

4   *Up-Front Investments*: initial financial investments in design, construction, land acquisition, etc. may be necessary to protect residents from flood impacts. This approach is far preferable to chronic and repetitive costs typical of a recovery-based system for coping with floods. Generally, the return on investment is higher over the long term when investments in the most effective mitigation measures are made at the outset.

5   *Diversity, Redundancy, and Connectivity*: these characteristics are the cornerstones of a resilient approach to flood risk reduction (Gunderson and Holling 2001; Walker and Salt 2006). Diversity entails the implementation of a multi-mitigation approach to addressing flood-related problems; it focuses on keeping all of the working parts of the social-ecological system functioning so it can better withstand a flood-based disturbance in a way that is inclusive of the most vulnerable human populations (Berkes et al. 2003). Redundancy ensures that multiple back-up measures and lines of defense will effectively avoid losses at the system level. This characteristic could be structurally based, to protect populations even when one measure fails. It could also be related to the way organizational functions are replicated to ensure that the system as a whole is operational in the face of a disaster (Paton and Johnston 2006). Connectivity is a place-based concept that enables the flow of people, biotic, and abiotic components throughout a social-ecological system. This circuitry helps ecosystems to continue to provide critical ecological services and better integrates humans into flood-prone landscapes. Connectivity is also a social-organizational characteristic in which communicative links across neighborhoods and organizations enhance the flow of information and knowledge throughout a system. These connections act as the foundation for storm-related social cohesion, preparation, and collective action (Adger 2003).

6   *Lower Risk Threshold*: one of the most significant shortcomings of the NFIP, and more broadly the U.S. approach to flood risk reduction, is the selection of the 100-year floodplain as the primary standard of risk. The boundaries delineating the 1 percent probability of flooding each year drives insurance purchases and other household protective actions, permitting for development, and planning at the community level. For example, the Flood Insurance Rate Map, which contains the boundaries of the regulatory 1 percent flood (often referred to as the "100-year flood" or base flood), determines whether a homeowner must purchase insurance and sets the premium rate the policyholder must pay. FEMA also uses the 100-year floodplain maps as a basis for enforcing local mitigation requirements (e.g., minimum building elevation) and compliance is accomplished through the use of a permitting program (e.g., building permits, subdivision regulations). Communities must also enact a floodplain management ordinance that meets NFIP minimum standards based on the 1 percent flood. Even outside the NFIP, communities use the 100-year floodplain as the basis for plans and programs that seek to reduce the adverse effects of floods.

However, empirical evidence suggests that the 100-year floodplain is a poor metric for measuring flood threats to human communities because it sets the risk threshold far too high. For example, in an analysis of insured flood claims across the Clear Creek Watershed south of Houston, TX, Brody et al. (2013) determined that 55 percent of these losses were located outside the 100-year floodplain and that the average claim distance from this boundary was approximately a quarter of a mile. The large amount of property loss occurring outside the floodplain is not necessarily a result of poor data or mapping techniques, but a fundamental limitation of using this boundary to delineate risk posed by damaging floods. It is important to note that none of the rainfall events causing property losses during the study period exceeded the 100-year frequency. Development permitting, community planning, and household preparedness, particularly in low-lying coastal areas, thus appear to be triggered by an insufficient threshold for flood risk. In sum, there is no solid evidence to justify a default 1 percent annual chance (100-year) design level for flood reduction (NRC 2014).

In recognition of these problems, other countries have set their risk thresholds to much lower tolerance levels. For example, in the Netherlands decision makers protect residents from flood risks ranging from 1 in 1,250–10,000 years. By comparison, U.S. risk tolerance seems unacceptably high given the amount of financial impact accruing each year.

Conclusion

Insured losses from floods in the U.S. amount to approximately $3 billion per year. These losses will continue to mount unless a national strategy based on protection is adopted. The current insurance-based system is predicated on recovery in which policy makers expect damage to structures, families, and economies. Mitigation activities focus on rebuilding after a storm rather than avoiding adverse impacts altogether. In contrast, we propose a more protective approach to flood risk reduction that focuses on eliminating the threat at the outset and integrating contingencies if flood damage were to occur. This shift in overall policy does not accept failure and places an emphasis on protecting residents ahead of a flood. Above all, a protection-based approach is more in line with the idea of developing flood-resilient communities over the long term. These concepts are strongly supported by the recently released NRC (2014) report—*Reducing Coastal Risks on the East and Gulf Coasts*, which calls for a comprehensive national approach to coastal risk reduction based on the implementation of proactive mitigation strategies that protect coastal localities across the U.S.

References

Abell, R.A. (1999) *Freshwater Ecoregions of North America: A Conservation Assessment*, Washington DC: Island Press.

Adger, W.N. (2003) "Social Capital, Collective Action, and Adaptation to Climate Change," *Economic Geography* 79(4): 387–404.

Alexander, D. (1993) *Natural Disasters*, New York: Chapman & Hall.

Birkland, T.A., R.J. Burby, D. Conrad, H. Cortner, and W.K. Michener. (2003) "River Ecology and Flood Hazard Mitigation," *Natural Hazards Review* 4: 46–54.

Beatley, T. (2009) *Planning for Coastal Resilience: Best Practices for Calamitous Times*, Washington, DC: Island Press.

Berkes, F., J. Colding, and C. Folke, eds. (2003) *Navigating Social Ecological Systems: Building Resilience for Complexity and Change*, Cambridge: Cambridge University Press.

Brody, S. D. (2013) "The Characteristics, Causes, and Consequences of Sprawling Development Patterns in the United States," *Nature Education Knowledge* 4(5): 2.

Brody, S.D., R. Blessing, A. Sebastian, P. Bedient. (2013) "Delineating the Reality of Flood Risk and Loss in Southeast, Texas," *Natural Hazards Review* 14(2): 89–97.

Brody, S.D., J. Gunn, W.E. Highfield, and W.G. Peacock. (2011a) "Examining the Influence of Development Patterns on Flood Damages along the Gulf of Mexico," *Journal of Planning and Education Research* 31(4): 438–448.

Brody, S.D., W.E. Highfield, and J.E. Kang. (2011b) *Rising Waters: Causes and Consequences of Flooding in the United States*, Cambridge: Cambridge University Press.

Brody, S.D., W.G. Peacock, and J. Gunn. (2012) "Ecological Indicators of Flood Risk Along the Gulf of Mexico," *Ecological Indicators* 18: 493–500.

Brody, S.D., S. Zahran, W.E. Highfield, H. Grover, and A. Vedlitz. (2008) "Identifying the Impact of the Built Environment on Flood Damage in Texas," *Disasters* 32(1): 1–18.

Brody, S. D., S. Zahran, P. Maghelal, H. Grover, and W. Highfield. (2007) "The Rising Costs of Floods: Examining the Impact of Planning and Development Decisions on Property Damage in Florida," *The Journal of the American Planning Association* 73(3): 330–345.

Burby, R.J. and L.C. Dalton. (1994) "Plans Can Matter! The Role of Land Use Plans and State Planning Mandates in Limiting the Development of Hazardous Areas," *Planning Administration Review* 54(3): 229–238.

Burby, R.J., S.P. French, B. Cigler, E.J. Kaiser, D. Moreau, and B. Stiftel. (1985) *Flood Plain Land Use Management: A National Assessment*, Boulder CO: Westview Press.

Federal Emergency Management Agency. (2007) *National Flood Insurance Program Community Rating System Coordinator's Manual*, Washington DC: Author.

Federal Emergency Management Agency. (2012) Community Rating System Fact Sheet. Available at: www.fema.gov/media-library/assets/documents/9998. Accessed September 23, 2013.

Federal Emergency Management Agency. (2013) National Flood Insurance Program Community Rating System Coordinator's Manual. FIA-15/2013. FEMA. www.fema.gov/media-library/assets/documents/8768?id=2434 Accessed September 23, 2013.

Fontini, F., G. Umgiesser, and L. Vergano. (2010) "The Role of Ambiguity in the Evaluation of the Net Benefits of the MOSE System in the Venice Lagoon," *Ecological Economics* 69(10): 1964–1972.

Gunderson, L.H. and C.S. Holling, eds. (2001) *Panarchy: Understanding Transformations in Human and Natural Systems*, Washington DC: Island Press.

Holway, J. and R. Burby. (1990) "The Effects of Floodplain Development Controls on Residential Land Values," *Land Economics* 66(3): 259–271.

Highfield, W. and S. D. Brody. (2013) "Evaluating the Effectiveness of Local Mitigation Activities in Reducing Flood Losses," *Natural Hazards Review* 14(4): 229–236.

Highfield, W., S.D. Brody, R. Blessing. (2014) "Measuring the Impact of Mitigation Activities on Flood Loss Reduction at the Parcel Level: The Case of the Clear Creek Watershed along the Upper Texas Coast," *Natural Hazards* 72(2): 687–704.

Highfield, W.E., Norman, S., Brody, S.D. (2012) "Examining the 100-Year Floodplain as a Metric of Risk, Loss, and Household Adjustment," *Risk Analysis* 33(2): 186–191.

Keane, J. (2013) "Saving Venice: The MOSE Project," *Industry Tap*. Retrieved June 8, 2014, from www.industrytap.com/saving-venice-the-mose-project/471.

King, R. O. (2011) "National Flood Insurance Program: Background, Challenges, and Financial Status," Washington, DC: UNT Digital Library. http://digital.library.unt.edu/ark:/67531/metadc40078/. Accessed September 23, 2013.

Kousky, C. and H. Kunreuther. (2013) "Addressing Affordability in the National Flood Insurance Program," *Washington, DC: Resources for the Future and the Wharton Risk Management and Decision Processes Center, Issue Brief*, 13–02.

Larson, L. and D. Pasencia. (2001) "No Adverse Impact: New Direction in Floodplain Management Policy," *Natural Hazards Review* 2(4): 167–181.

Michel-Kerjan, E. and C. Kousky. (2010) "Come Rain or Shine: Evidence on Flood Insurance Purchases in Florida," *Journal of Risk and Insurance* 77(2): 369–397.

National Research Council. (2014) *Reducing Coastal Risks on the East and Gulf Coasts*, Washington DC: National Academies Press.

Netherlands Ministry of Infrastructure and Environment. (2014) *Water Innovations in the Netherlands: A Brief Overview*, The Hague, Netherlands: Author.

Pasterick, E.T. (1998) "The National Flood Insurance Program," 125–154, in H. Kunreuther and R. J. Roth (eds.) *Paying the Price*, Washington DC: Joseph Henry Press.

Paton, D. and D. Johnston, eds. (2006) *Disaster Resilience: An Integrated Approach*, Springfield IL: Charles C. Thomas.

Poirier, L. (6 March 2014) "New Orleans Storm Risk Reduction System Receives FEMA Accreditation," *Engineering News-Record*. Retrieved June 14, 2014 from http://texas.construction.com.

Schuster, W.D., J. Bonta, H. Thurston, F. Warnemuende, and D.R. Smith. (2005) "Impacts of Impervious Surface on Water-Shed Hydrology: A Review," *Urban Water Journal* 2(4): 263–275.

Shilling, D. R. (2012) "Sinking Cities, Rising Waters, the Solution is Biblical," retrieved June 10, 2014, from www.industrytap.com/sinking-cities-rising-waters-the-solution-is-biblical/.

Squires, N. (6 December 2008) "'Moses Project' to Secure Future of Venice," *The Telegraph*. Retrieved June 14, 2014 from www.telegraph.co.uk.

Stein, J., P. Moreno, D. Conrad, and S. Ellis. (2000) *Troubled Waters: Congress, the Corps of Engineers, and Wasteful Water Projects*, Washington DC: Taxpayers for Common Sense and National Wildlife Federation.

United States Army Corps of Engineers (USACE). (2002) "Services to the Public: Flood Damage Reduction," available online at: www.corpsresults.us/docs/VTNFloodRiskMgmtBro_loresprd.pdf. Accessed July 7, 2005.

United States Interagency Floodplain Management Review Committee. Federal Interagency Floodplain Management Task Force. (1994) *Sharing the Challenge: Floodplain Management into the 21st Century: Report of the Interagency Floodplain Management Review Committee to the Administration Floodplain Management Task Force*, Washington DC: U.S. Government Printing Office.

Walker, B. and D. Salt. (2006) *Resilience Thinking: Sustaining Ecosystems and People in a Changing World.* Washington DC: Island Press.

White, G.F., W.A.R. Brinkmann, H.C. Cochrane, and N.J. Ericksen. (1975) *Flood Hazard in the United States: A Research Assessment*, Boulder CO: University of Colorado Institute of Behavioral Science.

9

HAZARD MITIGATION AND CLIMATE CHANGE ADAPTATION

Himanshu Grover

Introduction

Hazard mitigation and climate change adaptation are both concerned with reducing causalities and losses from future hazard events. Hazard mitigation consists of coordinated processes, activities and actions that are "directed toward eliminating the causes of a disaster, reducing the likelihood of its occurrence, or limiting the magnitude of its impacts if it does occur" (Lindell et al. 2006: 18). Climate change adaptation also focuses on minimizing the negative impacts of climate change as a result of changes in the global climatic system (global warming). As the scientific understanding of climate change and associated impacts has advanced in the past few decades, increasing frequency and intensity of extreme environmental events has become a critical concern for adaptation policy. To this end both hazard mitigation policy and climate change adaptations share the common goal of minimizing risks, causalities, and losses from future hazard events.

A number of communities are already starting to experience increasing losses from the influence of climate change on weather-related hazards. As per a recent report released by Munich Re (2012), weather-related losses between 1980 and 2011 nearly quintupled in North America. Increasing sea levels are likely to directly impact millions of people and the global supply chain, which will have crippling consequences for a number of nations (Rowley et al. 2007). Direct observations of significant deviations in normal weather patterns in increasing number of communities across North America, which are consistent with the scientific climate change simulations, confirm that the consequences of changing climatic conditions are beginning to occur (Gillett 2014). Thus, the link between weather-related hazards and climate change is no longer just a theory but is now underway as documented by scientific research. Communities can no longer continue to ignore the influence of changing climatic conditions on their future risks from weather-related hazards.

To this end, it is the right time to push for an integrated framework for hazard mitigation and climate change adaptation. With the common goal of safeguarding human life and property, as well as ensuring sustainability of ecosystem services, there is certainly an allure in adopting an integrated agenda to promote community resilience. This approach is in line with the international climate change policy recommendations to incorporate future climate change scenario impacts into the framing of local development policies and for directing development away from areas likely to face severe extreme consequences of climate change (IPCC 2014). A similar view is echoed by the hazards researchers who have expressed the need for incorporating climate change sensitivity in planning and implementation of hazard mitigation activities (Gall et al. 2011; Helmer and Hilhorst 2006; O'Brien et al. 2006; Solecki et al. 2011; Venton and Trobe 2008).

This chapter explores synergistic opportunities between hazard mitigation planning and climate change adaptation, and outlines an integrated conceptual framework to facilitate convergence of policies in these two areas. First, key aspects of contemporary hazard mitigation planning and climate change adaptation are discussed, followed by examination of vulnerability reduction as the common agenda of both frameworks. The second section discusses the concept of community resilience as the bridge between the two streams of research and practice. The final section presents an integrated conceptual framework. It is expected that the proposed approach will address local coping and adaptive capacities to minimize causalities and losses from future hazard events and climate change impacts.

Hazard Mitigation Planning Framework

Hazard mitigation planning includes planned anticipatory measures undertaken to eliminate cause and the likelihood of a disaster. And in case disasters do occur, hazard mitigation policies are likely to limit the negative impacts to human life and property. The primary goal of hazard mitigation planning is to implement proactive actions that reduce community vulnerability to environmental hazards. Hazard mitigation policy in the United States has primarily been guided by federal financial incentive-based legislations including the National Flood Insurance Act of 1968, the Robert T. Stafford Disaster Relief and Emergency Assistance Act of 1988, and the Disaster Mitigation Act (DMA) of 2000 (Schwab 2011). Although the Stafford Act was the first to provide a comprehensive package for the four disaster management functions of mitigation, preparedness, response, and recovery, DMA pushed hazard mitigation to the forefront of the disaster management approach. DMA amended the Stafford Act by requiring states and localities to prepare multi-hazard mitigation plans as a precondition for receipt of federal mitigation grants and established a competitive program for predisaster mitigation planning and project grants.

Hazard mitigation encompasses a range of proactive measures whose key components are identifying hazards, analyzing vulnerabilities, and developing strategies to mitigate the hazard threat. The outcome of this exercise is a detailed Hazard Mitigation Plan that includes planning, policies, programs, projects, and activities for reducing hazards risks to the community. The Federal Emergency Management agency (FEMA) published an Interim Final Rule (2002) that provides specific requirements that a Hazard Mitigation Plan must include:

1 Documentation of the planning process that was followed to develop the plan.
2 Risk assessment that serves as the factual basis for the mitigation strategies proposed in the plan. The risk assessment should include: (i) detailed description of all natural hazards that affect a jurisdiction, and (ii) vulnerability assessment of the jurisdiction.
3 A mitigation strategy that provides the jurisdiction's blueprint for reducing the potential losses identified in the risk assessment.
4 A plan maintenance process that describes procedures for reviewing progress, monitoring implementation, and assessment of achievements of goals of implemented mitigation measures.
5 Documentation of formal adoption of the plan.

In order to assist local communities in their preparation of hazard mitigation plans, FEMA published the Local Mitigation Planning Handbook (FEMA 2013), which details the necessary tasks required for development of a hazard mitigation plan. The first step involves pre-planning activities that include tasks designed to facilitate plan making. Task 1 of pre-planning is to identify the planning area and the resources available for plan development. In general, the planning area refers to the administrative jurisdiction for which the plan is being made, such as a city, township, or county. Task 2 of pre-planning is to assemble a planning team of representatives from each agency or sub-jurisdiction and other stakeholders from private sector organizations and non-governmental organizations (NGOs) that have the expertise to develop the plan or the authority to implement the mitigation strategies developed through this planning process. Task 3 is to design an effective outreach strategy with the help of the planning team. Tasks 4–6 focus on creating the key constituents of an effective Hazard Mitigation Plan document. Task 4 is to recognize and take advantage of local capabilities that currently reduce disaster losses or can be used to reduce future disaster losses. Task 5 is to conduct a detailed risk assessment that assesses the potential impacts of hazards to people, the local economy, and the natural and built environment. Task 6 involves development of policies and actions that will provide long-term benefits of risk reduction. These can be structural or nonstructural activities. Structural approaches include building works to safeguard existing development such as seawalls, dikes, or levees, and strengthening of existing structures to enhance resistance to hazard impacts. Most structural approaches are hazard-specific and often need to be supplemented with nonstructural programs to ensure desired outcomes. Nonstructural programs include adoption and implementation of activities that direct development away from identified risk zones, and might sometimes require relocating existing development out of the risk zones. Common nonstructural hazard mitigation policies include land use planning and zoning policies to control development in flood plains and other risk zones, and building codes that require increased resistance to the destructive forces of hazards such as wind, water, and seismic shaking. The final outcome of this step is a draft hazard mitigation plan document that is ready for review and formal adoption. Subsequent tasks 7–9 included in the handbook provide guidance to keep the plan current, review, adoption, and implementation of the plan document.

Climate Change Adaptation Planning Framework

Climate change adaptation actions target specific systems that are vulnerable to anticipated impacts of climate change with the goal of reducing harm or taking advantage of the new opportunities likely to be created. Adaptation to climate change can be classified as reactive or anticipatory (Smith 1997). Reactive adaptation denotes actions that are implemented after or in response to climate change impacts. In comparison, anticipatory adaptation measures are undertaken in advance of climate change impacts. Planned anticipatory adaptation to climate change denotes actions and policies implemented to reduce risks from changing climatic conditions and capitalize on opportunities associated with these changes (Füssel 2007). A number of climate change policy researchers have analyzed a variety of adaptation policy options being implemented across the world (Aalst et al.2008; Hallegatte 2009; McGray and Hammill 2007; Smit et al. 2000; Smith and Lenhart 1996). Klein and Tol (1997) identify three key approaches to successful anticipatory adaptation climate change: (i) increasing flexibility of potentially vulnerable subsystems in human communities; (ii) reversing trends that increase vulnerability; and, (iii) improving public awareness and preparedness. Each of these approaches to planned adaptation is also relevant to hazard mitigation. The emphasis is on addressing factors that increase community vulnerability to hazard events. Füssel and Klein (2006) contend that, given the similarity of approaches in risk reduction, there is no distinguishable boundary between adapting to climate change and reducing risks from weather-related hazards.

Unlike hazard mitigation planning, preparing a climate change response strategy is only a voluntary policy pursuit because communities are not subject to federal sanctions for ignoring climate change. Despite this, a number of communities have adopted climate change adaptation plans, and more are in the process of doing so. Most frameworks for climate change adaptation planning have been proposed by research and non-governmental agencies. A few of the popular adaptation frameworks include: (i) *Preparing for Climate Change*, proposed by the Climate Impacts Group and partners (Snover et al 2007); (ii) *Cities Preparing for Climate Change*, proposed by the Clear Partnership (Ligeti et al. 2010); (iii) *Surviving Climate Change on Small Islands*, from the Tyndall Center for Climate Change Research (Tompkins et al. 2005); (iv) *Handbook on Methods for Climate Change Impact Assessment and Adaptation Strategies*, from the United Nations Environment Program (Eenstra et al. 1998); (v) *Coastal Hazards and Climate Change*, from the New Zealand Climate Change Office (Wratt et al. 2004); (vi) *Climate Adaptation Guidebook*, from the Center for Climate Strategies (Thomas et. al. 2011); (vii) *Adapting to Climate Change – A Planning Guide for State Coastal Managers*, from the National Oceanographic and Atmospheric Administration Office of Ocean and Coastal Resource Management (NOAA 2010); and, (ix) *A Guide to Community Drought Preparedness*, from Drought Ready Communities (2011). Each of these planning frameworks can be utilized for climate change adaptation planning by local governments (see Perkins et al. 2007, for detailed description of each framework).

Among these, the climate change adaptation planning framework proposed by the International Council for Local Environmental Initiatives (ICLEI) and partners has been the most popular among the North American communities. This framework reflects ICLEI's five milestones for climate adaptation. As per this framework, the steps involved in creating a climate change adaptation plan are: (i) conducting a climate vulnerability assessment, (ii) setting preparedness goals, (iii) developing a climate preparedness plan, (iv) publishing and implementing the preparedness plan, and (v) monitoring and re-evaluation. Two more specialized frameworks that seem useful for adoption for community-based climate change adaptation planning include those proposed by the Federal Highway Administration (FHWA), and the Wildlife Conservation Society for North America (WCS). The FHWA framework was developed specifically for evaluating transportation projects, but has significant potential for community-based climate change adaptation planning. This framework recommends a three-step methodology starting with scoping, followed by detailed vulnerability assessment, and finally, integrated decision making (FHWA 2014). The *Adaptation for Conservation Targets* (ACT) framework proposed by WCS was initially developed for identifying climate change adaptation priorities for wildlife conservation (Cross et al. 2012). The key component of this six-step methodology is a targeted risk assessment based on anticipated climate change impacts and the sensitivity of the system likely to be impacted. Overall, this approach presents a useful framework that incorporates necessary elements of natural resource planning—use of local knowledge, conceptual modeling, and adaptive management—into a logical process suitable for climate change adaptation at the local level.

Notably, each of these climate change adaptation frameworks adopts a risk management approach that relies on scientific data analysis for risk assessment, and aspires to reduce the negative impacts of anticipated climatic changes through adjustments in the community and its built environment. Figure 9.1 provides an overview of the contemporary climate change adaptation planning framework. The primary goal of climate change adaptation planning framework is to promote development of a safe community that is driven by values of safety and risk reduction (Aalst et al. 2008; Burton et al. 2002; Füssel 2007). The first step is risk analysis based on comprehensive analysis of existing and future conditions. This is a scientific and deterministic process that includes hazard identification and vulnerability assessment based on existing climate variability, as well as anticipated changes in the climatic system. Identification and assessment of risks from climate change is often best addressed by sector-specific analysis due to their specialized competencies. Therefore, engagement with representatives of each sector and municipal line department is critical for realistic hazard identification.

Although it would be ideal to promote zero risk development, this is never a feasible or a practical option. Therefore, decisions have to be made as regards the need for development with

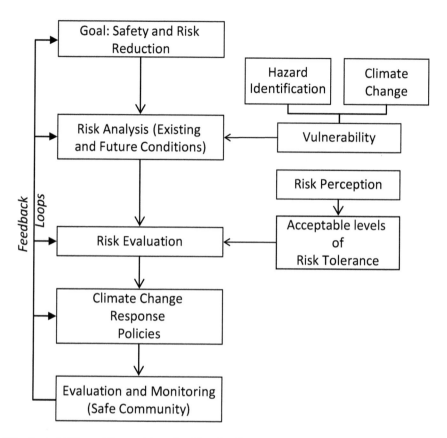

Figure 9.1 Typical Climate Change Adaptation Planning Process

respect to the level of risk that a community is willing to take. Risk evaluation is a methodologi-cal approach to make such policy decisions based on value analysis of acceptable levels of risk to the actors involved in the policy development process (Smith 2013). It is the process of making judgments about the need to address certain risks by balancing the assessed level of threat and sys-tem vulnerability (from the risk analysis) with respect to stakeholders' levels of risk tolerance. It is important to note that these decisions may not be truly objective, as the levels of risk tolerance are often based on perception of risk rather than the objective analysis of risk (Amendola 2002; Brecher 1997; French et al. 2004; Sjoberg 2003). Further, perceptions of risks, both at the level of individual actors and institutions influence such decisions (Douglas and Wildavsky 1982). Finally, based on the preceding analysis, proposed risk reduction policies are developed and implemented in order to ensure that future growth and development is safe and occurs within the acceptable levels of risk. The outcomes of these policies are monitored and evaluated in a timely manner to assure that the resulting development is safe and meets the adaptation goals. New development patterns and evolving climatic conditions will continue to influence future vulnerabilities, so it is important to incorporate ongoing feedback and adjustment loops in the planning process. Given the uncertainties associated with climate change, and scientific limitations of hazard risk assessment, climate change adaptation planning process is partly subjective. Therefore, extensive public participation is critical at each stage of the planning process and is, as such, implied in our discussion of the typical climate change adaptation planning process.

Vulnerability Reduction: The Common Agenda

It is evident that the primary goal of both hazard mitigation and climate change adaptation is to undertake anticipatory actions that reduce future risks to human life, the built environment, and other aspects of a community that are critical to its functioning. Hazard researchers identify risk to a community as a function of the frequency of the hazard event, its intensity, and vulnerability (Blaikie et al. 1994). Community vulnerability itself results from complex interactions among the residents, policies, institutional systems, and a host of other variables that directly or indirectly influence their exposure to hazards as well as the severity of the resultant impacts (Cutter et al. 2003; Kasperson et al. 1988; Timmerman 1981). Based on their extensive review of the disaster literature Wisner et al. (2003) offer this working definition of vulnerability, "the characteristics of a person or group and their situation that influence their capacity to anticipate, to cope with, resist, and recover from the impact of a natural hazard" (Wisner et al. 2003: 11).

Most disaster researchers agree that addressing community vulnerability is the key to limiting the impacts of natural hazards on society and economy. Community vulnerability is a social construct that arises out of everyday social and economic circumstances (Morrow 1999). The ultimate impact of a natural hazard event in any community is determined to a large extent by a complex set of inter-acting conditions, some influenced by location and others by socioeconomic conditions—including access to information and resources (Bolin and Stanford 1991; Lindell and Prater 2003; Morrow and Peacock 1997; Oliver-Smith 1988; Quarantelli 1998; Varley 1994). Decades of disaster research confirm that unequal access to resources and exposure to hazards risks result from a variety of socio-economic and political factors that will need to be addressed in order to effectively mitigate risks to a community (Cannon 1994; Mileti 1999).

Discussions of vulnerability in the climate change adaptation literature tend to fall into two categories—those focused on the damage caused or likely to be caused by the climate-related event (Jones and Boer 2004), and those addressing the pre-existing state of the system (Allen 2006). The former view has sought to analyze impacts as a function of the exposure to hazard event, the likeli-hood or the frequency of occurrence of the impact, the extent of human exposure to the impact, and sensitivity of the system to the impact. In this research approach, the social component asso-ciated with the system that acts to amplify or reduce the damage resulting from the impacts has often been addressed in the conceptualization even if it was not formally integrated into the model (Brooks 2003; Kates et al.1987). Although this conceptualization may seem similar to the current understanding in hazard research that the focus on the threat and disturbances alone is insufficient for understanding the impacts and responses to a hazard event, there is a significant point of depar-ture in terms of the analytical approach and the resulting policy response. This line of research has focused on analyzing community vulnerability using indicators of impacts such as casualties, dam-age, and monetary costs, rather than the indicators of preimpact state of the system (Jones and Boer 2004). Further, this analytical approach has not adequately distinguished between the variations within the exposed systems that lead to significant variations in the hazard impacts (Kasperson et al. 1988; Mitchell et al. 1989). And, finally, this approach has ignored the role of the political economy, including the social structures and institutions that shape the differential vulnerabilities and outcomes within a community (Hewitt 2014; Wisner and Luce 1993).

The second approach to understanding vulnerability recognizes that risk is explicitly defined as a function of the stressor, the hazard, and the vulnerability of the exposed system (Adger 2010). This line of research focuses its attention to the preimpact conditions that make the community unsafe, which leads to variations in vulnerability within the community that produce differential impacts across population segments and economic sectors. This conceptualization is similar to the social vulnerability approach commonly found in hazards research. Imbedded in this approach are two critical concepts of entitlement and coping capacity that influence the ultimate impacts of climate change related hazard events (Turner et al. 2003). Entitlement refers to legal and customary rights for exercising control over various resources. These entitlements vary among individuals and groups within a community, contributing to the overall sensitivity of the system to external stressors and hazards. Coping capacities refer to the abilities of individuals and groups that enable them to avert the potential harm as well as respond to, and recover from, harm caused by a hazard event. Both are linked through a complex system of institutions and networks (both formal and informal) within a community and ultimately influence the resultant losses from hazard impact.

Thus, both hazards and climate change adaptation researchers seem to agree that risk to a community is a function of both the hazard and the community's vulnerability to that hazard. The starting point in limiting losses from hazard events is to understand the nature of the hazard and the current levels of vulnerability so that appropriate policies for mitigation of harm can be formulated. Similarly, limiting the negative impacts of climate change on a community requires understanding the linkages between global climatic changes and local weather-related hazards, as well as the factors that are likely to exacerbate the resultant impacts of climate change on those hazards. In both approaches, environmental hazard prediction is based on the present level of scientific accuracy in modeling natural system, and thus implies a degree of uncertainty that policy makers must contend with. At the same time, community vulnerability appears to be the main factor that influences the variation in impact, and can be addressed effectively independent (to some extent) from the nature of the environmental threat. Additionally, researchers in related areas of scholarship have highlighted that focusing on community vulnerability can also help achieve broader objectives of sustainable development and social equity (see Berke 1995; Berke et al. 1993; Mileti and Gailus 2005; Schneider 2002).

This focus on vulnerability in hazards research tends to emphasize the importance of societal and individual outcomes, in which the contextual environmental stressor serves as backdrop to the complex integrations that take place in the human subsystems. Most hazards researchers agree that attention to societal outcomes provides an effective way to reduce community vulnerability to weather-related hazards whether or not their frequency and severity is exacerbated by climate change. In contrast, limited attention is given to environmental tradeoffs (such as impacts on water quality, carbon emissions, and deforestation) in implementing mitigation and adaptation strategies employed to counter vulnerability (Berkes 2007; Buckle et al. 2000). Consequently, vulnerability reduction policies tend to focus on threats to, sensitivity to, and adaptive capacity of specific community subgroups, independent of environmental threats from climate change or a combination of other environmental and sociopolitical factors (Turner 2010).

Community Resilience: The Bridging Framework

In contrast to vulnerability, the concept of community resilience, which originated from ecological science (see Holling 1973), focuses on understanding the internal processes of a community in the context of its environmental hazards. This approach views human communities as complex systems characterized by adaptive evolution, which results in continued emergence of new properties of social and behavioral interactions that ensure system survival and maintenance of critical system functions necessary for stress response and subsequent recovery (Dooley 1997; Holland 1992; Lansing 2003; Levin 1998). In hazards research, resilience is increasingly applied to the analysis of behavioral response, institutional response, community processes, and economies (Berkes and Ross 2013; Klein et al. 2003). Some resilience researchers highlight the importance of pre-event capacity to anticipate and prevent damage and post-event strategies to minimize impacts and promote recovery (Bruneau et al. 2003; Cutter et al. 2008; Tierney and Bruneau 2007). Others emphasize the capacity of the community to address not only preimpact issues but also the local capacity to respond to, and recover from hazard impacts (Berkes 2007). Berke and Campanella (2006) identify resilience as the ability to cope with and survive a disaster with minimal damages and loss of functional capacity. Inherent in understanding resilience in the context of hazards research is the capacity to withstand the impacts, and reduce the exposure to risk from anticipated hazards (Buckle et al. 2000; Manyena 2006). Although researchers continue to improve their understanding of what truly constitutes resilience, and how it can be applied to hazard mitigation, there is a general agreement that it is a forward-looking concept that can help address multiple hazard risks (Tompkins and Adger 2004). The resilience approach in hazard mitigation is consistent with the existing trend of an all-hazards approach in the broader hazards literature (Hewitt 2004). This approach also provides a framework to plan for future hazards that may not have been experienced before (Folke et al. 2002). This approach to hazard mitigation is increasingly gaining popularity among researchers to analyze community processes and their interactions with the surrounding environment that allow a community to absorb hazard impacts, and reorganize itself to retain critical functions after a hazard impact (Walker et al. 2004).

Although the general idea of enhancing community resilience is widely accepted by hazards researchers, it has been operationalized in a variety of ways. Two commonly adopted conceptualizations of the resilience framework are discussed here. The first perspective is the understanding of resilience as a community characteristic that enhances its ability to resist and mitigate hazard impacts. An example of this approach is the *five capitals* approach that interprets resilience as a composite indicator of performance measures of five community capitals—social, economic, human, physical, and natural capital (Mayunga 2007). Each of these capitals is evaluated in a predisaster community against benchmarks of resilience indicators. From this perspective, for example, a community with greater human capital in the form of education, health, skills, knowledge, and information is likely to have a higher degree of resilience. The second perspective focuses on postdisaster recovery outcomes, as in the *disaster resilience of place* (DROP) model (Cutter et al. 2008). That framework interprets community recovery as a consequence of antecedent conditions and the specific characteristics of the event and immediate effects. The DROP model identifies coping responses and adaptive capacity as critical factors that influence the degree of recovery. Similar to the five capitals approach, community resilience is measured as against resilience benchmarks across dimensions of ecological, social, economic, institutional, infrastructural, and community competence.

Although both these frameworks have been successfully utilized in empirical research on community resilience, they do share a few common issues. First, in both these frameworks, resilience is primarily linked to lack of vulnerability. Mayunga (2007) does not necessarily address the relationship between vulnerability and resilience (although it is implied), but Cutter and her colleagues (2008) do elaborate their perspective on this issue. The DROP model identifies vulnerability and resilience as related concepts with only a limited degree of overlap. The areas of overlap are evident in the discussion of the DROP model, but areas outside of the overlap have not been discussed. Further, the selection of resilience benchmarks is primarily based on measures of vulnerability indicators that were popular in the vulnerability research (Cutter et al. 2003; Cutter and Emrich 2006; Schmidtlein et al. 2008). Thus, the DROP model's measures of resilience do not adequately capture the dynamic nature of the concept. Another concern with respect to the measures suggested by both perspectives is equivalent treatment of each type of capacity (referred to as dimensions in DROP model); the lack of any differentiation thereby implies substitutability in operationalization of resilience. Further, these resilience frameworks do not address community adaptive capacity in much detail. The five capitals approach does address this issue directly but seems to incorporate the idea into the determination of resilience benchmarks. By contrast, the DROP model treats adaptive capacity as a postimpact process that influences community recovery, highlighting the distinction between the process and the lessons learned. Both these approaches thus address adaptive capacity, a critical component of resilience framework, in a limited and static manner.

In the context of climate change adaptation, researchers have also adopted the resilience framework to facilitate comprehensive analysis of interdependence between human and environmental systems in determining the condition and nature of disturbances in either subsystem or of the system as a whole (see Liu et al. 2007; Turner et al. 2003). Resilience frameworks adopted in climate change adaptation policy emphasize the two-way relationship between weather-related hazard events and community impacts. These researchers acknowledge the role of increasing greenhouse gas emissions, primarily from urban activities, in exacerbating existing weather related hazards (Adger 2000; Bahadur et al. 2010; Boelee et al. 2012; Brown 2013; Côté and Darling 2010; Côté and Nightingale 2012; Folke et al. 2010; Gallopín 2006; Hoffmann and Sgrò 2011; Hulme 2005; Nelson et al. 2007; Price 2004; Sgrò et al. 2011; Walker et al. 2004). At the same time, most researchers also acknowledge the critical role of enhancing local capacity to cope with hazard impacts. Coping capacity denotes the range of actions available, individually and collectively, to respond to perceived climate change risks and to avoid its potential impacts (Kelly and Adger 2000; Yohe and Tol 2002). Climate change adaptation policy research literature distinguishes coping capacity from adaptive capacity, primarily in terms of temporal and institutional aspects. For example, Lemos and Tompkins (2008) refer to coping capacity as the ability to reduce the worst and most acute hazard impacts, and adaptive capacity as long-term ability shaped over time by institutions and underlying causes of vulnerability. Similarly, Birkmann (2006) identifies coping strategies as those undertaken within existing institutional settings, and adaptive strategies as long-term actions that may require institutional changes. Thus, coping capacity is dependent on the ability of the community to draw upon available skills, resources, and experiences in immediate response to a hazard impact, and adaptive capacity is dependent of the ability to adjust existing community functions to prepare in advance for the changes and hazard impacts likely to happen in the future (Berman et al. 2012; Engle 2011; Smit and Wandel 2006).

Another key aspect of the resilience approach in climate change policy is the acceptance of uncertainty. All frameworks acknowledge that social and ecological systems are too complex and cannot be truly modeled, limiting our ability to predict scenarios (Hallegatte 2009; Hoffmann and Sgrò 2011; Kriegler et al. 2014; Pindyck 2012; Shove 2010; Smith and Lenhart 1996; Tompkins and Adger 2005). Response to climate change is particularly challenging because, even after decades of increasing scientific research, we are still unable to predict the exact influence of climate change on hazard events. The scientific community is still struggling to identify critical climatic conditions likely to trigger threshold effects that would result in stability shifts across socioecological systems (Walkers and Meyers 2004). Threshold effects are an increasing concern for climate change researchers as they signify irreversible changes in both ecological and social systems such that coping capacities are likely to be overwhelmed. Consequently, while scientific efforts are underway to decrease the degree of uncertainty regarding the dynamics of climate change impact, it is equally important to adopt policy frameworks that will enable communities to respond to and recover from extreme events.

In response this challenge of climate change uncertainties, there is an emerging perspective, influenced to a great extent by research in complex ecological systems, that views resilience as a process (Folke 2006; Mayena 2006) rather than an outcome. This conceptual approach promotes resilience thinking in all policy development frameworks that manage system functions. From this perspective, the characteristics of a resilient community include anticipating risks, resisting shocks, and recovering more quickly, as well as adapting effectively to increase the ability to anticipate, resist, and recover from future hazard impacts. This perspective also shifts the emphasis from risk mitigation to adaptive learning supplemented with local knowledge and culture. This perspective thus seems to be better suited in the context of uncertainties associated with hazard risks influenced by climate change. Further, this approach also incorporates the sustainability perspective, which has become an accepted cornerstone of local policy making in most communities (Cumming 2011; Folke and Gunderson 2010; Folke et al. 2010; Ludwig et al. 1997; Perrings 1998; Tobin 1999).

Characteristics of Resilience

Dovers and Handmer (1992) have proposed a typology of societal resilience actions that links resilience to planning for and adapting to hazards. They distinguished between reactive, marginal, and proactive resilience, which they referred to as Type I, Type II, and Type III (p. 270). The first type is characterized by efforts to strengthen the status quo, which commits the community to spending enormous resources. Consequently, such communities are poorly equipped and ill-prepared to deal with shocks that overwhelm the local resistant capacity. The second type, which is characterized by incremental changes, does not challenge the status quo but may lead to changes at the margins. This usually serves the interests of the elite few and not the environment or general public. The third type is characterized by efforts to adapt or change existing policies and institutional structures to ensure the community's functional continuity. This approach seeks to build community resilience through greater flexibility in policy formulation and implementation.

Current climate change policy response appears to be locked into the marginal resilience policy response wherein near-term economic and political concerns seem to outweigh long-term sustainable response. This type of resilience response continues to promote some degree of socioeconomic stability while promoting token actions that are unlikely to yield substantial benefits. A major concern with this type of response is that the perceived stability of the society comes with a much higher potential of risk, and could lead to collapse (through threshold events) as the society is unable to make social, economic, and political changes necessary for coping with, responding to, and recovering from climate-influenced extreme events. Proponents of greater hazard mitigation actions express similar sentiments related to continued disregard to environmental hazards in favor of local growth priorities (Burby 1998; Burby and Deyle 2000; Dalton and Burby 1994; Godschalk 2003). Recurring and ever-increasing losses from extreme natural events indicate that communities tend to favor short-term economic and social benefits at the expense of long-term hazard mitigation actions.

In order to formulate an integrated conceptual framework for hazard mitigation and climate change adaptation it is important to identify key characteristics of a resilient community. First is the ability to undergo change and still retain essential community functions. Inherent in this attribute is the ability of a community to live with uncertainty (Hewitt 2014). This requires building a societal and institutional memory of past events and increasing the capacity to learn from crisis. Extreme hazard events, including those exacerbated by climate change, will undoubtedly be damaging and will result in changes to the existing system. A resilient community will embrace the change while retaining the necessary elements for renewal and reorganization to ensure societal continuity (Folke et al. 2005). Each of the hazard events then becomes an opportunity for renewal and societal transformation into a better state. There is growing literature on the potential of combining local knowledge with scientific information to cope with environmental changes that threaten existing way of life (Gadgil et al. 2003; Mackinson 2001; Mercer et al. 2007; Riedlinger and Berkes 2001).

Second, is the ability and capacity to self-organize. Self-organization refers to the emergence of new patterns of system management or policy frameworks in response to a hazard threat or event experienced by the community. A number of hazards researchers have documented the emergence of self-organized networks in response to unmet needs during disaster response (Drabek and McEntire 2003; Quarantelli 1986; Turner 1976). In a resilient community, a number of these networks continue to evolve long after the immediate response needs are met to promote policy innovation during the window of opportunity (see Berke et al. 1993; Birkmann et al. 2010; Pelling and Dill 2006) in the recovery phase. Gladwell (2006) stresses the important roles of mavens—altruistic individuals with social skills who serve as knowledge brokers—in developing the new connections necessary for self-organization in human communities. Diversity has also been identified as an important ingredient necessary for promoting self-organization. In socioecological systems, diversity is not limited to the ecological diversity of genetic pools, species, and landscape diversity that are necessary for natural resource dependent economies, but also includes the diversity of stakeholders, organizations, and other social actors that promote a variety of perspectives for addressing the common issues that face the community. Diversity of constituencies that bring in a wide variety of views, considerations, and policy options has been identified as an important aspect of community resilience (Berkes 2007). Effective hazard mitigation, too, relies on a diversity of participants and perspectives to reduce risks in the policy development process. Increased participation of the community is also likely to result in more effective adoption and implementation of proposed policy actions.

Third, increased capacity for learning and adaptation is also critical to enhance community resilience. Learning in the context of resilience refers to institutional and social learning in a community, also known as adaptive management (Lee 1993). Such an ability is developed through continuous testing, monitoring and reevaluation of responses that acknowledge the inherent uncertainty associated with complex systems. Flexible institutions that share responsibility and powers connected to user groups, government agencies, and other non-governmental actors can promote social capacity for learning and adaptation (Tompkins and Adger 2004). Adaptive management can be operated through co-management of resources, and partnerships in policy formulation that incorporate dynamic learning characteristics combined with multilevel linkages among the co-management partners (Olsson et al. 2004). Communities that are able to develop such systems of co-management are better prepared to deal with external shocks that may arise from natural hazard threats or climate change impacts.

An Integrated Approach

Hazard mitigation and climate change are both concerned with reduction of risk to human communities. There exists a great degree of thematic overlap between hazard mitigation and climate change adaptation research (Figure 9.2). These areas of thematic overlap include threat analysis, focus on internal capacity, and emphasis on local policies to build resilience to external threats. Effective hazard mitigation relies on accurate and reliable understanding of the hazard threat. Most communities utilize past events as a guide to delineate risk areas (such as the 100-year floodplain) and formulate appropriate policies to manage them. This analysis of hazard risk is also referred to as the biophysical hazard. However, changing climatic conditions are likely to affect the existing distribution, frequency, and intensity of weather-related hazard threats (McBean and Ajibade 2009). Consequently, the present understanding of hazard risks will no longer be a reliable indicator of future risks. Threat analysis for mitigation of weather-related hazards, therefore, will need to consider the influence of climate change in policy formulation. In the context of climate change response, policy adaptation actions focused on understanding how anticipated changes in the climatic system are likely to influence weather-related hazards. Researchers are increasingly concerned with how present increase in intensity of hazard events will be influenced by climate change and subsequent weather-related hazards.

In recent years, both research communities have emphasized the role of internal community characteristics in producing disaster impacts. Hazard mitigation researchers agree that the internal capacity of a community to avoid, or at least minimize, losses due to hazard impacts are to a large extent dependent on local coping capacities, both individual and collective. This has also been referred to as social vulnerability in hazards research literature. Climate change adaptation researchers also highlight the importance of adaptive capacity in helping communities respond to and recover from hazard events. This assessment of adaptive capacity employs similar frameworks of analysis and measurements as those proposed for social vulnerability by hazards researchers. In terms of policy priorities, the hazard mitigation literature stresses the need to adopt an all-hazards approach by implementing policies that address causes of increased social vulnerability in the community.

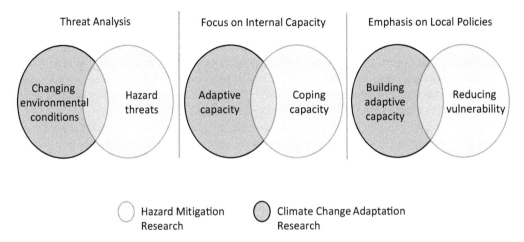

Figure 9.2 Locus of Traditional Research in Hazard Mitigation and Climate Change Adaptation

These policies focus on undertaking structural adjustments in the community to identify and reduce vulnerability of individuals, households, and the community as a whole. Climate change adaptation researchers, too, have recommended formulation and adoption of policies that enhance local adaptive capacities, both individual and collective. These policies also focus on addressing socioeconomic and political issues in a community that contribute towards differential impacts of hazard events.

Resilient Communities Framework

It is clear that hazard mitigation and climate change adaptation research are converging towards similar areas of thematic focus, so the framework described below is an initial attempt at an integrated conceptual model for connecting the two research disciplines. The proposed framework adopts the principles of *panarchy* (Gunderson and Holling 2001) in the context of the interrelated challenges of hazard mitigation and climate change adaptation. This framework is described as a continuous cycle of adaptive changes characterized in both ecological and social systems. The cycle of panarchy proceeds through forward loop stages of innovation, growth, exploitation, consolidation, predictability, and conservation, followed by back loop stages of instability, release, collapse, experimentation, novel recombination, and reorganization. This framework recognizes that adaptive cycles operate simultaneously on multiple scales—both spatial and temporal. The key to a stable system is not only the actors themselves but also the relationships between various actors and the nature of influence they exert upon each other and the system as a whole. The panarchial framework incorporates the dynamism of socioecological processes and transferability of adaptive learning wherein the learned benefits do not necessarily stay in the place where they were created, but can be transposed to flourish elsewhere (Holling 2005). Another important aspect of this framework is the role of memory that facilitates learning from past experiences and contributes towards the evolution of a subsequent stable state with enhanced adaptive capacity. Although this concept has achieved increasing attention in a number of disciplines that focus on complex systems, a number of authors have cited important concerns. Karkkainen (2005) notes that, although the concept is a general description of cycles of adaptive change, the proposed pattern of system change seems deterministic. Another concern relates to the seeming sense of futility implied by the inevitable failure of all forward loop endeavors—leading to back loop destabilization, and creative destruction. Researchers have also raised concerns regarding the lack of adequate attention to the role of human systems that now dominate the complex socioecological systems in the panarchial framework (Blythe 2015). Despite these conceptual limitations, the panarchial framework is well suited for understanding community resilience in the context of hazard mitigation and climate change adaptation.

The proposed Integrated Resilient Communities framework (Figure 9.3) emphasizes the cyclical relationship between community functions and the weather-related hazard events. In the pre-event community (Box A), various community activities such as urban development, utilization of natural resources, energy consumption, and travel patterns are the primary cause of greenhouse gas emission in Box B that are known to be the drivers of climate change. Thus, unsustainable community activities lead to increased emissions, contributing to the growing threat of climate change. Subsequently, these climatic changes influence the nature, distribution, and intensity of weather-related hazards such as floods, hurricanes, droughts, wildfires, and coastal inundation. Thus, human settlements both contribute to the problem of climate change and are likely to be at increasing risk from the resultant impacts of climatic changes. It is important to note that this relationship is not proportional. Because of the common pool nature of the climate system, a community's risk to a settlement is not proportional to the amount of greenhouse gas emissions it produces and the benefits are not exclusionary. Consequently, it is challenging to implement local policies that control a community's greenhouse gas emissions.

This framework also emphasizes the importance of coping capacity and adaptive capacity in influencing the functional state of the community on a continuous scale ranging from vulnerable to resilient (see the lower portion of Box A). In the context of this framework it is important to acknowledge that these capacities though linked are clearly distinct from one another. Coping capacity refers to the ability of the community to undertake responses within the existing structural constraints, whereas adaptive capacity is concerned with the ability to transform structurally and functionally when faced by hazard impacts that overwhelm the existing coping capacity. Consequently, a higher degree of coping and adaptive capacities of a community will make a community less vulnerable to hazard risks, and more resilient. In this framework vulnerability and resilience are identified as extreme ends of a continuous scale of community functions.

The hazard impacts, specifically related to weather-related hazards, are themselves influenced by climatic changes. A hazard impact on a community will result in some level of dysfunction (Box B). It is important to accept that no level of resilience or total lack of vulnerability precludes dysfunction. This degree of dysfunction following a hazard impact depends on the existing state of vulnerability in a community. Lower vulnerability and higher resilience is likely to result in transient dysfunction whereas greater vulnerability will result in more persistent states of dysfunction. This state of dysfunction will ultimately lead to a post-event stable state (Box C). This new stable state may be very similar to the pre-event community if the hazard impact is within the coping capacity of the community, but may also evolve into an emergent state as a result of adaptive learning. Coping and adaptive capacities will also change along with the evolution of community functions during this community recovery process. This stable state of community function continues on until the next hazard impact (Box D). This progressive cycle of evolution, impact, and adaptation continues indefinitely until the hazard impacts overwhelm both the coping capacity and the adaptive capacity of a community resulting in a catastrophe that pushes the community beyond the tipping point where the community continues to be in an endless state of persistent dysfunction. Historically, a number of communities, and at times large civilizations, have been known to be wiped out by hazard events. In present times, island communities such as that of Tuvalu face a similar fate as a result of the changing climate conditions (Farbotko and Lazrus, 2012).

Although this conceptual framework presents vulnerability and resilience as the opposing ends of the spectrum of community functions, in practice not all policies that reduce vulnerability are likely

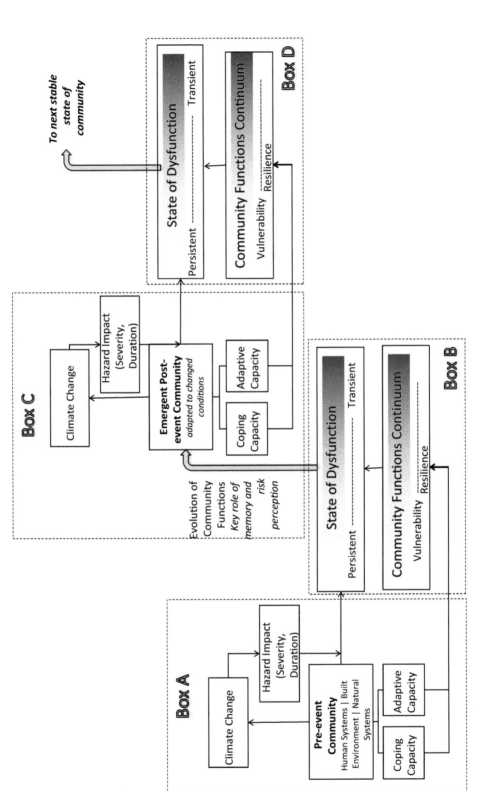

Figure 9.3 Integrated Resilient Communities Framework

to enhance resilience and vice versa. For example, provision of low-price electricity to provide relief during heat waves may be counterproductive in the long run as it is likely to contribute to increased greenhouse emissions that contribute to the problem of climate change. However, a number of policies that reduce community vulnerability are likely to promote local coping capacity that enhances community resilience to hazard impacts, providing climate change adaptation benefits.

Concluding Thoughts

This framework is a preliminary attempt to create an integrated conceptual framework to facilitate convergence of hazard mitigation and climate change adaptation scholarship and practice. It integrates the evolutionary nature of the panarchial framework from the socioecological perspective with the emphasis on coping capacity from the vulnerability perspective, and adaptive capacity from climate change policy response perspective. This integrated framework is a simplistic representation of community evolution from one stable state to another, mediated by hazard events that act as stimuli for adaptation. This framework, in essence, represents the core spirit of resilience—the ability to anticipate, respond to, recover from, cope with, and adapt to hazards. Focusing on a community's coping and adaptive capacities is critical in the context of increasing uncertainties about future weather-related hazard risks as a result of changing climatic conditions. With every evolution of community functions, adaptive actions will contribute towards increased coping capacities, and these coping capacities will be in turn leveraged into adaptive capacities following the next hazard impact. Consequently, a community's ability to undergo numerous cycles of adaptive evolution depends on its ability to learn from previous experiences and become better prepared for future hazard impacts. As documented in hazards research, disaster memory, and risk perception—both at the individual and collective scales—will play a critical role in enhancement of the coping and adaptive capacities of a community. Communities that fail to learn from past experiences or fail to appreciate increasing risks from climate change are likely to be overwhelmed by future hazard impacts leading to critical failure of community functions.

The logical next step in this research will be to identify measures of various concepts in this framework. The starting point will be to utilize the measures already operationalized by other researchers elsewhere for the various concepts utilized in this framework. From the perspective of this framework, it is critical to evaluate these measures over time rather than as static measures of resilience. This will require longitudinal studies that so far have been limited both in hazards and climate change adaptation research. This integrated framework is only a starting point in convergence of hazard mitigation and climate change adaptation research, and will likely (indeed, should) undergo further refinement similar to the evolutionary nature of community functions that it represents.

References

Aalst, M.K., T. Cannon, and I. Burton. (2008) "Community Level Adaptation to Climate Change: The Potential Role of Participatory Community Risk Assessment," *Global Environment Change* 18(1): 165–179

Adger, W.N. (2000) "Social and Ecological Resilience: Are They Related?" *Progress in Human Geography* 24(3): 347–364. doi:10.1191/030913200701540465.

Adger, W.N. (2010) "Social Capital, Collective Action, and Adaptation to Climate Change," 327–345, in M. Voss (ed.) *Der Klimawandel: Sozialwissenschaftliche Perspektiven*, Wiesbaden: VS Verlag für Sozialwissenschaften.

Allen, K.M. (2006) "Community-based Disaster Preparedness and Climate Adaptation: Local Capacity-building in the Philippines," *Disasters* 30(1): 81–101.

Amendola, A. (2002) "Recent Paradigms for Risk Informed Decision Making," *Safety Science* 40(1–4): 17–30.

Bahadur, A.V., M. Ibrahim, and T. Tanner. (2010) *The Resilience Renaissance? Unpacking of Resilience for Tackling Climate Change and Disasters, SCR Discussion Paper 1*, Brighton, UK: University of Sussex Institute of Development Studies.

Berke, P.R. (1995) "Natural-Hazard Reduction and Sustainable Development: A Global Assessment," *Journal of Planning Literature* 9(4): 383–395.

Berke, P.R. and T.J. Campanella. (2006). "Planning for Postdisaster Resiliency," *The Annals of the American Academy of Political and Social Science* 604(1): 192–207.

Berke, P.R, J. Kartez, and D. Wenger. (1993) "Recovery after Disaster: Achieving Sustainable Development, Mitigation and Equity," *Disasters* 17(2): 93–109.

Berkes, F. (2007) "Understanding Uncertainty and Reducing Vulnerability: Lessons from Resilience Thinking," *Natural Hazards* 41(2): 283–295.

Berkes, F. and H. Ross. (2013) "Community Resilience: Toward an Integrated Approach," *Society & Natural Resources* 26(1): 1–16. doi:10.1080/08941920.2012.736605.

Berman, R., C. Quinn, and J. Paavola. (2012) "The Role of Institutions in the Transformation of Coping Capacity to Sustainable Adaptive Capacity," *Environmental Development* 2(2): 86–100.

Birkmann, J. (2006) "Measuring vulnerability to promote disaster-resilient societies: Conceptual frameworks and definitions," in J. Birkmann (ed.), *Measuring Vulnerability to Natural Hazards: Towards Disaster Resilient Societies*, Tokyo: United Nations University Press.

Birkmann, J., P. Buckle, J. Jaeger, and M. Pelling. (2010) "Extreme Events and Disasters: A Window of Opportunity for Change? Analysis of Organizational, Institutional and Political Changes, Formal and Informal Responses after Mega-Disasters," *Natural Hazards* 55(3): 637–655.

Blaikie, P., T. Cannon, I. Davis, and B. Wisner. (1994) *At Risk: Natural Hazards, People's Vulnerability and Disasters.* London: Routledge.

Blythe, J. L. 2015. "Resilience and Social Thresholds in Small-Scale Fishing Communities," *Sustainability Science.* https://doi.org/10.1007/s11625-014-02539.

Boelee, E., M. Yohannes, J.-N. Poda, M. McCartney, P. Cecchi, S. Kibret, F. Hagos, and H. Laamrani. (2012) "Options for Water Storage and Rainwater Harvesting to Improve Health and Resilience against Climate Change in Africa," *Regional Environmental Change* 13(3): 509–519. doi:10.1007/s10113-012-0287-4.

Bolin, R., and L. Stanford. (1991) "Shelter, Housing and Recovery: A Comparison of U.S. Disasters," *Disasters* 15(1): 24–34.

Brecher, R.W. (1997) "Risk Assessment," *Toxicologic Pathology* 25(1): 23–26.

Brooks, N. (2003) *Vulnerability, Risk and Adaptation: A Conceptual Framework.* Norwich, UK: Tyndall Centre for Climate Change Research and Centre for Social and Economic Research on the Global Environment.

Brown, K. (2013) "Global Environmental Change I: A Social Turn for Resilience?" *Progress in Human Geography* 38(1): 107–117. doi:10.1177/0309132513498837.

Bruneau, M., S.E. Chang, R.T. Eguchi, G.C. Lee, T.D. O'Rourke, A.M. Reinhorn, M. Shinozuka, K. Tierney, W.A. Wallace, and D. von Winterfeldt. (2003) "A Framework to Quantitatively Assess and Enhance the Seismic Resilience of Communities," *Earthquake Spectra* 19(4): 737–738.

Buckle P., G. Mars, and R.S. Smale (2000) "New Approaches to Assessing Vulnerability and Resilience," *Australian Journal of Emergency Management* 15(2): 8–14.

Burby, R. J. (1998) *Cooperating with Nature: Confronting Natural Hazards with Land-Use Planning for Sustainable Communities.* Washington DC: Joseph Henry Press.

Burby, R.J., and R.E. Deyle. (2000) "Creating Hazard Resilient Communities Through Land-Use Planning," *Natural Hazards Review* 1(2): 99–106.

Burton, I., S. Huq, B. Lim, O. Pilifosova and E.L. Schipper (2002) "From Impacts Assessment to Adaptation Priorities: The Shaping of Adaptation Policy," Climate Policy 2(2–3): 145–159.

Cannon, T. (1994) "Vulnerability Analysis and the Explanation of 'Natural' Disasters," 13–30, in A. Varley et al. (eds.) *Disasters, Development and Environment*, Chichester: John Wiley.

Côté, I. M., and E. S. Darling. (2010) "Rethinking Ecosystem Resilience in the Face of Climate Change," *PLoS Biology* 8(7) e1000438. doi:10.1371/journal.pbio.1000438.

Côté, M., and A. J. Nightingale. (2012) "Resilience Thinking Meets Social Theory: Situating Social Change in Socio-Ecological Systems (SES) Research," *Progress in Human Geography* 36(4): 475–489.

Cross, M.S., E.S. Zavaleta, and D. Bachelet. (2012) "The Adaptation for Conservation Targets (ACT) Framework: A Tool for Incorporating Climate Change into Natural Resource Management," *Environmental Management* 50(3): 341–351.

Cumming, G. S. (2011) "Spatial Resilience: Integrating Landscape Ecology, Resilience, and Sustainability," *Landscape Ecology* 26(7): 899–909. doi:10.1007/s10980-011-9623-1.

Cutter, S. L., L. Barnes, M. Berry, C. Burton, E. Evans, E. Tate, and J. Webb. (2008) "A Place-Based Model for Understanding Community Resilience to Natural Disasters," *Global Environmental Change* 18(4): 598–606.

Cutter, S. L., B. J. Boruff, and W. L. Shirley. (2003) "Social Vulnerability to Environmental Hazards," *Social Science Quarterly* 84(2): 242–261.

Cutter, S. L., and C. T. Emrich. (2006) "Moral Hazard, Social Catastrophe: The Changing Face of Vulnerability along the Hurricane Coasts," *Annals of the American Academy of Political and Social Science* 604(1): 102–112.

Dalton, L. C., and R. J. Burby. (1994) "Mandates, Plans, and Planners: Building Local Commitment to Development Management," *Journal of the American Planning Association* 60(4): 444–461.

Dooley, K.J. (1997) "A Complex Adaptive Systems Model of Organization Change," *Nonlinear Dynamics, Psychology, and Life Sciences* 1(1): 69–97.

Douglas, M., and A. Wildavsky. (1982) *Risk and Culture: An Essay on the Selection of Technical and Environmental Dangers.* Oakland CA: University of California Press.

Dovers, S.R. and J.W. Handmer (1992) "Uncertainty, Sustainability and Change," *Global Environmental Change* 2(4): 262–276.

Drabek, T.E., and D.A. McEntire. (2003) "Emergent Phenomena and the Sociology of Disaster: Lessons, Trends and Opportunities from the Research Literature," *Disaster Prevention and Management* 12(2): 97–112.

Drought Ready Communities. (2011) *Guide to Community Drought Preparedness,* Lincoln NE: National Drought Mitigation Center.

Eenstra, J.G., I. Burton, J. B. Smith, and R. S. J. Tol. (1998) *Handbook on Methods for Climate Change Impact Assessment and Adaptation Strategies,* United Nations Environment Program, Nairobi and Institute for Environmental Studies, Free University, Amsterdam.

Engle, N. L. (2011) "Adaptive Capacity and Its Assessment," *Global Environmental Change* 21(2): 647–656.

Farbotko, C. and H. Lazrus. (2012) "The First Climate Refugees? Contesting Global Narratives of Climate Change in Tuvalu," *Global Environmental Change* 22(2): 382–390.

Federal Emergency Management Agency Interim Final Rule (2002) 44 CFR §201.6

FHWA. (2014) "Adaptation Framework," www.fhwa.dot.gov/environment/climate_change/adaptation/adaptation_framework/ [Accessed on 15 Jul. 2015].

Folke, C. 2006. "Resilience: The Emergence of a Perspective for Social–Ecological Systems Analyses," *Global Environmental Change* 16(3): 253–267. www.sciencedirect.com/science/article/pii/S0959378006000379.

Folke, C., S. Carpenter, T. Elmqvist, L. Gunderson, C.S. Holling, and B. Walker. (2002) "Resilience and sustainable development: building adaptive capacity in a world of transformations," *AMBIO: A Journal of the Human Environment* 31(5): 437–441.

Folke, C., S.R. Carpenter, B. Walker, M. Scheffer, T. Chapin, and J. Rockström. (2010) "Resilience Thinking: Integrating Resilience, Adaptability and Transformability," *Ecology and Society* 15(4). doi:10.1038/nnano.2011.191.

Folke, C., and L. Gunderson. (2010) "Resilience and Global Sustainability," *Ecology and Society.* 15(4): Article 43.

Folke, C., T. Hahn, P. Olsson, and J. Norberg. (2005). "Adaptive governance of social-ecological systems." *Annual Review of Environment and Resources* 30: 441–473.

French, D.P., T.M. Marteau, S. Sutton, and A.L. Kinmonth. (2004) "Different Measures of Risk Perceptions Yield Different Patterns of Interaction for Combinations of Hazards: Smoking, Family History and Cardiac Events," *Journal of Behavioral Decision Making* 17(5): 381–393.

Füssel, H.M. (2007) "Adaptation Planning for Climate Change: Concepts, Assessment Approaches, and Key Lessons," *Sustainability Science* 2(2): 265–275.

Füssel, H.M., and R.J.T. Klein. (2006) "Climate Change Vulnerability Assessments: An Evolution of Conceptual Thinking," *Climatic Change* 75(3): 301–329.

Gadgil, M., P. Olsson, F. Berkes, and C. Folke. (2003) "Exploring the Role of Local Ecological Knowledge in Ecosystem Management: Three Case Studies," 189–209, in F. Berkes, J. Colding, and C. Folke (eds.) *Navigating Social-Ecological Systems: Building Resilience for Complexity and Change,* Cambridge, UK: Cambridge University Press.

Gall, M., K.A. Borden, C.T. Emrich, and S.L. Cutter. (2011) "The Unsustainable Trend of Natural Hazard Losses in the United States," *Sustainability* 3(11): 2157–2181.

Gallopín, G. C. (2006) "Linkages between Vulnerability, Resilience, and Adaptive Capacity," *Global Environmental Change* 16(3): 293–303.

Gillett, N. P. (2014) "Detection and Attribution of Climate Change in North America," 251–267, in G. Ohring (ed.) *Climate Change in North America,* Heidelberg, Germany: Springer.

Gladwell, M. (2006) *The Tipping Point: How Little Things Can Make a Big Difference,* New York: Little, Brown and Company.

Godschalk, D.R. (2003) "Urban Hazard Mitigation: Creating Resilient Cities," *Natural Hazards Review* 4(3): 136–143.

Gunderson, L.H. and Holling, C.S. (2001). *Panarchy: Understanding Transformations in Human and Natural Systems,* Washington DC: Island Press.

Hallegatte, S. (2009) "Strategies to Adapt to an Uncertain Climate Change," *Global Environmental Change* 19(2): 240–247.

Helmer, M., and D. Hilhorst. (2006) "Natural Disasters and Climate Change," *Disasters* 30(1): 1–4.

Hewitt, K. (2004) A Synthesis of the Symposium and Reflection on Reducing Risk Through Partnerships. Paper presented at the conference of the Canadian Risk and Hazards Network (CRHNet), November 2004, Winnipeg.

Hewitt, K. (2014) *Regions of Risk: A Geographical Introduction to Disasters*, New York: Routledge.

Hoffmann, A.A., and C.M. Sgrò. (2011) "Climate Change and Evolutionary Adaptation," *Nature* 470(7335): 479–485. doi:10.1038/nature09670.

Holland, J.H. (1992) "Complex Adaptive Systems," *Daedalus* 121(1): 17–30.

Holling, C.S. (1973) "Resilience and stability of ecological systems," *Annual Review of Ecology and Systematics*, 4(1): 1–23.

Holling, C.S. (2005) "From Complex Regions to Complex Worlds," *Ecology and Society* 9(1): 11.

Hulme, P.E. (2005) "Adapting to Climate Change: Is There Scope for Ecological Management in the Face of a Global Threat?" *Journal of Applied Ecology* 42(5): 784–794.

Jones, R., and R. Boer. (2004) "Assessing Current Climate Risks," 93–117, in B. Lim and E. Spanger-Siegfried (eds.) *Adaptation Policy Frameworks for Climate Change: Developing Strategies, Policies and Measures*. United Nations Development Programme and Cambridge University Press, New York. Accessed 18 June 2018 at http://adaptation-undp.org/resources/training-tools/adaptation-policy-frameworks.

Karkkainen, B.C. (2005) "Panarchy and Adaptive Change: Around the Loop and Back Again," *Minnesota Journal of Law, Science and Technology* 7(1): 59–77.

Kasperson, R.E., O. Renn, P. Slovic, and H.S. Brown. (1988) "The Social Amplification of Risk: A Conceptual Framework," *Risk Analysis* 8(2): 177–187.

Kates, R.W., J.H. Ausubel, and M. Berberian. (1987) *Climate Impact Assessment: Studies of the Interaction of Climate and Society*, ICSU/SCOPE Report No. 27, New York: John Wiley.

Kelly, P.M. and W.N. Adger. (2000) "Theory and Practice in Assessing Vulnerability to Climate Change and Facilitating Adaptation," *Climatic Change* 47(4): 325–352. https://doi.org/10.1023/A:1005627828199.

Klein, R.J.T. and R.S.J. Tol (1997) *Adaptation to Climate Change: Options and Technologies: An Overview Paper*. *Technical Paper FCCC* TP/1997/3. Bonn, Germany: UNFCCC Secretariat. Retrieved from www. unfccc. int/resource/docs/tp/tp3.pdf.

Klein, R.J.T., R.J. Nicholls, and F. Thomalla. (2003) "Resilience to Natural Hazards: How Useful Is This Concept?" *Environmental Hazards* 5(1–2): 35–45.

Kriegler, E., J. Edmonds, S. Hallegatte, K.L. Ebi, T. Kram, K. Riahi, H. Winkler, and D.P. van Vuuren. (2014) "A New Scenario Framework for Climate Change Research: The Concept of Shared Climate Policy Assumptions," *Climatic Change* 122(3): 401–414.

Lansing, J.S. (2003) "Complex Adaptive Systems," *Annual Review of Anthropology* 32: 183–204.

Lee, K.N. (1993) *Compass and Gyroscope: Integrating Science and Politics for the Environment*. Washington DC: Island Press.

Lemos M.C. and E.L. Tompkins. (2008) "Creating Less Disastrous Disasters," *IDS Bulletin* 39(4): 60–66.

Levin, S.A. (1998) "Ecosystems and the Biosphere as Complex Adaptive Systems," *Ecosystems* 1(5): 431–436.

Ligeti, E., I. Wieditz, and J. Penney. (2010) *Cities Preparing for Climate Change: A Study of Six Urban Regions*, Toronto, Canada: Clean Air Partnership.

Lindell, M.K., and C.S. Prater. (2003) "Assessing Community Impacts of Natural Disasters," *Natural Hazards Review* 4(4): 176–185.

Lindell, M.K., C.S. Prater, and R.W. Perry. (2006) *Fundamentals of Emergency Management*, Emmitsburg MD: Federal Emergency Management Agency Emergency Management Institute. Available at www.training. fema.gov/EMYWeb/edu/fem.asp.

Liu, J., T. Dietz, S.R. Carpenter, and C. Folke. (2007) "Coupled Human and Natural Systems," *AMBIO* 36(8): 639–649.

Ludwig, D., B. Walker, and C.S. Holling. (1997) "Sustainability, Stability, and Resilience," *Ecology and Society* 1(1): 7. Accessed January 16, 2018 at www.consecol.org/vol1/iss1/art7/.

Mackinson, S. (2001) "Integrating Local and Scientific Knowledge: An Example in Fisheries Science," *Environmental Management* 27(4): 533–545.

Manyena, S.B. (2006) "The Concept of Resilience Revisited," *Disasters* 30(4): 434–450.

Mayunga, J.S. (2007) "Understanding and applying the concept of community disaster resilience: a capital-based approach," *Summer Academy for Social Vulnerability and Resilience Building* 1(1): 1–16.

McBean, G., and I. Ajibade. (2009) "Climate Change, Related Hazards and Human Settlements," *Current Opinion in Environmental Sustainability* 1(2):179–186.

McGray, H., and A. Hammill. (2007) *Weathering the Storm: Options for Framing Adaptation and Development*, Washington DC: World Resources Institute.

Mercer, J., D. Dominey-Howes, I. Kelman, and K. Lloyd. (2007) "The Potential for Combining Indigenous and Western Knowledge in Reducing Vulnerability to Environmental Hazards in Small Island Developing States," *Environmental Hazards* 7(4): 245–256.

Mileti, D.S. (1999) *Disasters by Design: A Reassessment of Natural Hazards in the United States*, Washington DC: National Academy Press.

Mileti, D.S., and J.L. Gailus. (2005) "Sustainable Development and Hazards Mitigation in the United States: Disasters by Design Revisited," *Mitigation and Adaptation Strategies for Global Change* 10: 491–504.

Mitchell, J.K., N. Devine, and K. Jagger. (1989) "A Contextual Model of Natural Hazard," *Geographical Review* 79(4): 391–409.

Morrow, B.H. (1999) "Identifying and Mapping Community Vulnerability," *Disasters* 23(1): 1–18.

Morrow, B.H., and W.G. Peacock. (1997) "Disasters and Social Change: Hurricane Andrew and the Reshaping of Miami," 226–242 in W.G. Peacock, B.H. Morrow, and H. Gladwin (eds.) *Hurricane Andrew: Ethnicity, Gender, and the Sociology of Disasters*, New York: Routledge.

Munich Re. (2012) *Severe Weather in North America: Perils, Risks, Insurance*. Munich, Germany: Author

National Oceanic and Atmospheric Administration (NOAA). (2010) *Adapting to Climate Change: A Planning Guide for State Coastal Managers*, NOAA Office of Ocean and Coastal Resource Management.

Nelson, D.R., W.N. Adger, and K. Brown. (2007) "Adaptation to Environmental Change: Contributions of a Resilience Framework," *Annual Review of Environment and Resources* 32(1): 395–419.

O'Brien, G., P. O'Keefe, J. Rose, and B. Wisner. (2006) "Climate Change and Disaster Management," *Disasters* 30(1): 64–80.

Oliver-Smith, A. (1988) "Natural Disasters and Cultural Responses," 1–34, in *Studies in Third World Societies* No. 36. Williamsburg VA: College of William and Mary.

Olsson, P., C. Folke, and F. Berkes. (2004) "Adaptive Comanagement for Building Resilience in Social–Ecological Systems," *Environmental Management* 34(1): 75–90.

Pelling, M., and K. Dill. (2006) "Natural Disasters as Catalysts of Political Action," *Media Development* 53(4): 7–10.

Perkins, B., D. Ojima, and R. Corell. (2007) *A Survey of Climate Change Adaptation Planning*, Washington DC: The H. John Heinz III Center for Science, Economics and the Environment.

Perrings, C. (1998) "Introduction: Resilience and Sustainability," *Environment and Development Economics* 3(2): 221–262.

Pindyck, R. S. (2012) "Uncertain Outcomes and Climate Change Policy," *Journal of Environmental Economics and Management* 63(3): 289–303.

Price, M. F. (2004) *Navigating Social–Ecological Systems: Building Resilience for Complexity and Change*, New York: Cambridge University Press.

Quarantelli, E.L. (1986) *Disaster Crisis Management*, Newark DE: University of Delaware Disaster Research Center.

Quarantelli, E.L. (1998) *What Is a Disaster? Perspectives on the Question*. London: Routledge.

Riedlinger, D., and F. Berkes. (2001) "Contributions of Traditional Knowledge to Understanding Climate Change in the Canadian Arctic," *Polar Record* 37(203): 315–328.

Rowley, R.J., J.C. Kostelnick, D. Braaten, X. Li, and J. Meisel. (2007) "Risk of Rising Sea Level to Population and Land Area," *Eos, Transactions American Geophysical Union* 88(9): 105–107.

Schmidtlein, M.C., R.C. Deutsch, W.W. Piegorsch, and S.L. Cutter. (2008) "A Sensitivity Analysis of the Social Vulnerability Index," *Risk Analysis* 28(4): 1099–1114.

Schneider, R.O. (2002) "Hazard Mitigation and Sustainable Community Development," *Disaster Prevention and Management* 11(2): 141–147.

Schwab, J. (2011) *Hazard Mitigation: Integrating Best Practices into Planning*, Chicago: American Planning Association.

Sgrò, C.M., A.J. Lowe, and A.A. Hoffmann. (2011) "Building Evolutionary Resilience for Conserving Biodiversity Under Climate Change," *Evolutionary Applications* 4(2): 326–337.

Shove, E. (2010) "Beyond the ABC: Climate Change Policy and Theories of Social Change," *Environment and Planning A* 42(6): 1273–1285.

Sjoberg, L. (2003) "Distal Factors in Risk Perception," *Journal of Risk Research* 6 (3): 187–211.

Smit, B., I. Burton, R.J.T. Klein, and J. Wandel. (2000) "An Anatomy of Adaptation to Climate Change and Variability," *Climatic Change* 45(1): 223–251.

Smit, B., and J. Wandel. (2006) "Adaptation, Adaptive Capacity and Vulnerability," *Global Environmental Change* 16(3): 282–292.

Smith, J.B. (1997) "Setting Priorities for Adapting to Climate Change," *Global Environmental Change* 7(3): 251–264.

Smith, J.B, and S.S. Lenhart. (1996) "Climate Change Adaptation Policy Options," *Climate Research* 6(2): 193–201.

Smith, K. (2013) *Environmental Hazards: Assessing Risk and Reducing Disaster*, New York: Routledge.

Snover, A.K., L.C. Whitely Binder, J. Lopez, E. Willmott, J.E. Kay, D. Howell, J. Simmonds. (2007) *Preparing for Climate Change: A Guidebook for Local, Regional, and State Governments*, University of Washington Climate Impacts Group and King County, Washington, in association with and published by ICLEI – Local Governments for Sustainability, Oakland, CA.

Solecki, W., R. Leichenko, and K. O'Brien. (2011) "Climate Change Adaptation Strategies and Disaster Risk Reduction in Cities: Connections, Contentions, and Synergies," *Current Opinion in Environmental Sustainability* 3(3): 135–141.

Thomas D.P., M. Wyman, G. Flora, W. Dougherty, J. Smith, S. Saunders, S. Chester, and T. Looby. (2011) *Adaptation Guidebook*, Washington DC: The Center for Climate Strategies.

Tierney, K., and M. Bruneau. (2007) "Conceptualizing and Measuring Resilience: A Key to Disaster Loss Reduction," *TR News*, 250(5): 14–15, 17.

Timmerman, P. (1981) "Vulnerability Resilience and Collapse of Society," *A Review of Models and Possible Climatic Applications*, Toronto, Canada: University of Toronto.

Tobin, Graham A. (1999) "Sustainability and Community Resilience: The Holy Grail of Hazards Planning?" *Environmental Hazards* 1(1): 13–25.

Tompkins, E.L., and W.N. Adger. (2004) "Does Adaptive Management of Natural Resources Enhance Resilience to Climate Change?" *Ecology and Society* 9(2): Article 10.

Tompkins, E.L., and W.N. Adger. (2005) "Defining Response Capacity to Enhance Climate Change Policy," *Environmental Science and Policy* 8(6): 562–571.

Tompkins, E., S. Nicholson-Cole, E. Boyd, L. Hurlston, G. Brooks Hodge, J. Clarke, N. Trotz, G. Gray, and L. Varlack. (2005) *Surviving Climate Change in Small Islands: A Guidebook*, University of East Anglia, Norwich: Tyndall Centre for Climate Change Research, School of Environmental Sciences.

Turner, B.A. (1976) "The Organizational and Interorganizational Development of Disasters," *Administrative Science Quarterly* 21(3): 378–397.

Turner, B.L. (2010) "Vulnerability and Resilience: Coalescing or Paralleling Approaches for Sustainability Science?" *Global Environmental Change* 20(4): 570–576.

Turner, B.L., R.E. Kasperson, P.A. Matson, J.J. McCarthy, R.W. Corell, L. Christensen, N. Eckley, et al. (2003) "A Framework for Vulnerability Analysis in Sustainability Science," *Proceedings of the National Academy of Sciences of the United States of America* 100(14): 8074–8079.

Varley, A. (1994) "The Exceptional and the Everyday: Vulnerability Analysis in the International Decade for Natural Disaster Reduction," 1–11, in A. Varley (ed.) *Disasters, Development and Environment*, Chichester: John Wiley.

Venton, P., and S. La Trobe. (2008) *Linking Climate Change Adaptation and Disaster Risk Reduction*. Teddington UK: Institute of Development Studies. www.preventionweb.net/publications/view/3007.

Walker, B., C.S. Holling, S.R. Carpenter, and A. Kinzig. (2004) "Resilience, Adaptability and Transformability in Social-Ecological Systems," *Ecology and Society* 9(2): 5.

Walker, B. and J.A. Meyers. (2004) "Thresholds in Ecological and Social-Ecological Systems: A Developing Database," *Ecology and Society* 9(2): 3.

Wisner, B., P. Blaike, T. Cannon, and I. Davis. (2003) *At Risk: Natural Hazards, People's Vulnerability and Disasters*, 2nd ed. London, UK: Routledge.

Wisner, B., and H.R. Luce. (1993) "Disaster Vulnerability: Scale, Power and Daily Life," *GeoJournal* 30(2): 127–140.

Wratt, D., B. Mullan, J. Salinger, S. Allan, T. Morgan, and G. Kenny. (2004) *Coastal Hazards and Climate Change. A Guidance Manual for Local Government in New Zealand*, New Zealand Climate Change Office.

Yohe, G. and R.S.J. Tol. (2002) "Indicators for Social and Economic Coping Capacity: Moving toward a Working Definition of Adaptive Capacity," *Global Environmental Change*. https://doi.org/10.1016/S0959-3780(01)00026-7.

PART III

Contributions of Emergency Response Planning to Community Resilience

10

EMERGENCY PREPAREDNESS AND RESPONSE PLANNING

Jennifer A. Horney and Garett Sansom

Introduction

The ways in which we prepare for and respond to emergencies has changed to meet the emerging challenges associated with urbanization, demographic shifts, and global climate change. A new, more comprehensive preparedness and response systems approach can effectively respond to threats from natural disasters, industrial risks, and terrorist activities. Research on risk communication, community resilience, evacuation planning and implementation, and global response can provide best practices for responders to emulate and improve upon. The following will briefly review the development of emergency preparedness and response, discuss how research has influenced emergency preparedness and response planning, and highlight areas still in need of improvement.

Historical Overview of Emergency Preparedness and Response in the United States

For most of United States history, emergency preparedness and response has existed in some form. For example, in 1803 the passage of the Congressional Fire Disaster Relief Legislation released federal funds to state and local governments to assist in the response to a series of fires in the Northeast (Drabek 1991). With this action, the government set an expectation of federal assistance in emergencies throughout the country.

This type of support for individual events continued for the next 150 years. However, these reactive legislative efforts lacked flexibility and offered little in the way of improving community preparedness. By 1916, the establishment of the Council of National Defense moved disaster response duties to the military and began to consider the role of preparedness (Miskel 2008). It was not until the 1950s that the federal government took a more proactive role in providing disaster relief, and most US residents certainly did not expect relief from the federal government until after this time.

The Federal Disaster Relief Act of 1950 was passed to provide funds to alleviate suffering and damage resulting from major disasters (PL 81-875 in Platt 1999). While the federal government remained limited in its ability to respond, it was firmly committed to an official role in disaster prevention and response. The responsibility for this program changed often in the years to follow, but it was relatively successful in responding to disasters. It was not until the early 1970s that immense disasters—such as Hurricane Agnes (Figure 10.1) and Three Mile Island—forced a reevaluation of the federal role (National Hurricane Center website; Miskel 2008). President Carter created the Federal Management Agency (FEMA) in 1979 in order to consolidate preparedness and response efforts.

Figure 10.1 Motorboating Down Main Street. Farmville, Virginia. Photo Credit: Virginia Governor's Negative Collection, Courtesy of the Library of Virginia

FEMA combined many agencies under a single framework to provide protection by preparing for and responding to all forms of disasters (Executive Orders 12127, 12148 1979). With increased knowledge, computational power, and new fields of study engaging in the study of disasters, the field of disaster research began to take entirely new directions. As responses to hurricanes, flooding, fires, droughts, and terrorist events were studied, many best practices were developed. New areas of inquiry, such as public and global health, community-based preparedness and planning, and risk communication were utilized to further enhance emergency planning and response.

Actors in Disaster Preparedness and Response in the United States

The role of government in preparedness for and response to disasters is critical for communities. For example, in the wake of a disaster, federal financial assistance can be released following the approval by the President of a formal request from the affected state's governor. The governor makes this request following a preliminary disaster report that identifies that the scope and type of damage goes beyond the ability of the state or local government to address on its own.

Following a Presidential Disaster Declaration (PDD), three types of assistance may be offered. First, individual assistance may be given to families, businesses, or individuals in the form of grants, loans, tax relief, or unemployment assistance. Second, public assistance can be released to help rebuild infrastructure, communities, or nonprofit groups. Last, matching mitigation funds can be provided to states and local communities for projects to reduce vulnerability.

The first phase accompanying a PDD is response. Working in partnership with other organizations to meet the needs of the "whole community," FEMA operates within the emergency management system, implementing programs to meet the unique response needs of each disaster. This phase offers immediate aid and relief to those affected, often in the form of emergency personnel, food, water, and equipment. This is accomplished through the National Response Framework (Brown and Magary 1998).

Following the response phase, the recovery phase begins the work of rebuilding the community (Brown and Magary 1998). This can last weeks, months, or even years. To be successful, disaster recovery must not simply bring a community back to where it was before the disaster; it must rebuild to be more resilient to future disasters. Recovery must also facilitate and strengthen institutional and community relationships that can aid in the coordination of ongoing recovery (Inam 1999; Olshansky et al. 2008). Recovery often encounters major challenges, including aligning policies and goals, effectively engaging the public, and setting priorities that guide the community towards a more disaster resilient future (Olshansky and Johnson 2010; Schwab et al. 1998; Smith 2012).

Responding to and preparing for disasters typically involves four distinct groups that work together: the public, private, and non-profit sectors, as well as individual citizens. It is only with proper coordination, communication, and expertise of the whole community that these functions can be successful.

Public Actors: Federal

The federal government plays a primary role in preparedness and response to major disasters. Following the 2003 creation of the Department of Homeland Security (DHS), FEMA became an agency of that department. In disaster response, FEMA follows a procedural framework for disaster preparedness and response with the National Response Framework, identifies key players, and supports the coordination of federal, state, and local resources. Ten FEMA regional administrators are responsible for operational aspects for all disaster response and recovery, mitigation, and preparedness activities throughout their regions.

Recognizing that there are shared responsibilities with regard to preparedness and response, there are many different federal departments involved. For example, the Department of Energy (DOE) is responsible for directing and restoring energy systems following a disaster. Furthermore, the DOE and the Nuclear Regulatory Commission (NRC) provide assistance and expertise on energy issues involving hazardous materials from nuclear facilities ranging from power plants to medical and industrial X-ray machines. The Department of Health and Human Services (HHS) provides medical care management assistance during and following a disaster. The Department of Housing and Urban Development (HUD) provides housing, both temporary and permanent, after disasters to those affected. The US Environmental Protection Agency (EPA) plays a major role with the cleanup efforts of hazardous materials and works directly with local efforts in planning and response.

Public Actors: State

State-level agencies and programs can often deal more directly with issues surrounding disaster preparedness and response than the federal government. A State Emergency Management Office works directly with their governors to meet requests for resources, as well as channel monetary funds from the federal government to local entities. A State Department of Public Health assists in locating and responding to disease outbreaks, terrorist attacks that utilize biological weaponry, and incidence reporting. Post-disaster, the State Department of Transportation, the State Department of Housing, and the State's National Guard may assist with infrastructure repair of roads and bridges, housing assistance and relocation services, and assistance with security and engineering.

Public: Local

Local government has the immediate responsibility for addressing concerns over a disaster. It is typically the local government that initially assesses damages, declares an emergency, requests assistance, and mobilizes resources. Each local jurisdiction within the United States has different strengths and weaknesses, priorities, and management structures; however, there are some similarities among locations. Elected executives, such as mayors and city managers, oversee the initial management of a disaster. Police, fire, and emergency medical services departments act in emergency management functions with the support of an emergency manager. There are also many unique private and non-profit local organizations, such as the American Red Cross, that play vital roles in disaster preparedness and response. Individuals acting as spontaneous volunteers, sometimes called to action by social media, may provide search and rescue, transportation, or distribution or relief supplies (Twigg and Mosel 2017)

How Has Research Influenced Disaster Preparedness?

Research has informed emergency preparedness and response on many levels. While traditional hazards research approaches focused more on physical sciences and engineering, more recent approaches—with input from fields such as sociology, planning, and public health, for example—have often taken on a more participatory approach to risk analysis and reduction. A key audience for this research is policymakers who decide how to structure preparedness and response to future disasters. This transition in focus has allowed for a more holistic approach that better prepares communities and governments to adequately react when the inevitable disaster strikes. Furthermore, this has had the effect of encouraging communities to participate in the planning process and legislative action (Mercer et al. 2008).

Specifically, public health professionals have become more integrated into preparedness and response following the September 11, 2001, terrorist attacks and the Anthrax attacks in October of that year. Their involvement has a potential impact on disaster preparedness and response through the introduction of new research methods and strategies such as participatory research, which may assist in creating a dialogue between communities, relevant stakeholders, and policy-makers. It also may play a role in creating disaster-ready communities that are well-informed, willing to perform grassroots campaigning, and ultimately can change the face of disaster research approaches (Mercer et al. 2008). The following will briefly explore recent research on risk communication, community service and evacuation planning, and emergency response in developing countries, as well as discuss how this research has impacted the way those involved in disaster preparedness and response approach their responsibilities.

Risk Communication

The field of risk communication is multi-faceted and draws upon many fields of study, such as social, cognitive, and economic psychology—as well as sociology, political science, anthropology, and communication research. Due to the fact that an individual's responses to dangerous situations can vary dramatically and oftentimes involve an array of emotional and cognitive responses, it is challenging to properly communicate during disasters. In addition, the risks themselves are fraught with uncertainty and complexity. Effective communication needs to juggle scientific knowledge, practical advice, and a sense of urgency, as well as promote logical responses from communities and individuals. There are, unfortunately, many examples where communication has failed. An example of this is global climate change, where risk communication must address an issue of global complexity with unknown uncertainties.

In the aftermath of disasters, it is common for researchers and government agencies to conduct After Action Reviews to discover what went well and what went poorly. After Hurricane Katrina, it was quickly noted that there was a failure to properly communicate with the community before and after the storm made landfall. Risk communication had failed to focus on the process and partners necessary to reach those who needed information to take action. For example, one study showed that predominantly low income and minority communities had stronger social and familial networks that both facilitated and hindered evacuation decisions. By putting more emphasis on channeling communications about risk and evacuation through these networks, official messages could have been more in sync with personal messages and potentially could have improved response. This is just one small example of the greater need to individualize messages to the target population and form relationships before a disaster strikes (Eisenman et al. 2007).

Situational, individual, and community level characteristics all play a role in successfully gaining compliance through the use of communication methods. Research has consistently shown that disaster warnings that are culturally and locally relevant increase the willingness of the community to comply. Fortunately, this aspect of disaster preparedness and response has demonstrated some success in the last decade, especially with cooperation from and coordination between responders, media outlets, and governing bodies (Glick 2007). However, for the best outcomes, it has been shown that each community should be individually addressed, which requires continuing research into best practices and approaches for risk communication in the future. Risk communication must consider the unique impacts on local populations, even when faced with challenges on a global scale.

Unmet Needs and Community Service Planning for Disasters

When communities work together to prepare and plan for disasters, it can positively impact how they respond and recover. Research into planning often suggests that disaster risk reduction can be accomplished by increasing community resilience through education, infrastructure improvement, and involvement. This approach can meet dual goals: addressing disparities in socio-demographic vulnerabilities to disaster and addressing the environmental challenges that can cause them. Monitoring and documenting the unmet needs of communities during non-disaster times can help prepare for the demand during disasters (Sansom et al. 2016).

In the aftermath of Hurricanes Katrina and Rita, the US Gulf Coast was faced with the obvious conclusion that existing systems had failed many of the populations impacted. In one study, the unmet needs of the population were analyzed by assessing real-time utilization of the 2-1-1 system for assistance. Based on 2-1-1 data, reporting of unmet needs for housing, food and water, transportation, and safety concerns surged during evacuations from Hurricanes Katrina and Rita. Not only did this research demonstrate what the population lacked, it also offered insight into future monitoring and surveillance programs that could utilize the 2-1-1 system to improve systems for better future response (Bame et al. 2012).

Shifts such as these—away from a top-down approach and toward one that focuses on the whole community—have several potential advantages. First, a focus on what the community can do for themselves may build resilience to future disasters and other crises. An essential role for disaster management is to coordinate services and resources. The "whole community" approach supports a focus on meeting the needs of vulnerable populations and potentially creates a long-lasting involvement in the community. For example, nonprofits and voluntary organizations are an invaluable part of this effort due to their ability to facilitate access to resources. Finally, it has been shown to be better at meeting the needs of populations than more traditional methods (Twigg 2009).

Evacuation Planning

The impact of hurricanes in the future will be influenced by many factors, including population growth and urbanization in coastal areas and global climate change. The US Atlantic and Gulf coasts, where hurricanes most often make landfall, are currently home to over 50 percent of the nation's population, and are projected to exceed 55 percent by 2015 (Adamo and de Sherbinin 2011). One way to reduce the impacts of hurricanes and other natural and technological disasters on human health is to successfully evacuate residents temporarily to a safer location. This requires improving our understanding of the hurricane evacuation decision-making process, particularly when large scale evacuations are called for.

People decide to evacuate—or not—from hurricanes based on a number of interrelated factors, including perception of risk, prior disaster experience and preparatory actions, sources of information, and household characteristics (Lindell and Perry 2012). Yet understanding the factors that influence the evacuation decision is only one part of the study of evacuation planning because many of the proposed variables are not modifiable through public policy intervention. A more contextual approach could be used by policy-makers and planners in the future to develop public health and safety interventions that facilitate evacuation among targeted sub-groups who are known to be either at heightened risk for hurricane-associated morbidity and mortality or who are thought to be less likely to evacuate.

For future research, there are several issues that need further consideration. These include the development of risk communication programs, improvements in evacuation modeling, and adoption of long-term hazard adjustments (Lindell and Perry 2012). More study of re-entry after evacuation is needed to understand resident expectations and concerns and to devise more effective messages. Additionally, comparative or longitudinal studies across multiple disaster events or geographic places are needed in order to understand trends and develop best practices. In the meantime, meta-analyses and other methods can be used to combine results from prior evacuation studies in new ways that may help practitioners and policy-makers draw additional conclusions from the published research (Huang et al. 2016).

Community Capacity and Rapid Response to Disasters: An International Perspective

Disasters know no boundaries. Over the last decade, China, the US, the Philippines, India, and Indonesia have been the nations most frequently impacted by natural disasters, and 70 percent of disaster mortality has been in low- and lower-middle income counties (CRED 2013). However, many US-based practitioners and students of emergency management and related fields are not aware of the different approaches to disaster preparedness and response in other countries. In many cases, a better understanding of global differences in resources, hazard vulnerability, infrastructure, and governance is needed in order to provide effective advice or assistance or to learn lessons from disasters affecting other countries. One way of facilitating this understanding is the study of disaster preparedness and response in the global context, as well as the sharing of information about approaches towards disaster preparedness and response in different settings.

Research conducted after the 2004 Indian Ocean tsunami highlights several examples of best practices in emergency response in developing counties that may be applicable in other settings impacted by major disasters. For example, international non-governmental organizations (NGOs) may play a large role in response, particularly in areas where the public sector is unable to prepare shelters for displaced residents and support response logistics. This post-disaster involvement of NGOs may open up new avenues for partnership with governments around social issues and other development issues that may have the synergistic effect of reducing vulnerability to future disasters (Arlikatti et al. 2012). In another example, new measures for adaptive capacity and collective efficacy—concepts studied almost exclusively with Western populations—were proposed to predict the capacity of members of a collective society to confront the consequences of the tsunami in Thailand (Paton et al. 2008). Findings from four case studies of international disasters will be presented and used to highlight potential translational research as well as gaps to be addressed in future research.

Conclusions

While only the concluding chapter of this section explicitly addresses the global nature of disasters, it is clear that globalization touches every aspect of emergency preparedness and response planning highlighted here. Global climate change raises challenges for the development of locally relevant risk communications messages in a setting of high uncertainty. The "whole community" approach stresses the unique ability of locally based nonprofits and other groups and agencies who routinely address unmet needs in their communities to coordinate services and resources for disaster preparedness and response. Finally, as disasters and emergencies—both natural and technological and including newly emerging and reemerging infectious diseases—are larger is size and scope, evacuees from one city, county, state, or nation may become members of new communities very rapidly. The linkages that allow us to benefit from the lessons of others also make us vulnerable to new threats, for which we much carefully prepare.

References

Adamo, S. B. and A. de Sherbinin. (2011) "The Impact of Climate Change on the Spatial Distribution of Populations and Migration," 161–195, in United Nations Department of Economic and Social Affairs, *Population Distribution, Urbanization, Internal Migration and Development: An International Perspective*, New York: United Nations.

Arlikatti, S., K. C. Bezboruah, and L. Long. (2012) "Role of Voluntary Sector Organizations in Post-Tsunami Relief: Compensatory or Complementary?" *Social Development Issues* 34(3): 64–80.

Bame, S. I., K. Parker, J. Y. Lee, A. Norman, D. Finley, A. Desai, A. Grover, C. Payne, A. Garza, A. Shaw, R. Bell-Shaw, T. Davis, E. Harrison, R. Dunn, P. Mhatre, F. Shaw, and C. Robinson. (2012) "Monitoring Unmet Needs Using 2-1-1 During Natural Disasters," *American Journal of Preventive Medicine* 43(6, supplement 5): S435–S442.

Brown, C. M. and C. Magary. (1998) *The Disaster Handbook*, Gainesville FL: University of Florida Press.

CRED—Center for Research on the Epidemiology of Disasters. (2013) *Annual Disaster Statistical Review 2012: The Numbers and Trends*, Louvain-la-Neuve, Belgium: Ciaco Imprimerie. Online version available at: http://reliefweb.int/report/world/annual-disaster-statistical-review-2012-numbers-and-trends. Accessed September 2, 2014.

Drabek, T. E. (1991) "The Evolution of Emergency Management," 3–29, in T. E. Drabek and G. J. Hoetmer (eds.), *Emergency Management: Principles and Practice for Local Government*, Washington DC: International City Managers Association.

Eisenman, D. C., K. M. Cordasco, S. Asch, J. Golden, and D. Glik. (2007) "Disaster Planning and Risk Communication with Vulnerable Communities: Lessons from Hurricane Katrina," *American Journal of Public Health* 97(S1): S109–S115.

Executive Order 12127, issued March 31, 1979 activated FEMA, effective April 1, 1979, and provided for the transfer of FIA, USFA and EBS functions.

Executive Order 12148, dated July 20, 1979 implemented the remaining transfer of functions to FEMA (DCPA, FPA and FDAA). (1979, 43239–43245)

Glick, D. (2007) "Risk Communication for Public Health Emergencies," *Annual Review Public Health* 28: 33–54.

Huang, S.-K., Lindell, M. K. and Prater, C. S. (2016) "Who Leaves and Who Stays? A Review and Statistical Meta-Analysis of Hurricane Evacuation Studies," *Environment and Behavior* 48(8): 991–1029.

Inam, A. (1999) "Institutions, Routines, and Crises: Post-Earthquake Housing Recovery in Mexico City and Los Angeles," *Cities* 16(6): 391–407.

Lindell, M. K. and R. W. Perry. (2012) "The Protective Action Decision Model: Theoretical Modifications and Additional Evidence," *Risk Analysis* 32(4): 616–632.

Mercer, J., I. Kelman, K. Lloyd, and S. Suchet-Pearson. (2008) "Reflections on Use of Participatory Research for Disaster Risk Reduction," *Area* 40(2): 172–183.

Miskel, J. (2008) *Disaster Response and Homeland Security: What Works, What Doesn't*, Stanford CA: Stanford University Press.

National Hurricane Center, Hurricane Preparedness Site, www.nhc.noaa.gov/outreach/history/#agnes. Accessed April 15, 2017.

Olshansky, R. B. and L. A. Johnson. (2010) *Clear as Mud: Planning for the Rebuilding of New Orleans*, Chicago: Planners Press.

Olshansky, R. B., L. Johnson, J. Horne, and B. Nee. (2008) "Planning for the Rebuilding of New Orleans," *Journal of the American Planning Association* 74(3): 273–287.

Paton, D., C. E. Gregg, B. F. Houghton, R. Lachman, J. Lachman, D. M. Johnston, and S. Wongbusarakum. (2008) "The Impact of the 2004 Tsunami on Coastal Thai Communities: Assessing Adaptive Capacity," *Disasters* 32(1): 106–119.

Platt, R. H. (1999) *Disasters and Democracy: The Politics of Extreme Natural Events*, Washington DC: Island Press.

Sansom, G., P. Berke, T. McDonald, E. Shipp, and J. A. Horney. (2016) "Confirming the Environmental Concerns of Community Members Utilizing Participatory-Based Research in the Houston Neighborhood of Manchester," *International Journal of Environmental Research and Public Health*, 13: 839. doi:10.3390/ijerph13090839.

Schwab, J., K. Topping, C. Eadie, R. Deyle, and R. Smith. (1998) "Planning for Post-Disaster Recovery and Reconstruction," Planning Advisory Service, Report No. 483/484, Chicago IL: American Planning Association.

Smith, G. P. (2012) *Planning for Post-Disaster Recovery: A Review of the U.S. Disaster Assistance Framework*, Washington DC: Island Press.

Twigg, J. (2009) *Characteristics of a Disaster-Resilient Community: A Guidance Note* (version 2), London: University of College, London.

Twigg, J. and I. Mosel. (2017) "Emergent Groups and Spontaneous Volunteers in Urban Disaster Response," *Environment and Urbanization*, 29(2): 443–458.

U.S. Code Cong. And Admin. Legis. Hist. for PL 81–875 (1950), 4024 as in R. H. Platt, *Disasters and Democracy: The Politics of Extreme Natural Disasters*, Washington DC: Island Press.

11

UNMET NEEDS AND COMMUNITY SERVICE PLANNING FOR DISASTERS

Sherry I. Bame and Sudha Arlikatti

Introduction

In the US, the notion that "all disasters are local" is widely recognized and emphasized. Established local government organizations such as fire, police, emergency medical services, and emergency management are the first to respond to routine incidents, as well as natural and human-induced disasters in a community. Also playing a vital role in providing community relief services are: a) private sector organizations such as utility companies, retail stores, media, banks, and hotels; b) nonprofit and volunteer organizations that are affiliated locally, nationally, or internationally to provide social support services such as Red Cross, United Way, Volunteers of America; and c) faith-based organizations providing resources and services through churches, temples, synagogues, and mosques (Arlikatti et al. 2012; Kapucu 2007; Waugh and Streib 2006). In the case of a devastating disaster where local community resources are inadequate to handle citizens' needs, the state and federal government entities step in through a more formalized process (Gajewski et al. 2011; Gazley 2013). This collaborative incident management system is articulated in the National Response Framework (NRF) document (U.S. DHS 2013).

The terrorist attacks of September 11, 2001 on U.S. soil brought to light the lack of coordination between multiple stakeholders (local, state, federal, and nonprofits) and the limited ability to track funding when responding to domestic threats (Government Accountability Office [GAO] 2002). Consequently, eleven days after the September 11, 2001 terrorist attacks in the U.S., the Department of Homeland Security (DHS) was established. Under the auspices of the DHS, the National Response Plan (NRP) was developed in 2004, "integrating all levels of government, the private sector, and non-governmental organizations into a common incident management framework" (U.S. DHS 2013: 3). In 2008, the NRP was replaced by the first National Response Framework (NRF) document to integrate lessons learned from responding to Hurricane Katrina and other major disasters. The NRF served as a comprehensive national guidance document that elaborated the roles, responsibilities and activities of partnering organizations involved in disaster and emergency response and short-term recovery.

In 2013, the NRF was revised (U.S. DHS 2013). This second edition of the NRF set the new requirements and terminology of the National Preparedness System mandated in Presidential Policy Directive (PPD) 8: National Preparedness (U.S. DHS 2013), aimed at strengthening the security and resilience of the United States. For the very first time it called attention to the importance of engaging the *"whole community"* to succeed in achieving resilience and national preparedness (FEMA 2011). *Whole community* is defined in the National Preparedness Goal as a focus on enabling the participation in national preparedness activities of a wider range of players from the private and nonprofit sectors, including nongovernmental organizations and the general public, in conjunction with the participation of federal, state, and local governmental partners in order to foster better coordination and working relationships.

(U.S. DHS 2013: 2)

A more generic understanding of "community" includes groups that share goals, values, and institutions, not always bound by geographic boundaries or political divisions. Instead, they may be faith-based organizations, neighborhood partnerships, advocacy groups, and social, volunteer, or professional associations. Whether clearing debris, providing shelter and food, or coordinating donations, community groups mobilize to reduce barriers in order to meet needs in their community, neighborhood, or group. Community groups may be *ad hoc* or based in existing community support networks or organizations. They are an essential component of successful disaster recovery and disaster planning for community resiliency.

Community-based Nonprofit Service Systems in the U.S.

The nonprofit sector plays a role in the provision of goods and services, and the legitimacy and distinctiveness of this sector can be aptly demonstrated by investigating its role in disaster response, relief and recovery. More often than not, voluntary agencies are the first to arrive and the last to leave the site of a disaster. Their ability to rapidly access and mobilize social capital is crucial for the success of the disaster recovery phase. Nonprofit organizations provide a wide variety of services, many of which are distinct from government and for-profit agencies, such as serving the marginal and vulnerable populations; focusing on capacity building; emphasizing local participation, leadership, and decision-making; and providing critical links between stakeholders (Arlikatti et al. 2012).

A perspective of community structure is that the nonprofit organizations provide social support and emergency services because the other sectors—namely, the government and the private for-profit sectors—do not provide them. Some scholars attribute these service gaps to "government failure" and "market failure" (Lohmann 1989; Powell and Steinberg 2006). Within democratic governance, the government failure occurs due to majoritarian constraint, size, scope and the median voter phenomenon (Weisbrod 1975). The majoritarian constraint, or the 51 percent rule for providing government services, is one of the preeminent reasons that the nonprofit sector is necessary. Without nonprofit involvement, the interests of the remaining 49 percent may not get addressed. In other words, the nonprofit sector addresses the interests of the minority. The government failure theory thus explains why nonprofit organizations provide public goods and services on a voluntary basis, even when government organizations have been identified to provide those services. It highlights the limitations that government agencies face in funding—or continuing to fund—programs after a period of time and how nonprofit organizations can fill the niche left unserved by government.

The theory of "market failure," also known as "contract failure" (Hansmann 1980, 1987), occurs because private companies prioritize services that are financially profitable. Therefore, service domains that offer low or negative profit margins will not likely be provided by the private sector. Nonprofits not only fill this gap, but also balance the scale regarding information asymmetries within the market. For example, nonprofits may educate marginalized low-income populations (e.g., single mothers) regarding funding options within the market (e.g., micro-financing opportunities). This is especially evidenced immediately after a disaster when local government agencies are scrambling to coordinate multiple efforts including the provisions of relief, shelter, and aid; avoiding epidemics; and ushering in normalcy. The complexity of the challenges associated with these responses make it necessary to involve the private for-profit and the voluntary sectors.

Both these sectors play important roles in disaster relief ranging from short-term relief provisions to long-term community development functions.

In the U.S., some nongovernmental organizations (NGOs) are officially designated as support elements to national response capabilities. These are the American Red Cross (ARC), the National Voluntary Organizations Active in Disaster (NVOAD), and Volunteers and Donations (FEMA 2012). The ARC has a unique relationship with the federal government. It is an independent entity that is organized and exists as a nonprofit, tax-exempt, charitable institution pursuant to a charter granted by the United States Congress. With the legal status of "a federal instrumentality," ARC carries out a number of responsibilities for the federal government, one of which is "to main-tain a system of domestic and international disaster relief, including mandated responsibilities under the National Response Framework coordinated by the Federal Emergency Management Agency (FEMA)" (ARC n.d.). Under the NRF and the Emergency Support Function 6 (ESF-6) the Red Cross and DHS/FEMA work together to coordinate federal mass care assistance for disaster victims (FEMA 2013). This includes sheltering, feeding operations, emergency first aid, bulk distribution of emergency items, and collecting and providing information on victims to family members.

The National Voluntary Organizations Active in Disaster (National VOAD) founded in 1970 has grown to consortium of approximately 100 organizations including 50 national organizations and territorial- and state-equivalent nonprofit, nonpartisan, membership-based voluntary organizations active in disaster management (FEMA 2012; NVOAD n.d.). VOADs serve as a forum for sharing knowledge, resources, and coordinating services throughout the disaster cycle to help survivors and their communities. The state/territory VOADs have a service-oriented mission that engages the whole community in building disaster resiliency. Voluntary organizations can range from large international charities such as Oxfam, CARE, and World Vision to national and smaller grass-roots or community-based voluntary groups such as United Way, and Salvation Army. When incident response operations exceed the resources of local government organizations, volunteers and donors support response efforts in many ways. For example, the Citizen Corps brings together local leaders from government and civic leaders from NGOs and the private sector to prepare for and respond to incidents.

Additionally, local Health and Human Service (HHS) social support programs extend exist-ing services to vulnerable groups such as special needs populations, undocumented residents, and ethnic and racial minorities. Regardless of voter support, fiscal priorities, or profit feasibil-ity, non-governmental HHS programs fill a gap in community support systems tailored to meet their communities' specific unmet needs. In collaboration with public services, HHSs, non-profit organizations, voluntary programs, and faith-based groups play a vital role in providing community support services during disaster response and recovery as well as preparation for resiliency.

Community Services and Access Barriers

Despite the guidance provided in the 2013 NRF document and the call for the *whole community* to participate in disaster preparedness and response efforts, a disaster does not impact everyone equally. Previous studies in the health care sector have regarded vulnerable populations as those at greater risk for poor health status and health care access. Vulnerable groups have typically been charac-terized by disease categories, age groups, and demographics (Aday 2001; Shi and Stevens 2005). Social scientists and disaster scholars posit that social vulnerability "derives from the activity and circumstances of everyday life or its transformers" (Hewitt 1997: 26). Regardless of how they are categorized, vulnerable populations generally include racial and ethnic minorities, low socioeco-nomic status populations, and those without adequate potential access to care (e.g., the uninsured or those without a regular source of care). Government and nonprofit sector initiatives to reduce access barriers created by social vulnerability recognize the common overlap of risk factors (e.g., low income, ethnic minority, female head of household) and continue to examine the combined influences of multiple risks on service gaps.

In disasters, low-income households, racial and ethnic minorities, and recent immigrants (Peacock et al. 1997); female-headed households (Enarson 1998; Enarson and Morrow 1998; Fothergill 1996); and people with disabilities and the elderly (McGuire et al. 2007) are affected to a higher degree. Highly vulnerable populations also have fewer resources to take protective measures such as purchasing insurance and living in safe, reinforced housing (Arlikatti and Andrew 2012). This segment of the population also suffers inequities in service availability even during normal times as they may be geographically segregated by race, age, and income; residing in cheaper and more hazardous areas (Phillips 1993); and served by fewer organizations due to their low tax base (Zakour and Gillespie 1998).

Access barriers are inevitable no matter how carefully support systems and preventive measures are planned and implemented for disaster management and community resiliency. These barriers interfere with individuals reaching the support services and resources they need, either delaying needs assessment or failing to mobilize help in a timely way. If those in need are fragile, these delays may be critical or lethal.

Access barriers may be systematic or random. Systematic barriers, whether intentional or not, consistently inhibit socially vulnerable subgroups from reaching support services they need—through discrimination by financial, age, racial, religious, or ethnic criteria. Organizations located in low-income neighborhoods may be expected to have fewer resources to exchange that would limit their links with more broad-based resource-rich organizations (Zakour and Gillespie 1998). This lack of linkage impedes the redistribution of goods and services among different geographic areas (Peacock et al. 1997). Systematic barriers may be reduced in contingency planning for disasters or addressed as discovered during the evolution of each unique disaster experience.

Random, haphazard barriers may be due to unplanned gaps in community services or to incident-specific conditions, such as a neighborhood blocked from rescue because a bridge was washed out or individuals were unable to reach distribution centers because of blocked roadways. Task forces and stakeholder groups may improvise decision-making to reduce or overcome these unexpected barriers. Given the uniqueness of each disaster's impacts and of each community's resiliency at a point in time, the nature of access barriers would vary as well as the community's ability to predict or overcome these barriers.

Extensive research examining access to health services has followed a model disaggregating access into multiple dimensions (Donabedian 1973; Penchansky and Thomas 1981; Wyszewianski and McLaughlin 2002). Based on this earlier work within health services research, four dimensions are presented here to examine the framework of access barriers to community health and human support services during disasters, easily remembered as "4 A's": (1) awareness, (2) availability, (3) affordability, and (4) acceptability. Applying this classification of access barrier dimensions can serve as a framework to better determine the types of access barriers occurring during disasters and to plan strategies for overcoming them so that all members of a community can use the social services and support systems provided for successful disaster response and recovery.

Awareness Barriers

The first stage of access to disaster support services is awareness or knowledge of appropriate services to meet the needs. Routine community support services may be known to subsets of the population who have needed these services in the past. However, eligibility criteria or types of services may have altered, especially during disasters. Residents suddenly homeless due to fire, flood, or evacuation may not know where to start looking for help or what kinds of help they may need tomorrow or next week.

Disaster survivors evacuating to another community may not be aware of services or resources in this new location to help meet their immediate needs as well as other short-term needs until they can return to their own communities to begin disaster recovery. Gajewski et al. (2011) found that the evacuation of Katrina survivors to cities across the country illustrated the weaknesses of devolution of government services to the local level and the increasing reliance on the non-profit sector. For example, the majority of the 800 Katrina households that went to Austin, Texas, found the city to be more expensive, less accessible, and culturally different from their home communities. "In addition, survivors experienced the disruption of all their social networks and previously learned strategies for surviving on a low income" (Gajewski et al. 2011: 392).

Another group often forgotten in this discourse are tourists that are inadvertently caught in an emergency or disaster. They are often unfamiliar with the threat, what protective actions to take, or which community resources may be available. The problem may be further exacerbated if they have difficulty seeking help due to language barriers (Burby and Wagner 1996). Visitors such as those attending conventions, festivals, sports events, concerts or other entertainment venues would need special directions for immediate help during a disaster until they can safely return home.

Overcoming awareness barriers involves a two-pronged approach to communicate the information about resources and services not only to the end-users (i.e., residents, evacuees, tourists, visitors) but also to the community stakeholders and support organizations. Planned multi-media communication "push" of information is developed and coordinated for rapid response. This would involve traditional media (i.e., newspapers, television, radio, telephone), as well as "social media" via smart phones, e-mail, and text alerts (Arlikatti et al. 2014). Awareness of these new channels of information about changes in services and resources is essential for emergency management and community support service organizations. Providers need updated information so they can refer requests for assistance—such as sheltering, emergency medical treatment, or other types of assistance—to the appropriate services or programs when encountering those in need.

Because support system services and resources in the U.S. are specialized and fragmented and the nature and types of disaster support services change so rapidly (Gajewski et al. 2011), the ability to refer to a clearinghouse of current information for all types of needs would be most effective and efficient. Smart phone applications and computer search engines help to locate local resources and programs; however, these are rarely updated as frequently as needed during disasters. Use of N-1-1 telephone numbers is developing into a national resource for such a clearinghouse of information and referral. Access to the 9-1-1 phone number throughout the U.S. provides this coordination and triage for emergency needs. In many communities in the U.S. and Canada, the 2-1-1 phone number can provide a clearinghouse of information for non-emergency needs regarding community services and resources that can be easily and frequently updated. Information about local governmental services may be available via 3-1-1, with this information often incorporated into the 2-1-1 referral database in many cities (Bame 2013; Arlikatti et al. 2010).

Availability Barriers

Individuals may be aware of what services and resources they need but those services may not be available locally. If routine community resources have been damaged or relocated during a disaster, then the community would need to be informed of current changes to the normal resources and alternatives to those resources. An example is distribution of food and water if grocery stores or food pantries are damaged or without electricity. Notifications of available Points of Distribution (PODs) for water and nonperishable foods are established by disaster management and volunteer programs throughout a community. However, if roads are blocked by fallen trees or electrical wires or high water levels, neighborhoods may not have access regardless of how near the POD is located. Updated information about alternative available resources would be needed.

Evacuees, tourists, and residents new to a community may expect certain types of services and resources that are not available locally. This would be expected if urban residents evacuated to rural communities or small towns were unable to find needed pharmaceuticals or medical equipment. Access to banking resources may not be available to non-residents of that community. Rural evacuation locations may encounter depleted fuel or infant supplies. As in the case of Hurricane Katrina evacuees, even moving from one urban area to another can prove challenging due to limited resources. Despite the enormous influx of funding following Hurricane Katrina, two basic needs remained unmet that could not be addressed adequately by any of the federal programs or NGO services: long-term affordable housing and transportation (Bame et al. 2012; Joh et al. 2015). For many Katrina survivors who had lived in tight-knit New Orleans communities, owning a car had been unnecessary. But when they were evacuated to larger cities such as Austin, Texas, and placed in vacant housing at the edge of town with limited access to public transportation, they were faced with numerous challenges in finding a job, going to a grocery, or accessing a health clinic (Gajewski et al. 2011).

Strategies to overcome availability barriers include rapid assessment of calls and requests for help to access existing services and resources. In addition, an information clearinghouse of reports by providers can update changes in status of service operations and availability constraints. Spatial analysis of blocked transit or depleted resources can be used to redirect deployment of resources and services. Temporary relocation may be established for distribution of services and resources, taking advantage of empty mall spaces or donated office or warehouse spaces. Increasingly, mobile venues (e.g., trucks, vans, mobile homes) are being developed to distribute services and resources to broad geographic areas affected by disasters. Electronic roadside signs are flexible sources of information and notification about changes for those passing. Media venues are needed for pushing information to the public and to providers about changes in resource availability. Telephone N-1-1 clearinghouse services are valuable both to enable providers to post changing availability and to those needing help to ascertain the current status of available resources and services.

Affordability Barriers

Although individuals may be aware of what types of resources they need and of services available, they may not be able to afford them. Purchasing shelter, food, fuel, and other support services in the private sector may exceed the financial capabilities of many households. Individuals may be turned away from available services because they do not meet eligibility criteria, such as age, demographic, disability, or nationality status. In spring 2007, the American Red Cross launched the *Means to Recovery* program to provide up to $10,000 toward client self-sufficiency as an additional resource for Hurricane Katrina survivors. Many survivors who had affordability barriers tried to purchase cars through this program, but few succeeded due to limited funds available and the bureaucracy of paperwork. Moreover, the elderly and people with disabilities who could not meet this program's guidelines of self-sufficiency were not even allowed to apply for these funds, compounding affordability barriers by age and disability status (Dewan and Strom 2007).

Other households normally able to afford support services may be without income for extended evacuation or recovery periods because their employment was interrupted by the disaster, hence limiting their ability to purchase resources and services. Interestingly, even those who could afford to purchase goods and services may be "cash poor" and not able to access credit because of power failures; thus, not able to pay for fuel or shelter or food because they did not carry enough cash with them.

Local public sector resources may not be sufficient for the demand of those who cannot afford commercial services. Since state and federal resources are often slower to mobilize, voluntary and faith-based organizations are relied on to fill the gaps. However, their resources and services are variable over time and between communities. Further, multiple disasters within a region may exhaust these sources of affordable or free support services. For example, hotels may fill up rapidly for evacuees who can afford those costs. Local shelter beds in schools and community centers also fill up quickly for the first-come-first-served surge of evacuees who could not access hotels or could not afford that option. If authorized by the Red Cross and disaster managers, churches, synagogues, and mosques open to shelter evacuees. When these have filled, local families can take evacuees into their homes to expand local sheltering capacity. If demand continued, evacuees would eventually be directed to other communities farther away from the disaster location, compounding evacuees' out-of-pocket transit costs as well as their stress and fatigue.

Strategies to reduce affordability barriers depend on the type of resources needed. For-profit stores donate vouchers for food, shelter, clothing, toiletries, infant care supplies and other goods. This is not only deductible for the stores' tax calculations but also provides positive advertising of their helpful role in the community. Other authorized commercial services supplement public sector resources and are eventually reimbursed from disaster management funds. Volunteer and faith-based organizations have been remarkably successful at collecting and distributing donations of goods, services, and funds to help disaster victims locally and at a distance. Thus, disaster planning strategies include formalizing commitments of local merchants as well as coordinating volunteer efforts to supplement the local public sector resources and social support services.

Acceptability Barriers

Support services and resources may be known, available, and affordable, but not acceptable. There may be language barriers with minimal or no access to interpreters that make support services difficult or unacceptable (Phillips and Morrow 2007). Another type of acceptability barrier is due to cultural or religious differences among providers and other clientele that may limit use by those with different characteristics or beliefs. Although their needs may be acute, some may not accept available and affordable services and resources. For example, some cultures and religions believe that nonrelated women and men should not be mixed in the same shelter. Volunteer and faith-based support services may be able to accommodate these acceptability differences. Alternatively, racial or ethnic biases and prejudices may keep others from using existing support services depending on the demographic mix of providers and clientele.

Language barriers may be anticipated by closely monitoring Census data and shifts in migration (both in and out) to be cognizant of the different languages spoken in a community. Subsequently, plans can be made to include various multi-lingual signage and computerized or telephone translation support. Various types of disabilities may also make access to services or resources unacceptable. Contingencies should be made to plan for blind, deaf, mental or neurological conditions, speech or motor disabilities in order to adapt services and resources for populations with special needs (Eldar 1992; McGuire et al. 2007; Vogt 1991). In addition, acceptable alternative shelter, food, transit, and support services would need to be established for those with acute and terminal illnesses and those with fragile health including the very young and very old. Accommodations for their caregivers would also have to be made. Acceptable modifications to shelters and food supplies would be needed for those with chronic diseases requiring careful hygiene, medication, or significant diet restrictions (e.g., those with asthma, diabetes, autoimmune diseases, special gastrointestinal or cardiovascular conditions). Disaster managers and stakeholders should determine policies *a priori* about separating resources to accommodate segregation or special needs.

Unmet Needs as a Measure of Access Barriers

Individuals encountering any type of access barrier to community health and human services experience unmet needs until they receive the required support services and resources. Moreover, their needs remain unmet if they receive inadequate or inappropriate support. Using the analogy of an iceberg, unmet needs remain below the surface and are difficult to measure. Once these needs surface and are addressed and served, they can be measured in several ways (e.g., counts and records of resources and services used, after-action debriefing and reports, financial records of claims and receipts for evacuation, response, and recovery). Determining the scope and nature of unmet needs is critical to identifying access barriers that limit community resiliency. But what about those needs that do not surface and receive resources that are counted and reported? Three approaches to measuring vulnerable groups with unmet needs are briefly summarized: risk analysis of "gap" subgroups, surveys of target disaster communities or evacuees, and analysis of N-1-1 tertiary datasets.

One strategy to measure unmet disaster needs has been to estimate gaps in a target population's use of disaster services compared to expected rates based on Census data within the disaster impact jurisdiction. The proportion of a subgroup below an expected rate would be assumed to be at risk for unmet needs. In order to calculate gaps from expected rates, available counts and socio-demographic data on those who used disaster services must be reliable and geographic identifiers must be valid. For example, number of shelter beds or number of meals provided could be counted and subtracted from the total population who evacuated, assuming that estimate was reliable and valid. Discrepancies in age, gender, and race/ethnicity percentages of those who used these services then may be compared to the evacuee baseline population's demographic proportions. The proportional differences, i.e., gaps in evacuation population actually served versus the total possible Census baseline data of evacuees, have been used to reflect those with unmet needs. However, the assumption that these non-users had unmet needs may not be valid. It would simply be unknown whether their needs remained unmet or if they purchased commercial resources or were supported by family and friends.

A second strategy to assess unmet needs and experiences with access barriers would be to collect primary data by surveying the target population in a disaster area or evacuation destination. This approach may obtain valuable insights to better understand disaster management of access barriers and effective planning to address unmet needs. However, sampling populations that have evacuated may be challenging or biased depending on the sampling protocol adopted, in turn, affecting the validity and utility of the findings. Reports must be explicit in discussing limitations in the scope and sampling of the target population. They must also account for response bias of the study population and interviewer bias of the researchers. Unknown recall biases may limit the validity and reliability of responses about disaster experiences and perspectives. Evacuees who sheltered in hotels or with friends or family may not be included, and hence, limit the ability to generalize the findings to all evacuees. Nevertheless, even with these constraints, survey findings can add important contributions to strategies to minimize access barriers and to enhance community resiliency.

A third approach is to analyze tertiary data from databases of telephone callers using three-digit N-1-1 numbers (i.e., 9-1-1, 2-1-1, and 3-1-1). These may serve as an inexpensive source of "real-time" data measuring unmet needs. If individuals already had access to services, they would not need to call the N-1-1 number(s). Thus, it is assumed that N-1-1 callers have unmet needs by encountering one or more of the access barrier dimensions. The caller data are collected in "real time," eliminating recall bias. The data are collected continuously enabling analyses over any period of time or aggregating into weekly or monthly measures. Unmet needs are documented by trained staff using either a standardized taxonomy or recorded verbatim or closely paraphrased, hence reducing interviewer bias and language barriers. All callers can be included, thus representing a broad geographic scope and potentially the total population within the N-1-1's jurisdiction. A major limitation, however, is the extent that the target population knows about the N-1-1 number. The 9-1-1 number seems universally known in the U.S., but 2-1-1 and 3-1-1 are evolving and have had limited aware-ness among the general population.

Statistical analysis of N-1-1 databases is relatively new, with challenges to merge multiple juris-dictions due to constraints and variations in caller-intake software and data collection and coding practices between call centers. Another limitation is that minimal socio-demographic information of callers is collected, which precludes determining how representative the caller population is com-pared to the general Census population. If the N-1-1 number is widely known, as with 9-1-1, then one could assume caller unmet needs are representative of those in the community. However, 2-1-1 and 3-1-1 call centers are still being developed and unfortunately are not consistently recognized or used throughout U.S. communities. It is simply not known how extensively different demographic segments in a community are familiar with 2-1-1 or 3-1-1 information and referral services. Hence, the representativeness of unmet needs of these callers remains unknown. Nevertheless, a description of unmet needs of callers who are aware of and use this resource does provide important real-time descriptions of access barriers over time and across locations in order to reflect patterns of unmet needs during all phases of a disaster as well as hot-spot high risk locations in broad-scale geographic coverage that includes evacuation routes and destinations as well as the breadth of a disaster area.

The remainder of this chapter describes findings from a nationwide case study of 2-1-1 programs throughout the U.S. to identify best practices in successfully managing a clearinghouse of informa-tion and referral for communities during a variety of disaster types in diverse regions of urban and rural communities. This study was funded by DHS's Division of Science and Technology, in col-laboration with 2-1-1 leadership in the U.S. (i.e., Alliance of Information and Referral Systems (AIRS n.d.), 211US, and United Way Worldwide). The findings from 2-1-1 data of unmet needs during Hurricanes Katrina and Rita over time and location are presented as a template of how N-1-1 data could be used in real-time to support efforts to identify and target vulnerable populations and evaluate interventions during disaster evacuation and recovery.

Case Study of 2-1-1 and Disaster Management Nationwide

The easy to remember 3-digit 2-1-1 telephone number for non-emergency information and refer-ral was approved by the U.S. Federal Communications Commission in 2000. The rapid growth of 2-1-1s since that time has helped to transform access to health and human services in communities throughout the U.S. and Canada. As of February 2014, 2-1-1s were available to almost 91 percent of the U.S., over 285 million people in all 50 states (211US 2014). In 2013, more than 60 percent of Canadians had access to 2-1-1 services. The most recent map of 211 coverage in the U.S. is shown in Figure 11.1. In 2013, except for four states with service coverage for less than 60 percent of their population and eight states with more than 60 percent but less than full coverage, the remaining states, as well as D.C. and Puerto Rico, had 2-1-1 services available to 100 percent of their population.

Beginning in 2001 in the aftermath of terrorist attacks on September 11th, 2-1-1s began to play key roles in disasters. The established Atlanta and Connecticut 2-1-1s fielded many thousands of calls from stranded travelers when airports grounded all flights. Other 2-1-1s throughout the U.S. received a high volume of calls both reaching out to offer help as well as seeking resources for concerns about recovery of victims' and responders' families. As more 2-1-1s developed since that time, many have collaborated closely with their regions' disaster management leaders to coordinate a clearinghouse of information and referral communication about non-emergency health and social services that could be updated rapidly as disasters evolve. These relations have proven so successful that 2-1-1s have been given a "seat" in many local and statewide emergency operations centers.

The 2-1-1 programs have maintained a "grass-roots" focus, with ongoing community relation-ships to maintain an up-to-date comprehensive database of referral information of health and human

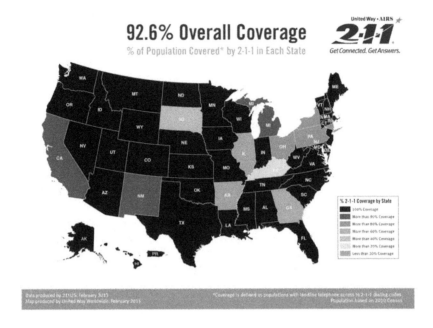

Figure 11.1   2-1-1 Service Coverage of U.S. Population by State, 2013

services agencies and programs in the non-profit and public sectors (211US 2014). Funding for their operating expenses is a mix of local and state non-profit and public sources that enable them to maintain services at no cost to the callers requesting help and no charges to the referral agencies and programs listed. Using their local familiarity of resources and ability to update information on an ongoing basis, 2-1-1s are well suited for social support during local disaster response and recovery.

Interestingly, 2-1-1's telephonic and software technology enables them to serve communities at a distance while maintaining local database information of resources and services. The Texas state-wide 2-1-1 network of 25 regional autonomous agencies switches into "roll-over" mode when a disaster is declared. If a local call is not answered by the nearest local 2-1-1 within a short time, the first available phone line in any other 2-1-1 across the state will answer seamlessly using that caller's local referral database. For example, a person in Houston calling for hurricane help may receive local Houston information from 2-1-1 staff located anywhere else in the state.

During Hurricane Gustav in 2008, the San Diego 2-1-1 developed a system that enabled them to help callers in Louisiana get services and resources in local communities or along evacuation routes using Louisiana's local resource databases. In other disasters when 2-1-1s themselves needed to evacuate, they could maintain service by working at a safe distance using "cloud technology" to access their localized referral database. In contrast, there are many communities who have not yet incorporated 2-1-1 into their disaster management plans, along with many living in disaster communities unfamiliar with 2-1-1 services (Gazley 2013). Thus, as 2-1-1s develop and expand, their roles during disasters and recovery evolve as well.

DHS Study of 2-1-1 Roles in Disasters

In 2010, DHS funded a study of 2-1-1s involvement in disasters (Bame 2013). Five regions in the U.S. were selected so that experiences during diverse types of disasters would be represented, building towards an "all hazards" model. National 2-1-1 leadership recommended 11 sites that

Table 11.1 Regional Distribution of DHS Project's Participating 2-1-1 Centers

U.S. REGION	2-1-1 CENTERS
East	New York City and New Jersey's statewide program
Southeast	Orlando and Tampa
Southwest	Houston, College Station, and Texas' statewide program
Midwest	Cedar Rapids plus eastern Iowa and Nebraska's statewide program
West	San Diego and Los Angeles

represented successful management during recent disasters as well as program diversity by region and agency size. The 2-1-1 call centers selected are listed in Table 11.1.

These study sites varied considerably in number of staff, size of jurisdiction area, and population served (Bame 2013; see Table 11.2). In 2010, New York City's 2-1-1 operated along with the city's 3-1-1. They employed 168 full-time equivalents (FTE) with jurisdiction over several boroughs that included approximately 9 million people. In contrast, Cedar Rapids' 2-1-1 employed only 1 FTE, contracting with other crisis center services for additional staff support. Cedar Rapids' service area of eastern Iowa included approximately 834,000 residents. The 2-1-1 service area and number of staff did not correlate with the proportion of population actually served in 2010. New York City's combined 2-1-1 and 3-1-1 averaged 2.5 calls per person within its jurisdiction, serving nearly 250 percent of its population. The Texas regional 2-1-1s each served approximately 13 percent of the residents within their jurisdictions; followed by Tampa's 2-1-1 handling calls from 11 percent of its population. The other study sites served less than 8 percent of their respective populations.

The number and types of disasters to which these study sites have responded are remarkable. Prior to 2012, these 2-1-1 sites had disaster management responsibilities during Hurricanes Charley, Dean, Dolly, Edouard, Emily, Katrina, Fay, Gustav, Humberto, Ike, Irene, Ivan, Rita, and Wilma. Other disasters they helped to manage included 9/11 terrorist attacks, blizzards, chemical spills, earthquakes,

Table 11.2 2-1-1 Study Site Variations by Size, 2010

2-1-1 Program	Total Calls 2010	Population in Jurisdiction	Calls per Population Served	Square Miles in Jurisdiction	Full-time Equivalent Staff
New Jersey	94,000	8,500,000	1.11%	8,722	25.0
NYC, NY	22,200,000	9,000,000	246.67%	305	168.0
Orlando, FL	196,151	2,465,278	7.96%	7,639	30.0
Tampa, FL	138,066	1,245,870	11.08%	1,072	20.8
Los Angeles, CA	508,427	10,000,000	5.08%	4,083	50.0
San Diego, CA	246,641	3,100,000	7.96%	4,200	64.0
TIRN – Texas	*	25,245,561	*	268,580	10.0
Houston, TX	802,599	6,000,000	13.38%	12,500	55.5
College Station, TX	40,780	301,358	13.53%	514	7 (+2 contract)
Cedar Rapids, IA	53,116	834,000	6.37%	*	1 (+7 contract)
Nebraska	65,347	1,976,690	3.31%	81,590	10.0

Note: * Missing

floods, firestorms, H1N1, hazardous materials incidents, ice storms, the "Miracle on the Hudson" air disaster, mudslides, power outages, sinkholes, tornadoes, tsunami, and wildfires.

Disaster planning and management practices were relatively similar across the study sites regardless of size, location, or types of disasters. The 2-1-1 sites reported frequent collaboration in planning for and responding to disasters with local and state Emergency Operations Centers (EOCs), local and statewide VOADs, and Red Cross, United Way, and other voluntary and faith-based organizations with designated key disaster management roles. Coordination with local and regional 9-1-1 and 3-1-1 services was common, both during disasters and routinely. These 2-1-1 personnel participated regularly in community and statewide disaster exercises, with 2-1-1 representatives playing key roles in regional and national disaster scenarios.

Analysis of 2-1-1 Data as a Case Study Template to Monitor Unmet Needs

The adoption of advances in telephone and software technology has varied widely among 2-1-1 call centers (Bame 2013). Their caller and referral database software structure and capability have differed greatly, with different sites using different software vendors. Each vendor has developed their own evaluation program and output. When database structure and output become consistent across 2-1-1s, an interface could be developed with disaster management software (e.g., WebEOC) in order for disaster responders and managers to monitor volume, types and locations of unmet needs in real-time from any location in the U.S. These analyses would help to determine timing and location of access barriers according to the nature of unmet need. With this real-time information, community support systems could mobilize resources to assist vulnerable populations in the "hot-spot," high-risk areas of communities.

The complex demands of multiple major disasters occurring simultaneously or in immediate succession may become less problematic if one could examine patterns of vulnerability for unmet needs as they unfold. To illustrate evaluations that could be used for real-time monitoring of unmet needs and access barriers, a database was created by merging all caller data from the 25, 2-1-1 call centers across Texas over a five-month study period. It encompassed a one-month baseline phase, the acute disaster phase comprising approximately three weeks post-Katrina and one week post-Rita (9/24/05), and early recovery that began in early October and was continued through the end of December.

Although there were hundreds of specific unmet needs coded for the 635,983 callers during the study period, two types of unmet disaster basic needs trumped all other needs. Of all callers, 28 percent had housing/shelter needs, and 4 percent had transportation/fuel-related unmet needs. It should be noted that these unmet needs were not mutually exclusive, with some families experiencing multiple kinds of unmet needs that compounded the severity of their condition. Each type of unmet basic need was analyzed over time and by location. Urban/rural biases were adjusted by controlling for population size, revealing the strain on resources in smaller communities compounded with additional evacuee needs, as well as the overwhelming volume of demand in large cities for help in disaster evacuation, response, and recovery.

Unmet Basic Disaster Needs over Time

The volume of 2-1-1 calls per day for each type of basic need was graphed over the five-month study period (Figure 11.2). The 2-1-1 personnel interpreted the obvious weekly patterns of low call volume during weekends and holidays to imply that callers simply wait until the support services are open and operating before seeking referral information. Calls peaking early in the week seemed to reflect pent-up demand from the weekend, decreasing as the week progressed. This weekly pattern

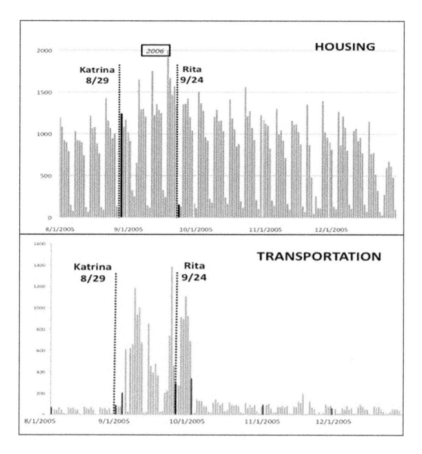

Figure 11.2   Daily Volume of Unmet Disaster Needs Over Time, Fall 2005

of call volume changed only during acute disaster phases and holidays, and was reported by all study sites nationwide regardless of their "24/7" hours of operation.

Patterns of the volume of 2-1-1 calls over time differed considerably by type of unmet need. Housing types of unmet disaster needs included temporary shelter, temporary housing, and housing repair and rehabilitation during recovery. Routine unmet housing needs were also present throughout the study period, including availability barriers of locator and realtor services, affordability barriers for rent, mortgage, utilities, and routine maintenance, and legal help needed for purchasing and for rental contract violations. Utility unmet needs (i.e., electricity, gas, water, telephone, and cable) included both disaster-related and routine problems.

Unmet housing needs increased slightly during the baseline phase, perhaps reflecting relocation needs for the school year. Following relocation of Hurricane Katrina evacuees into Texas, unmet needs for shelter and temporary housing increased greatly then surged upward for early evacuation from Hurricane Rita. High levels of unmet shelter needs continued immediately following Hurricane Rita, with slight decreases in volume of unmet needs for temporary housing or repair and rehabilitation of damages for over two months into recovery phase. Then the volume of these housing needs decreased to baseline levels and lower during the holiday season.

Unmet transportation needs were minimal until evacuees arrived in Texas after Hurricane Katrina. These evacuees needed transit to access other services and resources and to find temporary housing as the shelters began to close. Because mass transit is limited in Texas, evacuees needed to rely on disaster responders and volunteers to help navigate the maze of social services. Then transportation unmet needs soared when evacuating from Hurricane Rita. Personal transit was limited or failed on the congested highways. Fuel supplies were depleted as travelers stuck in traffic jams idled their engines to maintain their cars' air conditioning in the extreme heat and humidity. Transportation unmet needs remained high for approximately a week post-Rita until evacuees could return to their homes or find temporary housing. Then the pattern returned approximately to the low baseline levels.

Thus, routine transportation system support services and resources seemed to function well, with very minimal unmet needs until evacuees arrived without available transit resources. These evacuees' needs were compounded by needing to evacuate yet again from Hurricane Rita only three weeks later. Although the graphic portrait of unmet transportation needs was not surprising, it would have been helpful to have real-time monitoring of the patterns in order to address these problems more quickly. Following the after-action debriefing, Texas transportation emergency services revised their practices and resource allocation for successful transportation management during the five hurricanes affecting Texas in 2008.

Unmet Basic Disaster Needs by Location

Mapping the volume of 2-1-1 calls by type of need enables visualizing where and how much demand occurs for resources to meet unmet basic needs. The patterns shown in Figure 11.3 were unmet needs aggregated by county (N = 254 Texas counties) and ranked by volume over the five-month study period. The 2-1-1 data may also be displayed by city or zip code for more focused evaluations. The mapping may be done in real-time for ongoing portraits of where to move resources according to the volume of demand for vulnerable populations in order to overcome access barriers.

The patterns were similar for the volume of requests to meet housing and transportation needs as demonstrated in the grayscale map. The variability in need ranged from counties with the highest need colored "black" and those with no need colored "white." The volume of calls showed more widespread demand for housing than for transportation unmet needs. As seen in black, Texas' major metropolitan areas (Houston, Dallas, Ft. Worth, Austin, San Antonio) had the highest demand for each type of unmet need. Other urban areas that had also served as evacuation destinations had higher volume of unmet needs as well. As expected, demand for unmet needs was high where Hurricane Rita made landfall. The counties in 'gray' near the coast also contained major evacuation routes. Although these counties had less volume of demand, they also had less population. Counties farther

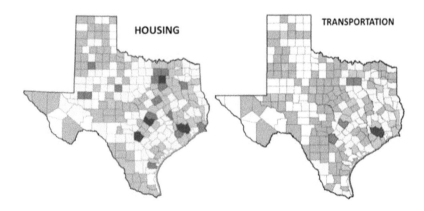

Figure 11.3   Volume of Unmet Disaster Needs by Location, Fall 2005

inland and seen in 'medium gray' had moderate demand for unmet needs, typically reflecting their routine volume of 2-1-1 calls. Counties in 'light gray' had low volume of calls among their sparse population, with no calls for that specific type of need in counties portrayed in white.

Thus, the volume of unmet needs followed expected patterns where greater demand corresponded to greater population in evacuation destinations, and with moderately high rates along evacuation routes. It was surprising to note higher call volume for unmet needs in the distant west-Texas communities where Katrina evacuees were bussed or flown when Hurricane Rita threatened (i.e., Abilene, Midland-Odessa, Lubbock, Amarillo, and El Paso). Without adequate disaster support resources, the needs of these remote evacuees exacerbated access barriers for residents seeking routine housing and transportation services as well. Less populated rural counties, mostly in west Texas, had few calls for help with unmet basic disaster needs.

"Hot Spots" At-Risk for Unmet Disaster Needs

Although the number of 2-1-1 calls about unmet needs is helpful to evaluate volume and location of resources needed, that mapping would not help disaster managers prioritize the degree of risk by location. To control for strong urban/rural bias, the number of calls per county was adjusted by population size measured as the number of households per county (Figure 11.4). These rates were calculated per 100 households then ranked according to a 1 percent expected average rate of 2-1-1 calls per specific type of unmet need. The counties colored black were "hot spots" of highest risk, reflecting a call rate more than ten times greater than would be expected for that type of unmet need, adjusting for population size. Counties colored in gray were between 5–10 times greater than expected. Counties in medium gray were within expected ranges, whereas counties colored light gray were much lower than expected call rates.

The patterns of counties "at-risk" differed remarkably according to type of unmet need after adjusting for urban/rural bias. There were few locations during the study period with slightly greater than expected levels at-risk for transportation unmet needs (colored medium gray). Although significant transportation problems were encountered during evacuation from Hurricane Rita, these were time-limited, with the level of unmet transportation needs dropping considerably and remaining low after the acute disaster phase (see Figure 11.2). Locations with mild risk were jurisdictions in the path of Hurricane Rita; Houston and Austin evacuation destinations; and scattered rural counties with routine types of transportation unmet needs. The remainder of the state had fewer than expected (colored light gray) or no reported (colored white) unmet transportation needs during the study period. Thus, it was likely that disaster management as well as local and highway safety services and

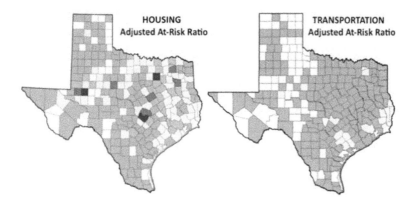

Figure 11.4 "Hot Spots" At-Risk for Unmet Disaster Needs, Adjusting for Population Size, Fall 2005

resources kept transportation access barriers to a minimum and responded to vulnerable population transit needs in a timely way. Depleted fuel supplies were restocked and personal transit capabilities were restored rapidly.

In contrast, "hot-spots" (colored black) for unmet housing needs included some of the urban evacuation destinations, especially those at considerable distance from the Gulf Coast. It was reported that, although evacuees had been sent to these inland locations, disaster housing support services and resources had not mobilized within these cities in a timely way or adequate amount. Hence, unmet shelter and housing needs for evacuees compounded local access barriers to routine housing services and resources. Slightly less at-risk "warm-spots" (colored gray) were identified in rural counties along evacuation routes toward those cities at highest risk. On one hand, perhaps this reflected geographic overflow of evacuees' demand for shelter and temporary housing. On the other hand, these surrounding communities may have experienced considerable competition by evacuees for local housing resources and support services, hence raising access barriers and increasing risk of their community's routine unmet housing needs.

Mapping and graphing unmet needs over time and location during Hurricanes Katrina and Rita illustrate a case study template in using tertiary N-1-1 caller data to monitor population needs and to identify where and when subgroups are vulnerable to access barriers. An emphasis is on supporting "grass-roots" resources used for meeting routine community needs, with the ability to target "hot-spots" where additional services and resources can be mobilized for disaster support. The types of access barriers may be identified and strategies devised to increase a community's resiliency to minimize damage and more efficiently and effectively manage disaster recovery.

Summary and Conclusions

The emphasis on *whole community* involvement to manage disasters and plan for more resilient communities is becoming more pervasive in the U.S. This approach encompasses the perspective that "all disasters are local," encouraging support of local and regional disaster support services as much as possible and supplementing with outside resources as needed. An essential role for disaster management is to coordinate services and resources from the commercial sector, nonprofit and volunteer organizations, and public agencies, not only to serve residents, but also evacuees, visitors, and tourists. Nonprofits and voluntary organizations seem invaluable in communities because of their flexibility to accommodate to the dynamic disaster situations as well as to fill gaps in the market and public sectors. However, a key element needed in communities is a clearinghouse for information to facilitate access to the maze of resources as well as to enable timely posting of changes in availability. The N-1-1 phone numbers are configured for this, with 9-1-1 used to access emergency services, 3-1-1 for public services, and 2-1-1 for nonprofit, voluntary, as well as public services. The 9-1-1 services are well known and available throughout the U.S. The 2-1-1 (and some 3-1-1) services are evolving into an important partner in disaster planning, management, and recovery, particularly in communities that have experienced large-scale disasters.

Access barriers are inevitable during disasters regardless of careful and coordinated planning or preparation. The N-1-1 telephone clearinghouse database may also serve as a means to monitor real-time problems encountered in access to disaster services and resources. The types of access barriers may be identified for focused strategies to reduce risk to vulnerable subgroups. Awareness barriers may be addressed with a "push" of information via traditional and social media. Availability barriers may be diagnosed using the N-1-1 caller database and updated clearinghouse of resources and services. Affordability barriers require mobilizing resources from public, market, and nonprofit sectors as well as requesting state- and federal-level support beyond what the local community resources can manage. Acceptability barriers may be less obvious, but nevertheless, endanger subgroups of the population.

Whether language, race, ethnicity, or religious beliefs inhibit access to disaster support, local policies need to determine if and how disaster services will accommodate these different perspectives.

DHS's study of 2-1-1's involvement in disaster management identified key roles played by 2-1-1 call centers throughout the U.S. Although 2-1-1s varied considerably in size (number of staff, geographic jurisdiction, and population served), they performed successfully as an information clearinghouse throughout a diverse set of natural and human-made disasters. They maintain a "grass-roots" focus as they play central roles in their community VOADs along with collaborating with emergency responders, 9-1-1s, local and state EOCs, and regional FEMA programs.

A template for analyzing 2-1-1 data was developed to determine unmet community needs measured in real-time as a disaster unfolds. Merging databases from the 25 regional 2-1-1s in Texas to study unmet disaster basic needs (e.g., housing/shelter and transportation/fuel) during and after Hurricanes Katrina and Rita revealed how these needs changed over time and varied by location. In the aftermath, local and state disaster managers confirmed the timing and locations of surges and "hot-spots" in unmet housing and transportation needs. They recommended developing this capability to monitor real-time patterns of vulnerable populations' unmet needs capable of being displayed along with emergency responder information (e.g., WebEOC). It remains to be seen what resources may be allocated for this and what developments 2-1-1 leadership promotes and implements. Following the disaster response phase, services and resources can be mobilized for recovery not only in disaster sites but also in communities where evacuees have depleted resources for meeting routine community needs.

Contextual factors that affect use and quality of social support services vary greatly from community to community. Although FEMA's NRF provides standard guidelines for organizations and programs to coordinate for community resiliency during disasters, each community needs to assess its strengths and gaps in services and resources to cope with expected threats and response scenarios. Regardless, unexpected risks will occur and access barriers will threaten vulnerable populations. Integration of communication among disaster managers and community stakeholders about social support services and resources will enable greater flexibility to meet community needs, identify types and clusters of unmet needs, and reduce access barriers for more equitable disaster resiliency.

References

211US. (2014) Nationwide Status, retrieved on December 15, 2017 from www.211us.org/status.htm.

Aday, L. (2001) *At Risk in America: The Health and Health Care Needs of Vulnerable Populations in the United States*, 2nd edition, San Francisco CA: Jossey-Bass.

Alliance of Information and Referral Systems (AIRS). (n.d.). "Who we are." Retrieved on June 18, 2019 from https://www.airs.org/i4a/pages/index.cfm?pageid=1.

Arlikatti, S. and S. Andrew. (2012) "Housing Design and Long-term Recovery Processes in the Aftermath of the 2004 Indian Ocean Tsunami," *Natural Hazards Review* 13(1): 34–44.

Arlikatti, S., K. Bezboruah, and L. Long. (2012) "Role of Voluntary Sector Organizations in Posttsunami Relief: Compensatory or Complementary?" *Journal of Social Development Issues* 34(3): 64–80.

Arlikatti, S., A. M. Pulido, H. Slater, and A. Kwarteng. (2010) "Role of Spanish Language Media in Promoting Disaster Resiliency: Public Service Announcements and 2-1-1 Program," *The Journal of Spanish Language Media* 3: 78–89.

Arlikatti, S., H. Taibah, and S. A. Andrew. (2014) "How Do You Warn Them If They Speak Only Spanish? Challenges for Organizations in Communicating Risk to Colonias Residents in Texas, USA," *Disaster Prevention and Management* 23(5): 533–550.

American Red Cross (ARC). (n.d.) "Our Federal Charter," retrieved on August 12, 2014 from www.redcross.org/about-us/history/federal-charter.

Bame, S. (2013) *2-1-1 Disaster Data Management System*, Washington DC: Department of Homeland Security, Science and Technology Directorate, Resilient Systems Division.

Bame, S., K. Parker, J. Y. Lee, et al. (2012) "Monitoring Unmet Needs: Using 2-1-1 during Natural Disasters," *American Journal of Preventive Medicine* 43(655): S435–442.

Burby, R. J. and F. Wagner. (1996) "Protecting Tourists from Death and Injury in Coastal Storms," *Disasters* 20: 49–60. doi: 10.1111/j.1467-7717.1996.tb00514.x.

Dewan, S. and S. Strom. (August 10, 2007) "Red Cross Faces Criticism over Aid Program for Hurricane Victims," *New York Times*. Retrieved August 7, 2014 from www.nytimes.com/2007/08/10/us/10redcross. html?pagewanted=all&module=Search&mabReward=relbias%3Ar&_r=0.

Donabedian, A. (1973) *Aspects of Medical Care Administration: Specifying Requirements for Health Care*, Cambridge MA: Harvard University Press.

Eldar, R. (1992) "The Needs of Elderly Persons in Natural Disasters: Observations and Recommendations," *Disasters* 16(4): 355–357.

Enarson, E. (1998) "Through Women's Eyes: A Gendered Research Agenda for Disaster Social Science," *Disasters* 22(2): 157–173.

Enarson E. and B. Morrow. (1998) *The Gendered Terrain of Disaster*, New York: Praeger.

Federal Emergency Management Agency (FEMA). (2011) A Whole Community Approach to Emergency Management: Principles, Themes and Pathways for Action. Retrieved on August 12, 2014 from www.fema. gov/media-library-data/20130726-1813-25045-0649/whole_community_dec2011__2_.pdf.

Federal Emergency Management Agency (FEMA). (2012) Voluntary Organizations Active in Disaster. Retrieved on August 12, 2012 from www.ready.gov/voluntary-organizations-active-disaster.

Federal Emergency Management Agency (FEMA). (2013) Emergency Support Function #6 – Mass Care, Emergency Assistance, Temporary Housing, and Human Services. Retrieved on August 12, 2014 from www.fema.gov/media-library-data/20130726-1913-25045-4194/final_esf_6_mass_care_20130501.pdf.

Fothergill, A. (1996) "Gender, Risk and Disaster," *International Journal of Mass Emergencies and Disasters* 14(1): 33–56.

Gajewski, S., H. Bell, L. Lein, and R. J. Angel. (2011) "Complexity and Instability: The Response of Nongovernmental Organizations to the Recovery of Hurricane Katrina Survivors in a Host Community," *Nonprofit and Voluntary Sector Quarterly* 40(2): 389–403.

Gazley, B. (2013) "Building Collaborative Capacity for Disaster Resiliency," 84–98, in N. Kapucu, C. V. Hawkins, and F. I. Rivera (eds.), *Disaster Resiliency: Interdisciplinary Perspectives*, New York: Routledge.

Government Accountability Office (GAO). (2002) "September 11: Interim Report on the Response of Charities." Retrieved on June 15, 2014, from www.gao.gov/new.items/d021037.pdf.

Hansmann, H. (1980) "The Role of Nonprofit Enterprise," *Yale Law Journal* 89(5): 835–901.

Hansmann, H. (1987) "Economic Theories of Nonprofit Organizations," 27–42, in W. W. Powell (Ed.) *The Nonprofit Sector: A Research Handbook*, New Haven CT: Yale University.

Hewitt, K. (1997) *Regions of Risk: A Geographical Introduction to Disasters*, Harlow: Longman.

Joh, K., A. Norman, S. I. Bame. (2015) "A Spatial and Longitudinal Analysis of Unmet Transportation Needs During Hurricanes Katrina and Rita," *Journal of Homeland Security & Emergency Management* 12(2): 387–406.

Kapucu, N. (2007) "Nonprofit Response to Catastrophic Disasters," *Disaster Prevention and Management: An International Journal* 16(4): 551–561.

Lohmann, R. A. (1989) "And Lettuce Is Nonanimal: Towards a Positive Economics of Voluntary Action," *Nonprofit and Voluntary Sector Quarterly* 18(4): 367–383.

McGuire, L. C., E. S. Ford, and C. A. Okoro. (2007) "Natural Disasters and Older US Adults with Disabilities: Implications for Evacuation." *Disasters* 31(1): 49–56.

National Voluntary Organizations Active in Disaster (NVOAD). (n.d.) "National Organization Members." Retrieved on November 10, 2017 from www.nvoad.org/voad-members/national-members/.

Peacock, W. G., B. H. Morrow, and H. Gladwin (eds.). (1997) *Hurricane Andrew: Ethnicity, Gender and the Sociology of Disaster*, New York: Routledge.

Penchansky, R. and J. W. Thomas. (1981) "The Concept of Access: Definition and Relationship to Consumer Satisfaction," *Medical Care* 19(2): 127–140.

Phillips, B. D. (1993) "Cultural Diversity in Disasters: Sheltering, Housing and Long-term Recovery," *International Journal of Mass Emergencies and Disasters* 11(1): 99–110.

Phillips, B. D. and B. H. Morrow. (2007) "Social Science Research Needs: Focus on Vulnerable Populations, Forecasting, and Warnings," *Natural Hazards Review* 8(3): 61–68.

Powell, W. W. and R. Steinberg (eds.). (2006) *The Nonprofit Sector: A Research Handbook*, 2nd ed., London: Yale University Press.

Shi, L. and G. D. Stevens. (2005) "Vulnerability and Unmet Health Care Needs," *Journal of General Internal Medicine* 20(2): 148–154.

U.S. Department of Homeland Security (DHS). (2011) Presidential Policy Directive / PPD-8: National Preparedness. Retrieved on August 12, 2014 from www.dhs.gov/presidential-policy-directive-8-national-preparedness.

U.S. Department of Homeland Security (DHS). (2013) *National Response Framework*, 2nd Edition. Accessed on August 4, 2014 from https://s3-us-gov-west-1.amazonaws.com/dam-production/uploads/20130726-1914-25045-1246/final_national_response_framework_20130501.pdf.

Vogt, B. M. (1991) "Issues in Nursing Home Evacuations," *International Journal of Mass Emergencies and Disasters* 9(2): 247–265.

Waugh, W. L. and G. Streib. (2006) "Collaboration and Leadership for Effective Emergency Management," *Public Administration Review* 66(s1): 131–140.

Weisbrod, B. A. (1975) "Towards a Theory of the Voluntary Nonprofit Sector in a Three Sector Economy," 171–195, in E. Phelps (ed.), *Altruism, Morality and Economic Theory*, New York: Russell Sage Foundation.

Wyszewianski, L. and C. McLaughlin. (2002) "Access to Care: Remembering Old Lessons," *Health Services Research* 37(6): 1441–1443.

Zakour, M. J. and D. F. Gillespie. (1998) "Effects of Organizational Type and Localism on Volunteerism and Resource Sharing during Disasters," *Nonprofit and Voluntary Sector Quarterly* 27(1): 49–65.

12

EVACUATION PLANNING

Hao-Che Wu, Shih-Kai Huang, Michael K. Lindell

Introduction

Evacuation, which can be defined as people's temporary movement away from a hazardous location to a safer place, has long been recognized as one of the most effective approaches to protect people against a wide range of natural or technological hazards. Sometimes evacuation can be improvised in small-scale emergencies such as building fires or hazardous materials spills involving a few city blocks. However, large-scale evacuations of thousands to millions of people from areas that are tens to thousands of square miles in size must be planned rather than improvised. The large scope of impact makes pedestrian evacuation infeasible and therefore necessitates vehicular evacuation. It also frequently implies a large population at risk and thus the evacuation of many vehicles. Finally, this large traffic demand usually exceeds the capacity of the evacuation route system (ERS), which has only been designed to meet normal traffic conditions. In response to these challenges, the following sections will describe the planning process for mass evacuations when facing threats from large-scale environmental hazards.

The evacuation plan is usually a significant part of a community's emergency operations plan (EOP), Perry and Lindell (2007). As will be discussed in the following sections, a pre-impact evacuation plan typically presumes that there is a forecast system that can detect the hazard and disseminate warnings before impact and that the impact area can be identified with enough precision to let people know which areas are dangerous and which are safe. Indeed, the National Weather Service recommends that emergency managers avoid considering evacuation as a protective action recommendation (PAR) for tornadoes.

An evacuation plan should be based upon accurate assumptions about the forecast/warning system for a given hazard as well as an understanding of how people respond in disasters. In addition, the evacuation plan should address three issues. First, it should be based upon a hazard/vulnerability analysis that clearly identifies the geographical areas that might be affected by environmental hazards. Second, it should be based upon an evacuation analysis that identifies the ERS, estimates evacuation demand (the number of vehicles evacuating from different locations over time), and identifies evacuees' destinations and the routes they are likely to take. Third, data from the evacuation analysis should be used as input for a model of traffic behavior. Moreover, the evacuation plan should indicate what traffic management strategies will be used to guide evacuees as effectively and efficiently as possible. Last, the evacuation plan should describe the procedures for determining when to begin and how to facilitate evacuees' re-entry into the evacuation zone after it has been determined to be safe to reoccupy.

Forecasting and Warning

Successful implementation of pre-impact evacuations requires relatively accurate detection and forecasting techniques. That is, detection authorities must detect signs of an imminent disaster; forecast its probability of occurrence, magnitude of impact, and time of arrival; and notify local emergency management authorities so they can transmit warnings to those in the risk area. Among the large-scale geophysical hazards, there is no capability for forecasting earthquakes with enough forewarning to support large-scale evacuations but volcanic eruptions and tsunamis can sometimes be forecast in time for people to evacuate. Among the large-scale meteorological hazards, there is usually adequate forecast technology for floods (although not necessarily for flash floods), wildfires, and hurricanes. Finally, evacuations for hazardous materials releases during transportation incidents must often be improvised but there is often adequate forecast technology for releases at fixed site facilities such as nuclear power plants and toxic chemical facilities.

Among all hazards, hurricanes have perhaps the best forecast and warning technology and thus the most extensively studied evacuation processes. Tropical cyclones (hurricanes and typhoons) have relatively long forewarning because weather services agencies are able to monitor their formation and passage over the open ocean many days before landfall. The U.S. National Hurricane Center (NHC) and local Weather Forecast Offices (WFOs) issue hurricane forecast advisories as well as watch and warning messages to the general public in addition to providing other information directly to state and local emergency managers. The forecast advisories describe hurricane features such as forecast track, forward movement speed, intensity (wind speed), and hurricane wind radius (see the archive of past hurricane forecast advisories at www.nhc.noaa.gov/archive/). In addition, the NHC provides an uncertainty cone, which is a graphical display that shows the hurricane's most probable deviation from its forecast track. The NHC also provides data that is used to generate graphic displays in HURREVAC (www.hurrevac.com/), a storm tracking and evacuation decision support tool.

There has been little research on local officials' ability to process different types of hurricane information displays. One experiment showed that the participants understood that hurricane tracks are quite uncertain and that hurricanes might even reverse their directions (Wu et al. 2014). Surprisingly, there was no difference in strike probability judgments for participants who were shown a forecast track only, an uncertainty cone only, or a forecast track and an uncertainty cone. Another experiment that tested people's search for different types of hurricane information during a simulated six-day hurricane scenario found that the participants mostly searched for hurricane forecast track (graphic), uncertainty cone (graphic), and intensity (numeric) information (Wu et al. 2015a) and used that information to make judgments about strike probabilities and PARs (Wu et al. 2015b). In addition, participants' information search patterns were more stable on the fourth hurricane scenario than on the first one (Wu et al. 2015a). This finding is important because they had already read a training manual, the *Local Official's Guide To Making Hurricane Evacuation Decisions* (Lindell and Prater 2009), that is a refinement of the one that the state of Texas provides to local officials. Thus, the changes in the participants' search patterns from the first to the fourth hurricane indicate that presentation of abstract principles in the *Official's Guide* was not sufficient for them to learn how to track hurricanes effectively. However, they were able to significantly improve their search efficiency with a modest amount (roughly an hour) of practice with an actual tracking task. Although these results are probably not typical of what could be expected from professional emergency managers, they are likely to generalize to local elected officials—few of whom have received training in hurricane evacuation decision making. Thus, the results of these experiments need to be replicated in future studies with other samples, but experiments testing responses to more complex hurricanes, such as those with gradually curving tracks or radical 90 degree changes in direction, are likely to increase the need to supplement the *Official's Guide* with practice on tracking tasks.

Unlike the limited amount of research on the protective action decision making by local officials, there is a substantial literature on households—especially on hurricane evacuation decisions. There are two major theoretical models that have been used to summarize existing research and guide new research on people's disaster responses (Tierney et al. 2001; Trainor et al. 2013). Mileti and Sorensen (1990) proposed the *Warning Response Model* to describe how a recipient responds as: (1) hearing the warning message, (2) understanding its contents, (3) believing the warning, (4) confirming the information, (5) personalizing the threat, and (6) responding with a protective action. Although these stages describe people's behavior in general, some people might skip one or more steps and proceed directly to the final response stage.

Lindell and Perry (1992, 2004, 2012) proposed the *Protective Action Decision Model* (PADM, see Figure 12-1) as an extension of their earlier evacuation research (Houts et al. 1984; Perry et al. 1981). The PADM notes the importance of social warnings that originate from one or more original information sources and pass as messages through different channels to intermediaries and ultimate receivers (Lindell et al. 2007a). However, it also notes that people receive information about environmental threats by observing social and environmental cues. Similar to Mileti and Sorensen's (1990) Warning Response Model, the PADM posits a pre-decision process in which those at risk (or who think they are at risk) are exposed to, heed, and interpret the available information, but they use this information to form three core perceptions: perceptions of the threat, perceptions of alternative protective actions, and perceptions of different social stakeholders (especially information sources). The three core perceptions determine whether people seek additional information or take protective response actions, both of which are promoted by situational facilitators and inhibited by situational impediments. If people are uncertain about the need for, effectiveness of, or feasibility of their chosen course of action (either information-seeking or protection), they return to the information sources at the beginning of the model to obtain additional information.

The PADM has been used to guide post-storm surveys that collected data on risk area residents' recollections of their perceptions, decisions, and the actions they took in Hurricanes Lili, Katrina, Rita, and Ike (Huang et al. 2016; Huang et al. 2012; Lindell et al. 2005; Lindell et al. 2011; Wei et al. 2014; Wu et al. 2012; Wu et al. 2013). In addition, the PADM has been used to study response to the 2009 tsunami in American Samoa (Lindell et al 2015; Lindell et al. 2013), tsunami threat from the 2011 Christchurch and Tohoku earthquakes (Wei et al., 2017), and the 2010 Boston water contamination incident (Lindell et al., 2017).

Patterned Human Behaviors in Evacuation Expectations

In addition to relying on the results of research on past evacuations, emergency managers should conduct surveys that use hypothetical scenarios to examine residents' evacuation expectations in order to support their evacuation analyses and evacuation plans. These evacuation expectations questionnaires almost invariably ask at what storm category the respondent would evacuate, but they also include questions about the number of vehicles people plan to take, their expected evacuation routes and destinations, and their level of emergency preparedness (e.g., Lindell et al. 2001, 2013). Some evacuation planners have placed these questionnaires as inserts in local papers and asked people to return them, but this procedure is extremely unlikely to produce accurate results because it will not produce the representative sample of the risk area population that is needed.

Some analysts have expressed concern that, even if evacuation planners use appropriate sampling procedures, there will still be some differences between local residents' behavioral expectations (what they think they will do in an evacuation) and protective action adoption (what they actually do in emergencies). That is, the question is sometimes raised how people could know how they will respond to an evacuation order if they have never experienced one before. Although more research on this issue is needed, the available data indicate that people can provide quite accurate answers to

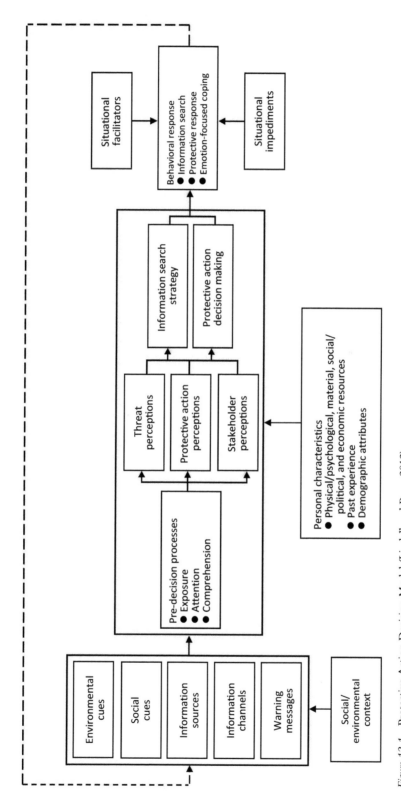

Figure 12.1 Protective Action Decision Model (Lindell and Perry 2012)

some questions and less accurate answers about others (Kang et al. 2007). For example, Kang and her colleagues found that coastal residents had accurate insight into the conditions under which they would decide to evacuate and many of the actions they would take while implementing that decision. They also had accurate expectations about their information sources, evacuation transportation modes, number of vehicles, evacuation accommodation types. However, expectations were less accurate regarding the number of trailers taken and evacuation destinations—the latter perhaps because the places they had hoped to stay were taken by others before they arrived.

The conclusions from Kang et al. (2007) are consistent with results from a recent statistical meta analysis of 38 studies on people's hurricane evacuation behavior (Huang et al. 2016). This comprehensive analysis found non-significant differences between actual and hypothetical evacuation studies in the effect sizes for 11 out of 17 demographic (age, female gender, white ethnicity, household size, education, income, homeownership), location (risk area, mobile home residence, coastal tenure), hurricane experience, information source (official warning, environmental cues), and expectations (expected storm intensity, expected rapid onset, expected flood risk, expected wind risk) variables. Indeed, there were only four variables for which hypothetical evacuation studies had significantly smaller (or more negative) effect sizes than the actual evacuation studies: female gender, official warning, mobile home residence, and expected wind risk. In addition, there were two variables that had larger (or more positive) effect sizes for hypothetical evacuation studies than the actual evacuation studies: homeownership and expected storm intensity. The most important of these findings are that people are likely to underestimate the effect of official warning on their evacuation decisions and, conversely, that they are likely to overestimate the effect of expected storm intensity on their evacuation decisions. However, official warnings are based substantially on expected storm intensity so these effects are likely to offset each other. In summary, the available evidence suggests that hypothetical evacuation studies can provide reasonably accurate information for evacuation analyses and evacuation plans, although more research is needed to examine the correspondence of the results from actual evacuations and evacuation expectations studies.

Identify the Risk Area (Potential Evacuation Zones)

A risk area is the geographic area that is exposed to the physical impacts of a given hazard and thus is an area from which people will need to evacuate. One major objective of risk area identification is to communicate risk information to residents so they will understand whether they are vulnerable to extreme events that could threaten their safety, health, property, or normal activities (Lindell and Perry 2004; Perry and Lindell 2007). That is, those who are at risk should be prepared to take protective action and those who are not at risk should be reassured that they do not need to do so.

The risk area map should subdivide the entire impact area according to the level of threat for each subdivision. For example, hurricane risk areas are often divided into separate areas for minor and major hurricanes. Similarly, Vulnerable Zones for toxic chemical facilities and Emergency Planning Zones for nuclear power plants can be divided into separate areas for minor and major releases. These subdivisions allow emergency planners to recommend evacuations that are appropriate to the level of threat in each part of the risk area. The boundaries between subdivisions and of the overall risk area need to be very clearly defined because previous research has found that many people find it difficult to identify their locations on risk area maps (Arlikatti et al. 2006; Zhang et al. 2004). Subdivision boundaries can be defined by postal code boundaries, political boundaries (city/county/state lines), geographical features (e.g., rivers), and major roads. However, it is essential that emergency planners consider geographical impediments to evacuation (e.g., water bodies, barrier islands, or mountains) when defining risk area subdivisions. This is especially important in evacuation route analysis because those geographical impediments might reduce the traffic capacity or limit evacuation speed.

In addition, risk area maps distributed to the community before an incident occurs should be accompanied by information about how to prepare for evacuation, how people will be warned when an emergency occurs, what the different levels of emergency are (e.g., watch and warning), what protective actions they should be prepared to take, which routes they should use, and how they can obtain evacuation assistance and additional information. Risk area maps that are displayed during an emergency on agency websites or by the news media (e.g., in newspapers and TV broadcasts) should clearly indicate what PARs—e.g., evacuate, shelter in-place, or monitor the situation—are advised for different subdivisions of the risk area. It will also be important to communicate that people within a risk area are generally advised to take the same protective action, although there are exceptions. For example, occupants of mobile homes might be advised to evacuate from areas in which occupants of site-built houses and multi-family dwellings are advised to remain.

Evacuation Analysis

Emergency planners should use their risk area maps as the foundation for their evacuation analyses, specifically estimating ERS capacity and evacuation demand. There are two major objectives of evacuation analysis, the first of which is to compute evacuation time estimates (ETEs) for different evacuation scenarios. These ETEs allow emergency managers to determine how long they can wait for additional information before initiating an evacuation. For example, knowing that the ETE for their jurisdiction is 36 hours tells them that they must initiate an evacuation no later than 36 hours before the arrival of Tropical Storm force wind if they are to clear the risk area before evacuation becomes hazardous for high profile vehicles such as recreational vehicles and buses (Lindell and Prater 2007a; Wu et al. 2015b). ETEs also allow them to assess the effectiveness of different traffic management strategies and determine which one allows them to clear the evacuation zone more rapidly and completely. For example, they can identify locations that are likely to experience bottlenecks and divert evacuation traffic onto routes that are likely to be less congested (Murray-Tuite and Wolshon 2013).

The following sections will introduce some factors that must be considered in computing ETEs. These factors include identifying the ERS and the capacities (in vehicles per hour) of its individual highway segments (links), forecasting traffic demand, modeling traffic behavior, and identifying strategies for traffic management.

Identify the ERS and Assess its Capacity

Managing a mass evacuation before the actual disaster threat arrives is complicated by the fact that drivers' behavior during an evacuation differs in several ways from their behavior as daily commuters. First, drivers' departures from their homes are more similar during evacuations than during normal commutes to work. In Hurricanes Katrina (2005), Rita (2005), and Ike (2008), for example, about half of the evacuees evacuated the day that the National Hurricane Center (NHC) issued a hurricane warning, which was two days before the hurricane struck (Wu et al. 2012, 2013). Moreover, most of these people left in the morning, with decreasing percentages of evacuees leaving as the day progressed. Thus, most vehicles tend to enter the ERS at essentially the same time.

Second, evacuation is almost exclusively one-way (outbound) traffic, so the inbound lanes of the ERS are usually empty. In hurricane evacuations, people move from coastal areas to inland areas. In riverine floods, this problem is somewhat less severe because people moving away from the river can generally travel in two directions. Hazardous materials incidents are potentially even less problematic in this respect because people can generally travel radially away from the hazard. Of course, the direction of travel is constrained by the characteristics of the ERS. For example, the Maine Yankee nuclear power plant (now decommissioned) was located at the base of a peninsula so evacuees from Boothbay Harbor at the peninsula's tip would have needed to travel 10 miles toward the plant before moving away from it. Emergency managers and traffic personnel need to direct risk area residents through stop signs and traffic signals so they can evacuate with minimal interruptions (Lindell 2013). Also, in the case of hurricanes, making the unused inbound lanes available to evacuees—a strategy known as contraflow—will facilitate evacuation (Wolshon 2001).

There are three steps to estimate ERS capacity (Lindell 2013). First, evacuation analysts must identify the routes that evacuees in different risk areas should take when evacuating. In practice, evacuation analysts focus on major highways because there are insufficient police and transportation department personnel to manage all routes out of the evacuation zone. Second, evacuation analysts must consider the number of lanes, road width, and speed limit for each link (section of roadway) and the characteristics of any intersections when choosing evacuation routes (Dotson and Jones 2005; TRB 2010). When assessing link capacity it is important to recognize that roadway characteristics sometimes change; for example, temporary road repairs might cause a link to change from four lanes to two lanes with a resulting reduction in its capacity in vehicles per hour (vph).

Third, evacuation analysts must consider the configuration of the links in the ERS. *Parallel links* (roads that do not intersect) have additive capacities. For example, the combined capacity of parallel links that can handle 800 vph and 600 vph, respectively, is 1400 vph. By contrast, *converging links* are limited by the capacity of the downstream link. For example, if the two links in the previous example (with a combined capacity of 1400 vph) converge into a link with 1000 vph, the capacity of this evacuation route would be 1000 vph at most. Moreover, the process of merging traffic from these two routes—even if managed by a traffic patrol officer—will further decrease capacity. Thus, evacuation analysts should expect bottlenecks to form at convergence points that will cause queues extending upstream and delaying departure from the risk area. The converse problem arises with *diverging links*, which will have unused capacity because the traffic flow to them is limited by the capacity of the upstream link. For example, if a link with 1000 vph feeds two links with capacities of 800 vph and 600 vph, this evacuation route will only transport 1000 vph. There is 400 vph (800 + 600 − 1000 vph) of illusory capacity that cannot be used. However, diverging traffic is much less likely than converging traffic to experience bottlenecks that would further decrease capacity.

Forecast Evacuation Demand

Evacuation analysts need to recognize that risk area populations are heterogeneous with respect to car ownership, so many transit dependent residents will need to have buses dispatched to pick them up—probably at their neighborhood schools. In addition, there are households having special needs that might require other assistance such as ambulances. These data on different population segments at risk need to be combined with estimates of the number of evacuating households and the number of evacuating vehicles per household in order to estimate the number of evacuating vehicles. It is important for evacuation analysts to recognize that the size of each population segment within the risk area varies over time. There is variation by time of day as people travel to work, school, and other daytime activities and return home at night—especially during the work week.

Vehicle Owning Households

Data on evacuees' car utilization during Hurricanes Lili, Katrina, Rita, and Ike found that the average number of registered vehicles per household was 1.70 in Lili, 2.15 in Katrina and Rita, and 2.03 in Ike (Lindell et al. 2011; Wu et al. 2012, 2013). However, the numbers of vehicles actually taken in evacuation were significantly less than the number of registered vehicles: 1.42 (Katrina and Rita) and 1.25 (Ike) vehicles taken during evacuation. Nonetheless, this decreased number of vehicles was offset by boat and travel trailers—0.12 (Katrina and Rita) and 0.69 (Ike)—so there were more than 1.5 vehicle-equivalents that each household took when evacuating. These results show that a very large number of vehicles will be traveling on the ERS during hurricane evacuations in major urban areas (e.g., the National Hurricane Center 2006, estimated that more than two million people evacuated the Houston area during Hurricane Rita). Therefore, assuming roughly 2.5 persons per household (data for individual jurisdictions can be found at www.factfinder.census.gov), this means there were roughly 1.2 million vehicles traveling during Hurricane Rita evacuation. To ensure that demand is equitably distributed across evacuation routes, warning messages should include evacuation route suggestions for different risk areas to avoid overuse of some evacuation routes and underuse of others. Nonetheless, the Hurricane Katrina and Rita survey results showed that some evacuees avoided major highways and other routes recommended by authorities because they wanted to avoid traffic congestion (Wu et al. 2012).

Transit Dependent Households

Some risk area residents lack vehicles that are reliable enough to travel long distances or simply do not have a car. Indeed, the carless population can reach 40 percent of the risk area population in some urban neighborhoods. Unsurprisingly, dependence on public transit is not randomly distributed throughout the population. Recent surveys have shown that disadvantaged population segments such as ethnic minorities, women, the unmarried, and those with lower education and income levels are less likely to own a vehicle (Lindell et al. 2011; Wu et al. 2012, 2013). To some extent, this lack of evacuation transportation is solved by rides from peers (friends, relatives, neighbors, and coworkers)—with recent hurricane evacuation surveys reporting that 9percent of evacuees from Hurricane Lili, 8 percent of those from Hurricanes Katrina and Rita, and 10 percent of those from Hurricane Ike were able to obtain such assistance. Less than 1 percent of evacuees from Katrina, Rita, and Ike reported having taken public transportation.[1] However, rides from peers are unlikely to be forthcoming if those peers also lack their own vehicles. Consequently, public authorities need to sign contracts with agencies (e.g., public transit and school districts) from which they can obtain buses, identify and train bus drivers, work with bus drivers to ensure their families can evacuate if needed, select pickup points (e.g., local elementary schools) for transit-dependent population segments, and publicize the transportation assistance procedures and policies (especially pickup point locations, pickup schedules, allowable number of bags, and provisions for pets). For example, evacuation bus stops in Miami Beach, Florida, are clearly identified on the streets so local residents and travelers can readily find them during a hurricane evacuation (Miami Beach Hurricane Information Center, 2014).

Households with Special Needs

Another evacuation issue that emergency planners should anticipate is the need to provide transportation for evacuees with special needs. The special needs population includes people with physical and mental disabilities, households with pets, school children, tourists with language/communication problems, and prisoners (Perry and Lindell 2007). Those groups have difficulties that either require more time to prepare to evacuate or prevent them from leaving at all. Thus, emergency planners need to identify the locations of these individuals and prepare to support them during an evacuation. Some of these special needs population segments will be concentrated in facilities (e.g., schools, hospitals, nursing homes, and jails), whereas many others will dispersed throughout the community. For those in special facilities, much of the evacuation planning effort can be delegated to facility personnel, although it is important to review all facility plans to ensure that they are not relying on the same buses or ambulances. For the dispersed special needs population, community organizations such as home health care organizations can be used to identify the locations and transportation/ medical requirements of their clients. Emergency planners should prioritize the order of evacuation for these special needs populations and, in many cases, begin to transport them to safety before the rest of the risk area population begins to evacuate. In some cases, door-to-door contacts might be required to ensure that special needs populations are completely evacuated from impact areas (Litman 2006).

Transient Populations

Transients include individuals visiting the area on business as well as households vacationing in the area. The size of the transient population varies by time of year, day of week (especially weekday vs. weekend), time of day, and also due to special events such as festivals and athletic events. Thus, evacuation planners must address this variation in their analyses. Transients potentially pose a problem for evacuation analyses because they are less likely to receive informal warnings from peers, which can be an important source of warnings in emergencies (Lindell et al. 2007b). In turn, the lower level of informal warnings may delay their departure times. However, they have no need to protect property and need less time to pack, which tends to decrease their departure times. In most cases, transients have their own vehicles but some arrive by public transportation, especially commercial airlines, and might have difficulty in rescheduling their departures.

Noncompliance and Shadow Evacuation

Noncompliance is the failure of those advised to evacuate to do so whereas shadow evacuation refers to evacuation by people whom the authorities do not consider to be at risk (Perry and Lindell 2007; Zeigler et al. 1981). Noncompliance not only risks the lives of those who remain in the disaster impact area, but also the lives of emergency responders who are attempt to rescue these people. Shadow evacuation is an important issue because the evacuation of people from areas that are not at risk can delay the evacuation of those who are located in areas that are expected to experience disaster impacts. These problems of noncompliance and shadow evacuation are illustrated by Lindell and Prater's (2007b) data on Texas coastal residents' expectations of evacuating from each of five risk areas (RA1 is on the coast and RA5 is farthest inland) for hurricanes in each of the five Saffir-Simpson hurricane categories (CAT1 is 74–95 mph and CAT5 is over 156 mph). Table 12.1 shows that, for a CAT1 hurricane, only 45.9 percent of those in RA1 would evacuate as they should (a problem of non compliance) but 35.9 percent of those in RA2 would also evacuate (a problem of evacuation shadow). Indeed, 26.5 percent of those in RA5 expected to evacuate from a CAT1 hurricane.

Table 12.1 Percentages of Households Expecting to Evacuate for Hurricanes in CAT1-CAT5, by Risk Area (RA)

RA	CAT1	CAT2	CAT3	CAT4	CAT5
1	45.9	63.7	87.8	98.2	100.0
2	35.9	53.7	77.8	88.2	91.4
3	31.1	48.9	73.0	83.4	86.6
4	28.2	46.0	70.1	80.5	83.7
5	26.5	44.3	68.4	78.8	82.0

The table also shows that there would be similar problems of noncompliance and evacuation shadow for other hurricane categories as well. This general pattern of noncompliance and evacuation shadow was later confirmed in evacuations from Hurricane Ike (Huang et al. 2012) even though the specific percentages differed. Indeed, the Hurricane Rita evacuation had a substantial level of evacuation shadow from areas much farther inland than RA5, although this appears to be quite unusual. Thus, emergency planners should anticipate noncompliance and evacuation shadow in any evacuation, and they should conduct their own evacuation expectations surveys, not just inside the official risk areas, but also in areas adjacent to the official risk areas. Once the rates of compliance and evacuation shadow have been identified, analysts can use census data and Geographical Information System (GIS) analyses to convert these rates to the projected number of households evacuating from each subdivision of the risk area. In turn, they can estimate the number of vehicles evacuating from each subdivision of the risk area by multiplying the number of evacuating households by the number of evacuating vehicles per household—usually about 1.5 vehicles per household or about 75 percent of all registered vehicles (Baker 2000).

Estimate Departure Time Distributions

After forecasting the number of evacuation vehicles, emergency planners should estimate when evacuees will start their evacuation trips and enter the ERS. Previous studies have concluded that households' departure timing depends on the anticipated arrival of hazard impact, the wording—especially the perceived urgency—of evacuation warnings, and the household's proximity to the point of impact (i.e., location in or near the risk area). Moreover, people are less likely to leave before authorities issue official evacuation warnings and are less likely to depart at night unless they believe they would not be safe to delay until daylight (Dixit et al. 2008; Fu et al. 2007; Huang et al. 2012; Lindell et al. 2005).

One common assumption emergency planners have adopted in evacuation models is an S-shaped curve for the cumulative departure time distribution. Under this assumption, the rate of entry into the ERS increases slowly during the beginning of an evacuation but increases rapidly as time progresses. Finally, the rate of entry into the ERS decreases as the time of impact approaches (Dixit et al. 2008; Lindell et al. 2005; Sorensen 2000). The simple S-shaped distribution only occurs for evacuations that are completed within a single day. When evacuations take place over multiple days, as is the case for hurricanes striking urban areas, there is S-shaped distribution each day, in which the evacuation rate is low in the early morning, increases through the late morning and afternoon, and declines after that.

Figure 12.2 depicts households' departure timing in Hurricanes Katrina, Rita, and Ike. Although some households sought to avoid traffic jams by evacuating before an NHC Hurricane Warning was issued, most used this NHC announcement and their local jurisdiction's official warning as an indication that it was time to leave. Generally, less than 10 percent households evacuated in the first 12 hours after a hurricane watch (Hurricane Katrina was an exception) was issued. This is because households require preparation time for evaluating their risk, making their evacuation decisions, preparing for leave work, gathering household members, packing travel items, and securing their property (e.g., turning off utilities, installing storm shutters) before they leave (Lindell et al. 2007b). In addition, many households make short trips to get money; buy gas, water, and groceries; and pick up others who need rides (Noltenius and Ralston 2009; Yin et al. 2014).

Households' perceptions of the urgency for responding to the threat determine the slope of the departure curve in the following phases of the evacuation process. People left quickly in Hurricane Katrina because the storm made its landfall less than 48 hours after the hurricane watch was issued and, especially, because the New Orleans mayor issued an evacuation order less than 24 hours before

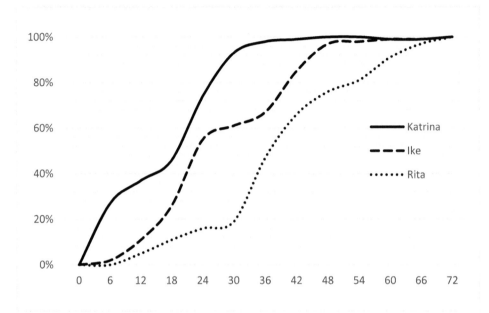

Figure 12.2 Cumulative Departure Timing in Hurricanes Katrina, Rita, and Ike

landfall. By contrast, people had much longer time to prepare for evacuation in Hurricane Rita. As expected, all three of the hurricanes show that the cumulative departure time curve is a sequence of daily S-curves in which departure rates are very low in the early morning (before 6am), increase during the later morning hours (6am–noon), decrease during the afternoon hours (noon–6pm), and virtually disappear at night (6pm–midnight) because people want to finish their driving and settle in by dark (Huang et al. 2017; Lindell et al. 2005).

Forecast Evacuation Logistics

In addition to studying household evacuation decisions, researchers have also examined the logistics of household travel behavior during hurricane evacuations. These studies have collected data on important variables such as the amount of time people take to prepare to evacuate, evacuation routes and destinations, types of accommodations they use (commercial facilities, peers' homes, public shelters), evacuation distance and duration, and evacuation costs. The average evacuation distance has been around 150–250 miles from the evacuees' homes in some recent hurricanes (Lili: 192.50 miles; Katrina: 266.42 miles; Rita: 199.17 miles; and Ike: 156.92 miles). More than 50 percent of the evacuees stayed with friends and relatives during all four hurricanes; less than 5 percent of them stayed in public shelters while away from home. These studies also found that evacuees tended to use interstate highways to evacuate during the Katrina and Rita evacuations (Figure 12.3), although evacuees from some Texas risk areas also took other routes such as state and county highways to avoid the traffic on the interstate highways. These studies have found that people who live closer to the coastal areas have tended to evacuate farther inland and spent more time away from their homes (Lindell et al. 2011; Wu et al. 2012) and people who evacuated farther inland preferred to stay at hotels/motels while they were gone (Lindell et al. 2011; Wu et al. 2012, 2013).

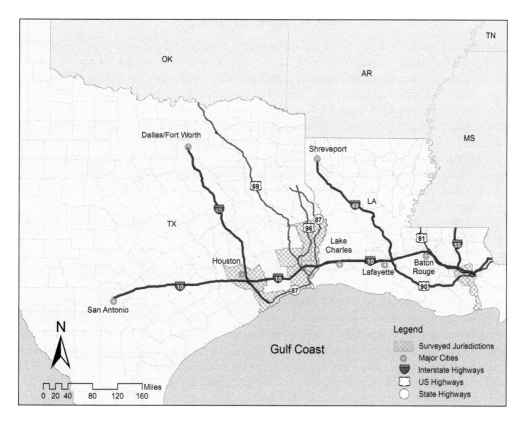

Figure 12.3 Cumulative Departure Timing in Hurricanes Katrina, Rita, and Ike

Model Traffic Behavior

In a very simple case, the time required to evacuate a risk area could be estimated by dividing the number of evacuating vehicles by the roadway capacity (in vehicles per hour). Unfortunately, evacuations frequently produce demand that exceeds the available roadway capacity, thus producing queues of vehicles waiting for access to the roadway. The need to incorporate queuing time into the calculations means that modeling traffic behavior for any area other than a small neighborhood (low demand) with an extensive road network (high capacity) requires the use of computer programs. Only these computer programs can adequately model the behavior of drivers as they leave their points of origin, drive to one of the principal evacuation routes, queue for access to the evacuation route, travel along that route to a proximal destination at the edge of the risk area, and travel from there to the household's ultimate destination where they will seek shelter. There are many traffic models available (Hardy and Wunderlich 2007; Lindell et al. 2019; Pel et al. 2012), but all of them require analysts to input the data from the previous stages of the evacuation analysis discussed in this section—identifying risk areas/sectors; identifying the ERS and its capacity; and forecasting evacuation demand. These traffic modeling programs can generate data on total traffic volume; average vehicle travel time; risk area clearance time; hourly volume, total volume, and average speed at each exit from the risk area; and queue lengths at selected intersections (Jones et al. 2011). It is important to conduct sensitivity analyses using a range of plausible values for each of the major input parameters to see how the model's output data, especially ETEs, are affected (Lindell 2008).

Traffic Management Strategies

Major evacuations generate a level of traffic that greatly exceeds ERS capacity, thus causing severe traffic jams. However, the longer that evacuees wait on evacuation routes, the greater the risk that they will be overtaken by hazard impact. To minimize the time required to clear the ERS, evacuation planners can either manage traffic demand (the number of vehicles entering the ERS at any given time) or increase ERS capacity. They can manage traffic demand in three ways. First, evacuation planners can provide credible pre-impact risk communication and emergency warnings that minimize the amount of evacuation shadow. That is, they need to assure people outside the official risk areas that it will be safe if they remain at home rather than evacuating. Second, emergency planners can flatten the peaks in evacuation demand by phasing the evacuation—with those at greatest risk being advised to leave first and those farther away being asked to wait until those at greatest risk have cleared the ERS. Third, evacuation planners can encourage people to take fewer vehicles.

There are two principal strategies for maintaining/increasing ERS capacity: modified traffic control and ERS lane augmentation (Abdelgawad and Abdulhai 2009; Chiu et al. 2008; Murray-Tuite and Wolshon 2013; Parr and Kaisar 2011). Modified traffic control refers to a simple set of traffic management strategies for maintaining the capacity of existing traffic lanes. One traffic impediment is cross-traffic that is attempting to move perpendicular to the flow of evacuation traffic, which can be reduced by eliminating all cross-traffic at minor intersections. In this case, any vehicles attempting to reach destinations on the other side of the evacuation routes must join the evacuation flow until they exit the evacuation zone, at which point they can continue in their original direction. Another traffic impediment is the delay that occurs between the time that a signal changes to green and the vehicle queue reaches full speed. This delay can be reduced by increasing signal length so that many more vehicles can pass through an intersection on each green signal. Merging traffic also causes delays, so another traffic control strategy is to close some freeway on-ramps. A fourth evacuation impediment is the difference in the capabilities of different types of vehicles (e.g., differences in acceleration rates and stopping distances for cars and buses). This problem can be addressed by segregating different types of vehicles onto different evacuation routes (e.g., establishing exclusive bus routes).

ERS lane augmentation increases capacity by increasing the number of lanes being used for evacuation traffic. The simplest version of this strategy is to allow motorists to use the shoulders of evacuation routes (Cova and Johnson 2003; Wolshon and Lambert 2004). A more complex version is to use contraflow, which reassigns the normally inbound lanes of an evacuation route to outbound traffic (Chiu et al. 2008; U.S. Department of Transportation and U.S. Department of Homeland Security 2006). Although it might seem that contraflow will double ERS capacity and, thus cut evacuation clearance times in half, this is not the case. A simulation in Texas coastal areas found that contraflow operation would only reduce evacuation clearance time by 14 percent (Chiu et al. 2008). However, clearance time would decrease by 30 percent if the contraflow operation strategy were implemented in conjunction with a phased evacuation strategy (i.e., by managing demand in addition to managing capacity).

Re-entry Plan

As soon as they think the danger has passed, most people want to return immediately to their homes to resume their normal lives. However, local officials and emergency managers sometimes find it difficult to decide when to allow re-entry into evacuated areas. For most geophysical hazards, scientists cannot say with certainty when the danger is over. For example, earthquake aftershocks can continue unpredictably for weeks or months and volcanic eruptive sequences can continue for years. The resulting conflicts between local residents who are trying to resume their normal activities by returning to their homes and authorities who are trying to protect people's safety by keeping them out of the evacuation zone can be very contentious (Tobin and Whiteford 2002).

In the case of meteorological hazards such as tornadoes, hurricanes, and floods, there is usually a clear end to the danger, so it is easier for local officials to decide when to allow evacuees to return to their homes, although there may still be dangers from unstable buildings and lack of infrastructure. Even when local authorities have formulated a schedule for a safe reentry, it is often very difficult to contact evacuees and provide them information about the appropriate time and procedures for their return. As indicated above, most evacuees stay at friends and relatives' homes, so hurricane evacuees are often scattered over thousands of square miles (Lin et al. 2014). Consequently, many people attempt to re-enter the evacuation zone before authorities are willing to permit them to do so (Dash and Morrow 2000). Indeed, Siebeneck and Cova (2007) reported that only 46 percent of their respondents complied with official re-entry plans by returning home on or after their scheduled return date.

Some of the reasons for early returns can be found in a study of re-entry after Hurricane Ike. First, respondents who believed there was a higher risk of returning home were more likely to seek the official reentry plan. Second, respondents who paid more attention to local news media and who had greater reliance on local authorities were more likely to receive and comply with the official reentry plan. Third, although the Internet has been identified as an easy way to obtain information, local news, and peers were the major information sources for reentry information during the Ike reentry process (Lin et al. 2014). Other data from Hurricane Ike show that people were quite concerned about physical or structural issues such as protecting their homes from further damage or from looting but were also concerned about traffic congestion during the return trip and the availability of utilities such as electric power and water when they arrived home (Siebeneck et al. 2013). One particularly interesting finding from this study is that people actually faced less trouble when they returned than they expected. This suggests that emergency managers could reduce the amount of premature re-entry by publicizing the areas in which utilities are (and, especially, are *not*) available and by reassuring people about the presence of security patrols in the evacuation zone to protect property.

Conclusions

An evacuation and re-entry plan is a very important part of a community's EOP. The effectiveness of an evacuation plan depends upon the collaboration of all stakeholders (households, emergency managers, transportation officials, police, etc.) before a disaster strikes. Emergency managers and planners should base their evacuation plans on research findings from previous evacuations, behavioral expectations surveys, and census data.

Evacuation planners need to begin by clearly identifying the areas that are exposed to environmental threats so they can reduce residents' confusion about whether or not they are at risk. In addition, they need to identify the ERS, forecast evacuation demand (in terms of both level and timing), and forecast evacuation logistics. Once these steps have been completed, they can identify the traffic management strategies that can be implemented during an evacuation and the re-entry plan for returning people to their homes when it is safe to do so.

Finally, they need to recognize that some elements of an evacuation plan for one hazard agent (e.g., a hurricane) might not be suitable for another (e.g., a toxic chemical plant). If emergency managers consider these issues when developing their evacuation plans, they will be better able to protect the public health and safety while minimizing social disruption, economic cost, and psychological distress.

Acknowledgment

This chapter is based, in part, on work supported by the National Science Foundation (IIS-1540469 and CMMI-1760766). Any opinions, findings, and conclusions or recommendations expressed in this chapter are those of the authors and do not necessarily reflect the views of the National Science Foundation.

Note

1 These data probably underestimate carpooling and transit ridership because these survey respondents were somewhat more likely to be White, older, homeowners with higher education and income.

References

Abdelgawad, H., and B. Abdulhai. (2009) "Emergency Evacuation Planning as a Network Design Problem: A Critical Review," *Transportation Letters* 1(1): 41–58.

Arlikatti, S., M. K. Lindell, C. S. Prater, and Y. Zhang. (2006) "Risk Area Accuracy and Hurricane Evacuation Expectations of Coastal Residents," *Environment and Behavior* 38(2): 226–247.

Baker, E. J. (2000) "Hurricane Evacuation in the United States," 308–319, in R. Pielke, Jr. and R. Pielke, Sr. (eds.), *Storms*, Vol. 1, London: Routledge.

Chiu, Y.-C., H. Zheng, J. A. Villalobos, W. G. Peacock, and R. Henk. (2008) "Evaluating Regional Contra-flow and Phased Evacuation Strategies for Texas Using a Large-scale Dynamic Traffic Simulation and Assignment Approach," *Journal of Homeland Security and Emergency Management* 5(1): 1–29.

Cova, T. J. and J. P Johnson. (2003) "A Network Flow Model for Lane-based Evacuation Routing," *Transportation Research Part A: Policy and Practice* 37(7): 579–604.

Dash, N. and B. H. Morrow. (2000) "Return Delays and Evacuation Order Compliance: The Case of Hurricane Georges and the Florida Keys," *Global Environmental Change Part B: Environmental Hazards* 2(3): 119–128.

Dixit, V., A. Pande, E. Radwan, and M. Abdel-Aty. (2008). "Understanding the Impact of a Recent Hurricane on Mobilization Time During a Subsequent Hurricane," *Transportation Research Record: Journal of the Transportation Research Board* 2041: 49–57.

Dotson, L. J. and J. Jones. (2005) "Development of Evacuation Time Estimate Studies for Nuclear Power Plants" NUREG/CR-6863 SAND2004-5900. Washington DC: US Nuclear Regulatory Commission. Downloaded from www.nrc.gov/docs/ML0502/ML050250240.pdf, accessed 2/6, 2018.

Fu, H., C. G. Wilmot, H. Zhang, and E. J. Baker. (2007) "Modeling the Hurricane Evacuation Response Curve," *Transportation Research Record* 2022: 94–102.

Hardy, M. and K. Wunderlich. (2007) *Evacuation Management Operations (EMO) Modeling Assessment: Transportation Modeling Inventory*, Falls Church, VA: Noblis. Downloaded from www.its.dot.gov/its_publicsafety/evacuation.htm, accessed 2/6, 2018.

Houts, P. S., M. K. Lindell, T. W. Hu, P. D. Cleary, G. Tokuhata, and C. B. Flynn. (1984) "The Protective Action Decision Model Applied to Evacuation During the Three Mile Island Crisis," *International Journal of Mass Emergencies and Disasters* 2(1): 27–39.

Huang, S.-K., M. K. Lindell, and C. S. Prater. (2016). "Who Leaves and Who Stays? A Review and Statistical Meta-Analysis Of Hurricane Evacuation Studies," *Environment and Behavior* 48(8): 991–1029.

Huang, S.-K, M. K. Lindell, and C. S. Prater. (2017) "Multistage Model of Hurricane Evacuation Decision: Empirical Study of Hurricanes Katrina and Rita," *Natural Hazards Review* 18(3): 05016008.

Huang, S.-K., M. K. Lindell, C. S. Prater, H.-C. Wu, and L. Siebeneck. (2012) "Household Evacuation Decision Making in Response to Hurricane Ike," *Natural Hazards Review*, 13(4): 283–296.

Jones, J., F. Walton, and B. Wolshon. (2011) *Criteria for Development of Evacuation Time Estimate Studies*, NUREG/CR-7002 SAND2010-0016P. Washington DC: US Nuclear Regulatory Commission. Downloaded from www.nrc.gov/reading-rm/doc-collections/nuregs/, accessed 2/6, 2018.

Kang, J. E., M. K. Lindell, and C. S. Prater. (2007) "Hurricane Evacuation Expectations and Actual Behavior in Hurricane Lili," *Journal of Applied Social Psychology* 37(4): 881–897.

Lin, C. C., L. K. Siebeneck, M. K. Lindell, C. S. Prater, H-C. Wu, and S-K. Huang. (2014) "Evacuees' Information Sources and Reentry Decision Making in the Aftermath of Hurricane Ike," *Natural Hazards* 70(1): 865–882.

Lindell, M. K. (2008) "EMBLEM2: An Empirically Based Large-scale Evacuation Time Estimate Model," *Transportation Research A* 42(1): 140–154.

Lindell, M. K. (2013) "Evacuation Planning, Analysis and Management," 121–149, in A. B. Bariru and L. Racz (eds.), *Handbook of Emergency Response: A Human Factors and Systems Engineering Approach*, Boca Raton FL: CRC Press.

Lindell, M. K., S.-K. Huang, and C. S. Prater. (2017) "Predicting Residents' Responses to the May 1–4, 2010, Boston Water Contamination Incident," *International Journal of Mass Emergencies and Disasters* 35(1): 84–113.

Lindell, M. K., J. E. Kang, and C. S. Prater. (2011) "The Logistics of Household Hurricane Evacuation," *Natural Hazards* 58(3): 1093–1193.

Lindell, M. K., J. C. Lu, and C. S. Prater. (2005) "Household Decision Making and Evacuation in Response to Hurricane Lili," *Natural Hazards Review* 6(4): 171–179.

Lindell, M. K., P. Murray-Tuite, B. Wolshon, and E. J. Baker. (2019) *Large-Scale Evacuation*. New York: Routledge.

Lindell, M. K. and R. W. Perry. (1992) *Behavioral Foundations of Community Emergency Planning*, Washington DC: Hemisphere.

Lindell, M. K. and R. W. Perry. (2004) *Communicating Environmental Risk in Multiethnic Communities*, Thousand Oaks CA: Sage.

Lindell, M. K. and R. W. Perry. (2012) "The Protective Action Decision Model: Theoretical Modifications and Additional Evidence," *Risk Analysis* 32(4): 616–632.

Lindell, M. K. and C. S. Prater. (2007a) "A Hurricane Evacuation Management Decision Support System (EMDSS)," *Natural Hazards* 40(3): 627–634.

Lindell, M. K. and Prater, C. S. (2007b) "Critical Behavioral Assumptions in Evacuation Analysis for Private Vehicles: Examples from Hurricane Research and Planning," *Journal of Urban Planning and Development* 133(1): 18–29.

Lindell, M. K. and C. S. Prater. (2009) *Local Official's Guide to Making Hurricane Evacuation Decisions*, College Station TX: Texas A&M University Hazard Reduction & Recovery Center.

Lindell, M. K., C. S. Prater, C. E. Gregg, E. Apatu, S.-K. Huang, and H.-C. Wu. (2015) "Households' Immediate Responses to the 2009 Samoa Earthquake and Tsunami," *International Journal of Disaster Risk Reduction*, 12, 328–340.

Lindell, M. K., C. S. Prater, and W. G. Peacock. (2007a) "Organizational Communication and Decision Making in Hurricane Emergencies," *Natural Hazards Review* 8(3): 50–60.

Lindell, M. K., C. S. Prater, and R. W. Perry. (2007b) *Introduction to Emergency Management*. Hoboken NJ: Wiley.

Lindell, M. K., C. S. Prater, W. G. Sanderson, Jr., H. M. Lee, Y. Zhang, A. Mohite, and S. N. Hwang. (2001) *Texas Gulf Coast Residents' Expectations and Intentions Regarding Hurricane Evacuation*, College Station, TX: Texas A&M University Hazard Reduction & Recovery Center.

Lindell, M. K., D. S. Sutter, and J. E. Trainor. (2013) "Individual and Household Response to Tornadoes," *International Journal of Mass Emergencies and Disasters* 31(3): 373–383.

Litman, T. (2006) "Lessons from Katrina and Rita: What Major Disasters Can Teach Transportation Planners," *Journal of Transportation Engineering* 132(1): 11–18.

Miami Beach Hurricane Information Center. (2014) *Evacuation*, Retrieved December 30, 2014, from http://web.miamibeachfl.gov/publicsafety/scroll.aspx?ID=46694.

Mileti, D. and J. Sorensen. (1990) *Communication of Emergency Public Warnings, ORNL-6609*, Oak Ridge TN: Oak Ridge National Laboratory.

Murray-Tuite, P., and Wolshon, B. (2013) "Evacuation Transportation Modeling: An Overview of Research, Development, and Practice," *Transportation Research Part C: Emerging Technologies* 27: 25–45.

National Hurricane Center. (2006) *Tropical Cyclone Report Hurricane Rita*, Retrieved December 4, 2009, from www.nhc.noaa.gov/pdf/TCR-AL182005_Rita.pdf.

Noltenius, M., and B. A. Ralston. (2009). "Pre-evacuation Trip Behavior," 395–413, in P. S. Showalter and Y. Lu (eds.), *Geospatial Techniques in Urban Hazard and Disaster Analysis*, Dordrecht, Netherlands: Springer.

Parr, S. A. and E. Kaisar. (2011) "Critical Intersection Signal Optimization during Urban Evacuation Utilizing Dynamic Programming," *Journal of Transportation Safety and Security* 3(1): 59–76.

Pel, A. J., S. P. Hoogendoorn, and M. C. J. Bliemer. (2012) "A Review on Travel Behaviour Modelling in Dynamic Traffic Simulation Models for Evacuations," *Transportation* 39(1): 97–123.

Perry, R. W., M. K. Lindell, and M. R. Greene. (1981) *Evacuation Planning in Emergency Management*, Lexington MA: Heath Lexington Books.

Perry, R.W., and M. K. Lindell. (2007) *Emergency Planning*, Hoboken NJ: John Wiley.

Siebeneck, L. K. and T. J. Cova. (2007) "An Assessment of the Return-Entry Process of Hurricane Rita 2005," *International Journal of Mass Emergencies and Disasters* 26(2): 91–111.

Siebeneck, L. K., M. K. Lindell, C. S. Prater, T. H. Wu, and S.-K. Huang. (2013) "Evacuees' Reentry Concerns and Experiences in the Aftermath of Hurricane Ike," *Natural Hazards* 65(3): 2267–2286.

Sorensen, J. H. (2000). "Hazard Warning Systems: Review of 20 Years of Progress," *Natural Hazards Review* 1(2): 119–125.

Tierney, K. J., M. K. Lindell, and R. W. Perry. (2001) *Facing the Unexpected: Disaster Preparedness and Response in the United States*, Washington DC: Joseph Henry Press.

Tobin, G. A. and L. M. Whiteford. (2002) "Community Resilience and Volcano Hazard: The Eruption of Tungurahua and Evacuation of the Faldas in Ecuador," *Disasters* 26(1): 28–48.

Trainor, J. E., P. Murray-Tuite, P. Edara, S. Fallah-Fini, and K. Trantis. (2013) "Interdisciplinary Approach to Evacuation Modeling," *Natural Hazard Review* 14(3): 151–162.

TRB—Transportation Research Board. (2010) *Highway Capacity Manual*, Washington DC: Author.

U.S. Department of Transportation and U.S. Department of Homeland Security. (2006) *Catastrophic Hurricane Evacuation Plan Evacuation: A Report to Congress*, Washington DC: Author.

Wei, H.-L., M. K. Lindell, and C. S. Prater. (2014) "'Certain Death' from Storm Surge: A Comparative Study of Household Responses to Warnings about Hurricanes Rita and Ike," *Weather, Climate & Society* 6(4): 425–433.

Wei, H.-L., H.-C. Wu, M. K. Lindell, S.-K. Huang, H. Shiroshita, D. M. Johnston, and J. S. Becker. (2017) "Assessment of Households' Responses to the Tsunami Threat: A Comparative Study of Japan and New Zealand," *International Journal of Disaster Risk Reduction* 25: 274–282.

Wolshon, B. (2001) "'One-way-out': Contraflow Freeway Operation for Hurricane Operation," *Natural Hazards Review* 2(3): 105–112.

Wolshon, B. and L. Lambert. (2004) *Convertible Roadways and Lanes, NCHRP Synthesis of Highway Practice 340*, Washington DC: Transportation Research Board.

Wu, H.-C., M. K. Lindell, and C. S. Prater. (2012) "Logistics of Hurricane Evacuation in Hurricanes Katrina and Rita," *Transportation Research Part F: Traffic Psychology and Behavior* 15(4): 445–461.

Wu, H.-C., M. K. Lindell, and C. S. Prater. (2013) "The Logistics of Household Hurricane Evacuation," 127–140, in J. Cheung and H. Song (eds.), *Logistics: Perspectives, Approaches and Challenges*, Hauppauge NY: Nova Science Publishers.

Wu, H.-C., M. K. Lindell, and C. S. Prater. (2014) "Effects of Track and Threat Information on Judgments of Hurricane Strike Probability," *Risk Analysis* 34(6): 1025–1039.

Wu, H.-C., M. K. Lindell, and C. S. Prater. (2015a) "Process Tracing Analysis of Hurricane Information Displays," *Risk Analysis* 35(12): 2202–2220.

Wu, H.-C., M. K. Lindell, and C. S. Prater. (2015b) "Strike Probability Judgments and Protective Action Recommendations in a Dynamic Hurricane Tracking Task," *Natural Hazards* 79(1): 355–380.

Yin, W., P. Murray-Tuite, S. V. Ukkusuri, and H. Gladwin. (2014) "An Agent-Based Modeling System for Travel Demand Simulation for Hurricane Evacuation," *Transportation Research Part C: Emerging Technologies* 42: 44–59.

Zhang, Y., C. S. Prater, and M. K. Lindell. (2004) "Risk Area Accuracy and Evacuation from Hurricane Bret," *Natural Hazards Review* 5(3): 115–120.

Zeigler, D. J., S. D. Brunn, and J. H. Johnson, Jr. (1981) "Evacuation from a Nuclear Technological Disaster," *Geographical Review* 71(1): 1–16.

13

EMERGENCY PREPAREDNESS AND IMMEDIATE RESPONSE TO DISASTERS

An International Perspective

Sudha Arlikatti and Carla S. Prater

Introduction

Research on emergency management suffers from the same problems as much of the rest of social science research in that most of it has been done in English speaking countries and much of the remaining work has been done in former colonies of these nations (Heady 1998). Students of emergency management are thus exposed to a great deal of information on what is being done in the English speaking world, yet are often unaware of the different approaches to emergency management used in other regions. A few scholars have examined the applicability of emergency management principles developed in rich countries to other areas. They have concluded that the principles of an all-hazards, integrated, and comprehensive approach covering all phases of emergency management and integrating relevant agencies, together with a focus on building community resilience at the local level, are viable and useful in a wide variety of settings (Martin et al. 2001). However, resources, both human and technical, are frequently lacking for the development and monitoring of adequate programs (Arlikatti et al. 2018).

One reason to study policies and programs used in other countries is that, increasingly, countries are learning and borrowing policies from each other (Dolowitz and Marsh 2000). Globalization has had two important effects. First, it has exposed all countries to an increasingly competitive economic system. Second, advances in the technology of communications have made it possible for policy makers to communicate quickly and easily. Because of these changes, policy makers increasingly look beyond their national borders for ideas on how to address problems at home. Governments are not the only institutions looking for ideas abroad. Non-governmental organizations (NGOs) of all sorts are also engaged in learning from a wide range of institutions outside of their host countries' borders (Stone 2000). Indeed, many NGOs such as Greenpeace are multinational organizations that routinely engage in cross-national policy transfer. Corporations, independent policy institutes, and less well organized transnational social movements can also engage in cross-national policy transfer.

The transfer of policy across national borders is complicated by the presence of "national policy styles" (Howlett 1991). These policy styles are based on the different characteristics of the governments involved, as well as the ways in which the governments relate to civil society. As a result, policy transfers are not always successful. Dolowitz and Marsh (2000) identified three ways in which policy transfers can fail.

- *Uninformed* transfers fail because the borrowing country does not understand the policy it is adopting;
- *Incomplete* transfers fail because the borrowing country does not transfer the necessary elements of the policy; and
- *Inappropriate* transfers fail because there are too many differences between the social, political, and economic contexts of the two countries.

Factors that Cause Variation in Policy Choices

Countries can be compared on the basis of many characteristics including regime type (roughly on a continuum from totalitarian to democratic—the level of policy centralization), political culture (traditionalistic, moralistic, and individualistic), orientation toward modernity (Barber 1997; Inglehart 1997), and level of economic development (Gross Domestic Product per capita, Human Development Index, or Gini coefficient). Gross Domestic Product is a measure of the total value of goods and services produced within a territory. For purposes of comparison, it is usually expressed in per capita terms, to adjust for different numbers of people living within the territories in question. The Human Development Index is a combination of life expectancy at birth, literacy and school enrollment rates, and GDP/capita (detailed information on how to calculate HDI values is available at hdr.undp.org). The Gini Index measures inequality over the entire distribution of income or consumption, with 0 representing a perfectly equal distribution and 100 representing perfect inequality. In other words, in a country with a Gini index of 0 everyone would have the same amount of income or consume the same amount of goods (details on the Gini index can be found at hdr.undp.org). Regardless of countries differences in these other characteristics, administrative structures are relatively similar across a broad range of countries because the function of an administrative structure has an effect on the shape it takes (Peters 1996). In addition, organizational models are frequently shared among groups of countries. For example, former colonies frequently have administrative structures that closely resemble those of their former colonial masters.

Countries that share membership in a multinational organization such as the European Union (EU) frequently come to share forms of administrative organization in order to simplify cross-national cooperation. In disaster management, the influence of the UN through programs such as the United Nations International Decade for Natural Disaster Reduction (IDNDR) and its successor program, the United Nations International Strategy for Disaster Reduction (UNISDR) have contributed to the diffusion of common models, while encouraging countries to adapt these models to their own realities. Regional emergency management organizations such as the Asian Disaster Preparedness Center (ADPC) based in Bangkok, Thailand, have also influenced the evolution of emergency management across a wide variety of nation states.

When comparing emergency management policies, the most relevant dimensions of comparison must include the

> propensity for disaster, local and regional economic resources, organization of government, and availability of technological, academic, and human resources, . . . level of local responder training, resilience of infrastructure, public opinion of the government's ability to manage the crisis, and the availability of specialized assets.
>
> *(Haddow and Bullock 2003: 165–166)*

In other words, the most relevant dimensions of comparison arise from what emergency managers know as *hazard vulnerability* and *community capacity*. The rest of this chapter will consider the dimensions listed above, but will also examine the role of the military in society, the development of civil society, and other factors affecting the various programs mentioned throughout this chapter. These factors are linked in complex ways to produce a unique profile of hazard vulnerability, community capacity and rapid response to disasters in each country.

Hazard Vulnerability

Hazard vulnerability varies according to hazard type and level of exposure. Frequently, a country's types and levels of hazard exposure influence the organizational structure and the quality of its emergency management organizations. Countries with high levels of hazard exposure have often been found to have "disaster cultures" enabling them to adapt and respond to recurrent events. Their emergency management organizations can also show a high level of adaptation to particular hazards. For example, countries facing frequent typhoons have developed more sophisticated programs and policies for coastal evacuation than those that do not.

An example of the way in which hazard exposure and experience can shape national emergency management programs occurred in Asia after the 2004 Indian Ocean tsunami. Prior to the arrival of this devastating tsunami, most countries located along the Indian Ocean—including Indonesia, Sri Lanka, Thailand, Myanmar, and India—had devoted little attention to emergency management. They had established basic "civil defense" programs, usually associated with the military and concentrating on disaster response. However, there was little or no attention paid to the connections between threat identification and risk communication, urbanization, livelihoods and economic development programs and the production of hazard vulnerability. Following the devastation wrought by the 2004 Indian Ocean tsunami, countries worked together to set up a tsunami warning system for the Asian countries and national governments set up formalized institutions charged with disaster mitigation, preparedness, response and recovery to increase community capacity. For example, on December 23, 2005, within a year of the tsunami, the Government of India enacted the Disaster Management Act. This led to the creation of the National Disaster Management Authority (NDMA), spearheaded by the Prime Minister of India. The NDMA advocates for a holistic, integrated approach to disaster management and mitigation in India (Arlikatti 2009). Similarly, countries in Central America affected by Hurricane Mitch in 1998 changed their national development programs to a sustainable development paradigm. These new programs emphasized the linkages between social factors, environmental degradation, and hazard vulnerability (Lavell 2004).

Economic Resources

Emergency management is low on the priority list in poorer countries, as indeed it is elsewhere. Whole societies live on the brink of economic collapse, so other problems such as the provision of basic education and public health services seem to be of more immediate concern. Thus, emergency management is often an afterthought that rises on the public agenda only after a disaster occurs. In the meantime, poverty and uncontrolled rapid urbanization generate large concentrations of vulnerable populations in high-risk urban areas. Moreover, the lack of attention to the links between the environment and human settlements has delayed the development of a better understanding of sustainable development and increased the incidence of disasters in rich and poor countries.

The quality of emergency management in a country is related to the amount of resources available nationally for emergency management and to the amount of resources available from outside a country. Many poor countries struggle with high levels of foreign debt, often incurred by undemocratic regimes. To compound the problem, some of these countries have devoted much of their national budget to the military, leaving little money for improving the country's human capital through education and health care. Emergency management is often left far down the list of discretionary spending programs. Sometimes this situation has been exacerbated by structural adjustment programs imposed by multinational lending agencies such as the World Bank and the International Monetary Fund. Nonetheless, some countries have been able to use funds from donor nations and international lending agencies to improve their emergency management capabilities.

Hazard insurance availability varies from one country to the next. Few countries have systems with market penetration as widespread as the National Flood Insurance Program (NFIP) of the U.S. In part, this is because participation in the NFIP is a condition for getting a mortgage. Many foreign countries lack disaster insurance programs or, if they have one, premiums are too high for the majority of the population to afford them. In these countries, businesses might have disaster insurance, but few homeowners do. National governments might want homeowners to purchase hazard insurance but cannot require hazard insurance as a condition of mortgage approval because many fewer people in these countries borrow money to buy a house. In many cases, people first buy the land and then build each room of the house as they save enough money to purchase the construction materials.

Haddow and Bullock (2003) mention the availability of "specialized assets" as a factor affecting emergency management. These specialized assets may include items needed during response operations such as heavy equipment, trained Urban Search and Rescue (US&R) teams, hazardous materials (hazmat) response capabilities, technical expertise such as Geographical Information Systems (GIS), and training facilities. Such resources are not available universally, but often are shared regionally through organizations such as the Caribbean Community's Caribbean Disaster Emergency Management Agency. US&R teams, in particular, are eager to participate in emergency response efforts no matter where they occur. Such teams can provide valuable assistance, on-the-job training, and assistance with the psychologically difficult task of body retrieval, although it is rare that they are able to arrive quickly enough to accomplish rescues during the critical first hours. This is due to the logistics of moving large numbers of people and their equipment, as well as due to legal and political problems with such movements. In some cases, flyover rights have been denied to US&R teams, and the entry of search dogs without the normal quarantine or veterinary procedures frequently causes problems. Thus, there is increasing interest in local development of hazmat and US&R capabilities, using foreign teachers if necessary, or sending a first generation of practitioners to study abroad and bring back the needed knowledge for a train-the-trainers approach.

Organization of Government

One of the most important issues affecting emergency management is the degree of political centralization in a particular country. The control of policies, programs, and resources by the highest (national) level limits the ability of local governments to mount a rapid emergency response or to develop appropriate hazard mitigation programs. Emergency management is a service, like police and fire protection, that is delivered over a dispersed area. Consequently, it benefits from a significant degree of decentralization, allowing local governments to manage the service delivery (Peters 1996).

An excessive emphasis on major disasters tends to lead to over-centralization because it is assumed there will be a need for governmental coordination over the very large areas affected. Unfortunately, this also requires communication of information through multiple layers of government—which delays response and recovery operations. In reality, small frequent events may cause more deaths and economic losses than larger ones. When small events occur, local governments can respond with more agility and effectiveness if they do not need to await instructions and resources from the central government. Thus, empowering local governments and their populations to deal with small events is very effective, but is frequently resisted by the national governmental authority because it reduces central control. Frequent small events also point to the connections between patterns of development and hazard vulnerability, which can increase calls for more public participation in national goal setting. This also threatens a *status quo* that benefits entrenched elite groups.

The location of emergency management agencies in all levels of government is related to these agencies' effectiveness and also to the emphasis given to different aspects of hazard vulnerability. An agency that is charged with responding to disasters, but has a low status in government, will have difficulty in finding and delivering the needed resources in a timely manner. Emergency managers benefit when they receive input from scientific agencies, such as those who map the national territory or deliver weather forecasts, but they are often isolated from such valuable input by their location in government. The level of professionalization within the emergency management agency also varies a great deal across countries. When an agency has a high political profile and adequate resources, it is more able to attract and keep well-qualified and dedicated personnel. Few countries have an adequate supply of well-trained emergency management professionals, however. This situation is of great concern, and a number of countries are addressing the lack of personnel by beginning training programs at the university and postgraduate level. Istanbul Technical University in Turkey, the Autonomous University of Nicaragua, Asian Development Bank Institute in Thailand, and the Tata Institute of Social Sciences have developed multidisciplinary programs in emergency management that include both physical and social science components, and there are other examples worldwide.

Quality of the Built Environment

The quality of a country's infrastructure, housing, and business and industrial installations affects the level of its disaster exposure and the type of emergency management program required to meet its needs. For example, the better the quality of the construction, the less need there is for US&R techniques after earthquakes. Similarly, good roads make it easier to evacuate large numbers of people from flood and hurricane zones. Countries with large numbers of high-rise office and apartment buildings have an increased need for highly developed firefighting capabilities, as do those with large chemical manufacturing installations. However, it does not necessarily follow that countries with higher quality infrastructure and better building codes and land use planning practices have uniformly lower hazard vulnerability. In the US, the states of California and Florida have stricter building codes and code enforcement mechanisms for earthquakes and hurricanes, respectively, whereas poorer states like Louisiana face problems of corruption and citizens living in extreme poverty—resulting in poorer construction practices that increase vulnerability. Hurricane Katrina, which damaged over a million housing units in the Gulf Coast region (about half of them located in Louisiana), is a case in point (The Data Center, www.datacenterresearch.org/data-resources/katrina/facts-for-impact/).

Several factors tend to increase the vulnerability of the built environment in developing or underdeveloped countries of the world. For example, the coastal communities of south India that were destroyed by the 2004 Indian Ocean tsunami were mostly informal unplanned settlements, built over centuries by fishermen and farmers themselves, using weak construction materials. Pre-tsunami housing was a mix of brick or concrete one- and two-story structures and smaller one-story houses built entirely or partly of wood or bamboo with palm leaf panels for walls and palm thatch roofs. Only commercial and government buildings were masonry structures, and these withstood the tsunami surge.

Following the tsunami, the State Government of Tamil Nadu (GoTN) was mandated by the central government to enforce existing Coastal Regulation Zone (CRZ) ordinances while finding sites to relocate tsunami survivors. They also adopted better earthquake- and tsunami-resistant building codes following recommendations of the Bureau of Indian Standards Codes. On 30 March 2005, the GoTN issued G.O. Ms. 172, stating that all state government-sponsored houses would be constructed (with appropriate compensations) beyond 200 meters from the high-tide line (HTL) of the Indian Ocean coast and adopt earthquake- and tsunami-resistant building standards. All these activities reflect that disaster experience can be a powerful impetus for regional and local governments to commit to rural land use planning strategies and environmental protection for hazard mitigation purposes.

On the other hand there is also ample evidence to suggest "resilience building measures during recovery, disaster preparedness, and early warning system development, rarely address the underlying causes of vulnerability and trajectories of social inequality in disaster prone societies" (Larsen et al. 2011: 482). For example, the strict enforcement of setback lines from the high tide line of post- tsunami housing in south India has affected the fishing livelihoods of many tsunami survivors. Relocated households reported lower levels of recovery than those that rebuilt in-situ on their own plots of land (Arlikatti and Andrew 2012; Arlikatti and Andrew 2017). Also, those households that reported having improved roofing materials since the tsunami generally reported a lower level of recovery than those that did not make improvements in roofing because the concrete slab roofs and modern design and construction practices are poorly ventilated and difficult if not impossible to repair compared to thatched roofs seen in traditional rural housing designs along the south Indian coastline (Andrew et al. 2013).

In Thailand, new planning guidelines and building codes introduced for hotels in coastal tourism communities still lack enforcement and the problems of corruption continues (Cohen 2007). Moreover rapid urbanization in highly exposed because flat and low-lying coastal lands of Thailand has made a continually growing population susceptible to coastal hazards such as cyclones, floods, and tsunamis. Hence, the built environment can increase vulnerability if policymakers are not mindful of the sociopolitical climate, community capacity, and availability of traditional building materials and expertise in mitigation and reconstruction efforts to ensure sustainability.

Civil Society

Civil society is the aggregate of organizations that are independent of the government. It includes religious groups, civic clubs such as the Rotary International, political parties, and other groups organized around specific interests. In post-disaster situations, the government's ability to serve disaster survivors expeditiously and equitably is often questioned. The public and private sectors are limited in their roles and abilities to meet the specific needs of all disaster survivors, especially in developing countries with scarce resources, exploding populations, and stark inequalities in the educational and economic attainment of the populace. The opinion that people have of their government's abilities in general can affect the degree of trust they will place in government agencies' pronouncements about their emergency management efforts. When people are well informed and have strong beliefs in their rights as citizens, they are likely to demand more competence from government emergency responders and other emergency management agencies (Arlikatti et al. 2007). Non-governmental organizations (NGOs) and community-based organizations (CBOs) are ways in which civil society seeks to change governmental priorities or supplement weak governmental powers with its own capabilities. True civil society groups are not organized by government agencies, as is the case with the local neighborhood fire brigades in Taiwan, but are grass roots organizations that emerge independently of government to meet specific needs such as flood mitigation in a local watershed. They can exert substantial influence and can even contribute to processes of regime change, as the Communidades Eclesiais de Base in Brazil did during the 1980s. In addition, existing organizations may be strengthened as they extend their missions to take on new disaster-related tasks as observed after the 1985 earthquake in Mexico City (Dynes et al. 1990). For this and other reasons, governments might be wary of strengthening civil society.

Edwards and Hulme (1996) elaborate that NGOs have always been an integral part of disaster response and recovery efforts across the globe and have become "the magic bullet" for solving social problems when governments fail, due to limited resources (Arlikatti et al. 2012; Prater et al. 2006). Civil society is often strengthened during disaster response and recovery operations, especially when government agencies prove inadequate to the task and emergent organizations arise to take on seemingly intractable problems. Such organizations were created in Mexico City during the response to the 1985 earthquake (Velázquez 1986) and in Kobe, Japan during the response to the Great Hanshin earthquake of 1995 (Shaw and Goda 2004). In South India following the 2004 Indian Ocean tsunami, countless faith-based organizations (FBOs) like the Mata Amrithanandamayi Math and the Art of Living foundation offered mental health and counseling services, while other NGOs including Seva Bharathi, Sevalaya, Society of Nutrition, Education, Health and Action, and Phoenix Federation were focused on child and maternal health and education, women's livelihood and empowerment, financial planning, and hygiene services. These FBOs and NGOs continued working in the impacted communities long after the public sector representatives had left. Their focus was on serving the needs of the poorest of the poor, underprivileged or disadvantaged, including women, children, people with disabilities, and those belonging to the lower caste (Arlikatti et al. 2012).

The Role of the Military in Society and Emergency Management

The armed forces are involved in emergency management to some degree almost everywhere. The military has a high degree of organization and in some countries that is enough to differentiate it from other governmental agencies and NGOs. In addition, the armed forces usually have more resources of the type needed for disaster response. These resources include communications hardware, transportation equipment, fuel, generators, water, temporary shelters, health care, and food. Equally important is the organization into groups of large numbers of strong young people who are accustomed to taking orders.

There is another reason for the strong influence of the military in emergency management, however. This has to do with the nearly universal roots of emergency management in civil defense—that is, the organization, training, and equipping of nonmilitary personnel to repel invaders. In many countries, the military retains a strong influence on emergency management organizations. This usually leads to an overemphasis on disaster response using command-and-control models. The problem is not that there are clear lines of authority. Rather, it is that little or no opportunity exists for civilian input and little attention is given to long-term emergency management needs such as disaster recovery and hazard mitigation. In other countries, the military is part of the emergency management system but is under civilian control. In such cases, it cannot respond unless its presence has been requested. This option is preferred, if only to avoid overdependence on the military and risking a slide into an authoritarian government during a period of national weakness.

The Role of International Organizations

Countries vary widely in their approaches to foreign aid, including policies on when and how to send or receive help for disaster areas or become involved in mitigation projects. Some countries have adopted a reactive approach and confine themselves to offering assistance with search and rescue or post-disaster cleanup. By contrast, others have adopted a more developmentalist perspective and assist in the formation of intergovernmental institutions and programs. Major goals from this perspective are to reduce the incidence of disasters and to increase the capacity of poor countries in responding to emergencies.

Many regional and international institutions are devoted to promoting improved emergency management practices. Noteworthy among them are the United Nations and its UNISDR, the organization formed to carry on the goals of the U.N. International Decade for Disaster Reduction.

The United Nations Office of Disaster Risk Reduction, together with the Government of Japan, the World Meteorological Association, and the Asian Disaster Reduction Center, published *Living with Risk: A Global Review of Disaster Reduction Initiatives* online in 2004. This resource is a valuable compendium of information on emergency management worldwide.

In addition to the UN with its global range, there are regional organizations supporting emergency management programs. The Organization of American States (OAS) supports the development of disaster resistant schools, hospitals, and road networks through its Department of Sustainable Development the Centro de Coordenación para la Reducción de los Desastres Naturales en América Central (CEPREDENAC) is an intergovernmental organization dedicated to the development of emergency management in Central America and the Dominican Republic. La Red is a network of Latin American social scientists that publishes scholarly work on disasters in the region, highlighting the issues of social vulnerability and sustainable development. Pan American Health Organization (PAHO), the regional office of the World Health Organization, has emphasized retrofitting hospitals and strengthening public health programs to improve emergency management practices throughout the region. It also publishes an influential newsletter, *Disasters: Preparedness and Mitigation in the Americas*. Other regions around the world have similar organizations that facilitate regional discussions and technology transfer for disaster mitigation and management and provide mutual aid.

Country Case Studies Showcasing Community Emergency Response

Case studies discussed herein present findings gleaned by the authors from their U.S. National Science Foundation (NSF)-funded RAPID research projects: (1) Understanding individual behaviors following the 2011 NZ and Japan earthquakes, (2) Examining organizations, religion and cultural response to the great floods of 2011–2012 in rural, suburban, and urban provinces of Thailand, and (3) Examining warning systems and understanding household response during the earthquake and tsunami affecting American Samoa in 2009. While not directly addressing all of the factors discussed in the previous section, the cases illustrate both the varied improvised responses by communities and planned responses countries have developed to address the challenges of emergency management and the varied contexts faced by disaster researchers.

The 2011 Christchurch New Zealand and Tohoku Japan Earthquakes

In a period of about three weeks in 2011, significant earthquakes struck both the South Island of New Zealand and the northeast coast of Japan. The Mw 6.3 New Zealand earthquake struck at 12:51 pm local time on February 22 in Christchurch, causing 185 deaths, 5,000 injuries and US$11 billion in damage (Bannister and Gledhill 2012). The Christchurch earthquake was an aftershock of the Mw 7.1 Darfield earthquake that had occurred approximately six months prior on September 4, 2010. The epicenter of the Darfield earthquake was west of Christchurch city and thus produced peak ground accelerations in the city that were only about one quarter to one eighth of those experienced in the February 22 event. The Mw 9.0 Japan (Tōhoku) earthquake struck at 2:46 pm local time on March 11, 2011. The earthquake and resulting tsunami caused 15,854 deaths, 26,992 injuries and US$235 billion in damage throughout northeast Japan. Christchurch was the most severely affected community in New Zealand and Hitachi was selected as a Japanese equivalent because of its similar coastal location, population size, commercial/industrial base, and shaking intensity whereas Japan was subjected to three to five minutes of shaking (Lindell et al. 2016).

The immediate behavioral responses of Christchurch and Hitachi residents were relatively similar to each other. Our study found that overall only 20 percent of those at risk evacuated immediately and this was most common in Hitachi, where the duration of shaking was much longer than in Christchurch, with 12 percent seeking cover (see Lindell et al., 2016). This estimate of immediate evacuation is higher than the 2 percent reported in Arnold et al. (1982) and 6 percent reported in Bourque et al. (1993), but is lower than the 38 percent reported in Prati et al. (2012). About 12 percent of the respondents in both Hitachi and Christchurch took cover, similar to the estimates of Prati et al. (2012) but substantially lower than at Whittier Narrows, where 43 percent of those at home and 40 percent of those at work took cover. They are also somewhat lower than some groups in Loma Prieta, where the percentages seeking shelter ranged from 0–68 percent, with a median of 21 percent, depending upon the respondent's location. The present study found percentages of the respondents in Christchurch (38 percent) and Hitachi (31.8 percent) who froze in place that are comparable to the corresponding percentages in the Umbria/Marche earthquake (32 percent), but larger than the corresponding percentages in the Whittier Narrows earthquake (20 percent—Goltz et al. 1992) and the Loma Prieta earthquake (ranging from 8–48 percent, with a median of 27 percent—Bourque et al. 1993).

To explain these similarities and differences in immediate behavioral response to earthquakes, one would like to be able to identify the corresponding similarities and differences in the different studies' demographic composition; earthquake experience, information, and emergency preparedness; and physical, social, and household contexts when the shaking began. Indeed, there are some statistically significant correlations of demographic variables with immediate behavioral responses. In the study, age was consistently correlated with immediate behavioral responses, being related to four of the six variables, but the correlations are modest (median r = .13 in absolute value). But the study failed to replicate the Bourque et al. (1993) finding that gender correlated with freezing (which they found more likely among women) and immediate evacuation (which they found more likely among men). More generally, the findings of the study were consistent with the Bourque et al. (1993: B11) conclusion that there are "few differences in response behavior at the time of the earthquake by demographic characteristics" and also with Baker's (1991) similar conclusion about the role of demographic variables in hurricane evacuation.

The data also indicated that experience and emergency preparedness were positively related to vigilance and negatively related to shock and fear. Moreover, prior experience also was significantly negatively related to risk perception. Moreover, earthquake experience, information, and emergency preparedness had some significant correlations with immediate behavioral response, a finding that is consistent with Weisæth (1989) and with Prati et al. (2012), who found that the physical, social, and household contextual variables had relatively few and small correlations with emotional reactions, risk perceptions, and immediate behavioral response.

The finding in the present study that fear was positively related to immediate evacuation is consistent with past research and theorizing that fear does not necessarily produce loss of control or nonrational flight (Aguirre 2005; Mawson 2005; Quarantelli 1954). However, the data from this study reveal a broader pattern of relationships between emotional reactions and behavioral response. The immediate behavioral responses ranged from continuing previous activities, through freezing, trying to protect persons, taking cover, trying to protect property, and ending with immediate evacuation. This ordering is consistent with what one might expect to be the rank ordering of these actions with respect to emotional arousal—those with the least fear continue their previous activities and those with the most fear evacuate immediately.

Ratings of fear were much higher than ratings of shock and vigilance for all response actions. Thus, fear is a dominant emotional reaction no matter what people's behavioral response is to earthquake shaking. However, there was a relatively small increase in the level of fear over the six behavioral responses in the study (continue what you were doing, freeze in place, take cover, protect persons, protect property, evacuate immediately).

In summary, all of the respondents were rather frightened and not so much shocked or vigilant. In any event, the level of emotional arousal did not strongly determine which action people took; some people with high fear and shock were able to take appropriate protective action and others with low fear and shock were not. The reactions of people in both earthquakes were remarkably similar, and prompted by remarkably similar factors, mainly previous earthquake experience and risk perception. Fear and shock did not dominate people's reactions. In these widely varying sociopolitical contexts, human reactions to natural phenomena remained broadly consistent at the individual level. This seems to highlight the role of sociopolitical systems as factors promoting or impeding adaptive disaster response and recovery at the individual or household level.

The Great Thailand Floods of 2011–2012

On July 25, 2011 the remnants of Tropical Storm Nock-ten impacted Thailand, resulting in severe inland flooding in many parts of the country. Unlike previous floods, the floods of 2011–2012 were different as the flood waters continued to inundate communities for over six months (starting on July 25 through mid-January 2012). Three months of heavy monsoons in the northern and central regions led to storm water overflows along the Mekong and Chao Phraya river basins. Over 12.8 million people in 65 of the Thailand's 75 provinces were affected by the floodwaters. The floods also impacted industrial and agricultural production at the regional and global levels. Thailand is the world's largest manufacturer of rice, rubber, and computer hard drives, accounting for more than 25 percent of world production (PreventionWeb 2013). Honda, Mitsubishi, and Toyota factories in the northern regions of Thailand were forced to shut down or relocate. With an estimated $45.7 billion in economic losses, the World Bank (2012) ranked the flooding in Bangkok to be the fourth costliest disaster to date (after Hurricane Katrina, the 1995 Kobe Earthquake, and the 2011 Japanese Tsunami).

Although the floodwaters receded in many parts of the country by early 2012, the Thai media reported that more than two million residents faced flooding of their agricultural lands and millions more continued to struggle with post-flood recovery. This resulted in street protestors demanding expedited repairs and compensation from the government and foreign investors threatening to close their factories and relocate their operations to safer locales in the second and third quarter of 2012, if the Thai government failed to provide specific flood-prevention plans. Although the business sector expected the government to find flood prevention solutions almost immediately, local leaders and government officials expressed concerns regarding the threat of repeated flooding during the next monsoon season, which was only a few months away.

We conducted a study to identify and document the earliest processes, programs, and policies (both established and emergent) used to address the immediate and short-term needs of the flood-affected communities in Thailand (see Andrew et al. 2016). The capital city of Bangkok and the two provinces of Pathum Thani and Ayutthaya were selected as they were continually mentioned in news headlines even until January 2012. Their respective public, private, and non-governmental organizations came under heavy scrutiny from the media for their inability to help their communities bounce back and function in the short term and solve problems associated with business continuity (i.e., tourism sector, hotels, and restaurants), coordinate evacuations and sheltering and provide equitable relief, compensation, and recovery programs to all citizens.

We interviewed 44 contacts from multi-sector organizations including schools, universities, tourism offices, nonprofit civic organizations, Buddhist FBOs, the automobile sector, and others to gauge what contributed to their organizational resiliency and capacity to help their constituents recover from the floods. Respondents from Bangkok noted that resources were quickly available to them as they are located in Thailand's capital, the seat of government, economic, and military power, and also the tourism industry. Respondents from the rural province of Ayutthaya noted the strength of social networks and the help provided by monks from numerous temples and FBOs. Respondents from the suburban community of Pathum Thani said they were the worst off because of several factors including unfamiliarity and the lack of knowledge of how to deal with floods, the few and ambiguous warnings they had received from the central government, and the actions of the government that favored Bangkok and allowed the positioning of sandbags to protect the capital in such a way that their province flooded.

Respondents from all three provinces and all sectors were extremely appreciative of the outstanding coverage of the floods that the news media provided, especially TV Channel 3, which resulted in an outpouring of aid from other parts of the country and the international community. The media were able to direct the Thai Army to survivors while the Army provided transportation and delivered emergency supplies. The Army spent most of its time in the water to help with evacuations by boat. Everyone we spoke to noted that being Buddhist meant helping those in need and sharing and showing kindness as a way of life. This shared religious belief system seemed to have been the invisible impetus for neighbors helping neighbors, and even the for-profit private sector organizations volunteering services to help the flood victims despite the lack of formal agreements, national disaster response plans, or direction from the government.

Varying improvised decisions adopted by multi-sector organizations including local, provincial, and central government agencies, the media, universities and schools, local nonprofit agencies, and FBOs helped communities recover faster. However, all study interviewees' perceived Ayutthaya (the rural province) to have been most resilient because of the local population's familiarity with dealing with floods annually—what is popularly known as a "disaster sub-culture." Wenger and Weller (1973) described disaster subculture as a community segment whose previous disaster experience informs actions that aid the community's survival and preserve its culture. Cuny (1991), as described in Dynes (1991), discussed how adaptation to flooding in Thailand has developed throughout the centuries and that homes located in the floodplains are typically built on stilts, roads are made parallel to the river, and the floodwaters help sustain the growth of crops and cultivation of fish. Although these rural communities lack the economic capital that urban areas like Bangkok city have, they may not be less resilient to the impacts of disasters because of their social capital (Siebeneck et al. 2015).

This was also evidenced by Larsen et al. (2011) who found that the most prominent source of resilience following the impacts of the 2004 Indian Ocean tsunami, in the tourism-dependent coastal communities was from the newly established or previously existing social networks between the public and private sectors and citizens at large. However, one must be cognizant of the fact that if success is dependent on the power relationships between stakeholders, it may also mean that such social networks may benefit only those in the collective and isolate others without resources even more.

Thailand has had its share of problems in strengthening disaster management due to failures at all levels of government, in enforcing rules, regulations, and mandates. Corruption and lack or misuse of resources, both human and financial, exacerbated the problems (Siebeneck et al. 2015). Agencies failed to coordinate or cooperate with one another as they lacked the political will and backing. Shook (1997) noted that most vulnerability assessments did not clearly identify these management components of hazard risk which were in fact a fundamental issue needing review in the Thai context. The 2004 Indian Ocean tsunami that struck the Thai coast and claimed countless lives spearheaded post-tsunami disaster risk reduction efforts including a memorandum of agreement among the Thailand International Cooperation Agency, the Thailand National Disaster Warning Center, and the National Oceanic and Atmospheric Administration (NOAA) of the United States of

America for technical cooperation in effective tsunami system analysis (Larsen et al. 2011). This has improved the national warning system tremendously.

Furthermore, since the 2011 floods, the Thai government has worked to deploy a nationwide flood management scheme, to increase its resilience to similar disasters. Sixty million Bhat (approximately $1.9 billion US dollars) were earmarked for constructing water retention areas in the Chao Phraya River in order to protect the low-lying communities in the area. Additionally, the construction of 20 dams, floodways, and diversion channels as part of a comprehensive drainage system was promised. However, these projects were put on hold until public hearings mandated by the Thai court could be held in order to fully assess the impacts these projects would have on people, natural resources, and the environment (Associated Press 2013). In September 2013 civic and engineering groups protested the $12 billion plan for being poorly conceived with little public input and narrowly focused on improving the irrigation system to counter droughts, rather than holistically focused environmental conservation and flood mitigation (VOANews 2013).

Although there is very little data related to emergency management programs and operations in Thailand, the newly placed emphasis on emergency management, and in particular hazard mitigation and preparedness, provides an opportunity for future researchers to monitor the impacts the expanding emergency management program will have on increasing disaster resilience in Thailand (Siebeneck et al. 2015).

American Samoa Tsunami of 2009

On Tuesday September 29, 2009, at 6:48 am local time (17:48 UTC), a large Mw 8.1 submarine earthquake occurred about 200 km (125 mi) south of American Samoa (Fritz et al. 2011). This caused the twelfth American Samoa tsunami with one foot or more of run up in the past century but only the second substantial one during that time (USACE 2012). The tsunami struck the American Samoa shoreline within 15 minutes (only three minutes after the first tsunami warning was issued over the radio) and killed 189 people throughout the islands of American Samoa, Samoa, and Tonga. A survey of five villages, two located directly on the coast and the others located inland of these two, provided the data summarized here (Lindell et al. 2015).

Some of the respondents' demographic characteristics highlight how different the American Samoa population is from that of mainland U.S. First, respondents' community tenure (how long they have lived in the community) is a significant percentage of their ages (25.3/41.3 = 61%), suggesting that people have lived in their communities for such a large percentage of their lives that they have developed extensive social ties that are likely to be important sources of informal warnings. Second, household sizes are quite large (an average of 8.6 persons/household; standard deviation = 4.5), which could increase evacuation difficulties if there are many young children or very old adults. Third, although most respondents (57 percent) were in their own homes at the time of the earthquake, over a third of them had family members who were absent and either known to be in danger (4 percent) or whose safety was unknown (35 percent). These conditions have the potential for delaying evacuation because families are reluctant to evacuate unless all members are accounted for (Drabek 1986; Perry et al. 1981).

The results of this survey confirmed that environmental cues, especially the experience of earthquake shaking combined with knowledge that an earthquake can cause a tsunami, are a major source of first information about the tsunami, consistent with Iemura et al. (2006) and Gaillard et al. (2008). The difference between the level of tsunami knowledge in American Samoa and in the area of Indonesia studied by Iemura et al. (2006) can probably account for the difference in the percentage of the two samples that were caught in the two tsunamis—5 percent in American Samoa vs. 74 percent in Indonesia. The finding that village bells in American Samoa provided only one-third of the warnings from social sources and one sixth of all first indications of the potential for a tsunami is likely due to the fact that they, like sirens, provide an ambiguous signal that is susceptible to incorrect interpretation by those who do hear them (Lachman et al. 1961; Gregg et al. 2007) as well as attenuation by white noise generated by high wind and surf (Lindell and Prater 2010).

The results also confirmed broadcast media as the most common first sources of social warnings, by showing that 57 percent of the respondents had radio/TV as their source of first warning, followed by peers (26 percent) and authorities (11 percent). However, people were more likely to receive an evacuation advisory (as opposed to their first information about the tsunami) from peers (36 percent) than from authorities (32 percent) or the media (19 percent). Overall, the data from American Samoa, together with data from warnings of the 2004 Indian Ocean tsunami in Mauritius (Perry 2007), and the 1980 eruption of Mt. St. Helens (Perry and Greene 1983) suggest that radio and TV can reach a large percentage of the risk area population because their broadcast process can transmit messages to many people simultaneously whereas peers can reach a large percentage of the risk area population because their diffusion process involves so many people relaying messages through social networks (Lindell et al. 2007; Rogers and Sorensen 1988). Authorities are much more limited in their ability to directly warn the risk area population because they lack the broadcast capacity of radio and TV (unless they have a dense array of electronic sirens), and they lack the large number of staff members that would be needed to substitute for the social networks that transmit peer warnings. However, the news media can only function as an effective first source if there are radio and TV stations available to broadcast warning messages.

Phone access is lower in American Samoa than in the continental US even in normal conditions. The high level of face-to-face warnings is consistent with the high level of observations of people evacuating (34 percent) because onlookers could easily speak to those who were evacuating on foot. Moreover, roads in the residential areas are frequently unpaved and relatively narrow so onlookers could easily speak to those evacuating slowly in cars as well.

We expected to find that people would be more likely to evacuate to the homes of relatives and friends (compared to hotels and motels or public shelters) than in other disasters because of the limited number of commercial facilities. Indeed, none of the respondents reported evacuating to a hotel or motel, perhaps due to the limited supply of hotels on the island. Moreover, the low level of public shelter utilization (5.6 percent) is at the low end of the range reported by Mileti et al. (1992). As expected, the homes of relatives (31.1 percent) were common evacuation destinations but, unexpectedly, the homes of friends (11.7 percent) were much less common than parks (38.9 percent)—an evacuation destination that has not previously been reported in evacuation research (Lindell et al. 2011; Wu et al. 2012, 2013). The greater use of parks than peers' homes might be explained by the common use of these locations as gathering places in American Samoa because flat land is fairly limited on the island—where narrow coastal lowlands separate the ocean from steep mountainous areas. Furthermore, the potential tsunami inundation zone is relatively small (compared to hurricane surge zones) because the topography rises rapidly everywhere except the extreme northwest corner of the island. Indeed, only two of the five villages in the study area were even partially inundated by the tsunami. As it turned out, the others were not inundated by this tsunami—although it is important to note that the respondents had no way to know that at the time of the earthquake.

On average, people evacuated well before they expected the tsunami to arrive. Somewhat surprising is the finding that risk area residents responded rapidly even though 51 percent of them took the time to seek more information and perform some preparatory tasks (locate family members, warn others, help others, pack an emergency kit, and protect property) before evacuating. Since many respondents lived well above sea level and high ground was fairly easily accessible in a short distance, they may have delayed evacuation accordingly. However, tsunami hazard zones were not delineated by signage as they are, for example, in Oregon (Lindell and Prater 2010).

There are few multi-story structures on the island, so it is unsurprising that less than 1 percent of the respondents took advantage of them for evacuation. However, they do exist in the most densely populated areas, so local authorities can prepare for tsunamis by preparing a systematic inventory of these structures and publicizing their availability and suitability for vertical evacuation. Surprisingly, of those who evacuated farther inland, vehicular evacuation (82.1 percent) was almost five times as popular as pedestrian evacuation (17.9 percent). However, it is unclear why people chose to evacuate in cars, especially since high ground was fairly easily reached by foot. It may be that people are not in the habit of walking long distances and misjudged the distance needed to achieve personal safety from tsunami inundation. Whatever the reasons for taking cars, there was not a significant difference in the probability of being overtaken by the tsunami between those who evacuated in vehicles (3.4 percent) and those who did not (6.2 percent). This indicates that the vulnerability of vehicular evacuees should be addressed by conducting systematic site-specific evacuation analyses (CRTWFS 2010) of the likelihood of vehicles being overtaken by tsunami waves based on the demand for space on the evacuation route system in relation to that system's capacity (Lindell 2013; Murray-Tuite and Wolshon, 2013). In addition, a population's general physical fitness may be a factor in their ability to evacuate on foot, which is one more reason to promote healthy lifestyles through good nutrition and exercise programs.

The differences of the Samoan population from that of the mainland US make these results particularly interesting. Where the Samoa data match the findings from studies of mainland populations in other types of disasters, we can reasonably infer that the behaviors generalize. Where the Samoa data provide findings that have not been studied in mainland populations in other types of disasters, caution is required in making any generalizations, but further studies of more diverse populations should be undertaken to verify their applicability, especially studies using similar methods across cultures. (Lindell et al 2016; Paton et al 2013). Similarly, comparison of the American Samoa data with Perry's (2007) Mauritius data indicates some patterns of response to a remote-source tsunami have limited generalizability to a near-source tsunami. In turn, this suggests caution in trying to generalize from the findings of slow onset disasters such as hurricanes to rapid onset disasters such as flash floods.

Conclusions

This chapter discussed the varying approaches to emergency management adopted in different parts of the world. It identified how countries differ systematically in their vulnerability to hazards, economic resources, government organizations, quality of the built environment, civil society, role of the military, and role of international organizations, which thereby impacts their immediate response to environmental threats. The chapter then turned to four NSF-funded case studies to highlight key findings from the authors' research to capture the immediate responses of risk area residents to the 2011 earthquake in New Zealand and Japan, the 2011–2012 great floods in Thailand and the 2009 earthquake and tsunami in American Samoa.

The Hyogo Framework of Action 2005–2015 adopted by 168 governments at the World Conference on Disaster Reduction held in Kobe, Japan, in January 2005 re-emphasized the need for community-based disaster risk management—CBDRM (UNISDR 2005). CBDRM fosters the participation of threatened communities in both the evaluation of risk (including hazards, vulnerability, and capacities) and in the ways to reduce it. Understanding the process by which natural disasters produce community impacts is important for four reasons. First, information from this process is needed to identify the *pre-impact conditions* that make communities vulnerable to disaster impacts. Second, information about the disaster impact process can be used to identify *specific segments of each community* that will be affected disproportionately (e.g., low-income households, ethnic minorities, isolated populations, or specific types of businesses). Third, information about the disaster impact process can be used to identify the *event-specific conditions* that determine the level of disaster impact. Fourth, an understanding of disaster impact process allows planners to identify *suitable emergency management interventions*. These reasons are broadly generalizable to all national contexts and can be used to organize hazard analyses for any level of government. Sensitivity to local conditions must be the foundational element for any system of emergency management, yet lessons can be learned from widely varying contexts around the world. The business of policy makers is to apply such lessons without violating the principles of sound policy transfer.

The Sendai Framework (2015–2030), the successor instrument to the (HFA) 2005–2015, was signed by UN Member States on March 18, 2015 at the Third UN World Conference on Disaster Risk Reduction in Sendai City, Miyagi Prefecture, Japan. Its Four Priorities for Action— (1) Understanding risk, (2) Strengthening disaster risk governance to manage disaster risk, (3) Investing in disaster risk reduction for resilience, (4) Enhancing disaster preparedness for effective response—and to "Build Back Better" in recovery, rehabilitation, and reconstruction (UNISDR n.d. underscore the importance of joint global efforts to strengthen accountability in disaster risk reduction. Nation states have been called upon to work not only on their own national and local disaster risk reduction strategies but to substantially increase and extend international cooperation to developing countries to complement their efforts. It is envisioned that such support will help generate practical guidelines and ensure ownership by all stakeholders of the world's safety.

Acknowledgments

This research was funded by the U.S. National Science Foundation's grants SES 0527699, SBE 0838654, CMMI 1138612, CMMI 0900662, IMEE-113861 and RAPID 1242004. We would also like to acknowledge the contributions of our research team members from the US (Prof. Walter Gillis Peacock, Prof. Michael K. Lindell, Dr. Himanshu Grover, Dr. Simon A. Andrew, Dr. Laura Siebeneck, Dr. Christopher E. Gregg), research contributors in the countries studied (Dr. Kannappa Pongpurat in Thailand, Col. Sam Taylor in American Samoa, Hideyuki Shiroshita in Japan, Dr. David Johnston and Julia Becker in New Zealand), and our graduate research assistants on the projects (Emma J. I. Apatu, Shih-Kai Huang, Hao-Che Wu, and Jaikampan Pongpurat). None of the conclusions expressed here necessarily reflects views other than those of the authors.

References

Aguirre, B. E. (2005) "Emergency Evacuations, Panic, and Social Psychology," *Psychiatry* 68: 121–129.

Andrew, S., S. Arlikatti, L. Long, and J. Kendra. (2013) "The Effect of Housing Assistance Arrangements on Household Recovery: An Empirical Test of Donor-assisted and Owner-driven Approaches," *Journal of Housing and the Built Environment* 28(1): 17–34.

Andrew, S.A., S. Arlikatti, L. Siebeneck, K. Pongponrat, and J. Kraiwuth. (2016) "Sources of Organizational Resiliency during the Thailand Floods of 2011: A Test of Bonding and Bridging Hypotheses," *Disasters* 40(1): 65–84.

Arlikatti, S. (2009) "Terrorism Watch or Natural Hazard Mitigation, Why the Difference? The Context of Community Policing in India," 142–154, in S. Ekici, A. Ekici, D. A. McEntire, R. H. Ward, and S. Arlikatti (eds.), *Building Terrorism Resistant Communities: Together Against Terrorism*, NATO Science for Peace and Security Series – E. Human and Societal Dynamics, Vol. 55.

Arlikatti, S. and S. Andrew. (2012) "Housing Design and Long-term Recovery Processes in the Aftermath of the 2004 Indian Ocean Tsunami," *Natural Hazards Review* 13(1): 34–44. DOI: 10.1061/ (ASCE) NH.1527-6996.0000062.

Arlikatti, S. and S. A. Andrew. (2017) "Disaster Housing Recovery in Rural India: Lessons from 12 Years of Post-Tsunami Housing Efforts," 175–190, in A. Sapat and A. M. Esnard (eds.), *Coming Home after Disaster: Multiple Dimensions of Housing Recovery*, Boca Raton FL: CRC Press.

Arlikatti, S., K. Bezboruah, and L. Long. (2012) "Role of Voluntary Sector Organizations in Posttsunami Relief: Compensatory or Complementary?" *Journal of Social Development Issues* 34(3): 64–80.

Arlikatti, S., M. Lindell, and C. Prater. (2007) "Perceived Stakeholder Role Relationships and Adoption of Seismic Hazard Adjustments," *International Journal of Mass Emergencies and Disasters* 25(3): 218–256.

Arlikatti, S., P. Maghelal, N. Agnimitra, and V. Chatterjee. (2018) "Should I Stay or Should I Go? Mitigation Strategies for Flash Flooding in India," *International Journal of Disaster Risk Reduction* 27: 48–56.

Arnold, C., M. Durkin, R. Eisner, and D. Whitaker. (1982) "Occupant Behavior in a Six Story Office Building Following Severe Earthquake Damage," *Disasters* 6(3): 207–214.

Associated Press. (June 27, 2013) "Thai court delays start of flood prevention works. Retrieved on June 27, 2013 from www.wral.com/thai-courtdelays-start-of-flood-prevention-works/12601186/.

Baker, E. J. (1991) "Hurricane Evacuation Behavior," *International Journal of Mass Emergencies and Disasters* 9(2): 287–310.

Bannister, S. and K. Gledhill. (2012) "Evolution of the 2010–2012 Canterbury Earthquake Sequence," *New Zealand Journal of Geology and Geophysics* 55(3): 295–304.

Barber, B. R. (1997) "The New Telecommunications Technology: Endless Frontier or the End of Democracy?" *Constellations* 4(2): 208–228.

Bourque, L. B., L. A. Russell, and J. D. Goltz. (1993) "Human Behavior During and Immediately after the Earthquake," B3–B22 in P. A. Bolton (ed.) *The Loma Prieta, California, Earthquake of October 17, 1989: Public Response*, Professional Paper 1553-B, Washington DC: U. S. Geological Survey. Available at pubs.usgs.gov/pp/pp1553/pp1553b/.

Cohen, E. (2007) "Tsunami and Flash-floods: Contrasting Modes of Tourism-related Disasters in Thailand," *Tourism Recreation Research* 32(1): 21–39.

CRTWFS—Committee on the Review of the Tsunami Warning and Forecast System and Overview of the Nation's Tsunami Preparedness. (2010) *Tsunami Warning and Preparedness: An Assessment of the U.S. Tsunami Program and the Nation's Preparedness Efforts*, Washington DC: National Research Council.

Cuny, F. C. (1991) "Living with Floods: Alternatives for Riverine Flood Mitigation," 62–73, in A. Kreimer and M. Munsasinghe (eds.) *Managing Natural Disasters and the Environment*, Washington DC: The World Bank.

Dolowitz, D. P. and D. Marsh. (2000) "Learning from Abroad: The Role of Policy Transfer in Contemporary Policy-Making," *Governance* 13(1): 5–23.

Drabek, T. E. (1986) *Human System Responses to Disaster: An Inventory of Sociological Findings*, New York: Springer-Verlag.

Dynes, R. R. (1991) "Disaster Reduction: The Importance of Adequate Assumptions about Social Organization," Preliminary Paper #172, Newark DE: University of Delaware Disaster Research Center. http://udspace.udel.edu/handle/19716/547.

Dynes, R. R., E. L. Quarantelli, and D. Wenger. (1990) *Individual and Organizational Response to the 1985 Earthquake in Mexico City, Mexico*, Newark DE: Disaster Research Center.

Edwards, M. and D. Hulme. (1996) *Beyond the Magic Bullet: NGO Performance and Accountability in the Post-Cold War World*, Hartford CT: Kumarian Press.

Fritz, H. M., J. C. Borrero, C. E. Synolakis, E. A. Okal, R. Weiss, V. V. Titov, B. E. Jaffe, S. Foteinis, P. J. Lynett, I.-C. Chan, and P. L.-F. Liu. (2011) "Insights on the 2009 South Pacific Tsunami in Samoa and Tonga from Field Surveys and Numerical Simulations," *Earth Science Reviews* 107(1–2): 66–75.

Gaillard, J.-C., E. Clavé, O. Vibert, A. Dedi, J.-C. Denain, Y. Efendi, D. Grancher, C. C. Liamzon, D. R. Sari, and R. Setiawan. (2008) "Ethnic Groups' Response to the 26 December 2004 Earthquake and Tsunami in Aceh, Indonesia," *Natural Hazards* 47: 17–38.

Goltz, J. D., L. A. Russell, and L. B. Bourque. (1992) "Initial Behavioural Response to a Rapid Onset Disaster: A Case Study of the October 1, 1987 Whittier Narrows Earthquake," *International Journal of Mass Emergencies and Disasters* 10(1): 43–69.

Gregg, C. E., B. F. Houghton, D. Paton, D. M. Johnston, D. A. Swanson, and B. S. Yanagi. (2007) "Tsunami Warnings: Understanding in Hawai'i," *Natural Hazards* 40: 71–87.

Haddow, G. D. and J. A. Bullock. (2003) *Introduction to Emergency Management*, New York: Butterworth-Heinemann.

Heady, F. (1998) "Comparative and International Public Administration: Building Intellectual Bridges," *Public Administration Review* 58(1): 32–39.

Howlett, M. (1991) "Policy Instruments, Policy Styles, and Policy Implementation," *Policy Studies Journal* 19(2): 1–21.

Iemura, H., Y. Takahashi, M. P. Pradono, P. Sukamdo, and R. Kurniawan. (2006) "Earthquake and Tsunami Questionnaires in Banda Aceh and Surrounding Areas," *Disaster Prevention and Management* 15(1): 21–30.

Inglehart, R. (1997) *Modernization and Postmodernization: Cultural, Economic, and Political Change in 43 Societies*, Princeton NJ: Princeton University Press.

Lachman, R., M. Tatsuoka, and W. J. Bonk. (1961) "Human Behavior during the Tsunami of May 23, 1960," *Science* 133: 1405–1409.

Larsen, R. K., E. Calgaro, and F. Thomalla. (2011) "Governing Resilience Building in Thailand's Tourism-dependent Coastal Communities: Conceptualising Stakeholder Agency in Social Ecological Systems," *Global Environmental Change* 21(2): 481–491.

Lavell, A. (2004) "The Lower Lempa River Valley, El Salvador: Risk Reduction and Development Project," 67–82, in G. Bankoff, G. Frerks, and D. Hilhorst (eds.) *Mapping Vulnerability: Disasters, Development and People*, London: Earthscan.

Lindell, M. K. (2013) "Evacuation Planning, Analysis, and Management," 121–149, in A. B. Bariru and L. Racz (Eds.) *Handbook of Emergency Response: A Human Factors and Systems Engineering Approach*, Boca Raton FL: CRC Press.

Lindell, M. K., J. E. Kang and C. S. Prater. (2011) "The Logistics of Household Evacuation in Hurricane Lili," *Natural Hazards* 58(3): 1093–1109.

Lindell, M. K. and C. S. Prater. (2010) "Tsunami Preparedness on the Oregon and Washington Coast: Recommendations for Research," *Natural Hazards Review* 11(2): 69–81.

Lindell, M. K., C. S. Prater, C. E. Gregg, E. Apatu, S.-K. Huang, and H.-C. Wu. (2015) "Households' Immediate Responses to the 2009 Samoa Earthquake and Tsunami," *International Journal of Disaster Risk Reduction* 12: 328–340.

Lindell, M. K., C. S. Prater, and W. G. Peacock. (2007) "Organizational Communication and Decision Making in Hurricane Emergencies," *Natural Hazards Review* 8(3): 50–60.

Lindell, M. K., C. S. Prater, H.-C. Wu, S.-K. Huang, D. M. Johnston, J. S. Becker, and H. Shiroshita. (2016) "Immediate Behavioral Responses to Earthquakes in Christchurch New Zealand and Hitachi Japan," *Disasters* 40(1): 85–111.

Martin, B., M. Capra, G. van der Heide, M. Stoneham, and M. Lucas. (2001) "Are Disaster Management Concepts Relevant in Developing Countries?" *Australian Journal of Emergency Management* 16(4): 25–33.

Mawson, A. R. (2005) "Understanding Mass Panic and Other Collective Responses to Threat and Disaster," *Psychiatry* 68(2): 95–113.

Mileti, D. S., J. H. Sorensen, and P. W. O'Brien. (1992) "Toward an Explanation of Mass Care Shelter Use in Evacuations," *International Journal of Mass Emergencies and Disasters* 10(1): 25–42. Accessed February 26, 2013 at www.ijmed.org.

Murray-Tuite, P. and B. Wolshon. (2013) "Evacuation Transportation Modeling: An Overview of Research, Development, and Practice," *Transportation Research Part C* 27: 25–45.

Paton, D., Okada, N., and S. Sagala. (2013) "Understanding Preparedness for Natural Hazards: Cross Cultural Comparison," *Journal of Integrated Risk Management* 3(1): 18–35.

Perry, S. D. (2007) "Tsunami Warning Dissemination in Mauritius," *Journal of Applied Communication Research* 35(4): 399–417.

Perry, R. W. and M. Greene. (1983) *Citizen Response to Volcanic Eruptions*, New York: Irvington.

Perry, R. W., M. K. Lindell, and M. R. Greene. (1981) *Evacuation Planning in Emergency Management*, Lexington MA: Heath Lexington Books.

Peters, B. G. (1996) "Political Institutions, Old and New," 205–221, in R. E. Goodin and H.-D. Klingemann (eds.) *A New Handbook of Political Science*. New York: Oxford University Press.

Prater, C., W. G. Peacock, S. Arlikatti, and H. Grover. (2006) "Social Capacity in Nagapattinam, Tamil Nadu after the December 2004 Great Sumatra Earthquake and Tsunami," *Earthquake Spectra* 22(S3): 715S–729S.

Prati, G., V. Catufi, and L. Pietrantoni. (2012) "Emotional and Behavioural Reactions to 35 Tremors of the Umbria–Marche Earthquake," *Disasters* 36(2): 439–451.

PreventionWeb (2013). Thailand—Disaster statistics. www.preventionweb.net/english/countries/statistics/?cid=170. Accessed 11 July 2013.

Quarantelli, E. L. (1954) "The Nature and Conditions of Panic," *American Journal of Sociology* 60(3): 267–275.

Rogers, G. O. and J. H. Sorensen. (1988) "Diffusion of Emergency Warnings," *Environmental Professional* 10(4): 185–198.

Shook, G. (1997) "An Assessment of Disaster Risk and its Management in Thailand," *Disasters* 21(1): 77–88.

Shaw, R. and K. Goda. (2004) "From Disaster to Sustainable Civil Society: The Kobe Experience," *Disasters* 28(1): 16–40.

Siebeneck, L., S. Arlikatti, and S.A. Andrew. (2015) "Using Provincial Baseline Indicators to Model Geographic Variations of Disaster Resilience in Thailand," *Natural Hazards* 79(2): 955–975.

Stone, D. (2000) "Non-governmental Policy Transfer: The Strategies of Independent Policy Institutes," *Governance: An International Journal of Policy and Administration* 13(1): 45–62.

United Nations International Strategy for Disaster Reduction (UNISDR) 2005. Hyogo Framework for Action 2005–2015: Building the Resilience of Nations and Communities to Disasters. Retrieved on 22 May 2019 from www.unisdr.org/2005/wcdr/intergover/official-doc/L-docs/Hyogo-framework-for-action-english. pdf.

United Nations International Strategy for Disaster Reduction (UNISDR). (n.d.). Sendai Framework for Disaster Risk Reduction. Retrieved on 22 May 2019 from https://www.unisdr.org/we/coordinate/ sendai-framework.

USACE—US Army Corps of Engineers, Honolulu District. (2012) *American Samoa Tsunami Study*, Honolulu HI. Retrieved on November 11 from, http://americansamoarenewal.org/sites/default/files/resource_ documents/ASTS_Final_Report-031312.pdf.

Velázquez, D. R. (1986) "La Organización Popular Ante el Reto de la Reconstrucción," *Revista Mexicana de Ciências Políticas e Sociales* 123: 59–79.

VOA News. (September 26, 2013) "Thai Flood Prevention Dam Draws Criticism," Retrieved on 11 November from www.voanews.com/a/thai-flood-prevention-dam-draws-criticism/1757489.html.

Weisæth, L. (1989) "A Study of Behavioural Response to an Industrial Disaster," *Acta Psychiatrica Scandinavica* 80(s355): 13–24.

Wenger, D. E. and J. M. Weller. (1973) "Disaster Subcultures: The Cultural Residues of Community Disasters," Preliminary Paper #9, Newark DE: University of Delaware Disaster Research Center. http://udspace.udel. edu/handle/19716/399.

World Bank (2012). Thai Flood 2011. Rapid Assessment for Resilient Recovery and Reconstruction Planning. www.gfdrr.org/sites/gfdrr/files/publication/Thai_Flood_2011_2.pdf (last accessed on 4 February 2015).

Wu, H. C., M. K. Lindell, and C. S. Prater. (2012) "Logistics of Hurricane Evacuation in Hurricanes Katrina and Rita," *Transportation Research Part F: Traffic Psychology and Behaviour* 15(4): 445–461.

Wu, H. C., M. K. Lindell, C. S. Prater, and S.-K. Huang. (2013) "Logistics of Hurricane Evacuation in Hurricane Ike," 127–140, in J. Cheung and H. Song (eds.), *Logistics: Perspectives, Approaches and Challenges*, New York: Nova Science Publishers.

Websites

Asian Development Bank Institute (ADBI) www.adbi.org/

Asian Disaster Preparedness Center (ADPC) -www.adpc.net/igo/

Autonomous University of Nigaragua Universidad Nacional Autónoma de Nicaragua (UNAN) www.unan. edu.ni/

Caribbean Disaster Emergency Management Agency (CDEMA) www.cdema.org/

Centro de Coordinación para la Prevención de Desastres Naturales en América Central (CEPREDENAC) www.cepredenac.org/

Disasters: Preparedness and Mitigation in the Americas www.paho.org/disasters/newsletter/

International Monetary Fund (IMF) www.imf.org/external/index.htm

Istanbul Technical University (ITU) http://www.itu.edu.tr/en/

National Disaster Management Authority (NDMA) www.ndma.gov.in/en/

National Flood Insurance Program (NFIP) www.fema.gov/national-flood-insurance-program

Organization of American States (OAS) www.oas.org/usde/Working%20Documents/Naturaldesasterandland. htm

Pan American Health Organization (PAHO) www.paho.org

Tata Institute of Social Science (TISS) www.tiss.edu/

Urban Search and Rescue (US &R) www.fema.gov/urban-search-rescue

United Nations International Decade for Natural Disaster Reduction (UNIDNDR) www.unisdr.org/we/ inform/publications/31468

United Nations Office of Disaster Risk Reduction (UNISDR) www.unisdr.org/

PART IV

Contributions of Disaster Recovery Planning to Community Resilience

14

UNDERSTANDING DISASTER RECOVERY AND ADAPTATION

Michelle Annette Meyer

Introduction

Disaster recovery is "the differential process of restoring, rebuilding, and reshaping the physical, social, economic, and natural environment through pre-event planning and post-event actions" (Smith and Wenger 2007: 237). Recovery is one of the least understood stages of disaster and is also at least partially independent of the disaster incident itself. Instead, recovery is closely intertwined with regular community processes and activities (Ranous 2012). The majority of research on disaster recovery focuses on either broad economic recovery (Webb et al. 2002; Xiao and Peacock 2014; Xiao and Van Zandt 2012) or social and household recovery (Bates 1982; Comerio 1998; Highfield et al. 2014; Mileti 1999; Peacock et al. 1987; Peacock et al. 1997; Weber and Peek 2012).

Previous research shows conclusively that disaster recovery can be the longest stage of disasters and is often nonlinear. For example, Haas et al. (1977) identified stages within common recovery trajectories based on infrastructure and rebuilding: (1) restoration period of basic infrastructure (first few months), (2) replacement period (capital rebuilding for two years), and (3) commemorative, betterment, and development reconstruction period (large projects up to 10 years post-event). Recovery overlaps with response and immediately following an event, many decisions are made that have effects on community development and planning for decades. This extended timetable— and the opportunity to incorporate adaptation and resilience measures—makes disaster recovery one of the most important planning actions communities can undertake. Thus, understanding how communities and planners can develop practices and policies that turn disasters from devastation to opportunity is key to resilience across societies.

Disaster recovery is not progressively linear for communities or households; instead, it includes starts and stops and backtracking. At the community level, issues of inequality, coordination, and conflict can cause uneven recovery outcomes for different populations (National Research Council 2006). To understand individual and household recovery requires that it be placed within the context of community-level recovery decisions. The ability of those in the recovery process to garner resources for community projects affects individual resilience and recovery at the household *and* community level (Bolin and Stanford 1998; Brunsma et al. 2007; Peacock et al. 1997). For example, recovery of organizations and businesses is required for community residents to bounce back from a disaster (Abramson et al. 2010).

As social scientists, we recognize that community processes involve networks of social systems and social interaction among different organizations and systems (Bates and Pelanda 1994; Tierney and Oliver-Smith 2012). In disasters, complexity, competition, conflict, and coordination as well as heterogeneity and social inequality arise from the interaction between various social systems within one community (Peacock and Ragsdale 1997: 27). The focus for recovery research is thus on the nature of the social systems involved and the interactions between and among these systems that help determine and shape the social processes of disaster recovery. As Olshansky and Chang (2009: 200) described, disaster recovery and the corresponding management of reconstruction across various social systems remain important challenges for urban planners and scholars particularly because "time compresses, stakes increase, additional resources flow, and public interest is heightened." Poor recovery can become a "second disaster" (Erikson 1994) for individuals and communities that experience population loss, stagnant economies, and uneven and inequitable rebuilding and reconstruction. For example, Kamel and Loukaitou-Sideris (2004) found that recovery programs may not fully meet the housing needs of communities, nor be appropriately targeted based on damage or population need.

To counteract the negative possibilities of a poorly executed recovery, pre-disaster behavior and planning is needed because, as early disaster research highlighted in the concept of "principle of continuity," recovery outcomes and processes relate to pre-disaster behavior (Quarantelli and Dynes 1977; Wenger 1978). Recovery processes could perpetuate previous social and economic conditions that exacerbate inequality and inequitable development (Bolin 1985; Bolin and Stanford 1998; Dash et al. 1997; Finch et al. 2010; Geipel 1982; Pais and Elliot 2008; Peacock et al. 1997). On the other hand, well-designed recovery can provide opportunities for increased mitigation (Berke et al. 1993), incorporation of sustainability (Smith and Wenger 2007), and increased resiliency to all types of future hazards (Berke and Campanella 2006). Resource and development management and planning are strongly connected to the incorporation of mitigation into recovery operations (Reddy 2000). Thus, recovery, while a stage of the disaster life cycle, is highly dependent upon everyday planning practices and pre-disaster policies.

A challenge for communities, as we shall see in this section, is the tension between addressing needs quickly post-disaster and slowly working towards more thoughtful and often more resilient recovery strategies. For many communities, focusing on the immediate needs in recovery and working to rebuild as quickly as possible means that opportunities for incorporating resilience are largely lost. This contradiction between expediting recovery and thoughtfully working through recovery options can be addressed through pre-event planning for recovery, not just response (Berke and Campanella 2006; Boyd et al. 2014). Understanding how planning addresses these challenges and opportunities both pre- and post-disaster are areas of continuing research. Also, how planning can redirect community trajectories and reduce rather than exacerbate inequality are central for understanding disaster recovery in the years to come.

As noted in the following chapters, the many and varied processes that are critical for recovery are undertaken by a multitude of both governmental and nongovernmental organizations and emergent groups at multiple levels (Aldrich and Meyer 2015; Drabek 1987; Drabek and McEntire 2002; Smith and Birkland 2012). Individuals, nongovernmental organizations, businesses, and government agencies all have recovery activities that restore their own capacity to operate as well as have implications for the restoration of the entire community. The disaster literature describes how during the immediate response, organizations are searching for information, trying to meet needs quickly, and thus networks and collaborations change quickly as organizations attempt to fill the emergent needs. In these situations, it is not surprising that Gillespie and Colignon (1993) found that emergency management organizations are central to initial disaster response networks, and that these networks become more concentrated around these central organizations. Emergency management, with a history of paramilitary training, is often tasked with recovery planning and collaboration, which creates a potential disconnect between those tasked with long-term recovery planning and those best suited to engage in collaborative partnerships across communities. Waugh and Streib (2006) described the need for greater collaboration based on their assessment of poor communication during Hurricane Katrina, echoing similar findings from Averch and Dluhy (1997) after Hurricane Andrew that found emergency management agencies were changing from top-down authoritarian regimes to collaborative organizations. While recovery involves coordination across various sectors and parts of the community, the local government is assumed to be central to disaster recovery, but the leaders may not be emergency managers (Olshansky and Chang 2009). Successful features of local government-led recovery include pre-planning, mitigation, organizational coordination, leadership, availability of resources and distribution of those resources, linkage of recovery to on-going activities, and public participation (Berke and Wenger 1991; Lindell et al. 2006; Reddy 2000). Because of their training, skill-sets, and values (Boyd et al. 2014; Rabinovitz 1967; Schwab and Topping 2010) community planners can be well-suited to understand, communicate with, and advocate for stakeholders in their communities, yet they are frequently missing from emergency planning activities (Bierling 2012).

With the frequent outsourcing of government functions to nongovernmental entities (e.g., external planning consultants), the importance of understanding the capacities, functions, and limitations of nongovernmental and governmental organizations within the context of disaster activities and networking is increasing (Brudney and Gazley 2009). For example, local churches often respond with resources for the general community during disasters, and they can provide efficient disaster services when connected through organizational ties to emergency management (Phillips and Jenkins 2010). Thus, one of the most important aspects of recovery is *pre-disaster* planning. A lack of pre-event organizational coordination to develop recovery plans results in inefficient delivery of disaster services and resources, even when policies and plans for collaboration exist (Kapucu 2006a, 2006b; Kapucu et al. 2010). For example, Meyer (2013) found that pre-disaster networks of community organizations focused on disaster issues improved information sharing, organizational disaster planning, and planning for vulnerable populations, such as homeless and persons with access and functional needs. To increase the United States' resilience, FEMA (2011) encourages all local communities to collaborate with their "whole community" (including individuals, businesses, nonprofits, civic groups, recreational groups, and emergency management) to increase disaster resilience. In relatively recent large disasters, nongovernmental private sector initiatives such as Rebuild LA and We Will Rebuild emerged as major players in post-disaster planning and rebuilding (Morrow and Peacock 1997). We have also seen the protracted and conflict-ridden failures to address recovery planning needs after Katrina (Olshansky and Johnson 2010). How recovery is coordinated across different sectors and with various organizations and businesses (often those with limited disaster response roles) becomes a challenge for the future. Further, how planners and community planning processes generate the engagement and involvement of these sectors *pre-disaster* is an area of emergency management and community planning that is becoming increasingly important.

This section of the book focuses on this least theorized stage of disasters: long-term recovery and adaptation (Rubin 1985, 2009). With resilience implying the ability to bounce back *and* reduce future impacts, recovery is a link in the disaster chain that can connect returning the community to functioning with adapting and improving to meet the needs of future disasters in the community (Comerio 2014; Norris et al. 2008). Each disaster provides the opportunity for lessons to be learned and integrated into policy and planning practices across various scales of local government, nonprofit organizations, and businesses (Birkland 2006). The chapters in the section highlight the capacities and challenges for different sectors to recover and adapt, discuss differential challenges for recovery within and across communities with a focus on vulnerable populations, discuss adaptations to climate change which will affect both the impacts and capacity to recovery from future disasters, and also address the role of design in recovery.

Chapter 15, by John Cooper and Jaimie Masterson, lays the groundwork for the remaining chapters in this section by describing the history of disaster recovery efforts that led to the development of the National Disaster Recovery Framework (NDRF). Their review of the research literature examines the key stakeholders involved in the recovery process and the factors that the NDRF identifies as indicators of a successful recovery. They also identify neglected issues such as social vulnerability and climate change. Finally, they examine the NDRF's performance during Hurricane Sandy, HUD's recommendations based on this performance, and the legislation needed to improve recovery procedures.

In Chapter 16, Yang Zhang and William Drake describe housing recovery in the United States following three of the costliest storms in the country's history: Hurricanes Andrew (Florida 1992), Ike (Texas 2008), and Sandy (New Jersey and New York 2012). Housing recovery is a central tenet of most disaster recovery operations. Yang and Drake describe how ongoing housing inequities affect owners' decisions and abilities to return and rebuild. Across all three disasters, they show how higher income areas had the least storm damage and were the quickest to be rebuilt. These results relate to community planning in which low-income residents are more likely to live in substandard housing in more flood-prone and environmentally risky areas. Further, the authors highlight a common inequity in disaster recovery funding: rental versus owner-occupied. Rental recovery programs are less common and often proceed at a slower pace than owner-occupied recovery efforts. The authors conclude with a discussion on the role of housing recovery funding mechanisms in these three disasters.

Chapter 17 focuses on population displacement as one potential outcome of disaster. Ann-Margaret Esnard and Alka Sapat describe how displacement from environmental events has now outpaced displacement due to conflict, and that the United States is not immune to this displacement. This displacement, the long-term and potentially permanent movement of populations following an event, is a rising concern as climate change increases sea level rise and extreme events. Often, when planners and community leaders think of disaster recovery, the assumption is that populations will return once infrastructure and houses are repaired. As the authors discuss in this chapter, the issue of permanent displacement and population movement is one that should not be overlooked by planners of communities in hazard-prone areas *and* those neighboring communities that may become the new home of displaced residents. Preparing for increased population mobility due to disasters is an area of recovery that needs more research and policy attention.

In Chapter 18, we turn our attention to business and economic recovery and Yu Xiao discusses how business recovery is situated within the broader community recovery processes. Xiao offers a staggering statistic: 40 percent of small businesses fail after a disaster. She describes how this vulnerability relates to infrastructure, supply chain issues, labor, and consumer spending, which all can be affected in one disaster. Yet, as she describes, this is one side of the paradox of understanding economic and business recovery, because depending on the scale one analyzes, economies, on average, are resilient to disasters. Xiao provides more details on this issue of guiding current business recovery planning efforts and highlights the need for greater attention and research on this component of community recovery.

Finally, Chapter 19 grounds us in practical design. Jamie Masterson discusses the role that design plays in disaster recovery. Although we know that design of buildings, public places, and community amenities affect social interactions and use of these areas, Masterson describes how design can also address disaster resilience. She describes different projects that have considered disaster risk and recovery during the design and implementation phase. This chapter brings us full circle in placing disaster, a rare occurrence for many communities, within everyday planning activities that will improve their resilience to the unthinkable event that one day may come.

In the past decade, large, multi-state disasters such as Hurricanes Katrina (2005), Rita (2005), Ike (2008), Sandy (2012), Harvey (2017), Irma (2017), and Maria (2017), the California wildfires of 2017 and 2018, and smaller but still destructive disasters such as wildfires in Texas (2011) and Colorado (2013), the flooding of the Mississippi River (2011), and even technological disasters (West, Texas,

2013) have left numerous communities struggling with post-disaster planning, unequal and partial recovery outcomes, and recovery efforts that fail to reduce pre-disaster vulnerabilities. Increasing disaster frequencies and impacts mean more communities will struggle, often with little local experience in managing the difficult processes of achieving sustainable and resilient recovery. Helping understand the myriad of practices that communities use to recover and the outcomes for individuals, organizations, businesses, and communities as a whole are the central foci of disaster recovery research. By highlighting lessons from previous recoveries, best practices, challenges, and diversity of experiences in community recovery and adaptation, this section helps illuminate this understudied stage of disasters for the planning community, scholars, and community leaders.

References

Abramson, D., T. Stehling-Ariza, Y. S. Park, L. Walsh, and D. Culp. (2010) "Measuring Individual Disaster Recovery: A Socioecological Framework," *Disaster Medicine and Public Health Preparedness* 4(S1): S46–S54.

Aldrich, D. P. and M. A. Meyer. (2015) "Social Capital and Community Resilience." *American Behavioral Scientist* 59(2): 254–269.

Averch, H. and M. J. Dluhy. (1997) "Crisis Decision Making and Management," 75–91, in W. G. Peacock, B. H. Morrow, and H. Gladwin (eds.) *Hurricane Andrew: Ethnicity, Gender and the Sociology of Disasters*, New York: Routledge.

Bates, F. L. (1982) *Recovery Change and Development: A Longitudinal Study of the 1976 Earthquake*, Athens GA: University of Georgia Department of Sociology.

Bates, F. L. and C. Pelanda. (1994) "An Ecological Approach to Disasters," 145–159, in R. R. Dynes and K. J. Tierney (eds.), *Disasters, Collective Behavior, and Social Organization*, Newark DE: University of Delaware Press.

Berke, P. R. and T. J. Campanella. (2006) "Planning for Post-Disaster Resiliency," *Annals of the American Academy of Political and Social Science* 604(1): 192–207.

Berke, P. R., J. Kartez, and D. Wenger. (1993) "Recovery after Disaster: Achieving Sustainable Development, Mitigation and Equity," *Disasters* 17(2): 93–109.

Berke, P. R. and D. Wenger. (1991) *Linking Hurricane Disaster Recovery to Sustainable Development Strategies: Montserrat, West Indies*, College Station TX: Texas A&M University Hazard Reduction and Recovery Center.

Bierling, D. (2012) *Participants and Information Outcomes in Planning Organizations*, PhD Dissertation, College Station, TX: Texas A&M University.

Birkland, T. A. (2006) *Lessons of Disaster: Policy Change after Catastrophic Events*, Washington DC: Georgetown University Press.

Bolin, R. (1985) "Disasters and Long-Term Recovery Policy: A Focus on Housing and Families," *Review of Policy Research* 4(4): 709–715.

Bolin, R. and L. Stanford. (1998) "The Northridge Earthquake: Community-Based Approaches to Unmet Recovery Needs," *Disasters* 22(1): 21–38.

Boyd, A., J. B. Hokanson, L. A. Johnson, J. C. Schwab, and K. C. Topping. (2014) *Planning for Post-Disaster Recovery: Next Generation*, Chicago: American Planning Association.

Brudney, J. L. and B. Gazley. (2009) "Planning to Be Prepared: An Empirical Examination of the Role of Voluntary Organizations in County Government Emergency Planning," *Public Performance & Management Review* 32(3): 372–399.

Brunsma, D. L., D. Overfelt, and J. S. Picou. (2007) *The Sociology of Katrina: Perspectives on a Modern Catastrophe*, New York: Rowman.

Comerio, M. C. (1998) *Disaster Hits Home: New Policy for Urban Housing Recovery*, Berkeley CA: University of California Press.

Comerio, M. C. (2014) "Disaster Recovery and Community Renewal: Housing Approaches," *Cityscape* 16(2): 51–68.

Dash, N., W. G. Peacock, and B. H. Morrow. (1997) "And the Poor Get Poorer: A Neglected Black Community," 206–225, in W. G. Peacock, B. H. Morrow, and H. Gladwin (eds.), *Hurricane Andrew: Ethnicity, Gender and the Sociology of Disaster*, New York: Routledge.

Drabek, T. E. (1987) "Emergent Structures," 259–290 in R. R. Dynes, B. De Marchi, and C. Pelanda (eds.), *The Sociology of Disasters*, Milan, Italy: Franco Angeli Press.

Drabek, T. E. and D. A. McEntire. (2002) "Emergent Phenomena and Multiorganizational Coordination in Disasters: Lessons from the Research Literature," *International Journal of Mass Emergencies and Disasters* 20(2): 197–224.

Erikson, K. (1994) *A New Species of Trouble: The Human Experience of Modern Disasters*, New York: W. W. Norton.

Federal Emergency Management Agency (FEMA). (2011) *A Whole Community Approach to Emergency Management: Principles, Themes, and Pathways for Action*, Washington DC: Author.

Finch, C., C. Emrich, and S. Cutter. (2010) "Disaster Disparities and Differential Recovery in New Orleans," *Population and Environment* 31(4): 179–202.

Geipel, R. (1982) *Disaster and Reconstruction: The Friuli, Italy, Earthquakes of 1976*, London: Allen and Unwin.

Gillespie, D. F. and R. A. Colignon. (1993) "Structural Change in Disaster Preparedness Networks," *International Journal of Mass Emergencies and Disasters* 11(2): 142–162.

Haas, J. E., R. W. Kates, and M. J. Bowden, eds. (1977) *Reconstruction Following Disaster*, Cambridge MA: The MIT Press.

Highfield, W., W. G. Peacock, and S. Van Zandt. (2014) "Determinants of Damage to Single-Family Housing from Hurricane-induced Surge and Flooding: Why Hazard Exposure, Structural Vulnerability, and Social Vulnerability Matter in Mitigation Planning," *Journal of Planning Education & Research* 34(3): 287–300.

Kamel, N. M. and A. Loukaitou-Sideris. (2004) "Residential Assistance and Recovery Following the Northridge Earthquake," *Urban Studies* 41(3): 533–562.

Kapucu, N. (2006a) "Public-Nonprofit Partnerships for Collective Action in Dynamic Contexts of Emergencies," *Public Administration* 84(1): 205–220.

Kapucu, N. (2006b). "Interagency Communication Networks During Emergencies: Boundary Spanners in Multiagency Coordination," *The American Review of Public Administration* 36(2): 207–225.

Kapucu, N, T. Arslan, and M. L. Collins. (2010) "Examining Intergovernmental and Interorganizational Response to Catastrophic Disasters: Toward a Network-Centered Approach," *Administration & Society* 20(10): 1–26.

Lindell, M. K., C. S. Prater, and R. W. Perry. (2006) *Introduction to Emergency Management*, Hoboken NJ: John Wiley.

Meyer, M. A. (2013) "Social Capital and Collective Efficacy for Disaster Resilience: Connecting Individuals with Communities and Vulnerability with Resilience in Hurricane-Prone Communities in Florida," PhD Dissertation, Fort Collins CO: Colorado State University.

Mileti, D. (1999) *Disasters by Design: A Reassessment of Natural Hazards in the U.S.* Washington DC: Joseph Henry Press.

Morrow, B. H. and W. G. Peacock. (1997) "Disasters and Social Change: Hurricane Andrew and the Reshaping of Miami," 226–242, in W. G. Peacock, B. H. Morrow, and H. Gladwin (eds.), *Hurricane Andrew: Ethnicity, Gender and the Sociology of Disaster*, New York: Routledge.

National Research Council (NRC). (2006) *Facing Hazards and Disasters: Understanding Human Dimensions*, Washington DC: National Research Council.

Norris, F., S. P. Stevens, B. Pfefferbaum, K. F. Wyche, and R. L. Pfefferbaum. (2008) "Community Resilience as a Metaphor, Theory, Set of Capacities, and Strategy for Disaster Readiness," *American Journal of Community Psychology* 41(1): 127–150.

Olshansky, R. B., L. D. Hopkins, and L. A. Johnson. (2012) "Disaster and Recovery: Processes Compressed in Time," *Natural Hazards Review* 13(3): 173–178.

Olshansky, R. B. and L. Johnson. (2010) *Clear as Mud: Planning for the Rebuilding of New Orleans*, Chicago: The American Planning Association.

Olshansky, R. and S. Chang. (2009) "Planning for Disaster Recovery: Emerging Research Needs and Challenges," *Progress in Planning* 72(4): 200–209.

Pais, J. and J. R. Elliott. (2008) "Place as Recovery Machines: Vulnerability and Neighborhood Change after Major Hurricanes," *Social Forces* 86(4): 1415–1452.

Peacock, W. G., C. D. Killian, and F. L. Bates. (1987) "The Effects of Disaster Damage and Housing Aid on Household Recovery Following the 1976 Guatemalan Earthquake," *International Journal of Mass Emergencies and Disasters* 5(1): 63–88.

Peacock, W. G., B. H. Morrow, and H. Gladwin. (1997) *Hurricane Andrew: Ethnicity, Gender and the Sociology of Disaster*, New York: Routledge.

Peacock, W. G. and A. K. Ragsdale. (1997) "Social Systems, Ecological Networks and Disasters: Toward a Socio-Political Ecology of Disasters," 20–35, in W. G. Peacock, B. H. Morrow, and H. Gladwin (eds.), *Hurricane Andrew: Ethnicity, Gender and the Sociology of Disaster*, New York: Routledge.

Phillips, B. and P. Jenkins. (2010) "The Roles of Faith-Based Organizations after Hurricane Katrina," 215–238, in R. P. Kilmer, V. Gil-Rivas, R. G. Tedeschi, and L. G. Calhoun (eds.), *Helping Families and Communities Recover from Disaster: Lessons Learned from Hurricane Katrina and Its Aftermath*, Washington DC: American Psychological Association.

Quarantelli, E. L. and R. R. Dynes. (1977) "Response to Social Crisis and Disaster," *Annual Review of Sociology* 3(1): 23–49.

Rabinovitz, F. F. (1967) "Politics, Personality, and Planning," *Public Administration Review* 27(1): 18–24.

Ranous, R. (2012) *A Compendium of Best Practices and Lessons Learned for Improving Local Community Recovery from Disastrous Hazardous Materials Transportation Incidents*, Arlington VA: Transportation Research Board.

Reddy, S. D. (2000) "Factors Influencing the Incorporation of Hazard Mitigation During Recovery from Disaster," *Natural Hazards* 22(2): 185–201.

Rubin, C. B. (1985) "The Community Recovery Process in the United States after a Major Natural Disaster," *International Journal of Mass Emergencies and Disasters* 3(2): 9–28.

Rubin, C. B. (2009) "Long Term Recovery from Disasters—The Neglected Component of Emergency Management," *Journal of Homeland Security and Emergency Management* 6(1): 1–17.

Schwab, J. C. and K. C. Topping. (2010) "Hazard Mitigation: An Essential Role for Planners," 1–14, in J. C. Schwab (ed.), *Hazard Mitigation: Integrating Best Practices into Planning*, Planning Advisory Service Report 560, Chicago: American Planning Association.

Smith, G. P. and D. Wenger. (2007) "Sustainable Disaster Recovery: Operationalizing an Existing Agenda," 234–257, in H. D. Rodríguez, E. L. Quarantelli, and R. R. Dynes (eds.), *Handbook of Disaster Research*, New York: Springer.

Smith, G. P. and T. Birkland. (2012) "Building a Theory of Recovery: Institutional Dimensions," *International Journal of Mass Emergencies and Disasters* 30(2): 147–170.

Tierney, K. and A. Oliver-Smith. (2012) "Social Dimensions of Disaster Recovery," *International Journal of Mass Emergencies & Disasters* 30(2): 123–146.

Waugh, W. L. and G. Streib. (2006) "Collaboration and Leadership for Effective Emergency Management," *Public Administration Review* 66(s1): 131–140.

Webb, G. R., K. J. Tierney, and J. M. Dahlhamer. (2002) "Predicting Long-Term Business Recovery from Disaster: A Comparison of the Loma Prieta Earthquake and Hurricane Andrew," *Global Environmental Change Part B: Environmental Hazards* 4(2): 45–58.

Weber, L. and L. Peek (eds.). (2012) *Displaced: Life in the Katrina Diaspora*, Austin TX: University of Texas Press.

Wenger, D. E. (1978) "Community Response to Disaster: Functional and Structural Alterations," 18–47, in E. L. Quarantelli (ed.), *Disasters: Theory and Research*, London: Sage Publications.

Wenger, D. E. (1987) "Collective Behavior and Disasters Research," 213–238, in R. R. Dynes, B. De Marchi, and C. Pelanda (eds.), *The Sociology of Disasters*, Milan, Italy: Franco Angeli Press.

Xiao, Y. and W. G. Peacock. (2014) "Do Hazard Mitigation and Preparedness Reduce Physical Damage to Businesses in Disasters: The Critical Role of Business Disaster Planning," *Natural Hazards Review* 15(3): 04014007.

Xiao, Y. and S. Van Zandt. (2012) "Building Community Resiliency: Spatial Links between Households and Businesses in Post-Disaster Recovery," *Urban Studies* 49(11): 2523–2542.

15

THE NATIONAL DISASTER RECOVERY FRAMEWORK

John T. Cooper, Jr. and Jaimie Hicks Masterson

Introduction

Over the past two decades, we've seen changes in how recovery is handled at the federal level in the United States. As mentioned in previous chapters, recovery has long been considered the least understood among the four phases—mitigation, preparedness, response, and recovery (Berke and Beatley 1997; Haas et al. 1977; Mileti 1999; Olshansky 2005; Rubin et al. 1985). Quarantelli defines recovery as "bringing the post disaster situation to some level of acceptability [which] may or may not be the same as the pre-impact level" (1999: 2). Others have defined recovery in stages, short-term and long-term. Short-term recovery calls for temporary measures to get critical services and facilities up and running to a functional state as well as efforts to house affected populations and it can take days to weeks after the disaster (Haas et al. 1977; Masterson et al. 2014). Long-term recovery focuses on reconstruction and returning a community to a fully operational state, usually lasting months to years after the disaster (Haas et al. 1977; Masterson et al. 2014). Researchers have suggested that recovery look beyond restoring pre-impact conditions to the overall betterment of the community (Blaikie et al. 1994; Enarson and Morrow 1998; Godschalk 1991; Haas et al. 1977; Mileti 1999; Reddy 2000; Rubin 1991).

A community is considered resilient if it is robust enough to withstand disaster impacts, rapidly recover to pre-existing levels, and enhances community conditions to better than pre-existing conditions (Bruneau et al. 2003; Mileti 1999; Wisner et al. 2004). Community resilience is also the ability to withstand a disturbance and then to renew and reorganize through adaptive learning to avoid past mistakes (Gunderson and Holling 2002). Other definitions of resilience often mention the speed at which recovery takes place (Peacock et al. 2008; Bruneau et al. 2003). However, there is little discussion of what role the federal government plays in helping communities become more resilient during the disaster recovery phase.

In this chapter, we follow the evolution of federal disaster recovery programs and the "steadily growing federal role in post-disaster recovery policy and priority setting, as well as funding for recovery from large disasters" (Johnson and Olshansky 2017: 307). First, we describe the precursors of our current programs—the Federal Response Plan and the Stafford Act. Next, we explore the circumstances leading up to the creation of the National Disaster Recovery Framework (NDRF), the principles of the NDRF, and what is considered a "successful recovery." We then examine how the NDRF performed and changes in recovery policy post-Hurricane Sandy. Finally, we will look ahead to the future of disaster recovery at the federal level.

Background

Congress passed the Disaster Relief Act in 1974 to establish a process for federal assistance to disaster-stricken communities. The Federal Emergency Management Agency (FEMA) was established in 1979, bringing together many agencies and departments under one roof to handle emergency- and disaster-related issues in the United States. FEMA's original aims were to: 1) anticipate, prepare for, and respond to major civil emergencies; 2) use all available resources most efficiently; 3) be extensions of missions of current agencies, whenever possible; and 4) closely link hazard mitigation activities with emergency preparedness and response functions (AHSDR 2009). The focus was on effective *response* to emergencies and disasters, *preparation* for the response, and *mitigation* of hazards to ultimately reduce and eliminate the need for response. Recovery was not a part of the original focus, in and of itself.

In 1988, the Robert T. Stafford Disaster Relief and Emergency Assistance Act (Stafford Act) established federal disaster relief policies and procedures. Most federal disaster policies and practices today stem from this legislation, which described the need and procedures for a managerial framework of disaster response under a set of Emergency Support Function (ESF) annexes within the Federal Response Plan. Originally there were 12 annexes: ESF 1-Transportation; ESF 2-Communications; ESF 3-Public Works and Engineering; ESF 4-Firefighting; ESF 5-Information and Planning; ESF 6-Mass Care; ESF 7-Resource Support; ESF 8-Health and Medical Services; ESF 9-Urban Search and Rescue; ESF 10-Hazardous Materials; ESF 11-Food; and ESF 12-Energy. Each annex described agencies, departments, and organizations whose role and activation is based on the characteristics of the disaster and the needs of the public. This framework builds off the principle that disasters ultimately occur at the local level, where emergency responders are on the ground to provide support (Perry and Lindell 2007). The ESFs were intended to be a logical extension of the daily responsibilities of local emergency managers, police, and fire departments, and other response-oriented personnel (Quarantelli 1999). If the disaster is at a scale that exceeds the capacity of local agencies, the state's ESF agencies are activated to provide support. If the scale of the disaster exceeds the capacity of state resources, federal ESF agencies are activated to provide support. Generally, FEMA acts as a partner to states and tribal nations to facilitate coordination, not as a manager of state disaster decisions and processes (Mitchell 2006). However, recovery was not specifically addressed in the legislation.

Today it is still difficult to coordinate federal, state, and local organizations for recovery, in part because there has not been a recurrent necessity for coordination, as there has been with emergency response personnel (Quarantelli 1999). More recently, Johnson and Olshansky observe "the federal government has continually struggled to find a model for recovery that is both responsive to victims' needs and mindful of the nation's purse" (2017: 306). The lack of attention to recovery was evident in 1990 when only two trainings were available annually on mitigation and recovery through FEMA's emergency management training program (Rubin and Popkin 1990). However, although there seemed to be little evidence of attention to recovery in practice, FEMA has attempted to navigate the "interaction and decision making among a variety of groups and institutions, including households, organizations, businesses, the broader community and society" (Mileti 1999: 240). For example, ESF 6-Mass Care included sheltering and temporary housing for victims—short-term recovery components. With this support function, the agency began to handle the loss of available housing (AHSDR 2009).

Although the Stafford Act did not specifically address recovery, it established new funding streams to speed short-term recovery. The Public Assistance (PA) program made funds available for debris removal and critical infrastructure and facilities repair, such as sewage systems, water, schools, and government facilities. Cost-sharing between local or state and federal levels to rebuild infrastructure and public facilities was established after the Mt. St. Helens eruption where the state assisted in covering 25 percent of the costs (AHSDR 2009). Funds for recovery projects and grants for hazard mitigation and planning were also made available in the legislation.

Three more annexes were added to the Emergency Support Functions in 2004—ESF 13-Public Safety and Security; ESF 14-Long-Term Community Recovery, and ESF 15-External Affairs. The inclusion of ESF-14 marked a shift in FEMA's principles and scope, broadening recovery to "long-term community recovery" to help communities beyond immediate response and short-term recovery.

The 2005 and 2008 Hurricane Seasons

Less than a year after the addition of ESF-14, the impacts of Hurricanes Katrina and Rita quickly surpassed local and state capacity to handle response and recovery. In addition, it soon became evident that the federal government also could not handle catastrophes of this magnitude (AHSDR 2009). At the time, under the updated Stafford Act, FEMA was expected to handle temporary housing for victims (AHSDR 2009). After Katrina, for example, 150,000 trailers were ordered and still thousands of households were on wait lists. Each trailer cost roughly $59,000, totaling $5.5 billion in federal expenses (AHSDR 2009).

Trailers were the main solution following Katrina and Rita, due to legal interpretations of what FEMA could do under the Stafford Act. Specifically, FEMA leadership interpreted the law to mean that the agency could not provide funds for rental repairs, greatly limiting housing options, particularly for low-income households (AHSDR 2009). As a result, injustices and inequities permeated the entire recovery process. In other instances, FEMA delayed or denied assistance to qualified disaster victims, particularly to low-income households and minorities after Katrina (Hooks and Miller 2006). In reality, it seemed that FEMA's assistance was designed more for higher-income families who could rely on savings or commercial loans than for low-income families who were completely dependent on government assistance (Hooks and Miller 2006). This left many in the greatest need without options for recovery. When low-income families applicants did qualify, resources were slow to be delivered. Still other housing programs were used at the time, though to a lesser degree, but they also exposed major problems. For example, the Section 403 Hotel Program, a temporary housing solution for victims, created confusion and unpredictability as FEMA incrementally extended occupancy status. Consequently, tenants did not know if they would be allowed to continue living in their current situation from one month to the next. Likewise, the Rental Program, which provided vouchers to tenants, had several deadline changes, creating confusion and frustration among tenants and landlords alike. There were also flawed public assistance programs to help communities restore their normal functioning (AHSDR 2009). These factors contributed to slow recovery following Katrina and Rita and exposed FEMA's ill-equipped recovery process.

Prior to FEMA's creation, the Department of Housing and Urban Development (HUD) provided recovery assistance to communities. This was a logical assignment because HUD's mission is to "create strong, sustainable, inclusive communities and quality affordable homes for all" (HUD n.d.). Since FEMA's creation, that agency had largely taken on all roles pertaining to disasters, including housing recovery. Following Katrina, however, HUD was given authority to provide housing, but only to public housing clients affected by the hurricanes—a fraction of the total housing demand (AHSDR 2009). FEMA could have given more authority to HUD, but chose not to due to concerns that HUD could not meet the large demand (AHSDR 2009). That is, because HUD traditionally provided vouchers only for existing housing, many thought the limited housing choices available following a disaster would be insufficient to support the demand (AHSDR 2009).

From the criticisms following the 2005 hurricane season, FEMA and HUD agreed to work closely together to form the Disaster Housing Assistance Program. When Hurricane Ike struck the

Texas coast in 2008, many looked to see improvements in the programs and the recovery effort as a whole. After Hurricane Ike, unlike after Hurricanes Katrina and Rita, the Disaster Housing Assistance Program (DHAP-Ike) limited the use of funds to purchase mobile homes or trailers. Instead, housing vouchers were utilized. Unfortunately, because there was a shortage of rental housing (and housing in general), many residents were forced to find housing far from their pre-disaster homes. Prior to Hurricane Ike, only single-family homes received assistance. In December of 2008, a pilot program was created through the Federal Assistance to Individuals and Households (IHP) program to provide assistance to qualified multi-family properties.

A New Framework

Although the 2005 and 2008 hurricane seasons exposed deficiencies in the federal capacity to recover from disasters, these catastrophic events also created a "window of opportunity" (Birkland 1997) to improve the recovery process. To this end, in 2009 President Obama directed the U.S. Department of Homeland Security, which houses FEMA, and HUD, to develop a Long-Term Disaster Recovery (LTDR) working group to provide guidance on community recovery following a disaster. The LTDR working group released the *National Disaster Recovery Framework* (NDRF) in September 2011.

Today when a disaster exceeds the capacity of state, local, and tribal recovery programs, the federal government provides assistance through the *NDRF*, which was designed to be paired with the ESF annexes and the new *National Response Framework* (NRF). The NDRF specifies that FEMA is the federal agency responsible for disaster response and HUD is the federal agency responsible for long-term housing recovery. When a disaster occurs, NRF ESFs are to be activated. Once the disaster response begins to move to the recovery phase, responsibility transitions from the NRF to the NDRF.

Successful Recovery Factors with NDRF

The NDRF describes seven factors that contribute to a successful recovery process. The first is *Effective Decision Making and Coordination*, which occurs when leadership roles and responsibilities are well defined. Also, stakeholders should take the time to examine a number of alternatives for recovery in the community. This often occurs alongside other planning processes (Beatley 2009; Masterson et al. 2014). NDRF also describes the value of metrics to measure progress and hold key players accountable, including voluntary organizations active in disasters (VOADs). Developing an action-oriented implementation plan is considered a planning best-practice (Berke and Godschalk 2006).

Integration of Community Recovery Planning Processes is the second factor for successful recovery. Pre-disaster planning should take place in the community—for preparedness, response, mitigation, and *recovery*. Businesses and individuals should play a role in pre-disaster recovery planning, providing input that will ultimately help individuals, households, and the community prepare for recovery. The NDRF also describes how critical infrastructure should be addressed within this planning process, especially in terms of coordination across jurisdictions. Communities would then identify and prioritize actions for recovery. Such actions should be in line with, and incorporated into, comprehensive land use plans—something that is reiterated in the literature (Beatley 2009; Masterson et al. 2014). The framework also states that recovery plans should establish an organizational framework to manage recovery planning. Those involved in recovery planning should help to revise response plans to include recovery and mitigation. This also marks an important policy shift. Although the NDRF does not mandate production of pre-disaster recovery *plans*, it does advocate pre-disaster recovery planning—just as in response planning and mitigation planning—as a primary mechanism to reduce the duration of impact and improve overall outcomes. Generally, there has been weak support for planning from the federal government. Instead, costly funding for the construction of dams, levees, seawalls, and beach renourishment programs incentivize development that might not have occurred without these protections (Burby et al. 1999). Even mandated hazard mitigation plans, which can be vital tools to reduce long-term disaster costs, are generally weak and do not take advantage of the wide range of strategies a community can employ (Masterson et al. 2014). There are few incentives for local communities to plan for disaster recovery, so state and federal governments should provide incentives alongside the NDRF to encourage high-quality and consistent recovery plans (Berke and Campanella 2006). Such plans are critical tools for addressing interorganizational conflicts, gaps, and duplication (Quarantelli 1999).

A *Well-Managed Recovery* is the third factor in successful recovery as defined by the NDRF. A well-managed recovery starts with the relationships and networks established to accomplish the complex activities required for recovery. Such partnerships should be both vertical (among different levels of government) and horizontal (across local agencies, departments, organizations, etc.). Sound leadership and the development of both vertical and horizontal organizational ties has been found to yield better plans and quicker response and recovery (Berke and Campanella 2006; Briggs 2004; Perry and Lindell 2007). The NDRF says communities should be able to coordinate with outside sources, such as other jurisdictions, universities, foundations, etc. In addition, a well-managed recovery is directed by leaders who can guide the transition between response and recovery, complies with the building standards and guidelines of the comprehensive plan., and has strong support networks to facilitate outcomes (Berke and Campanella 2006; Bolin and Bolton 1986; Briggs 2004). Not only should the recovery planning process establish partnerships pre-disaster, but also provide equal opportunity for underserved and socially vulnerable populations. Grassroots organizing should be incorporated to promote inclusiveness to form new and renewed partnerships (Berke et al. 2011; Berke and Campanella 2006; Briggs 2004). Planning efforts performed by communities and organizations create more informed, engaged, adaptive and therefore resilient communities (Brody et al. 2011).

Proactive Community Engagement and Public Participation and Public Awareness is the fourth factor of a successful recovery, according to the NDRF. Collaboration should occur between varying stakeholders to provide needed resources for recovery. Inclusive planning practices have been touted in the planning field as a way to have authentic dialogue and engage citizens in civic and democratic processes and garner community support (Berke and Campanella 2006; Briggs 2004; Innes and Booher 2004). In reality, this community support may be more important than the hazards a community faces (Lindell et al. 1996). For decades, socially vulnerable populations have been studied as they relate to disaster, and repeatedly, these groups are marginalized and forgotten (Bryant and Mohai 1992; Pastor et al. 2006; Peacock et al. 1997). Taking particular care to include diverse populations, particularly socially vulnerable groups, will go a long way to effectively planning. Likewise, the NDRF states that recovery plans should be formed before a disaster to include local opinions and needs so they can be implemented immediately following a disaster. There are a number of strategies and inclusive techniques to reach all populations of varying languages and literacy. Effective communication during disaster events has long been a recommendation (Perry and Lindell 2007). Also, establishing a culture of unity with open and authentic dialogue builds community capacity (Innes and Booher 2004). Gaining trust in these ways can yield community efficacy and collective action (Meyer 2013).

The NDRF describes a *Well-Administered Financial Acquisition* as the fifth factor in successful disaster recovery. Stakeholders should be aware of a diverse array of funding sources for post-disaster recovery. The knowledge and ability to acquire and administer external funding and programs is also essential. Funders should have flexibility when administering such funds and resources, and financial monitoring should take place to prevent fraud. Federal dollars also help local business to speed overall community recovery (FEMA 2011). The speed at which victims and communities receive funds has long been a concern to the recovery process (Brown et al. 2013).

Organizational Flexibility is the sixth factor affecting a recovery within the NDRF. Scalability and adaptability of government systems, an emerging issue, ultimately allows for easier application of laws, regulations, and policies. Staffing also contributes to flexibility, in that the compression of activities immediately following a disaster may create a need for a substantial increase in staffing levels for some government functions such as building inspections. Additionally, appropriate training of staff plays a role in flexibility, allowing their roles and responsibilities adapt to post-disaster needs. Such adaptability requires partnerships to be established pre-disaster within the organizational framework to accomplish the wide variety of activities that need to take place. The malleability of staff roles and responsibilities allows for the most effective and productive use of taxpayer-funded programs, which are held accountable for post-disaster activities (FEMA 2011). Also, Berkes, Colding, and Folke explain that managing for sustainability and resilience in socio-economic systems "means not pushing the system to its limits but maintaining diversity and variability, leaving some slack and flexibility, and not trying to optimize some parts of the system but maintaining redundancy" (2003: 15).

Resilient Rebuilding is the NDRF's final factor for successful recovery. Rebuilding to mitigate future disaster impacts with the aim of ultimately eliminating risk is the key point in reconstruction efforts. Often we see a desire to build back quickly to previous conditions (Quarantelli 1999). We have learned however, that developing plans and policies to build back "smarter" and more resilient should be the focus (Blaikie et al. 1994; Brody et al. 2011; Enarson and Morrow 1998; Godschalk 1991; Haas et al. 1977; Masterson et al. 2014; Mileti 1999; Reddy 2000; Rubin 1991). NDRF recommends that building codes and land use ordinances should be utilized in communities—a point seen throughout the research literature (Berke and Campanella 2006; Masterson et al. 2014). NDRF suggests that structures exposed to hazards can be removed or retrofitted and that risk reduction should be a part of business and governmental operations and practices. We know that planning teams headed by emergency managers and city planners should work together to inform hazard mitigation policy (Godschalk 2003). If communities address hazard mitigation throughout their many plans, there is a greater likelihood they have thought through their hazard risks and can respond and recover effectively. For example, when a community has incorporated hazard mitigation into zoning and development codes and home association regulations, it has taken great strides to reduce vulnerabilities.

Key Players

Just as disaster response is scalable, the NDRF is intended to be scalable. Whether a disaster is presidentially declared or one that can be handled locally, the framework still applies. Federal assistance in disasters acts as a supplement to state and local resources, primarily because of state sovereignty and the federal principle that emergencies and disasters are best handled at the local level. In fact, only one percent of all disasters exceed state and local capacity and receive PDDs (Schwab 1998). Only when local and state governments do not have the resources does the federal government provide assistance.

Just as with the NRF, key players should be specified as coordinating agencies or partners at each scale. The NDRF establishes a Federal Disaster Recovery Coordinator, State or Tribal Disaster Recovery Coordinators, Local Disaster Recovery Managers, and Recovery Support Functions (RSFs). States and localities determine the coordinating agency or partner to work with federal coordinators. RSFs help activate key players to accomplish essential tasks and support efforts to recover. The RSFs include Recovery Planning and Capacity Building; Economic, Health and

Social Services; Housing; Infrastructure Systems; and Natural and Cultural Resources. These RSFs are different from the ESFs that guide emergency response. ESF timeframes last days to weeks following a disaster whereas RSF timeframes extend months to years following a disaster even though they might overlap with ESFs. Each ESF hands over responsibilities to RSFs once response efforts are completed. The following describes federal coordinating resources to support state and local governments.

RSF Community Planning and Capacity is coordinated by FEMA and primarily supported by FEMA and the Department of Health and Human Services (HHS), along with 13 other agencies. A primary goal of this RSF is to help organize, plan, manage, and implement recovery. Some of the key objectives of this RSF are to promote mitigation planning and to incorporate mitigation and recovery into local community plans and initiatives. Another important objective is to develop local leadership capacity through cross-training stakeholders such as emergency managers, city managers, planning staff, economic development staff, other local officials, and nonprofit and private sector partners. This RSF strives to utilize partnerships with extension programs, universities, national professional associations, and nongovernmental organizations to expand resources. The RSF also maintains communications among all partners in preparation for recovery.

RSF Economic is coordinated by the Department of Commerce (DOC) and primarily supported by FEMA, DOC, Department of Labor (DOL), Small Business Administration (SBA), U.S. Department of Treasury (TREAS), and U.S. Department of Agriculture (USDA), and with four other supporting agencies. For example, in Hurricane Sandy, Congress allocated $545 million to SBA to administer to businesses to ensure their return (Johnson and Olshansky 2017). The primary goal of this RSF is to rebuild businesses and employment for the return of economic and business activity.

RSF Health and Social Services is coordinated by the HHS) and primarily supported by the Corporation for National and Community Service (CNCS), FEMA, National Protection Program Directorate (NPPD), Office of Civil Rights and Civil Liberties (CRCL), Department of Interior (DOI), Department of Justice (DOJ), DOL, Department of Education (ED), Environmental Protection Agency (EPA), and Department of Veterans Affairs (VA). Other supporting agencies include the Department of Transportation (DOT), SBA, TREAS, USDA, American Red Cross (ARC), and National Volunteer Organizations Active in Disasters (NVOAD). This RSF supports local groups in restoring public health and health care and social services networks, and coordinates with ESFs 3, 6, 8, and 11 from the NRF. Objectives of the RSF include a focused attention on socially vulnerable populations, conducting needs assessments, providing technical assistance, and promoting clear communications across organizations.

RSF Housing is coordinated by HUD with primary agencies being FEMA, the Department of Justice, HUD, and the USDA. Supporting Organizations include CNCS, DOC, DOE, EPA, HHS, SBA, US Access Board, VA, ARC, NVOAD. The primary goal of this RSF is to "address pre- and post- disaster housing issues and coordinate and facilitate the delivery of. . .resources. . .in the rehabilitation and reconstruction of destroyed and damaged housing and to develop new accessible, permanent housing options" (FEMA 2011: 55). FEMA is the coordinating agency under NRF for ESF 6, now named Mass Care, Emergency Assistance, Temporary Housing, and Human Services. ESF 6 is able to move an individual or family from response (immediately after the disaster where the primary concerns are mass evacuations, sheltering, distribution of supplies, donations management, support for dependents and pets) to short-term and long-term recovery with temporary housing and repair loan assistance as well as non-housing functions such as crisis counseling, case management, unemployment services, legal services, and other service programs (FEMA 2011). The expanded ESF 6 is strongly linked to *RSF Housing* and *RSF Health and Social Services*. In a disaster, FEMA activates ESF 6 to respond to immediate needs of victims. As the ESF 6 role diminishes, HUD activates *RSF Housing*, which assumes these activities and roles. A part of the challenge is this period of transition from ESF 6 to *RSF Housing*.

RSF Infrastructure Systems is coordinated by DOD and USACE with primary agencies being FEMA, National Protection Programs Directorate (NPPD), U.S. Army Corps of Engineers (USACE), Department of Energy (DOE), and DOT. Supporting agencies include DHS, DOC, DOI, ED, EPA, Federal Communications Commission (FCC), General Services Administration (GSA), HHS, Nuclear Regulatory Commission (NRC), TREAS, USDA, and Tennessee Valley Authority (TVA). The focus of this support function is to provide support to state and local governments for infrastructure reconstruction following a disaster. For instance, in Hurricane Sandy DOT received $13 billion for road and rail repairs and USACE received $5.4 billion for coastal restoration from Congress. The goals of this RSF are the federal coordination for infrastructure priorities and action plan, and the leveraging of assets to reconstruct needed infrastructure.

RSF Natural and Cultural Resources is coordinated by DOI, with primary agencies being FEMA, DOI, and the Environmental Protection Agency (EPA). Supporting organizations include Advisory Council on Historic Preservation (ACHP), Corporation for National and Community Service (CNCS), Council on Environmental Quality (CEQ), DOC, Institute of Museum and Library Services (IMLS), Library of Congress (LOC), National Endowment for the Arts (NEA), National Endowment for the Humanities (NEH), USACE, USDA, and Heritage Preservation. The goal and primary function of this support function is to address the "long-term environmental and cultural resource recovery needs" (FEMA 2011: 61).

Social Vulnerability and Climate Change

As our understanding of the four disaster phases—mitigation, preparedness, response, and recovery—has become clearer, so has our understanding of the plethora of other issues within disaster planning. In particular, socially vulnerable populations and climate change are two emerging topics that have garnered significant attention. Understanding and identifying specific population segments that have additional needs is vital to an effective emergency response. Specifically, recovery efforts are not just for the recovery of the "average resident." Identifying and understanding the needs of particular population segments that may have a more difficult time coping with and recovering from disasters (i.e., socially vulnerable groups) is a key component to a more rapid recovery. In fact, through social vulnerability mapping, communities can actually predict levels of damage (Van Zandt et al. 2012). Oftentimes, socially vulnerable groups have been neglected because of systematic and institutionalized discrimination (Nigg and Tierney 1993). When the draft NDRF was published, there was only one instance in which it provided specific language on non-discrimination and even that was within a small checklist. Because discrimination against certain populations occurs systematically, the Brookings Institution recommends that awareness training on discrimination issues in response and recovery phases be a part of federal, state, and local requirements (Brookings-Bern Project 2010). Even when discrimination is unintentional, training can offer insight into socially vulnerable populations and how discrimination is manifested (Brookings-Bern Project 2010). Although there is little attention to specific discrimination concerns within the NDRF, there is an understanding of socially vulnerable populations (Cooper and Waddell 2010). At the time of that study, the NDRF was not yet complete, but there was a significant discussion of "underserved populations" and a more holistic approach to disasters. Cooper and Waddell (2010) gave the NDRF a strong rating because there was a statute or policy that addressed vulnerable populations.

Climate change is another topic that has emerged in disaster planning. Being a politically polarized issue, climate change has taken time to become visible within disaster planning practice. The NDRF does mention climate change and received a strong rating on its ability to address climate change (Cooper and Waddell 2010). Unfortunately, social vulnerability and climate change are inadequately discussed in federal and state disaster plans. Social mitigation and social vulnerability need to be a part of planning processes and must come at federal, state, and local levels (Cooper and Waddell 2010).

NDRF in Practice

Although NDRF is not a legislative statute, many states and localities have adopted and implemented the framework. The state of Louisiana began using NDRF in August 2013 with Hurricane Isaac, with St John Parish being one of the first localities to adopt it. Many within FEMA and NDRF hailed such local participation and grass roots efforts to plan for recovery. NDRF coordinator Wayne Rickard stated that local engagement in planning "Lays the foundation for successful recovery in any community. . .and can help jumpstart a community's cycle of success" (FEMA 2013).

The first major test in practice for NDRF came following Hurricane Sandy with a multi-state implementation. To facilitate coordination, President Obama asked FEMA to be the lead agency of the response (NRF) and recovery (NDRF). President Obama also asked the HUD secretary to closely work with FEMA to coordinate specific needs with New Jersey and New York governors, mayors, and other elected officials (Johnson and Olshansky 2017). Still there were areas for improvement in recovery, including the recognition for better federal coordination during the transition from the National Response Framework ESFs to National Disaster Recovery Framework RSFs. There is "strong evidence . . . since Sandy that the federal recovery framework is still overly complicated and potentially less effective because of the many disconnected pots of money" (Johnson and Olshansky 2017: 307). The FEMA Hazard Mitigation Grant Program must coordinate with HUDs Community Development Block Grant-Disaster Recovery (CDBG-DR) and FEMAs Public Assistance Program (Johnson and Olshansky 2017). Another recognized need was to use planning and analysis as the basis for decision-making (HUD 2013). In addition, HUD released six recommendations to guide federal investment in the region.

The first recommendation focused on infrastructure investments. Within NDRF, infrastructure is only mentioned twice, and both of those are in a single RSF—*Health and Social Services*. HUD recommends regional approaches to critical infrastructure such as electricity, water and wastewater, and communications networks. Mapping infrastructure interdependencies may be one way to understand and prioritize regional investment, which should also include an assessment of future conditions such as sea level rise and population growth (Guarnacci 2012; HUD 2013; Mayunga 2009). In practice, infrastructure should leverage alternative assets and be connected to other urban planning considerations and documents—something that is a core principle within the NDRF (Guarnacci 2012; HUD 2013; Mayunga 2009; Yi and Yang 2014).

Along with infrastructure investments addressing future conditions, overall urban development should take into consideration its own current and future risks. Assessing future risks is only mentioned three times within the NDRF, primarily through collecting and analyzing existing and future data within *RSF Community Planning and Capacity Building*. In order to assess such conditions, mapping tools are recommended for residents and communities to help visualize their vulnerabilities. Mapping, specifically, is not mentioned in the NDRF, however, prioritizing the engagement of vulnerable populations is an NDRF core principle (HUD 2013).

The third recommendation is for safe, affordable housing options. One way to do this is by helping victims stay in their homes with repair assistance, something that had not previously been provided. Within this recommendation is the need to retrofit structures and to create nationally consistent mortgage policies to reduce hazard exposures (HUD 2013). Within the NDRF, "affordable housing" is mentioned in *RSF Community Planning and Capacity Building* and in *RSF Housing*, though the word "affordable" is only mentioned thrice. First it is discussed briefly in regard to the recovery of local economies that are dependent on adequate housing, especially affordable housing. Second it is mentioned as an outcome of RSF Housing, or to rebuild permanent housing, including affordable housing. Third it is mentioned in reference to rental housing within a broader discussion of inclusive recovery efforts for the whole community.

The fourth recommendation is to support small businesses by making it easier for them to access federal contracts for rebuilding. Some recommendations include developing a one-stop shop online for business recovery. Improving access to recovery loans by increasing SBA's disaster loans limit and expediting disbursement of small loan amounts is also recommended. Another opportunity is in creating specialized skills training programs for rebuilding (HUD 2013). Within NDRF, businesses are discussed extensively, although "small business" is only mentioned twice. *RSF Economic*, the support function that specifically deals with rebuilding businesses and facilitating the return of economic activities to communities, specifically mentions leveraging federal resources to speed recovery.

The fifth recommendation focuses on insurance. HUD recommends property owners mitigate future risks, a component addressed within the NDRF. *RSF Economic* specifically discusses insurance, and the purchase of insurance is included within the Individuals and Families Pre-Disaster Checklist (FEMA 2011). HUD also recommends streamlining payouts at the federal level. Although the NDRF does not specifically discuss federal streamlining of funds, its Federal Government Post-Disaster Checklist does discuss adjusting federal assistance programs to provide timely resources to communities. However, streamlining funds is specifically mentioned within the State Government and Tribal Government Post-Disaster Planning Activities checklist. Finally, HUD recommended examining ways to address affordability challenges posed by Congressionally mandated reforms to the National Flood Insurance Program in the *Biggert-Waters Act* of 2012, which were intended to charge actuarial rates to households and businesses within the 100-year floodplain. However, this problem was subsequently obviated by the *Homeowner Flood Insurance Affordability Act* of 2014, which eliminated some of the more stringent flood insurance financing reforms (Schwab et al. 2014).

Lastly, HUD recommends building local government capacity by supporting regional planning efforts to create and implement locally created recovery and rebuilding plans and strategies. The NDRF mentions "recovery planning" extensively and reiterates the value of comprehensive and regional planning at all levels (FEMA 2011). Regional planning has been shown to take more holistic approaches to concerns and issues that do not necessarily fall within the boundaries of individual jurisdictions. Another recommendation is the need for a federally funded local Disaster Recovery Manager (HUD 2013). Currently, NDRF recommends a Local Disaster Recovery manager, but it does not provide funding mechanisms to support much needed local capacity. NDRF speaks of "local government capacity" within *RSF Community Planning and Capacity Building*, but does not give specific ways in which resources would be allocated to help increase capacity.

Sandy Recovery Improvement Act Changes

President Obama issued an executive order forming the Hurricane Sandy Rebuilding Task Force, which included representatives from all federal agencies a specified coordination with FEMA. Within 180 days of first meeting, the task force produced a long-term rebuilding plan with outcomes, goals, actions, recommended policies, and monitoring of progress. Among recommendations from HUD, the Hurricane Sandy Rebuilding Task Force released a Rebuilding Strategy in August 2013 to serve as a model for communities. There are 69 total policy recommendations throughout the report, making it clear there is room for significant improvement in disaster recovery and the implementation of NDRF (HSRTF 2013). The recommendations focused on strategies for government coordination and accountability, aligning funding with local needs, and a regional approach to resilience (HSRTF 2013; Johnson and Olshansky, 2017). Consequently, President Obama signed the Sandy Recovery Improvement Act (SRIA) in January 2013. The legislation is the most significant change to FEMA since the Stafford Act. Changes were made to tribal procedures, the individual assistance program, public assistance program, hazard mitigation program, dispute resolution, and federal assistance to individuals and households (FEMA 2014).

Tribal Procedures

SRIA 2013 amended Sections 401 and 501 of the Stafford Act on tribal procedures. Previously, tribes had to request a disaster declaration through state offices, treating tribal nations more like local governments as opposed to sovereign nations. Now tribes have the ability to directly request a disaster declaration from the federal government. The new policy seeks to better coordinate with tribal groups and to eliminate other procedures that may limit a tribe's sovereignty as a nation.

Individual and Family Assistance

SRIA mandated the review of factors that may contribute to the need for additional assistance for individuals and families—something that had not taken place since the 1988 legislation. The goal of this change is to speed the process by which individuals receive recovery resources. For instance, unsatisfied needs for childcare have long been seen as an impediment to housing and business recovery for families with children. SRIA expanded federal grant eligibility to private businesses that operate childcare services and facilities through the Other Needs Assistance (ONA) program. ONA previously only covered additional medical, dental, and funeral expenses during time of recovery. Now, personal property (clothing, furniture, etc.), transportation costs, and other expenses deemed necessary for recovery can qualify for financial assistance under ONA. A household can also receive support for 18 months and up to $31,900 with a 75 percent federal contribution and a 25 percent state contribution. Prior to 2008, there was no funding from the federal government for the repair and improvement of rental properties. SRIA has now made the multi-family housing assistance pilot program (established after Hurricane Ike) permanent by allowing rental housing to be used as temporary housing for victims and providing more housing choices for victims following a disaster. It also authorizes FEMA to enter into lease agreements with private owners of multi-unit apartment facilities (Brown et al. 2013).

Public Assistance

Qualified states and localities can apply for funds through the Public Assistance program (PA) to help lessen the burden of clearing debris and repairing and rebuilding public facilities. SRIA provides alternative procedures for Section 406 of the Stafford Act, which originally only allowed states and localities to be *reimbursed* for such costs. Now, FEMA can provide financial assistance before local jurisdictions pay to repair damaged public facilities, based on estimates of eligible costs. The goal is to speed recovery and incentivize cost-effective repairs. This is also true for debris removal activities and could ultimately encourage applicants to recycle the debris without reducing the grant amount. Reimbursements for overtime wages of specialists can also be approved. To reduce costs, localities can create a debris removal plan before a disaster for a 5 percent additional funding opportunity (Brown et al. 2013).

Another alternative for localities is an "in-lieu contribution" to repair, restore, or replace an alternate public facility. Prior to SRIA, FEMA was required by law to reduce eligible costs for in-lieu PA projects by 10 percent for public agencies and reduce by 25 percent for private non-profits. Since SRIA, there is no reduction. Instead, this program incentivizes localities to build back, not to replace existing exposed structures, but in less vulnerable areas and encourages more innovative uses of public facilities. For example, a school that was destroyed during a disaster may have been in a flood-plain in a community experiencing population loss. SRIA now allows that community to use funds for mitigation projects such as elevating police or fire facilities out of the floodplain. Also, these procedures will allow for multiple individual facilities to be a single PA project. For instance, funds for several different schools could be used to combine some schools or add additional uses to the facilities.

Hazard Mitigation Program

The Hazard Mitigation Program has traditionally been the slowest program of all disaster assistance programs. In order to expedite requests, SRIA will now allow multiple structures to be treated as a group so physically vulnerable structures could be packaged for collective buy-out instead of being processed individually. The program has also been given authority to provide 25 percent of the project grant funding as an advance (Brown et al. 2013).

Additional Changes

Other changes to the Stafford Act based on SRIA include the creation of a Dispute Resolution pilot program to resolve federal and state disagreements on costs and eligibility questions (Brown et al. 2013). SRIA has also authorized certain reimbursements of government employee wages during emergencies. This allows government employees to be paid for time that otherwise would have been unreimbursed. To speed the process of historical and environmental review, SRIA has created a joint and simultaneous process for recovery projects.

Looking Ahead

Going forward, there must be a greater focus on developing local pre-disaster recovery plans to guide short-term and long-term recovery decisions (Berke and Campanella 2006). As is the case with other community plans, recovery plans need a thorough fact basis and clear understanding of scientific evidence along with local knowledge of the community. Communities should utilize technical assistance, which should be provided by government (Johnson and Olshansky 2017), based on science-based tools and credible data (Johnson and Olshansky 2017) to "better anticipate community vulnerabilities for future disasters and to adopt measures that will reduce future human, economic, and environmental costs" (HSRTF 2013: 41). In fact, mapping software, "evidence-based information, risk-based analysis, and robust cost-benefit analyses" can help communities understand the complexities of disaster recovery and prioritize investments and decisions (HSRTF 2013: 41). As discussed previously, mapping vulnerabilities and hazard exposures can allow a community to predict damage and to subsequently mitigate and plan for recovery (Van Zandt et al. 2012). Recovery plans should promote pre-disaster mitigation planning as essential to recovery and integrate mitigation planning into other community plans, including comprehensive plans, land use plans, economic development plans, affordable housing plans, etc. (HSRTF 2013).

References

Ad Hoc Subcommittee on Disaster Recovery (AHSDR). (2009) "Far from Home: Deficiencies in Federal Disaster Housing Assistance After Hurricanes Katrina and Rita and Recommendations for Improvement," Washington DC: U.S. Government Printing Office.

Beatley, T. (2009) *Planning for Coastal Resilience: Best Practices for Calamitous Times*, Washington DC: Island Press.

Berke, P. and T. Beatley. (1997) *After the Hurricane: Linking Recovery to Sustainable Development in the Caribbean*, Baltimore MD: Johns Hopkins University Press.

Berke, P. and T. J. Campanella. (2006) "Planning for Postdisaster Resiliency," *Annals of the American Academy of Political and Social Science* 604(1): 192–207.

Berke, P., J.T. Cooper, D. Dalveson, D. Spurlock, and C. Raush. (2011) "Building Capacity for Disaster Resiliency in Six Disadvantaged Communities", *Sustainability* 3(1): 1–20.

Berke, P. and D. Godschalk. (2006) *Urban Land Use Planning*, Chicago IL: University of Illinois Press.

Berkes, F., J. Colding, and C. Folke. (2003) *Navigating Social Ecological Systems: Building Resilience for Complexity and Change*, Cambridge: Cambridge University Press.

Birkland, T. A. (1997) *After Disaster: Agenda Setting, Public Policy, and Focusing Events*, Washington DC: Georgetown University Press.

Blaikie, P., T. Cannon, I. Davis, and B. Wisner. (1994) *At Risk: Natural Hazards, People's Vulnerability, and Disasters*, London: Routledge.

Bolin, R. C. and P. A. Bolton. (1986) "Race, Religion, and Ethnicity in Disaster Recovery," *FMHI Publications*, Paper 88. http://scholarcommons.usf.edu/fmhi_pub/88

Briggs, X. D. (2004) "Social Capital: Easy Beauty or Meaningful Resouce?" *Journal of the American Planning Association* 70(2): 151–158.

Brody, S., W. Highfield, and J. E. Kand. (2011) *Rising Waters: Causes and Consequences of Flooding in the United States*, Cambridge: Cambridge University Press.

Brookings-Bern Project. (26 February 2010) "Comments on FEMA"s National Disaster Recovery Framework," *Brookings-Bern Project on Internal Displacement*.

Brown, J. T., F. X. McCarthy, and E. C. Liu. (2013) *Analysis of the Sandy Recovery Improvement Act of 2013*, Washington DC: Congressional Research Service.

Bruneau, M., S. Chang, R. Eguchi, G. Lee, T. D. O'Rourke, A. M. Reinhorm, . . . D. von Winderfeldt. (2003) "A Framework to Quantitatively Assess and Enhance the Seismic Resilience of Communities," *Earthquake Spectra* 19(4): 733–752.

Bryant, B. I. and P. Mohai. (1992) *Race and the Incidence of Environmental Hazards*, Boulder CO: Westview Press.

Burby, R., T. Beatley, P. Berke, R. Deyle, S. French, D. Godschalk, . . . and R. Platt. (1999) "Unleashing the Power of Planning to Create Disaster-Resistant Communties," *Journal of the American Planning Association* 65(3): 247–258.

Cooper, J. T. and J. Waddell. (2010) *Impact of Climate Change on Response Providers and Socially Vulnerable Communities in the US*, Boston MA: Oxfam America Research.

Enarson, E. and B. H. Morrow. (1998) *The Gendered Terrain of Disasters: Through Women's Eyes*, Miami FL: Florida International University Laboratory for Social and Behavioral Ressearch.

FEMA. (2011) *National Disaster Recovery Framework: Strengthening Disaster Recovery for the Nation*, Washington DC: Federal Emergency Management Agency.

FEMA (2013, April 13). *Parishes Take Charge of Recovery Using National Disaster Recovery Framework* [Press release]. Retrieved from www.fema.gov/news-release/2013/04/04/parishes-take-charge-recovery-using-national-disaster-recovery-framework.

FEMA. (2014) *Sandy Recovery Improvement Act Fact Sheet*, Washington DC: Federal Emergency Management Agency.

Godschalk, D. R. (1991) "Disaster Mitigation and Hazard Management," 131–160 in T. Drabek and G. Hoetmer (eds.), *Emergency Management: Principles and Practice for Local Government*, Washington DC: International City Management Association.

Godschalk, D. R. (2003) "Urban Hazard Mitigation: Creating Resilient Cities," *Natural Hazards Review* 4(3): 136–143.

Gotham, K. F. and M. Greenberg. (2008) "From 9/11 to 8/29: Post-Disaster Recovery and Rebuilding in New York and New Orleans," *Social Forces* 87(2): 1039–1062.

Guarnacci, U. (2012) "Governance for Sustainable Reconstruction after Disasters: Lessons from Nias, Indonesia," *Environmental Development* 2: 73–85.

Gunderson, L. H. and C. S. Holling (eds.). (2002) *Panarchy: Understanding Transformations in Human and Natural Systems*. Washington DC: Island Press.

Haas, J. E., R. W. Kates, and M. J. Bowden. (1977) *Reconstruction Following Disaster*, Cambridge: MIT Press.

Hooks, J. P. and T. B. Miller. (2006) "Continuing Storm: How Disaster Recovery Excludes Those Most in Need," *California Western Law Review* 43(1): 21–74.

HUD. (n. d.) Mission *U.S. Department of Housing and Urban Development Mission*. Retrieved from http://portal.hud.gov/hudportal/HUD?src=/about/mission.

HUD. (2013) Hurricane Sandy Rebuilding Task Force Fact Sheet: Progress to Date. Washington DC: U.S. Department of Housing and Urban Development.

Hurricane Sandy Rebuilding Task Force (HSRTF). (2013) *Hurricane Sandy Rebuilding Strategy: Stronger Communities, A Resilient Region*, Washington, DC: U.S. Department of Housing and Urban Development.

Innes, J. E. and D. E. Booher. (2004) "Reframing Public Participation: Strategies for the 21st Century," *Planning Theory and Practice* 5(4): 419–436.

Johnson, L. and R. Olshansky (2017). *After Great Disasters: An in-Depth Analysis of How Six Countries Managed Community Recovery*, Cambridge MA: Lincoln Institute of Land Policy.

Lindell, M. K., D. J. Whitney, C. J. Futch, and C. S. Clause. (1996). "The Local Emergency Planning Committee: A Better Way to Coordinate Disaster Planning." 234–249 in R. T. Silves and W. L. Waugh, Jr. (eds.), *Disaster Management in the U.S. and Canada: The Politics, Policymaking, Administration and Analysis of Emergency Management*, Springfield IL: Charles C. Thomas.

Masterson, J. H., W. G. Peacock, S. Van Zandt, H. Grover, L. F. Schwarz, and J. T. Cooper. (2014) *Planning for Community Resilience: A Handbook for Reducing Vulnerabilities to Disasters*, Washington DC: Island Press.

Mayunga, J. S. (2009) *Measuring the Measure: A Multi-dimensional Scale Model to Measure community Disaster Resilience in the U.S. Gulf Coast Region*, College Station TX: Texas A&M University.

Meyer, M. A. (2013) *Social Capital and Collective Efficacy for Disaster Resilience: Connecting Individuals with Communities and Vulnerability with Resilience in Hurricane-Prone Communities in Florida*, Fort Collins CP: Colorado State University.

Mileti, D. (1999) *Disasters by Design: A Reassessment of Natural Hazards in the United States*, Washington DC: John Henry Press.

Mitchell, J. K. (2006) "The Primacy of Partnership: Scoping a New National Disaster Recovery Policy," *The Annals of the American Academy of Policital and Social Science* 604(1): 228–255.

Nigg, J. and K. Tierney. (1993) *Disasters and Social Change: Consequences for Community Construct and Affect*, Newark DE: University of Delaware Disaster Research Center.

Olshansky, R. B. (2005) *How Do Communities Recover from Disaster? A Review of Current Knowledge and an Agenda for Future Research*. Paper presented at the 46th Annual Conference of the Association of Collegiate Schools of Planning, Kansas City.

Pastor, M., R. Bullard, J. K. Boyce, A. Fothergill, R. Morello-Frosch, and B. Wright. (2006) "Environment, Disaster, and Race after Katrina," *Race, Poverty, & the Environment* 13(1): 21–26.

Peacock, W. G., H. Kunreuther, W. H. Hooke, S. L. Cutter, S. E. Chang, and P. R. Berke. (2008) *Toward a Resiliency and Vulnerbaility Observatory Network: RAVON*, HRRC Report 08-02R. College Station TX: Texas A&M University Hazard Reduction & Recovery Center.

Peacock, W. G., B. H. Morrow, and H. Gladwin. (1997) *Hurricane Andrew: Ethnicity, Gender, and the Sociology of Disasters*, London: Routledge.

Perry, R. W. and M. K. Lindell. (2007) *Emergency Planning*, Hoboken NJ: John Wiley.

Quarantelli, E. L. (1999) *The Disaster Recovery Process: What We Know and Do Not Know from Research*, Newark DE: University of Delaware Disaster Research Center.

Reddy, S. D. (2000) "Factors Influencing the Incorporation of Hazard Mitigation During Recovery from Disaster," *Natural Hazards* 22(2): 185–200.

Rubin, C. B. (1991) "Recovery from Disaster," 224–259 in W. L. Waugh and K. Tierney (eds.), *Emergency Management: Principles and Practice for Local Governments*, Washington DC: International City Management Association.

Rubin, C. and R. Popkin. (1990) *Disaster Recovery After Hurricane Hugo in South Carolina*, Washington DC: Center for International Science, Technology and Public Policy.

Rubin, C. B., M. D. Saperstein, and D. G. Barbee. (1985) *Community Recovery from a Major Natural Disaster*. Paper presented at the Program on Environment and Behavior, University of Colorado, Boulder.

Schwab, J.C., (1998) *Planning for Post-Disaster Recovery and Reconstruction*, American Planning Association (APA), Planners Advisory Service (PAS) Report 483/484. Chicago IL: American Planning Association.

Schwab, J.C., K. Topping, C. Eadie, R. Deyle, and R. Smith. (2014) *Planning for Post-Disaster Recovery: Next Generation*, Federal Emergency Management Agency (FEMA) and American Planning Association (APA), Planners Advisory Service (PAS) Report 576. Chicago IL: American Planning Association.

Van Zandt, S., W. G. Peacock, D. W. Henry, H. Grover, W. E. Highfield, and S. Brody. (2012) "Mapping Social Vulnerability to Enhance Housing and Neighborhood Resilience," *Housing Policy Debate* 22(1): 29–55.

Wisner, B., P. Blaikie, T. Cannon, and I. Davis. (2004) *At Risk: Natural Hazards, People's Vulnerability and Disasters*, New York: Routledge.

Yi, H. and J. Yang. (2014) "Research Trends of Post Disaster Reconstruction: The Past and the Future," *Habitat International* 42: 21–29.

16
HOUSING RECOVERY AFTER DISASTERS

Yang Zhang and William Drake

Introduction

Decisions to rebuild housing units or relocate households after disasters have the potential to cause dramatic changes in the composition of a community, either redressing or exacerbating pre-existing socio-economic conditions (Peacock et al. 2014, 2017; Van Zandt et al. 2012; Zhang 2012, Zhang and Drake, 2016). The quality of community recovery after disasters is most clearly manifested through housing recovery within the impact area. The trajectories and distribution of housing recovery over time and space reflect both the speed and quality of disaster recovery and are therefore major indicators of the success or failure of the overall disaster recovery policies. This chapter examines variations of housing recovery following recent catastrophic disasters in the United States and subsequently synthesizes critical issues for managing post-disaster housing recovery.

Housing Recovery after Disasters: State of Knowledge

Disasters occur when natural or technological hazards interact with socio-ecological systems (Mileti 1999). Disaster impacts are due to interactions between hazard exposure, physical vulnerability, and social vulnerability. Hazard exposure refers to the geography of an area in terms of the probability of occurrence of extreme events (e.g., flooding, wind, surge, etc.) and their expected intensity distribution, whereas physical vulnerability refers to the level of resistance that the built environment has to protect itself from the physical forces of disasters. More recent perspectives have expanded the vulnerability construct to include social vulnerability, which refers to socioeconomic characteristics of subpopulations that create variability in disaster vulnerability (Blaikie et al. 1994; Cutter 1996; Cutter and Finch 2008; Peacock et al. 2008; Flanagan et al. 2011; Lindell and Prater 2003; NRC 2006; Van Zandt et al. 2012).

Hazard exposure and physical vulnerability are concerned with the damage a structure is likely to experience when exposed to the forces of a particular hazard agent. Disaster damage is a function of not only the hazard's physical interaction with the impacted area and the intensity of a particular event, but also the quality and nature of the built environment. Buildings of higher quality and built to stronger codes will fare better than others when facing similar destructive forces (Mileti 1999; Highfield et al. 2014). In this regard, improvements to building quality over time and space to better meet the potential hazard exposures can be critical for reducing hazards vulnerability. Furthermore, land use planning and development policies can be essential for keeping development out of highly vulnerable areas, or at least out of the relatively more disaster-prone areas in a community (Burby 1998; Daniels and Daniels 2003; Godschalk et al. 1999). These policies can also be critical for preserving natural resources such as wetlands, dunes, etc. that can act as important buffers to lessen the physical impacts and damage (Brody et al. 2011; Brody et al. 2007; Zahran et al. 2008).

Post-disaster recovery is often viewed as an opportunity to upgrade the built environment through improved building quality and land use as a means to lessen vulnerability (Birkland 2006). When a disaster destroys large concentrations of the built environment, the economic cost of implementing these hazard mitigation strategies is reduced and public support can be stronger (Burby 2006). However, the opportunities for post-disaster betterment are always complicated by the prolonged planning and negotiation process, which can frustrate victims who desire a quick return to normalcy. Despite the aspiration to "build back better" (Khasalamwa 2009), many recovery cases end up being a restoration that extends the impacted area's pre-event development trajectory (Kates et al. 2006, NRC 2006).

The social component of vulnerability analysis disputes the popular belief that natural disasters are impartial, indiscriminate events that equally impact everyone in their path and instead suggests that all natural disasters are socially constructed (Bates and Swan 2007; Hartman and Squires 2006; Peacock and Ragsdale 1997). This conceptualization asserts that both the physical vulnerability of a built environment and the affected population's ability to prepare for, mitigate, and recover from disaster impacts are closely related to the underlying social order of that community. In the US, the literature has provided ample evidence of social vulnerabilities related to race/ethnicity (Bolin 1986; Bolin and Stanford 1991; Fothergill et al. 1999; Lindell and Perry 2004; Peacock et al. 1997), income and poverty (Dash et al. 1997; Fothergill and Peek 2004; Peacock et al. 1997), and gender (Enarson and Morrow 1997, 1998;Fothergill 1999), as well as age, education, religion, social isolation, and others. Morrow (1999) notes that these factors are often present in combination (Black and poor, for example), which is likely to amplify their impact. Of course, these factors, alone or together, are not adequate to explain the unequal disaster experiences of victims observed in the literature. These markers have significant consequences in terms of determining where people live, their access to social and economic resources, and therefore their overall ability to cope with the impacts of natural disasters (Van Zandt et al. 2012). These differences, in terms of housing, are expressed in both the physical vulnerability of housing structures and differences in access to financial resources such as insurance and economic capital to address home repair costs and other recovery needs.

As discussed above, building quality is important for determining the built environment's physical vulnerability to hazards. In the housing sector, market theorists have suggested a sequential process of "filtering" in which lower-income households successively inhabit homes and neighborhoods as they physically deteriorate over time (Grigsby 1963; Myers 1975). Through this process, poor and minority residents are allocated to older and lower quality homes in less desirable neighborhoods with potentially higher hazard risk (Charles 2003; Foley 1980; Logan and Molotch 1987; Massey and Denton 1993; Phillips 1993; Phillips and Ephraim 1992; Van Zandt 2007). These homes are typically built to older, less stringent building codes, have lower-quality designs and construction materials, and are less well maintained (Bolin 1994; Bolin and Bolton 1983; Bolin and Stanford 1991; Girard and Peacock 1997). As a consequence, one of the most consistent findings in the disaster literature is that low-income and minority households tend to suffer disproportionately higher levels of damage in disasters (Blaikie et al. 1994; Bolin 1986, 1993; Highfield et al. 2014; Peacock and Girard 1997; Quarantelli 1982).

Both a cause and a consequence of physical vulnerability is the fact that socially vulnerable populations often lack insurance and access to financial resources. These inequalities in the housing market mean that some segments of disaster victims are less likely to have sufficient means to meet the financial burden of disaster recovery (Bolin and Stanford 1991; Peacock and Girard 1997). In addition, housing recovery public assistance policies in the United States have structurally favored middle-class homeowners. The recovery needs of renters and low-income homeowners have been insufficiently addressed (Fothergill and Peek 2004; Kamel and Loukaitou-Sideris, 2004; Mueller et al. 2011).

Because of the aforementioned reasons, natural disasters could both magnify and accelerate housing and development processes already occurring in communities (Kates et al. 2006; Olshansky et al. 2012). Consequently, preexisting conditions are important predictors of disaster impacts. For housing recovery, the disparities experienced by socially vulnerable populations, both in terms of the higher physical vulnerability of their homes and their difficulty in obtaining resources for recovery, may be expected to result in volatile and uneven recovery trajectories over time. In other words, the preexisting social process related to housing attainment will likely repeat itself after a disaster, and may even become accelerated (French et al. 2008). The preexisting inequalities, differences in damage, and differences in recovery trajectories could continue to distort overall recovery for households, thus exaggerating inequalities to become greater than those found in pre-disaster conditions. In the sections that follow, we present findings from our own empirical work from three catastrophic natural disasters: the 1992 Hurricane Andrew in Miami-Dade, Florida, the 2008 Hurricane Ike in Galveston, Texas, and the 2012 Hurricane Sandy in New Jersey, to assess and characterize long-term trends in housing recovery.

Data and Methods

The findings presented below are the results of three separate, comprehensive data collection efforts following hurricanes Andrew, Ike, and Sandy. Hurricane Andrew struck south Florida on August 24, 1992, as a Category 5 hurricane and generated an estimated $26.5 billion of damage. Hurricane Ike made landfall near Galveston, Texas, on September 13, 2008, as an uncharacteristically large Category 2 storm. Aggregate damage from Ike is estimated at $29.5 billion. Hurricane Sandy made landfall near Atlantic City, New Jersey, on October 29, 2012, as a Category 2 hurricane. The storm surge brought extensive damage to much of the New York-New Jersey-Connecticut coastal area. Total direct economic damage caused by Sandy is estimated at $68 billion, making it one of the costliest hurricanes in U.S. history.

Hurricanes Andrew and Ike

In the cases of hurricanes Andrew and Ike, housing damage and the subsequent recovery process was monitored using a longitudinal approach. We collected information in Miami-Dade, Florida, and Galveston, Texas, before and after the hurricanes. More specifically, our samples represented both physical and social vulnerability by including residential structures that were damaged to varying degrees (from unaffected to severely damaged), as well as houses from neighborhoods with a range of socio-demographic attributes (i.e., income, race, and ethnicity). In each case, we undertook detailed multivariate modeling of damage and trajectories of subsequent housing recovery.

Results of the Hurricane Andrew research effort consist of data analyses conducted on a dataset created by linking two major sources, the Miami-Dade County, Florida, tax appraisal data from 1992 to 1996 and the 1990 Census data at block group level. Yearly tax appraisal data were merged to provide data on each parcel throughout the five-year period and census data were linked spatially to each parcel using GIS (for detailed discussion about data collection and analytical procedures for the Andrew study, see Peacock et al. 2014; Zhang 2012; Zhang and Peacock 2010).

Findings from the Hurricane Ike research effort were the result of a combination of primary and secondary data sources. Following Ike, structural damage assessments and resident interviews were conducted annually over a three-year period on a random sample of just over 1,500 single-family residential structures on Galveston Island and Bolivar Peninsula. In addition, to track multiple forms of residential housing, a comprehensive data set was developed using the tax appraisal data for 2008–2012 for single family, duplexes, and apartment buildings for Galveston Island and Bolivar Peninsula (for detailed discussion about data collection and analytical procedures for the Ike study, see Highfield et al. 2014 and Peacock et al. 2014).

Hurricane Sandy

While the primary focus of the Andrew and Ike studies was to develop parallel models for trajectories of the housing recovery in respective locations, the Hurricane Sandy research effort focused on patterns of housing damage and the subsequent distribution of federal housing recovery funding in the State of New Jersey.

The housing data for this analysis were merged from three public secondary data sources. Predisaster socio-demographic information of the impact area was collected from the US Census Bureau American Community Survey 2011 5-year estimates. The damage variables were collected from the State of New Jersey Department of Community Affairs (NJ DCA 2014). Damage assessment was based on FEMA Individual Assistance Records, current as of March 13, 2013 (NJ DCA 2013: 13), which categorizes the damage to individual housing units into three levels: minimum (less than $8,000 worth of damage), major ($8,000 to $28,800 worth of damage), and severe (more than $28,000 worth of damage).

Data on housing recovery grant allocation was collected from the State of New Jersey Office of the State Comptroller. Its official website (http://nj.gov/comptroller/sandytransparency/) publishes Superstorm Sandy Community Development Block Grant Disaster Recovery (CDBG-DR) funds awarded within the State of New Jersey. To assist disaster-impacted states' recovery efforts, the federal government enacted the Disaster Relief Appropriations Act of 2013 (Public Law 113-2, approved January 29, 2013). The Act appropriates monies targeted for disaster recovery to various federal agencies. Among the monies appropriated, $16,000,000,000 were allocated in Community Development Block Grant Disaster Recovery (CDBG-DR) funds to be split among states that had declared natural disasters in 2011, 2012 or 2013. These CDBG-DR funds are administered by HUD and were used to address unmet disaster recovery needs— funding needs not satisfied by other public or private funding sources like FEMA Individual Assistance, SBA Disaster Loans or private insurance. Data were collected on CDBG-DR programs that target housing reconstruction and recovery in New Jersey. Reported data include the number of grants awarded, the amount awarded, and the amount spent. At the time of data collection (July 2014), all records were current as of April 1, 2014, for four separate housing programs. They include the Homeowner Resettlement Program (HRP), the Reconstruction, Rehabilitation, Elevation, and Mitigation program (RREM), the Landlord Incentive program, and the Neighborhood Enhancement program.

HRP and RREM are the two assistance programs targeting homeowners affected by Sandy (NJ DCA 2013: 46). The HRP intends to protect the local tax base in the heavily impacted counties by decreasing out-migration. The program provides $10,000 to homeowners to repair their primary residence that sustained major or severe damage. The $10,000 payment is intended to provide quick relief to homeowners trying to reoccupy their damaged homes, and in particular, the program is designed to provide relief for the expected cost increase of home insurance premiums. To become eligible for the HRP fund, homeowners are required to reside in the county for three years or else refund the money.

The RREM program provides grant awards for rehabilitation, reconstruction, elevation, and/or other mitigation activities to homeowners who sustained major or severe damage. The maximum award, $150,000, is based on the average rebuilding cost for the average size home in the coastal region of New Jersey. The RREM program puts explicit focus on using the opportunity provided by the disaster recovery to rebuild "safer and smarter" (NJ DCA 2013: 45). Grant recipients are required to follow the reconstruction and rehabilitation standards prescribed in the New Jersey Disaster Recovery Action Plan (NJ DCA 2013) based on the HUD guidelines for CDBG-DR fund allocations (US-HUD 2013). These standards incorporate hazards mitigation considerations, green building technology, and energy-efficient development.

The Landlord Incentive program encourages rental property owners to reduce the affordable housing shortfall that was made much direr by Superstorm Sandy. Landlords can receive a rental unit subsidy that matches the federal Section-8 project-based payment methodology. The grants are restricted to low- and medium-income households. The Neighborhood Enhancement program is designed to "stabilize threatened but viable neighborhoods through the creation of affordable housing" (NJ DCA 2013: 3–9, Amendment 7). The grant provides zero-interest loans to developers (non-profit and for-profit) to build affordable housing units. The maximum award is $250,000 per unit (up to $1.75 million per award) and the units must be affordable at 30 percent of the resident applicant's income.

To account for the wide variation in the size of municipalities when comparing damage and funding distribution across the state, we normalized the data before the analysis. We conducted our analysis at the municipality level because this is how the damage and grant progress are reported by the state. Census data were also collected at the municipality level in order to maintain a consistent scale of analysis. New Jersey does not have unincorporated areas. Instead, it consists of 565 municipalities that cover a wide range of geographic areas and population sizes. There are five types of municipal government—cities, boroughs, townships, towns, and villages. Each government type exists within a county and shares equal status with the others, with all being responsible for tax collection and the provision of common public services.

The damage data were converted into the damaged proportion of all housing units per municipality. Similarly, the data for the two homeowner grant programs were converted into the proportion of all owner-occupied units per municipality and the Landlord Incentive program data were converted into the proportion of all renter-occupied units per municipality. The Neighborhood Enhancement program targets both renter- and owner-occupied housing in blighted areas, so we converted the grant data into the proportion of all housing units (occupied and unoccupied) per municipality.

Findings

Inequality in Housing Damage

We consistently found that housing in higher-income neighborhoods appeared more resilient by experiencing less damage than the housing in lower-income areas. In both Hurricane Andrew and Hurricane Ike, houses in wealthier neighborhoods retained more of their pre-disaster value after impact, meaning that they did not suffer the same relative damage as houses in less-wealthy areas, net of other effects. Hurricane Sandy's damage data is more revealing. The owner-occupied homes' damage data (Table 16.1) indicates that homes in the lowest income strata (less than 80 percent of area median income [AMI]) had more damage in each of the three damage categories (minor, major, severe) than their counterparts in the medium and the top income strata. For example, 4,278 homes in the lowest income strata were severely damaged, whereas the numbers of severely damaged homes are 2,773 and 3,505, respectively, for the medium and the top income strata. Table 16.1 also reports the damage data grouped by flood zone designations. It shows that more homes in the lowest income strata were located in the high flood risk zone (NFIP Zone A) and the coastal high hazard zone (NFIP Zone V). The damage data for rental homes (Table 16.2) offer a more drastic

Table 16.1 Owner-Occupied Homes Damaged in New Jersey From Superstorm Sandy

Income Level	Amount of Damage			
	Minor	Major	Severe	Total
Less than 80% Area Median Income	**9,592**	**11,516**	**4,278**	**23,386**
NFIP Zone A	1,444	7,438	2,500	11,382
NFIP Zone V	221	2,820	1,457	4,498
*All Other Zones	7,927	1,258	321	9,506
80%—120% Area Median Income	**4,165**	**6,715**	**2,773**	**13,653**
NFIP Zone A	762	4,491	1,700	6,953
NFIP Zone V	149	1,664	894	2,707
*All Other Zones	3,254	560	179	3,993
Greater than 120% Area Median Income	**3,855**	**8,266**	**3,505**	**15,626**
NFIP Zone A	916	5,474	2,197	8,587
NFIP Zone V	185	2,146	1,045	3,376
*All Other Zones	2,754	646	263	3,363
No Income Data	**1,893**	**2,449**	**964**	**5,306**
NFIP Zone A	306	1,598	556	2,460
NFIP Zone V	74	610	330	1,014
*All Other Zones	1,513	241	78	1,832
All Homeowners	**19,505**	**28,946**	**11,520**	**59,971**

*Includes areas not yet determined by FEMA under ABFE.

Source: This table is reproduced based on NJ DCA 2013. Data source: FEMA Individual Assistance Data effective March 12, 2013 and FEMA ABFE maps.

contrast across different income levels. A total of 16,812 rental units in the lowest income strata were damaged by the storm, which is five times the number for the medium income strata (2,106) and the high income strata (1,270) combined. The flood zone data show that most damaged rental units in the lowest income strata were either in the high flood risk zone (NFIP Zone A) or in the coastal high hazard zone (NFIP Zone A).

A multivariate correlation analysis between home damage and several key socio-demographic indicators using the New Jersey statewide Hurricane Sandy dataset is reported in Table 16.3. Median household income has significant negative correlations with damage in all categories, ranging between −.094 and −.132. These correlations suggest that damage from the storm generally increased for homes in lower income neighborhoods. The correlations between damage and poverty, percent renter occupied, and percent African-American population demonstrate no clear association. The correlation coefficients for these three variables are close to zero and often not statistically significant. These results suggest that income had an overall inverse relationship with damage level from Sandy, but the relationship does not show clear evidence of uneven damage along racial lines. Although there are some lower-income communities on the north Jersey Shore that were hard hit, the location of the storm surge and the overall affluent socio-economic conditions of the New Jersey coastline (Figure 16.1) have not produced the type of socio-economic disparities in the disaster impact area that have been associated with some other disasters, such as hurricanes Andrew, Katrina, and Ike.

Table 16.2 Renters with Housing Damage in New Jersey From Hurricane Sandy

Income Level	Amount of Damage			
	Minor	Major	Severe	Total
Less than 80% Area Median Income	**5,230**	**9,435**	**2,147**	**16,812**
NFIP Zone A	1,882	6,433	1,335	9,650
NFIP Zone V	334	2,000	520	2,854
*All Other Zones	3,014	1,002	292	4,308
80%—120% Area Median Income	**366**	**1,347**	**393**	**2,106**
NFIP Zone A	202	931	262	1,395
NFIP Zone V	42	323	86	451
*All Other Zones	122	93	45	260
Greater than 120% Area Median Income	**200**	**788**	**282**	**1,270**
NFIP Zone A	115	557	213	885
NFIP Zone V	35	187	54	276
*All Other Zones	50	44	15	109
No Income Data	**493**	**974**	**245**	**1,712**
NFIP Zone A	197	647	164	1,008
NFIP Zone V	37	209	56	302
*All Other Zones	259	118	25	402
All Homeowners	**6,289**	**12,544**	**3,067**	**21,900**

*Includes areas not yet determined by FEMA under ABFE.

Source: This table is reproduced based on NJ DCA 2013. Data source: FEMA Individual Assistance Data effective March 12, 2013 and FEMA ABFE maps.

There are two important considerations that may have an impact of the magnitude of correlations reported in Table 16.3. First, the correlations were likely influenced by the fact that the analysis was conducted using the state-wide data instead of the coastal municipalities affected by the storm. The large number of data points with no damage could attenuate the overall association. Second, as noted earlier, the damage categories are based on absolute monetary damage, not as a percentage of the homes' total value. This may be reducing the association for these social vulnerability variables. A more appropriate categorization—damage hardship—based on damage proportional to a home's

Table 16.3 Correlations Between Damage and Key Social Vulnerability Indicators in New Jersey

	Minimum Damage	Major Damage	Severe Damage	Total Damage
Median Household Income	−.11***	−.13***	−.09**	−.14***
Poverty (%)	.08	.07	.03	.08
Renter Occupied (%)	.06	.02	.04	.05
Black Population (%)	−.05	−.07	−.09	−.08

N=563, **p < 0.05, ***p < 0.01

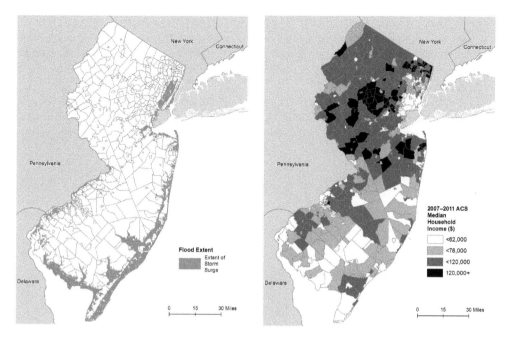

Figure 16.1   Hurricane Sandy Storm Surge Extent and Income Distribution in New Jersey

total value would likely result in a stronger positive association for the social vulnerability variables, but this information was not available for this analysis.

Inequalities in Housing Recovery

The literature suggests that social vulnerability markers such as income, race/ethnicity, and home tenure are not only associated with higher relative damage, but also slower recovery trajectories. Our results from the three cases suggest that the social vulnerability is indeed important in the recovery process; however, the nature of its influence is more nuanced than the literature would suggest. In both the Hurricane Andrew and the Hurricane Ike cases, housing in higher income areas was restored more quickly. With respect to race and ethnicity, very different pictures emerge. In Hurricane Andrew, the expected divergent impact and recovery patterns between minority and non-Hispanic White areas were clearly evident. However, in the case of Hurricane Ike, minority areas actually fared better than White dominant neighborhoods. Perhaps home tenure—one of the most overlooked elements of social vulnerability in the existing literature—was important for explaining this discrepancy between the two hurricanes. Figure 16.2 displays the impact and recovery trends for renter- and owner-occupied single-family housing based on the research of Peacock and his colleagues (2014). In Miami-Dade after Hurricane Andrew, there are considerable variations in housing recovery. By Year 2 (one year after the impact), owner-occupied housing is nearly at 88 percent of pre-impact value and surpasses restoration (1.0) levels by Year 3. Rental housing reaches 51 percent of pre-impact assessment by Year 2, 64percent by Year 3, and only reaches 71 percent by Year 4. The picture is perhaps even more sobering for Hurricane Ike, but a similar pattern emerges. First, there is a much greater differential in impact, with owner-occupied housing faring better than rental housing, which falls below 25 percent of its pre-impact assessment. In the subsequent years, owner-occupied housing climbs to over 50 percent by Year 2 and ultimately to 67.1 percent by Year 4. Rental housing languishes, climbing to only 30 percent in Year 2 and finishing at a miserable 36 percent by Year 4. Clearly, in comparison with the Andrew case in Miami-Dade, the picture is not good for either owner- or renter-occupied housing in Galveston—but is particularly dire for rental properties.

The Sandy study in New Jersey suggests a similar dichotomous recovery pattern between rental and owner-occupied homes. The CDBG-DR housing programs data (Table 16.4) shows that, although recovery funds were made available to owner-occupied homes and rental units, recovery

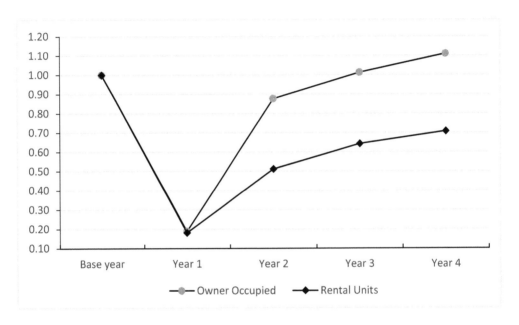

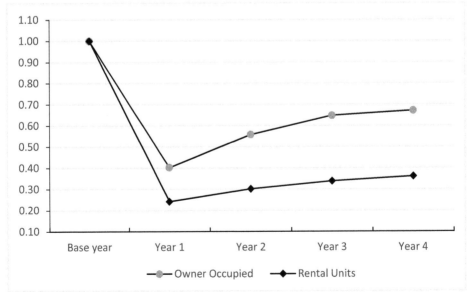

Figure 16.2 Recovery Trajectories for Owner and Renter Occupied Single Family Homes

Table 16.4 New Jersey Hurricane Sandy CDBG-DR Housing Program Summary (April 1, 2014)

	Homeowner Resettlement	RREM	Neighborhood Enhancement	Landlord Incentive
Grants Awarded	18,132	1,928	32	118
Funds Allocated	$181,320,000	$222,316,339	$25,519,905	$4,529,409
Funds Spent	$180,860,000	$47,093,217	$0	$278,513
Funds Remaining	$460,000	175,223,122	$25,519,905	$4,250,896
% Allocated Funds Spent	99.9%	21.2%	0%	6.1%
% Allocated Funds Spent (min.)	0%*	0%	0%	0%
% Allocated Funds Spent (max.)	100%	100%**	0%	41%

*A single award of $10,000 was unspent in one municipality. This is the only case of 0% spent.

**A single award of $83, 315 was fully spent in one municipality. This is the only case of 100% spent.

Source: This table is based on data obtained from New Jersey Department of Community Affairs (NJ DCA 2013).

for these two types of homes progressed at very different paces. At the time when the data was reported on April 1, 2014, about 18 months after Sandy, the HRP had spent 99.9 percent of the total obligated funding. By contrast, the RREM program, which also focused on owner-occupied homes, had only spent 21.2 percent of the total obligated funds. Although these two homeowners' programs had a drastically different performance (discussed in more detail below), they both achieved a much better pace of recovery than the two rental recovery programs. The Landlord Incentive program awarded 118 grants in the impact area worth a total of $4.5 million. However, only 6.1 percent of the funding was actually put into use. The Neighborhood Enhancement program, the other program that focuses primarily on rental properties, did not progress at all. Although it received more than $25 million, none of the funds had yet been spent as of April 1, 2014. A more detailed look at the distribution of the Landlord Incentive program funds across the state (Figure 16.3) shows that an average of 6.1 percent of allocated funds were spent across the state, but there was wide variation across municipalities. Some awardees had yet to use any of this money, while others used as much as 41 percent.

A multivariate correlation analysis between the CDBG-DR housing recovery funds and the social vulnerability variables sheds more light on the recovery challenges faced by socially vulnerable populations, especially renters (Table 16.5). There is a weak, negative association (though statistically significant) between median household income and the provision of recovery funds from all programs, including both homeowner programs and rental programs. This is evidence that the CDBG-DR funds allocation had some success targeting their awards to less-affluent residents who likely have greater unmet needs. However, the consistent negative correlations between income and outstanding funds in the RREM program and the two rental programs, defined as the difference between the obligated funds and the funds spent, suggest that recovery funding allocated to less affluent areas was spent at a slower pace. In other words, housing recovery in these areas progressed slower than in relatively wealthier areas. There is little evidence of significant associations between the two homeowner programs and percent poverty, percent rental, and percent African American. This is not surprising because the Homeowner Resettlement program and the RREM program are restricted to homeowners, and the areas occupied by homeowners are unlikely to have a concentration of families in poverty, renters, or African Americans. Both programs targeting rental homes, the Neighborhood Enhancement program and the Landlord Incentive program, demonstrate positive

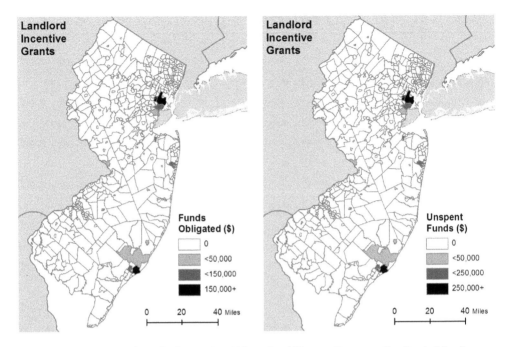

Figure 16.3 Comparison of Landlord Incentive Obligated and Unspent Recovery Funding in New Jersey

correlations with percent poverty, percent rental, and percent African American. But the outstanding funds in both programs have similar correlations with these variables. These results suggest that although more recovery funding for rental properties was allocated to areas with greater rental needs, the areas with greater needs for rental properties had more funding that remained unspent. In other words, rental properties are being reconstructed more slowly during the recovery process.

Table 16.5 Correlations Between CDBG-DR Housing Recovery Funds and Social Vulnerability Variables

	Median Household Income	Poverty Level (%)	Renter Occupied (%)	Black Population (%)
Total Funds Obligated	−.11**	.03	−.02	−.10
HR Funds Obligated	−.11***	.04	.04	−.08
HR Funds Outstanding	.01	.01	−.02	−.02
RREM Funds Obligated	−.11***	.06	.02	−.07
RREM Funds Outstanding	−.11***	.06	.02	−.07
NE Funds Obligated	−.16***	.19***	.19***	.21***
NE Funds Outstanding	−.16***	.19***	.19***	.21***
LI Funds Obligated	−.10**	.13***	.11**	.15***
LI Funds Outstanding	−.11**	.14***	.11***	.15***

N=563, **p < 0.05, ***p < 0.01

These findings suggests that the picture is much bleaker for rental housing, making it difficult for households that are dependent upon renting to find replacement housing following both disasters. Not surprisingly, renting is highly correlated with other social and physical vulnerabilities including lower household incomes, minority status, and poorer structural conditions (Kreimer 1980; Morrow 1999; Peacock et al. 2006). Renters have little to no control over the decision to rebuild, and are at much greater risk of temporary and permanent displacement (Burby et al. 2003). Traditionally, recovery funds have been targeted almost exclusively at property owners, with little if any attention given to the rights of renters to return (Rodriguez-Dod and Duhart 2007).

The other reason that rental properties—low-income housing in particular—experience slow recovery is due to post-disaster gentrification. Several public housing developments that primarily served African-American families in Galveston, Texas, were heavily damaged by Hurricane Ike in 2008 and were soon after demolished. As of summer 2013, five years after the storm, none of the more than 500 housing units had been replaced. The reconstruction of these units has become an intractable local and state political issue, with some island residents and political factions continuing to battle with each other and state officials over the disbursement of disaster recovery funds to the City. At issue is whether these funds should be conditional on the one-for-one replacement of the lost affordable housing units. Many long-time residents simply do not want to see public housing residents return to the island (Rice, April 17, 2013). Galveston's experience is not unusual—it mirrors similar gentrification controversies that have taken place along the Gulf Coast after Hurricane Katrina (Olshansky and Johnson 2010).

Accelerated Post-Disaster Neighborhood Decline

The Andrew data includes parcel level land use for the duration of the study period. This allowed us to track the occurrence of property vacancy and abandonment in the impact neighborhoods (for a detailed report, see Zhang 2012). Consistent with the literature that suggests natural disasters create disproportionate obstacles for marginalized households and neighborhoods, we found that Hurricane Andrew compressed in time the occurrence of housing vacancy and abandonment, whereas the distribution of vacancy and abandonment varied across space as defined by its pre-event characteristics. The number of vacant and abandoned residential parcels in the first year following Hurricane Andrew was 20 times greater than the number in the year prior to the storm. Though the intensity of vacancy and abandonment gradually attenuated, there were still more occurrences seven years after the storm than there were before the storm. The emergence of vacant and abandoned land after the hurricane was not random, but instead spatially selective. In general, properties in heavily damaged neighborhoods were more likely to become vacant or abandoned. When damage was comparable, vacant and abandoned properties tended to cluster in marginalized neighborhoods that were characterized by a high concentration of low-income households, rental units, and minorities. In a sense, the storm magnified the vulnerability and fragility of marginalized neighborhoods and compressed their deterioration into a shorter period of time, a process that would have been more gradual had it not been for the disaster. Although the Ike study did not directly examine home vacancy and abandonment in Galveston, a similar pattern of neighborhood decline in low-income areas where rental units cluster is also expected, at least during the study period. As we reported earlier, low-income housing and rental units had a bleak recovery experience in comparison to owner-occupied, wealthier homes.

The Andrew study empirically demonstrated the succession of neighborhood decline by showing that vacant and abandoned properties sprawled out over time in the study area. The existing vacant and abandoned properties were followed by more vacancy and abandonment in the same area at a later time. The temporal succession of vacant and abandoned properties was expected, given their well-documented consequences such as neighborhood destabilization, threatened business viability, lowered property values, and damage to the overall quality of life in a neighborhood (Accordino and Johnson 2000). Following the destruction caused by a natural disaster, these negative externalities take on added significance and prompt more property vacancy and abandonment. When a neighborhood is flattened by a disaster and many damaged properties are left unrepaired or abandoned, as seen in our data following Hurricane Andrew, it becomes very difficult for the remaining households to stay and keep investing in their properties with the hope that their neighborhood will soon become revitalized.

Managing Housing Recovery Funding

The existing literature suggests that lack of sufficient funding has been a major roadblock for disaster victims to achieve housing recovery, especially those socially vulnerable populations in the impact area. Our examination of the CDBG-DR housing recovery programs in New Jersey after Sandy reveals a more nuanced picture of the challenges associated with disaster recovery funding. The Disaster Relief Appropriations Act, Public Law 113-2, approved on January 29, 2013, made available housing recovery funding for different programs (Table 16.4). However, the performance of these programs, measured by the percentage of allocated funding spent by the time of data collection, varied to a great extent, as discussed above. For programs targeting homeowners, 99.9 percent of the HRP appropriation was put into use, but only 21.2 percent of the total RREM funds were spent 18 months after the storm. There was a wide variation across municipalities that received recovery funds. In some municipalities, none of the allocated RREM money was spent while in another location, RREM programs have finished—100 percent of the allocation was spent (Figure 16.4).

The correlation analysis provides more details about funding allocation and funding utilization across the state of New Jersey (Table 16.6). The positive associations between home damage level

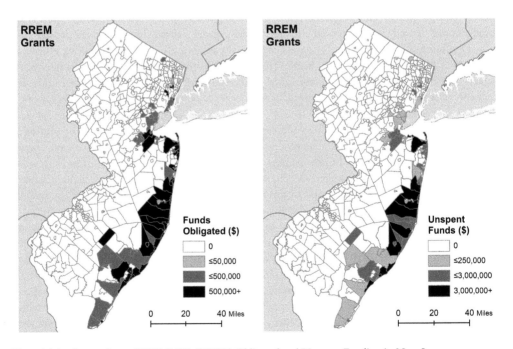

Figure 16.4 Comparison of CDBG-DR RREM Obligated and Unspent Funding in New Jersey

Table 16.6 Correlations Between Damage and CDBG-DR Housing Recovery Funds

	Minimum Damage	Major Damage	Severe Damage	Total Damage
Total Funds Obligated	.60***	.73***	.67***	.78***
HRP Funds Obligated	.56***	.75***	.86***	.87***
HRP Funds Outstanding	.01	.17***	.16***	.10
RREM Funds Obligated	.57***	.49***	.56***	.71***
RREM Funds Outstanding	.58***	.49***	.55***	.71***
NE Funds Obligated	.01	.00	−.01	.01
LI Funds Obligated	.06	.02	.01	.03

N=563, **p < 0.05, ***p < 0.01

and homeowner grant variables (i.e., HRP and RREM) are consistently significant. This is true for the total funds distributed, as well as for the individual grant programs. It is also worth noting that the strength of these positive associations becomes even stronger for major and severe damage levels. The only exception is the correlation between damage and the RREM funds obligated variable, which demonstrates a consistent strength of association across damage levels. This positive correlation between damage and grant variables suggest that the two homeowner grant programs successfully targeted funds to areas heavily damaged by the storm. The increase in the correlation strength as the level of damage rises would be expected, given that a stated goal of these two programs is to target major- and severely damaged homes.

The correlations between outstanding funds and damage show weak associations (though statistically significant) between outstanding HRP funding and major and severe damage levels (.167 and .157 respectively), suggesting that HRP funding was spent at a slower pace in municipalities with heavier damage, although overall this program has spent all obligated money. The magnitude of correlations between outstanding RREM funds and damage is much higher, .575 for minimum damage, .485 for major damage, and .552 for severe damage. These numbers are very troubling because they suggest that in municipalities that experienced more hurricane damage, and therefore had more recovery needs, the RREM grants had a more protracted implementation process.

The difference between HRP and RREM programs could be explained by how these two programs were designed and implemented. HRP is a straightforward cash grant program intending to provide $10,000 to homeowners to compensate their recovery cost. As noted earlier, all homeowners who had at least minimum damage were qualified to receive the grant, as long as they agreed to stay in the same county three years after receiving the money. In comparison, RREM was designed to provide grants for homeowners to restore their storm-damaged homes. Besides the damage criteria, the grant recipients from the RREM program would have to adhere to the new building standards when replacing or repairing their homes. These new standards (e.g., the 2009 Residential International Code and the ENERGY STAR ™ standards) are designed to improve the pre-existing built environment to become more disaster-resilient and more energy-efficient. An inevitable consequence of these requirements is that the planning, design, review, permit, and fund allocation process for the RREM projects involves additional steps and takes longer to complete. When reconstruction involves redesigning a home, elevating a structure that was not elevated before the storm, or relocating a home, the process can become complicated and controversial.

There have been additional problems that have plagued the implementation speed of the RREM program. The contracting firm originally hired to manage the program was terminated six months after the program launched amid claims of delays, poor performance, and improper rejection of aid applications (Hannah and Seidman 2014). A number of the administrative rules cited for causing delays, including in-person submission of application materials, the number of required inspections, home contractor eligibility requirements, and the payout schedule of grant funds to awardees, have recently been altered in hopes of improving the program's performance (Gurian 2014).

Turning to the two rental programs, the correlation analysis shows little association between damage and either the Landlord Incentive program or the Neighbor Enhancement program. The plausible explanations for this lack of associations relate to the nature of these two programs. The stated primary purpose of the Landlord Incentive program and the Neighborhood Enhancement program is to provide affordable housing. The Neighborhood Enhancement program is restricted to the nine coastal counties that received the most damage from the storm, but the Landlord Incentive program is available to the entire state. Neither program contains any housing damage eligibility criteria. Although the justification for these two grant programs speaks to the urgent need for affordable housing—a problem exacerbated by Sandy—neither specifically prioritizes municipalities that experienced significant damage, nor do they require a damage level threshold as part of the eligibility criteria.

A possible explanation for why the Landlord Incentive program has a weak, but significant correlation only for the minimum damage level may be related to the fact that the majority of the destruction from Sandy was concentrated along the coast of the state—an area that is largely affluent and features local economies that are heavily supported by a strong demand for beach tourism. The less-affluent, working-class communities—the type targeted by these two programs—are located further inland and also in the northern part of the state. Many of these municipalities were also impacted by Sandy, but they did not suffer the full brunt of the damage. As a result, the strongest correlation between these two programs and the damage variables exists at the minimum damage level variable and nowhere else.

Discussion and Policy Implications

Following a devastating natural disaster, housing recovery is a critical component of long-term community recovery. Major natural disasters are likely to be followed by uneven damage and housing recovery systemically related to the intersection of physical and social vulnerability. In other words, the most socially vulnerable populations, especially low-income households and renters, tend to live in the most physically vulnerable locations (although this is not always true for US coastal communities, see Lindell and Hwang, 2008). These differences in both physical and financial preexisting conditions place these populations at a disadvantage from the outset. After the disaster strikes and recovery is underway, they continue to face difficulties in achieving housing recovery. The consequences of both inequality in disaster damage and the inequality in recovery assets include a strong potential for redevelopment and neighborhood transition; permanent displacement of low-income households and renters; a loss of affordable housing stock; and ultimately, the worsening of preexisting inequalities. Our case studies suggest three approaches to overcome these problems associated with post-disaster housing recovery in the United States.

Improve Efficiency of Recovery Assistance Programs

Funding is always a significant barrier for post-disaster recovery; thus, it is important to make the best use of public housing recovery assistance programs to fill the market gap and to expedite reconstruction across all segments of disaster victims, especially low-income and renter families. From the local perspective, communities should promote coordination and concerted action among local, state, and federal recovery agencies. It is important to invest in and build long-term relationships with federal and state agencies involved in disaster recovery because it helps the communication after a disaster and, therefore, could expedite the procurement of federal recovery funding. Education and advocacy programs are critical to help local officials, property owners, and residents (especially among marginalized groups) to understand the various housing recovery programs and application procedures, as well as their rights with respect to insurance companies.

From the federal and state perspectives, it is obviously important to design recovery assistance programs that are sensitive to varying recovery needs in the impact communities, defined not only by damage, but also by financial conditions and home tenure status of the affected residents. More importantly, housing recovery assistance programs must be equipped with a mechanism that allows speedy allocation of funds for recovery projects on the ground. After Hurricane Sandy, while CDBG-DR recovery funds were obligated in the disaster areas in New Jersey, mostly aligning well with recovery needs, the use of these funds varied a great extent across recovery programs. In particular, the RREM program and the two programs targeting rental properties demonstrated a very poor record of getting obligated funds to storm victims. The frustration among coastal residents in New Jersey over the lack of efficiency of these programs led to the widely reported protest at Governor Christie's rally for the two-year anniversary of Sandy (Talev 2013). To this end, the funding agencies, primarily FEMA and HUD-CDBG in the United States, should do a better job to make funding programs work for their intended purposes. Categorically, this would include giving states, communities, and affected residents more flexibility in deciding how the money should be spent. Communities vary greatly in terms of their governance, location, planning capacity, and recovery needs. It would be very difficult for the state and federal agencies to maintain total control of funding application, allocation, waivers, and requirements while also trying to maintain funding efficiency.

Balance Hazard Mitigation and Recovery Speed

Although disaster recovery provides a "window of opportunity" (Birkland 2006) to rebuild better, smarter, and more resilient to future disasters, controversy and a prolonged post-disaster process to upgrade building standards and implement land use adjustments will delay housing recovery in the impact area. Such delay could also put the area in a long-term trajectory of decline. To prevent a natural disaster from becoming a catalyst for neighborhood decline, should local officials should make it a priority to attract displaced households to return and rebuild as soon as possible. Introducing building standards and land use adjustments for building back smarter and better *after* a disaster had little success in the literature (Kates et al. 2006; NRC 2006; Olshansky and Johnson 2010). Instead, hazards mitigation should be incorporated into communities' regular comprehensive planning *before* disaster strikes to accelerate housing recovery (Olshansky et al. 2012; Wu and Lindell 2004).

There are many benefits to incorporating mitigation into routine community planning activities. Existing research suggests that communities that incorporate hazard planning into their comprehensive planning efforts have better outcomes than those that separate these functions (Burby 2005; Brody et al. 2007). Proper land use planning can help limit development in high-hazard areas and can guide the adoption of building codes that strengthen structures in those areas as they are rebuilt. More importantly, comprehensive planning develops a common vision for change among residents in the community. Consequently, when a disaster opens the window of opportunity and outside recovery funds are available, a comprehensive plan will help the community to secure the recovery funds, alleviate the potential controversies pertaining to building standards upgrades and land use adjustment, and guide recovery in a way that is consistent with the vision laid out in the plan. All of these are important preconditions for a speedy housing recovery. A strong plan will also provide protection against the assertion of special interests (usually those backed by power and money) that often occurs after a disaster. If equity goals are strong in the plan, they provide the possibility of using the recovery and rebuilding period to actually lessen inequities for socially vulnerable populations rather than exacerbating them.

Make Recovery Assistance for Rental Properties Work

The findings from Hurricanes Andrew and Ike showing that post-disaster recovery is particularly slow for rental properties, coupled with the finding that CDBG-DR rental programs after Hurricane Sandy had a protracted implementation in New Jersey suggest that we must make recovery aid for rental housing work better. Planners must work actively to protect housing choices in their communities and guard the fair housing goals during disaster recovery.

In the US, the provision and spatial distribution of housing is primarily a market-driven process. Although the private-sector impediments for providing housing for marginalized populations are difficult to overcome, communities themselves often support these practices by employing exclusionary practices. Oftentimes, in response to pressure from affluent stakeholders, cities act to maintain neighborhood homogeneity, support affluent and fast-growing areas of the community, and allocate few resources to poor or minority neighborhoods. After a disaster, when resource competition is high, communities become more likely to succumb to powerful investors or special interests groups. Consequently, low-income housing and rental housing often become low-priority agenda items and may even face pressure for removal.

To overcome long-standing spatial inequities and improve the resilience of low-income housing and rental properties, federal and state funding agencies and communities should build financial incentives, remedial actions, and rewards for the compliance of private developers and landlords with fair housing goals during disaster recovery. In a fast-paced and conflict-ridden post-disaster environment, local governments must work as arbiters of equitable housing choice for the long-term benefits of their communities. Local governments must be proactive in facing NIMBY opposition head-on after a disaster and setting expectations among the development community that fair housing obligations and speed of restoration of rental development are to be taken seriously. To this end, communities can use zoning and other land use planning tools to prevent removal of rental units. They should work to channel recovery funds to ensure that investment in infrastructure and capital improvements is equitable.

Future Research

This study has its limitations, which should be considered when generalizing our findings and in future research. Primary among these is that, despite the overall consistent findings across three cases, we found a more nuanced picture of the relationship between key social vulnerability variables (e.g., race, income, tenure) and housing recovery trajectories than what the literature would suggest. More case studies in locations with varying socio-demographic compositions are needed. Second, it will be critical in future work to include more information on the household/owner/landlord's post-impact decision-making process to fully understand disaster impact and recovery. For example, a critical and important element not included in these models is the nature of household financial resources such as insurance, savings, loans, and grants that were available and employed in the recovery processes. Third, future research must grapple with the important issue of identifying the mechanisms producing differentials in impact, which are likely to be related to housing maintenance, building code differentials, and differentials in infrastructure investments. Although our data do not permit such assessments, differences in storm water and sewer provisions have the potential to contribute to damage and recovery after flood disasters. Lastly, more program evaluation research is needed to fully decipher the diverging performance of different CDBG-DR programs, particularly the challenges associated with the implementation of the RREM program and the rental housing programs.

Acknowledgment

This material is based upon work supported by the National Science Foundation under Grant No. CMMI 1029298 and the Peking University—Lincoln Institute Center for Urban Development and Land Policy under Grant No. PLC A204.

References

Accordino, J. and G. T. Johnson. (2000) "Addressing the Vacant and Abandoned Property Problem," *Journal of Urban Affairs* 22(3): 301–315.

Bates, K. A. and R. S. Swan (eds.). (2007) *Through the Eye of Katrina: Social Justice in the United States*, Durham NC: Carolina Academic Press.

Birkland, T. A. (2006) *Lessons of Disaster: Policy Change after Catastrophic Events*, Washington DC: Georgetown University Press.

Blaikie, P., T. Cannon, I. Davis, and B. Wisner. (1994) *At Risk: Natural Hazards, People's Vulnerability and Disasters*, London: Routledge.

Bolin, R. (1986) "Disaster Impact and Recovery: A Comparison of Black and White Victims," *International Journal of Mass Emergencies and Disasters* 4(1): 35–50.

Bolin, R. (1993) "Post-Earthquake Shelter and Housing: Research Findings and Policy Implications," 107–131, in K. J. Tierney and J. M. Nigg (eds.), *Socioeconomic Impacts*, Monograph 5 prepared for the 1993 National Earthquake Conference. Memphis TN: Central U.S. Earthquake Consortium.

Bolin, R. and P. Bolton. (1983) "Recovery in Nicaragua and the U.S.A.," *The International Journal of Mass Emergencies and Disasters* 1(1): 125–144.

Bolin, R. and L. Stanford. (1991) "Shelter, Housing and Recovery: A Comparison of U.S. Disasters," *Disasters* 15(1): 24–34.

Brody, S. D., W. E. Highfield, and J. E. Kang. (2011) *Rising Waters*, New York: Cambridge University Press.

Brody, S. D., S. Zahran, P. Maghelal, H. Grover, and W. E. Highfield. (2007) "The Rising Costs of Floods: Examining the Impact of Planning and Development Decisions on Property Damage in Florida," *Journal of the American Planning Association* 73(3): 330–345.

Burby, R. J. (1998) *Cooperating with Nature: Confronting Natural Hazards with Land-Use Planning for Sustainable Communities*, Washington DC: Joseph Henry Press.

Burby, R. J. (2005) "Have State Comprehensive Planning Mandates Reduced Insured Losses from Natural Disasters?" *Natural Hazards Review* 6(2): 67–81.

Burby, R. J. (2006) "Hurricane Katrina and the Paradoxes of Government Disaster Policy: Bringing About Wise Government Decisions for Hazardous Areas," *The Annals of the American Academy of Political and Social Sciences* 604(1): 171–191.

Burby, R. J., L. J. Steinberg, and V. Basolo. (2003) "The Tenure Trap the Vulnerability of Renters to Joint Natural and Technological Disasters," *Urban Affairs Review* 39(1): 32–58.

Charles, C. Z. (2003) "The Dynamics of Racial Residential Segregation," *Annual Review of Sociology* 29: 167–207.

Cutter, S. L. (1996) "Vulnerability to Environmental Hazards," *Progress in Human Geography* 20(4): 529–539.

Cutter, S. L. and C. Finch. (2008) "Temporal and Spatial Changes in Social Vulnerability to Natural Hazards," *Proceedings of the National Academy of Sciences* 105(7): 2301–2306.

Daniels, T. and K. Daniels. (2003) *The Environmental Planning Handbook for Sustainable Communities and Regions*. Chicago IL: American Planning Association.

Dash, N., W. G. Peacock, and B. Morrow. (1997) "And the Poor Get Poorer: A Neglected Black Community," 206–225, in W. G. Peacock, B. H. Morrow, and H. Gladwin (eds.), *Hurricane Andrew: Ethnicity, Gender and the Sociology of Disaster*, London: Routledge.

Enarson, E. and B. H. Morrow. (1997) "A Gendered Perspective: The Voices of Women," 116–140, in W. G. Peacock, B. H. Morrow, and H. Gladwin (eds.), *Hurricane Andrew: Ethnicity, Gender and the Sociology of Disaster*, London: Routledge.

Enarson, E. and B. H. Morrow. (1998) *The Gendered Terrain of Disaster*, Westport CT: Praeger.

Flanagan, B. E., E. W. Gregory, E. J. Hallisey, J. L. Heitgerd, and B. Lewis. (2011) "A Social Vulnerability Index for Disaster Management," *Journal of Homeland Security and Emergency Management* 8(1): 1–22.

Foley, D. L. (1980) "The Sociology of Housing," *Annual Review of Sociology* 6: 457–478.

Fothergill, A. (1999) "An Exploratory Study of Woman Battering in the Grand Forks Flood Disaster: Implications for Community Responses and Policies," *International Journal of Mass Emergencies and Disasters* 17(1): 79–98.

Fothergill, A., E. G. Maestas, and J. D. Darlington. (1999) "Race, Ethnicity and Disasters in the United States: A Review of the Literature," *Disasters* 23(2): 156–173.

Fothergill, A. and L. A. Peek. (2004) "Poverty and Disasters in the United States: A Review of Recent Sociological Findings," *Natural Hazards* 32(1): 89–110.

French, S. P., E. Feser, and W. G. Peacock. (2008) "Quantitative Models of the Social and Economic Consequences of Earthquakes and Other Natural Hazards," Project SE-2 Final report, Mid-America Earthquake Center Project. Atlanta GA: Georgia Institute of Technology.

Girard, C. and W. G. Peacock. (1997) "Ethnicity and Segregation: Post Hurricane Relocation," 191–205, in W. G. Peacock, B. H. Morrow, and H. Gladwin (eds.), *Hurricane Andrew: Ethnicity, Gender and the Sociology of Disaster*, London: Routledge.

Godschalk, D. R., T. Beatley, P. Berke, D. J. Brower, and E. J. Kaiser. (1999) *Natural Hazard Mitigation: Recasting Disaster Policy and Planning*, Washington DC: Island Press.

Grigsby, W. (1963) *Housing Markets and Public Policy*, Philadelphia PA: University of Pennsylvania Press.

Gurian, S. (2014) "Two Years after Hurricane Sandy, New Jersey's Recovery Trudges Along," *New Jersey Spotlight*, October 29. www.njspotlight.com/stories/14/10/29/two-years-after-hurricane-sandy-new-jersey-s-recovery-trudges-along/.

Hannah, M. and A. Seidman (2014) "Hurricane Sandy: Thousands Await Aid to Rebuild in New Jersey," *Emergency Management*, October 29. www.emergencymgmt.com/disaster/Hurricane-Sandy-Thousands-Await-Aid-Rebuild-New-Jersey.html.

Hartman, C. W. and G. D. Squires (eds.). (2006) *There is No Such Thing as a Natural Disaster: Race, Class, and Hurricane Katrina*, New York: Routledge.

Highfield, W., W. G. Peacock, and S. Van Zandt. (2014) "Mitigation Planning: Why Hazard Exposure, Structural Vulnerability, and Social Vulnerability Matter," *Journal of Planning Education and Research* 34(3): 287–300.

Kamel, N. M. O. and A. Loukaitou-sideris. (2004) "Residential Assistance and Recovery Following the Northridge Earthquake," *Urban Studies* 41(3): 533–562.

Kates, R. W., C. E. Colten, S. Laska, and S. P. Leatherman. (2006) "Reconstruction of New Orleans after Hurricane Katrina: A Research Perspective," *Proceedings of the National Academy of Sciences* 103(40): 14653–14660.

Khasalamwa, S. (2009) "Is 'Build Back Better' a Response to Vulnerability? Analysis of the Post-Tsunami Humanitarian Interventions in Sri Lanka," *Journal of Geography* 63(1): 73–88.

Kreimer, A. (1980) "Low-Income Housing under 'Normal' and Post-disaster Situations: Some Basic Continuities," *Habitat International* 4(3): 273–283.

Lindell, M. K. and S. N. Hwang (2008) "Households' Perceived Personal Risk and Responses in a Multihazard Environment," *Risk Analysis* 28(2): 539–556.

Lindell, M. K. and R. W. Perry. (2004) *Communicating Environmental Risk in Multiethnic Communities*, Thousand Oaks CA: Sage.

Lindell, M. K. and C. S. Prater. (2003) "Assessing Community Impacts of Natural Disasters," *Natural Hazards Review* 4(4): 176–185.

Logan, J. R. and H. L. Molotch. (1987) *Urban Fortunes: The Political Economy of Place*, Berkeley CA: University of California Press.

Massey, D. D. and N. A. Denton. (1993) *American Apartheid: Segregation and the Making of the Underclass*, Cambridge MA: Harvard University Press.

Mileti, D. (1999) *Disasters by Design: A Reassessment of Natural Hazards in the United States*, Washington DC: Joseph Henry Press.

Morrow, B. H. (1999) "Identifying and Mapping Community Vulnerability," *Disasters* 23(1): 1–18.

Mueller, E. J., H. Bell, B. B. Chang, and J. Henneberger. (2011) "Looking for Home after Katrina: Posdisaster Housing Policy and Low-income Survivors," *Journal of Planning Education and Research* 31(3): 291–307.

Myers, D. (1975) "Housing Allowances, Submarket Relationships and the Filtering Process," *Urban Affairs Quarterly* 11(2): 215–240.

National Research Council (NRC). (2006) *Facing Hazards and Disasters: Understanding Human Dimensions*, Washington DC: National Academy of Sciences Press.

New Jersey Department of Community Affairs (NJ DCA). (2013) *Community Development Block Grant Disaster Recovery Action Plan*, Trenton NJ: Author.

New Jersey Department of Community Affairs (NJ DCA). (2014) "Superstorm Sandy Recovery Division," retrieved July 2, 2014, from www.renewjerseystronger.org/.

Olshansky, R. B., L. D. Hopkins, and L. A. Johnson. (2012) "Disaster and Recovery: Processes Compressed in Time," *Natural Hazards Review* 13(3): 173–178.

Olshansky, R. B. and L. A. Johnson. (2010) *Clear as Mud: Planning for the Rebuilding of New Orleans*, Chicago IL: American Planning Association Press.

Peacock, W. G., N. Dash, and Y. Zhang. (2006) "Shelter and Housing Recovery following Disaster," 258–274, in H. Rodriguez, E. L. Quarantelli, and R. Dynes (eds.), *The Handbook of Disaster Research*, New York: Springer.

Peacock, W. G., N. Dash, Y. Zhang, and S. Van Zandt. (2017) "Post-Disaster Sheltering, Temporary Housing and Permanent Housing Recovery," 569–594, in H. Rodriguez, W. Donner and J. E. Trainor (eds.), *Handbook of Disaster Research*, 2nd ed., New York: Springer.

Peacock, W. G. and C. Girard. (1997) "Ethnic and Racial Inequalities in Hurricane Damage and Insurance Settlements," 171–190, in W. G. Peacock, B. H. Morrow, and H. Gladwin (eds.), *Hurricane Andrew: Ethnicity, Gender and the Sociology of Disaster*, London: Routledge.

Peacock, W. G., H. Kunreuther, W. H. Hooke, S. L. Cutter, S. E. Chang, and P. R. Berke. (2008) Toward a Resiliency and Vulnerability Observatory Network: RAVON. Final Report NSF Grant SES-08311115. Hazard Reduction and Recovery Center, Texas A&M University. http://archone.tamu.edu/hrrc/Publications/researchreports/RAVON.pdf.

Peacock, W. G., B. H. Morrow, and H. Gladwin (eds.). (1997) *Hurricane Andrew: Ethnicity, Gender and the Sociology of Disasters*, London: Routledge.

Peacock, W. G. and A. K. Ragsdale. (1997) "Social Systems, Ecological Networks and Disasters: Toward a Socio-Political Ecology of Disasters," 20–35, in W. G. Peacock, B. H. Morrow, and H. Gladwin (eds.), *Hurricane Andrew: Ethnicity, Gender and the Sociology of Disaster*, London: Routledge.

Peacock, W. G., S. Van Zandt, Y. Zhang, and W. Highfield. (2014) "Inequality in Long-Term Housing Recovery after Disasters," *Journal of American Planning Association*, 80(4), 356–371.

Phillips, B. D. (1993) "Culture Diversity in Disasters: Sheltering, Housing and Long-Term Recovery," *International Journal of Mass Emergencies and Disasters* 11(1): 99–110.

Phillips, B. and M. Ephraim. (1992) *Living in the Aftermath: Blaming Processes in the Loma Prieta Earthquake* (No. 80), Boulder CO: University of Colorado, Institute of Behavioral Science, Natural Hazards Research Applications Information Center.

Quarantelli, E. L. (1982) "General and Particular Observations on Sheltering and Housing in American Disasters," *Disasters* 6(4): 277–281.

Rice, H. (2013) "Galveston Ends Defiance on Housing," *Houston Chronicle* April 17, 2013. Accessed September 30, 2013 at www.chron.com/news/houston-texas/houston/article/Galveston-ends-defiance-on-housing-4443137.php.

Rodriguez-Dod, E. C. and O. Duhart. (2007) "Evaluating Katrina: A Snapshot of Renters' Rights Following Disasters," *Nova Law Review* 31(3): 467–485.

Talev, M. (2013) "Christie Heckler Does Not Want to 'Sit Down and Shut Up,'" *The Bloomberg News*, October 24, 2014. Retrieved on December 15, 2014 from www.bloomberg.com/politics/articles/2014-10-29/christie-heckler-does-not-want-to-sit-down-and-shut-up.

U.S. Department of Housing and Urban Development (HUD). (2013). Community Development Block Grant Disaster Recovery Program. Retrieved April 10, 2016, from www.hudexchange.info/programs/cdbg-dr/.

Van Zandt, S. (2007) "Racial/Ethnic Differences in Housing Outcomes for First-Time, Low-Income Home Buyers," *Housing Policy Debate* 18(2): 431–474.

Van Zandt, S., W. G. Peacock, D. Henry, H. Grover, W. Highfield, and S. Brody. (2012) "Mapping Social Vulnerability to Enhance Housing and Neighborhood Resilience," *Housing Policy Debate* 22(1): 29–55.

Wu, J. Y. and M. K. Lindell. (2004) "Housing Reconstruction After Two Major Earthquakes: The 1994 Northridge Earthquake in the United States and the 1999 Chi-Chi Earthquake in Taiwan," *Disasters* 28(1): 63–81.

Zahran, S., S. D. Brody, W. G. Peacock, H. Grover, A. Vedlitz. (2008) "Social Vulnerability and the Natural and Build Environment: A Model of Flood Casualties in Texas," *Disasters* 32(4): 537–560.

Zhang, Y. (2012) "Will Natural Disasters Accelerate Neighborhood Decline?" *Environment and Planning B* 39(6): 1084–1104.

Zhang, Y. and W. Drake. (2016) "Planning for Housing Recovery after the 2008 Wenchuan Earthquake in China," 191–208, in A. Sapat and A. M. Esnard (eds.), *Coming Home after Disaster: Multiple Dimensions of Housing Recovery*, New York: Routledge.

Zhang, Y. and W. G. Peacock. (2010) "Planning for Housing Recovery? Lessons Learned from Hurricane Andrew," *Journal of the American Planning Association* 76(1): 5–24.

17

POPULATION DISPLACEMENT

Ann-Margaret Esnard and Alka Sapat

Introduction

Displacement due to natural disasters is expected to increase in the coming decades due to increases in population, hazard exposure, and vulnerability to natural hazards, as well as increased frequency and intensity of extreme weather events. According to the Internal Displacement Monitoring Center, 24.2 million of the total 31.1 million new cases of internal displacement resulted from disasters triggered by sudden-onset hydro-meteorological, climatological, and geophysical hazards (IDMC 2017). Developing countries bear the brunt of disasters and displacement each year, but high-income countries are not immune to displacement.

In the United States, a number of natural disasters produced involuntary displacement in the 1990s: Hurricane Andrew in 1992 (Morrow 2005; Peacock et al. 1997; Sanders et al. 2004; Smith and McCarty 1996); the Mississippi River flood (Iowa, Illinois, Missouri) in 1993 (Changnon 1996); the Loma Prieta and Northridge earthquakes in 1989 and 1994, respectively (Eadie 1998; Olshansky et al. 2003); and Hurricane Floyd in 1999 (Dow and Cutter 2000; Maiolo et al. 2001). However, hurricane Katrina, which made landfall in southern Louisiana in August 2005, was the defining catastrophic disaster that brought population displacement and the resultant dilemmas and complexities of long-term recovery to the fore in planning and policy domains. More than 1 million people from throughout the Gulf Region including Louisiana, Mississippi, and Alabama were displaced and more than 1.2 million housing units were damaged in New Orleans (U.S. Senate 2009; Hori et al. 2009). Myers et al. (2008) reported high outmigration rates of disadvantaged populations, while other scholars reported on multiple and protracted displacement that resulted when these vulnerable households were forced to relocate several times after the initial displacement, primarily because of the difficulty of finding suitable and affordable housing (Meyer 2013; Weber and Peek 2012a, 2012b).

Since 2010, other notable disasters have produced involuntary displacement. In May 2011, an EF-5 tornado with wind speeds of 200 miles per hour affected Joplin, Missouri (NOAA 2011). The City of Joplin reported that the tornado affected more than 17,000 people, damaged an estimated 4,000 homes, and displaced more than 9,000 people (City of Joplin 2016). In 2012 Hurricane Sandy delivered a blow to 24 states, with major impacts felt in New Jersey and New York. As of 2013, Hurricane Sandy was the largest tropical cyclone on record in the Atlantic basin and the costliest storm disaster in U.S. history, with economic damage assessed at $71 billion (IDMC 2013). Power outages, flight cancelations, airport closures, and two consecutive days of suspended trading on Wall Street resulted in impacts being felt far beyond New York City (Malone 2012). According to IDMC

(2013), this storm forced more than three quarters of a million people to leave their homes. In September 2013, the Colorado floods affected 20 counties in Colorado, including the Boulder area, Northeast Colorado, and Central Colorado (Colorado United 2014). An unprecedented rainfall of 18.4 inches in September near the Boulder area (versus an average September rainfall of 1.6 inches) resulted in a 1,000-year rain, 100-year flood (Brennan and Aguilar 2013). This led to the destruction of 1,852 homes, damage to 28,363 homes, a $2.9 billion economic impact, and displacement of more than 100,000 persons (Colorado United 2014; IDMC 2014). As we write this chapter, communities, regions, and nations are still reeling from the 2017 and 2018 catastrophic hurricanes and wildfires that will no doubt lead to record levels of damage, destruction, and displacement in the U.S. and its territories. Hurricane Maria has accelerated the pace of migration from Puerto Rico to the U.S. mainland, as families seek housing, schools, jobs, and health care.

This pattern of intense weather events and resultant destruction and population displacement signals a wake-up call for planning and policy professionals who find themselves in disaster-stricken recovering communities, as well as for host communities oftentimes making short- and long-term decisions to address multi-faceted and complex housing, infrastructure and land use decisions. This chapter is an attempt to place population displacement in the context of planning for resilient communities. We begin by defining the concept of population displacement and discussing different definitions and measures for resilience. Next, we focus on how housing recovery affects displacement, and the dilemmas engendered by displacement for host communities. Planning challenges and opportunities that encompass issues such as host community pre-disaster planning efforts, planning for non-displaced populations, land re-development and gentrification, and vulnerability reduction are addressed in the final section. Parts of this book chapter draw from the authors' previously published work as indicated in the acknowledgments.

Population Displacement: Concepts and Definitions

Population movement encompasses several distinct concepts that refer to those who are evacuated and those who are displaced. Part of the conceptual confusion lies in the temporal dimensions of the evacuation continuum. Other aspects stem from different terms used by planners, policy-makers, and even within the research community to characterize voluntary and involuntary displacement. In its most basic form, evacuation is an instinctual flight response to danger in any form and has been generalized in the research literature as people's withdrawal actions from a specific area because of a real or anticipated threat or hazard (Sorenson and Vogt 2006). To deal with potential disasters, planning for evacuation became integral to disaster management processes and more current definitions of evacuation reflect a systematic planning component. As defined by the Department of Homeland Security, evacuation is "the organized, phased, and supervised withdrawal, dispersal, or removal of civilians from dangerous or potentially dangerous areas, and their reception and care in safe areas" (U.S. Department of Homeland Security 2008: 139). Whether systematic or spontaneous (Gerber 2010), evacuation has been historically understood to be temporary and short-term (usually up to a week). However, the growing intensity and scale of disasters has increasingly meant that evacuation is often not short-term, and evacuated residents and businesses can no longer return to the area (Quarantelli 1980; Sorenson and Vogt 2006). If placed on a temporal dimension, displacement is typically viewed as being of longer duration. Displacement has been described by Oliver-Smith (2006) as the uprooting of people in response to physical, economic, or environmental danger or harm, such as a natural hazard. In the U.S. context, past scholarly research has also referred to the mass movements of populations (Lin 2009) and households due to damage by natural hazards to housing and infrastructure as population dislocation (Van Zandt et al. 2009).

Population displacement can be conceptualized as arising from three different factors: a) displacement ensuing from people fleeing violent conflict or war (Belcher and Bates 1983); b) displacement due to development projects such as construction of dams (de Wet 2006; Oliver-Smith 2009); and c) displacement due to disasters and crises (Esnard and Sapat 2014; Weber and Peek 2012a), including, more recently, displacement induced by environmental degradation spurred by climate change (Oliver-Smith 2012; Oliver-Smith and Sherbinin, 2014; Singh 2012). Other facets of displacement include its voluntary/involuntary nature. The terms relocation and resettlement of populations are often used to characterize those displaced by infrastructure projects. Governments that deploy these projects usually depict resettlement processes as voluntary in nature, implying the acquiescence of populations in the migration that it entails, particularly in developing countries. However, as noted by scholars, most large-scale infrastructure projects compel the displacement and resettlement of millions of people. Although they aim to generate economic growth, these projects mostly leave local people displaced, disempowered, and destitute (Oliver-Smith 2009).

Most of the displacement occurring due to disasters is also seen as being voluntary in nature. However, there is a fine line between the voluntary and involuntary nature of displacement. For the displacees, there often is no choice of return due to physical devastation of an area or due to economic reasons. At times, what starts out as temporary evacuation can evolve into permanent displacement and eventual relocation. As Belcher and Bates (1983) note, the choice to relocate voluntarily depends on the interplay between push factors (e.g., unemployment and political conflict) in the home place and pull factors (e.g., employment prospects and potential for better quality of life) from elsewhere. For some, displacement is for a very long duration, resulting in protracted displacement; sometimes this occurs when displaced persons move back and forth between insecure locations, displaced many times in search of safety and livelihoods (IFRC 2012: 23). With protracted displacement, the prospect of finding a durable solution to displacement such as return or permanent resettlement stalls, as displacees continue to require assistance, often remaining marginalized in host societies and communities. Those in protracted displacement are also more vulnerable to repeated displacement and risk getting caught in further cycles of disaster and displacement.

Various factors affect the potential risk for displacement. Among them are levels of displacement vulnerability and equally, if not more importantly, levels of community resilience. To understand how resilience affects displacement, we briefly discuss different concepts of resilience and various metrics used in its measurement.

Resilience and Population Displacement

Building community resilience is increasingly seen as being vitally important to reduce vulnerability and to mitigate potential displacement. Over the years, different understandings have emphasized engineering and seismic resilience (Bruneau et al. 2003; Miles and Chang 2006), physical resilience (Bodin and Wiman 2004; Gordon 1978); ecological resilience (Gunderson 2000; Holling 1973; Longstaff 2005; Waller 2001), social resilience (Adger 2000; Godschalk 2003), community resilience (Coles and Buckle 2004; Cutter et al. 2008; Patel et al. 2017; Pfefferbaum et al. 2005), and individual resilience (Butler et al. 2007; Egeland et al. 1993; Masten et al. 1990). Although myriad definitions exist, the concept of disaster resilience, as Comerio (2014: 2) notes, "can be defined simply as the capacity to rebound from future disasters." Norris et al. (2008: 4) contend that resilience also needs to be understood not as an outcome but as "a process linking a set of adaptive capacities to a positive trajectory of functioning and adaptation after a disturbance." Based on this conceptualization, they discuss four sets of adaptive capacities encompassing economic development, social capital, information and communication, and community competence—that together provide a basis for disaster resilience.

Norris et al. (2008) also contend that some dimensions of resilience cannot be captured by a singular metric. They point out that the value of the resilience concept is its ability to describe the characteristics of, and interactions between, stressors (disasters), adaptive capacities, and wellness, and they maintain that capacities and disaster readiness can be enhanced through interventions and policies. Nevertheless, scholars and policy-makers have made important progress in measuring resilience, while acknowledging assumptions and limitations. Cutter et al. (2010) discuss baseline indicators and related social economic infrastructure along with institutional capacity (mitigation) and community competence variables. Peacock (2009) used a different conceptual model based on a combination of a community's capital resources (social, economic, physical, and human) and the four phases of a disaster: mitigation (perceptions and adjustments); preparedness (planning and warning); response (pre- and post-impact); and recovery (restoration and reconstruction).

Typically, higher levels of resilience are likely to be associated with lower numbers of households being displaced. However, the relationship between resilience levels and potential displacement is not always straightforward. Displacement risk is affected by levels of vulnerability; complex interactions among different variables that affect resilience, vulnerability, and displacement; and cumulative impacts of repeated disasters (Esnard et al. 2011). Also, resilience is a multi-dimensional concept and complementary and conflicting relationships may exist between different dimensions (Sapat 2012). For instance, social resilience may be linked to economic resilience at the individual level as those who have better social networks are likely to have better access to economic resources such as jobs (Norris et al. 2008). However, some dimensions of resilience may conflict with and reduce overall community resilience; for instance, strengthening economic resilience (in the form of greater industrialization or more development) can have a negative impact on ecological resilience, as seen in environmental impacts (Sapat 2012: 75). In developing a displacement risk index, Esnard et al. (2011) note that there are numerous variables that affect displacement risk, including social and physical vulnerability and levels of resilience. Levels of vulnerability can in some instances be offset by adaptive capacities increasing resilience. For instance, although urban areas might be more vulnerable to the impacts of natural hazards due to increasing levels of property and infrastructure that lie in harm's way, these areas may also have greater capacities, such as emergency medical services, housing, and non-governmental organizations that can offset the levels of vulnerability. However, as noted by Mayunga (2007: 4) "conceptualizing resilience [as the opposite of vulnerability] may not be desirable because it does not add much to our understanding."

In the various examples used throughout the remainder of this chapter (i.e., Hurricane Andrew; Hurricane Katrina, Joplin tornadoes, Hurricane Sandy, Colorado Floods), we see a wide variability in levels of involuntary displacement, damage to homes, and economic impacts that resulted from natural disasters. Sheltering and housing options and lengthening recovery periods are cross-cutting dilemmas with implications for household and community resilience. Such dilemmas are discussed in the following section.

Housing Recovery and Displacement

Housing damage and subsequent recovery is critical in determining the lengths and levels of displacement. Policy change and learning in post-disaster housing has been problematic despite repeated disasters; the dilemmas are rooted in pre-disaster vulnerabilities, shortages in affordable housing and rentals, inadequate vertical and horizontal coordination among multiple entities dealing with housing, the lack of effective and politically strong low-income housing advocacy groups and coalitions, and the "messy" nature of post-disaster housing policy (Sapat et al. 2011). Locating housing for displaced populations and in some countries, dealing with land tenure problems and the lack of clear land ownership issues also present challenges to effective housing recovery

(Sapat and Esnard 2017). After Hurricane Katrina, affordable housing was in short supply, as rents rose by as much as 46 percent three years after the hurricane (Brookings Institution 2008; Sapat et al. 2011). Following Hurricane Sandy, post-disaster housing complexities resulted from limited access to affordable housing options in New York City, one of the most difficult housing markets due to low vacancy rates and high rents (Conlin 2012; Schwirtz 2012; Soria 2013). Similar problems have cropped up in Houston and other parts of Texas impacted by Hurricane Harvey that are experiencing shortages in temporary and permanent affordable housing.

Problems also arise in post-disaster housing in the U.S. due to widely held assumptions that market mechanisms, both in insurance and property, can provide a solution in post-disaster situations, and that market solutions exist for renters to find alternate accommodations in post-disaster scenarios (Comerio 1998, 2014). However, as noted by Comerio (2014), economic downturns such as the 2008 recession can severely impact housing markets. Additionally, a number of homeowners do not have full or high-quality property and disaster insurance coverage (Bolin and Stanford 1998; Peacock and Girard 1997); personal savings and peer relief are usually inadequate, and there are problems with the National Flood Insurance Program (Lindell, et al. 2016). Many fail to apply for, or are not covered by Small Business Administration (SBA) loans (Dash et al. 2007) and typical FEMA individual assistance programs do not cover housing repair costs. Market mechanisms also do not supply housing to the most vulnerable. Comerio points out that it took 10 years to replace 75 percent of the affordable housing lost in the San Francisco Bay Area after the 1989 Loma Prieta earthquake, it took 4 years to rebuild middle-class apartments lost in Los Angeles after the 1994 Northridge earthquake, and it took 7 to 10 years to rebuild housing in Kobe, Japan, after the 1995 Hanshin-Awaji earthquake (Comerio 2014). Rental recovery in minority neighborhoods, particularly in areas with higher concentrations of Blacks, was also found to be much slower following Hurricane Andrew, leaving them worse-off in the long-term (Zhang and Peacock 2010). During recovery processes, a larger portion of the influx of governmental assistance and private insurance typically accrues to homeowners and more affluent residents, who are then better positioned than the less affluent to acquire available housing following displacement (Esnard and Sapat 2014).

Given these impediments, planning for post-disaster housing can be critical. As noted by Olshansky and his colleagues, the presence of a planning and recovery framework in New Orleans before Katrina could have aided in developing recovery scenarios and project prioritizations based on community consensus, would have helped reduce the long-term displacement of thousands of individuals by helping the city to quickly procure federal funds for reconstruction, and would have reduced the intense time pressure to quickly rebuild something after the storm (Olshansky 2006; Olshansky et al. 2008; Olshansky and Johnson 2010; Olshansky et al. 2012).

The problems following Katrina were the impetus leading to the adoption of the National Disaster Housing Strategy (NDHS) and some critical policy changes that were put into motion (Sapat et al. 2011). Under this framework, a number of states have developed disaster housing task forces. We turn to Florida as an example.

Since 2010 Florida has been developing the State Disaster Housing Planning Initiative (SDHPI), which has led to the development of a state disaster housing plan and a county disaster housing template. The state disaster housing task force consists of representatives from federal, state, and local agencies in the fields of emergency management, housing and neighborhood programs, land-use planning, human services, and engineering. It also includes representatives from nongovernmental organizations, faith-based and volunteer groups, and private sector partners.

In addition to the role played by state governments, what local governments do can be critical in post-disaster planning. Smith (2017: 287) notes that after Hurricane Fran hit Kinston, North Carolina, the city developed a pre-disaster recovery plan incorporating a number of hazard mitigation goals and identifying homes in flood-prone areas. When Hurricane Floyd struck, the city was ready with its application to the state and FEMA that received quick approval, substantially speeding up the housing recovery process (Smith 2011: 65). Local governments can take measures to develop local ordinances to facilitate the placement of temporary housing units on private property to house disaster survivors, as well as pre-identify potential community sites to expedite the placement process (Florida Department of Emergency Management 2012). Identifying temporary housing sites is one of the most daunting problems faced by FEMA (Aldrich and Crook 2013; Zensinger 2013). For instance, as Hurricane Katrina, Hurricane Ike, and other disasters in the past have shown, manufactured housing is limited in its use, cannot be placed in floodplain areas, is expensive and time consuming to install, and faces community opposition in the form of NIMBYism (Aldrich and Crook 2008). Community housing sites for temporary housing are often far removed from jobs, schools, and other social infrastructure (Zensinger 2013). Flood-resistant areas that decision-makers choose for housing are less willing to accept such projects than more flood-prone ones, and residents in more geographically suitable and safer locations may zealously and vociferously protect their areas from encroachment by those wishing to locate temporary housing there. This leads then to temporary housing being placed in areas that are geographically more vulnerable, increasing vulnerability to the next disaster (Aldrich and Crook 2013).

Displacement to nearby and far-flung communities is another important consideration in addressing long-term recovery, given that similar housing and land use problems persist. The next section documents some of the other challenges faced by displaced persons, as well as host community residents, agencies, and institutions.

Host Communities: Place and People Dilemmas

Host communities (or receiving communities) are communities that are not affected by the primary event, but are those to which displaced persons turn as safe havens. Host communities that provide refuge for displaced populations affected by natural disasters tend to be in close proximity to impacted communities if vacant housing is available (Smith and McCarty 1996). As documented in Welsh and Esnard (2009), Broward County in South Florida has had a unique experience historically as a receiving area in 1992 with Hurricane Andrew and as an area of direct impact from Hurricane Wilma in 2005. In 1992 Hurricane Andrew struck adjacent southern Miami-Dade County, destroying more than 40,000 homes (Peacock et al. 1997) causing more than 40 percent of households to leave their homes due to damage and infrastructure loss (Smith and McCarty 1996), and leaving more than 80,000 people unemployed (Hartwig 2002). The devastation from the 145-mile-per-hour storm instigated an unanticipated housing rush as displaced households migrated north to reestablish their lives. Just as southern Miami-Dade communities were unprepared to assist residents with the significant challenges delivered with the storm (Benedick 2002), Broward County was similarly unprepared with urban growth plans to receive these displaced households. The increase in Broward County's population over the next decade, driven in part by the migrations from Miami-Dade County, rapidly transformed sparsely developed, rural western Broward County into a sprawling suburb with little remaining vacant land (Welsh and Esnard 2009). Major spikes in home prices, driven by high demand and a shrinking supply of homes, accompanied the migration.

Weber and Peek (2012a) provide one of the most thorough accounts of experiences of Katrina displacees in some of the host communities around the country—Colorado (Denver), Georgia (Atlanta), Louisiana (Baker, Baton Rouge, Lafayette, New Orleans), Mississippi (Jackson), Missouri (Columbia), South Carolina (Columbia/West Columbia), and Texas (Austin, Dallas, Houston, Huntsville). One common story that emerged in the aftermath of Katrina was that urban evacuees were displaced to rural host communities and small cities lacking public transportation, infrastructure, and social services (Miller 2012a, 2012b). Additionally, some displaced persons faced economic hardships related to housing and jobs, whereas others faced widespread marginalization, prejudice, and stigmatization in their new locations (Weber and Peek 2012a, 2012b). What remains unclear, however, is how to address the spikes in demand for housing, schools, government services, health and human services, and infrastructure in these host communities. Transfer or sharing of the tangible and intangible costs and coordination between originating and hosting municipalities, housing insecurity for both residents and displaced persons in those areas, and decisions about the kinds of zoning and policy ordinances needed for placement of temporary manufactured homes and/or accommodation of displaced persons are some of the issues brought to the fore, especially after Hurricane Katrina (Gerber 2010; Gongol 2005; Levine et al. 2007; Mitchell et al. 2012).

Gerber's research on evacuation ingress management begins to address the important issue of governance and capacity needs. That research has led him to conclude that identifying basic relationships between local capacity and preparedness activities permits a better understanding of how well local governments can perform this important aspect of disaster management (Gerber 2010). He identified four critical tasks to address evacuation ingress management. These are: (1) establish reception areas/centers for evacuees in the host jurisdiction; (2) monitor traffic movement of self-evacuees into/through the area, and redirect as necessary; (3) coordinate with support agencies to provide short-term needs for evacuees, such as directions, information, shelter, medical care, and other assistance; and (4) plan, in coordination with social service agencies, for long-term support for evacuees (Gerber 2010).

Since displacement is a multi-faceted and complex phenomenon, building resilience to potential displacement involves the consideration of multiple factors (social, economic, political, and physical) and requires planning and policy interventions at different levels and across a spectrum of organizations. Some of the planning challenges and opportunities are discussed in the following section.

Planning: Challenges and Opportunities

As most disasters are as much man-made as they are natural, much more can be done in order to strengthen community-based and national resilience to prevent the worst impacts of natural hazards, and to better prepare for events that cannot be avoided.

(IDMC 2014: 6)

Some of the main challenges that arise in dealing with planning challenges relate to host community pre-disaster planning efforts, planning for non-displaced populations, land re-development and gentrification issues, and ongoing efforts at reducing vulnerability.

Pre-Disaster Planning for Post-Disaster Displacement

In addition to intra-county and inter-county movement after disasters, Hurricane Katrina highlighted that far-flung inter-state displacement is a new reality. Studies by Groen and Polivka (2008, 2010) demonstrate that in the first two years after the hurricane, displaced persons who had never married and those with lower levels of education were much less likely to return compared to those in other racial, marital, or educational groups. They also determined that compared to Alabama and Mississippi, Louisiana had the lowest numbers of people who returned to the area in which they were living prior to the storm, and that the metropolitan areas of Houston, New Orleans, Dallas, Baton Rouge, Atlanta, and Memphis received the largest number of evacuees who relocated outside their pre-Katrina counties (Groen and Polivka 2008).

One of the lingering questions by Gongol (2005)—should cities actively prepare for the next time by beginning the planning process for receiving displacees—is an important one with many implications for planning and community resiliency. We turn to Florida's Post-Disaster Redevelopment Planning (PDRP) initiative for an answer. Following the devastating hurricanes of 2004, the State of Florida mandated that coastal communities prepare and adopt "Post-Disaster Redevelopment Plans" (PDRPs, Florida Department of Community Affairs 2010). The program developed planning guidelines that were tested in six pilot counties, acknowledging the vulnerability of both coastal and inland communities and the importance of fostering multi-county discussions about interdependent regional infrastructure systems and evacuation. One of the pilot communities, Polk County, was the only inland county that is located between the two major urban areas of Orlando and Tampa. As such, Polk County's plan acknowledges the possibility that even if Polk County is not directly impacted by a disaster, an incident in either one of these two major urban areas will mean providing host services to displaced survivors. During the planning process, the existing disaster housing plan was reviewed and found by the Building, Housing, and Historic Preservation Workgroup to be insufficient if the county were directly impacted by an incident and needed to house its own residents, or if it were indirectly affected as a host county to long-term evacuees from other counties. These deficiencies might have been due to the fact that the current plan was not developed in consultation with land use planners to ensure consistency with future land use designations (Florida Department of Emergency Management 2010).

Gentrification and Land Redevelopment

In post-disaster rebuilding efforts, planners must continue to draw on lessons learned about linking recovery, reconstruction, and resiliency initiatives. Disasters potentially result in opportunities to redress mistaken historical patterns and redevelop in better ways. However, it is important to note that although redevelopment might produce positive impacts, post-disaster period and redevelopment projects can also entail redevelopment injustices for disaster survivors and displaced persons, which in turn reproduce the conditions for future risk (Derickson 2014; Gotham and Greenberg 2014; Greenberg 2014). The question of "redevelopment for whom" persists because recovery is often skewed towards wealthier sections of society and uneven redevelopment takes the form of gentrification of low-income neighborhoods destroyed by disaster. After Katrina, city officials in the cities of Gulfport and Biloxi, Mississippi, pronounced the storm as having brought forth new opportunities. However, these opportunities were mainly to use the destruction wrought by Katrina as an impetus for regional development that would negatively impact poor African American neighborhoods that had already been squeezed by urban development strategies prior to the disaster (Derickson 2014).

In an empirical study of changes after major hurricanes, using geographic information systems (GIS) data from the "billion dollar" storms of the early 1990s and demographic data from local census tracts, Pais and Elliott (2008) found that in the years following such disasters, pro-growth coalitions took advantage of new sources of material and symbolic capital to promote development that favored residential elites and their allies. The authors' findings substantiated previously researched case studies on increases in social inequality following disaster. They contend that these pro-growth coalitions act as "recovery machines," contributing to uneven and more segregated transformations of local neighborhoods across affected regions and exacerbating existing and future vulnerabilities to disasters (Pais and Elliott 2008). Moreover, they astutely point out that during recovery, funds available for redevelopment increase substantially via government disaster aid and private insurance claims. This provides additional fuel to recovery machines led by developers, residential elites, and their allies, allowing these groups to exercise disproportionate control over post-disaster resources and decisions.

Invisible Displacees

Planning for displaced populations is particularly difficult when the displaced, such as urban displacees and undocumented immigrants, cannot be "seen" or remain invisible to state actors. As Crisp et al. (2012) pointed out, refugees and internally displaced persons are increasingly not found among host communities or camps in rural areas, but are instead in towns and cities. In developing countries, cities absorb these populations, but they typically have no choice but to live in shanty towns or on the outskirts of urban settlements, mostly in sordid and destitute conditions, remaining undocumented and often facing settled residents' opposition to their presence. Although there is a higher percentage of urban displacees in developing and middle-income countries, there are also "shadow renters" (Comerio 2014) in cities within the U.S. and other high-income countries. Often these are undocumented immigrants who may be living with extended family or friends or doubling up with other renters. Typically, these populations are served by non-state actors such as nongovernmental organizations (NGOs) and religious groups. Diaspora-led community-based organizations (CBOs), such as Vietnamese-American and Haitian-American organizations, also serve immigrant populations in host communities and advocate on their behalf (Edwards 2010; Esnard and Sapat 2011; Sapat and Esnard 2012). Although NGOs and CBOs can be critical links to these populations, they also need to be included in planning and policy processes for disaster recovery. The challenges in reaching "invisible" populations include finding effective modes of communication, resolving issues related to trust, and addressing gaps with enumeration, metrics, and measurement.

Planning for Non-Displaced Populations

Although more attention is needed in planning for displaced populations, planners and policy-makers also need to prepare for those populations that cannot move due to factors such as the lack of resources and transportation. As noted in an empirical analysis of Miami-Dade County after Hurricane Andrew, households with higher socio-economic status showed a greater tendency to leave their homes and communities (Lin 2009). Typically, households with higher socioeconomic status often possess greater mobility as a result of their access to transportation, internal and external resources, and wider or better-resourced social networks (Morrow-Jones and Morrow-Jones 1991). The affluent can often bypass disaster assistance processes by moving to hotels or moving in with friends and families (Lindell et al. 2011; Wu et al. 2012, 2013). Given that the most marginalized sections of society might be unable to move, planners and policy-makers need to adopt a variety of approaches to address their needs, including the provision of safe temporary shelters.

A post-Hurricane Andrew study of one Miami-Dade working-class community immediately following the 1992 hurricane and then again a decade later (in 2003) highlighted the lingering "deep-seated impacts on many households" that lacked the resources to rebuild locally after the storm, as well as changes in the demographic profiles in the South Miami Heights community, especially a significant increase in the number of renters (Dash et al. 2007). The authors emphasized the importance of planning for a lengthy recovery period in such working-class communities. Similarly, rates of insurance and of business revival and returns are intricately linked to post-disaster housing and are important determinants of recovery (Xiao and Van Zandt 2012; Zhang and Peacock 2010).

Vulnerability Reduction

Ultimately, planners need to acknowledge that vulnerability reduction is a key aspect of community resilience. King (2006: 293) noted that planning is an "indirect reducer of vulnerability, through its role in developing services, facilities, infrastructure and access," but that most of the existing vulnerable locations are the results of historical decisions about which modern planners can do little until redevelopment. Colorado has had a long history of costly flood disasters, and the City of Boulder is known for its long tradition of floodplain mitigation and management pioneered by famous geographer, Dr. Gilbert White. Citing FEMA, Aguilar (2014) has noted that 24,000 individuals and businesses in Colorado had federal flood coverage at the time of the devastating flood. Recovery efforts will require ongoing education and outreach efforts to homeowners and business owners who are in high flood risk areas. New York City is also looking ahead; Hurricane Sandy presented a window of opportunity to address long-term recovery and mitigate the impact of future storms. At the time of the hurricane, the NYC Mayor's office convened a Special Initiative for Rebuilding and Resiliency (SIRR) to immediately begin analyzing Sandy's impact and to produce a comprehensive update to PlaNYC entitled *A Stronger, More Resilient New York*. Building on the successes of PlaNYC's 2007 and 2011 incarnations, a new plan set forth hundreds of recovery, mitigation, and long-term sustainable development strategies and tactics (SIRR 2013). Furthermore, New York City updated its Coastal Storm Plan to double the number of evacuation zones and work more closely with the NYC Housing Authority and other agencies to prioritize and protect vulnerable populations (Gibbs and Holloway 2013).

Plans need to rely on basic vulnerability data (pre-disaster) and magnitudes of population displacement (post-disaster). At the most basic level, it would be useful to know the numbers of individuals and families displaced so that authorities in both disaster-stricken and host communities can determine housing, school, health care, and other needs (Plyer et al. 2010). Like Crisp et al. (2012), we acknowledge that accounting for displaced persons (internal and transnational) and returnees is and will remain an inexact science. Currently, in the United States, there is no systematic approach, nor are there guidelines to keep track of the number of displaced persons, though the technology certainly exists. In the case of Hurricane Katrina, the Louisiana Recovery Authority (LRA) led the policy- and decision-making process after the hurricane, and was responsible for supervising the rebuilding and recovery processes. At the same time, the LRA led the necessary initiatives to estimate initial displacement in the area, and population declines and increases. Plyer et al. (2010) provide a useful summary of the pros and cons of using various administrative data sets (e.g., school enrollment, utility connections, United States Postal Service counts) as a basis for population estimates following Hurricane Katrina. It is encouraging, however, that the International Federation of Red Cross and Red Crescent Societies (IFRC 2014) has expanded the list of vulnerability indicators. That list incorporates displaced populations, migrants, and returnees. Widespread adoption of this expanded list can have implications for how we conceptualize and measure resiliency.

Conclusion

Being proactive in planning is critically important for smoother disaster recovery processes and reductions in the duration and extent of displacement. Local governments need to work in a coordinated fashion with state governments and the federal government to adopt, implement, and enforce state legislation and plans that aim to reduce potential displacement. Building and sustaining the collective capacity of networks of relevant stakeholders during the recovery planning process as well as developing actionable pre-disaster recovery plans that proactively tackle housing issues can catalyze housing recovery (Smith 2017). Additionally, ongoing dialogue is needed about the use of standard definitions and terminology for displaced persons in agencies and organizations. There are several terms that are employed in defining population movement and migration, as well as persons displaced by disasters (Esnard and Sapat 2014). Differences in the definitions of these terms are not merely semantic distinctions; rather they have social-legal and political implications (Sapat and Esnard 2012). As noted by Schneider and Ingram (1993), designations and terminologies reflect social constructions and underlying narratives. These play an important role in determining who benefits and who loses from policies. Particular designations and terms may determine certain rights and expectations for services. Planners need to be cognizant of the multiple "publics" that they serve, which include communities with differential vulnerabilities, capabilities, and capacities.

Finally, we need to pay greater attention to the transnational nature of disasters and catastrophes: those that extend beyond politically and socially constructed borders. As noted by Sapat and Esnard (2012), the 2010 Haiti earthquake and resultant population displacement had repercussions not just in Haiti, but transnationally in several countries, including the United States. Planners must have the necessary knowledge and skills to engage a broader spectrum of stakeholders or the "whole community." This whole community approach, according to FEMA (2011: 3), engages a broad spectrum of the community as "vital partners in enhancing resiliency" of the entire Nation, and "determines the best ways to organize and strengthen their assets, capacities, and interests." In addition to the role played by local governments, coordination with non-governmental actors and the private sector is vitally important for policy initiatives such as post-disaster housing, vulnerability reduction, and building resilience in both disaster-stricken and host communities. Education and outreach programs to these groups are needed for effective program implementation. Provision of information and knowledge to marginalized populations about insurance, housing recovery programs, and policy initiatives is also needed to improve recovery processes, which in turn helps reduce displacement locally or across city, county, state, and national borders.

Acknowledgment

The material in this chapter is based on research supported by the U.S. National Science Foundation Grants (NSF Grant Nos. CMMI-0726808; CMMI-1034667; and CMMI-1162438). Any opinions, findings, and conclusions or recommendations expressed in this material are those of the authors and do not necessarily reflect the views of the National Science Foundation. We also wish to acknowledge Routledge for granting us permission to reuse small portions of our book *Displaced by Disaster: Recovery and Resilience in a Globalizing World*.

References

Adger, W. (2000) "Social and Ecological Resilience: Are They Related?" *Progress in Human Geography* 24(3): 347–364.

Aguilar, J. (2014) "Colorado Re-emerging from $2.9 billion Flood Disaster a Year Later," *Denver Post* (September 7, 2014). Retrieved November 2014 from: www.denverpost.com/news/ci_26483288/colorado-re-emerging-from-2-9-billion-flood.

Aldrich, D. P. and K. Crook. (2008) "Strong Civil Society as a Double-Edged Sword: Siting Trailers in Post-Katrina New Orleans," *Political Research Quarterly* 61(3): 379–389. doi:10.1177/1065912907312983.

Aldrich, D. P. and K. Crook. (2013) "Taking the High Ground: FEMA Trailer Siting after Hurricane Katrina," *Public Administration Review* 73(4): 613–622.

Belcher, J. C. and F. L. Bates. (1983) "Aftermath of Natural Disasters: Coping through Residential Mobility," *Disasters* 7(2) 118–128.

Benedick, R. (2002) "Hurricane Andrew Left Legacy of Higher Housing Costs," *The Sun Sentinel*. Retrieved November 2014 from: www.sun-sentinel.com/news/weather/hurricane/sfl-sbuildaug20, 0,5553762.story?page=2.

Bodin, P. and B. Wiman. (2004) "Resilience and Other Stability Concepts in Ecology: Notes on their Origin, Validity, and Usefulness," *ESS Bulletin* 2(2): 33–43.

Bolin, R. and L. Stanford. (1998) *The Northridge Earthquake: Vulnerability and Disaster*, London: Routledge.

Brennan, C. and J. Aguilar. (2013) "Eight Days, 1,000-year Rain, 100-year Flood," *Daily Camera*. Retrieved November 2014 from: www.dailycamera.com/news/boulder-flood/ci_24148258/boulder-county-colorado-flood-2013-survival-100-rain-100-year-flood.

Brookings Institution and GNOCDC. (2008) *Anniversary Edition: Three Years after Katrina. New Orleans: Greater New Orleans Community Data Center*. Retrieved November 2014 from: www.gnocdc.org.

Bruneau, M., S. E. Chang, R. T. Eguchi, G. C. Lee, T. D. O'Rourke, A. M. Reinhorn, M. Shinozuka, K. T. Tierney, W. A. Wallace, and D. von Winterfeldt. (2003) "A Framework to Quantitatively Assess and Enhance the Seismic Resilience of Communities," *Earthquake Spectra* 19(4): 733–752.

Butler, L., L. Morland, and G. Leskin. (2007) "Psychological Resilience in the Face of Terrorism," 400–417, in B. Bongar, L. Brown, L. Beutler, J. Breckenridge, and P. Zimbardo (eds.), *Psychology of Terrorism*, New York: Oxford University Press.

Changnon, S. A. (1996) *Great Flood of 1993: Causes, Impacts, and Responses*, Boulder CO: Westview Press.

City of Joplin. (2016) Joplin, Missouri hit by EF-5 Tornado on May 22, 2011. Retrieved December 2017 from: www.joplinmo.org/DocumentCenter/View/1985.

Coles, E. and P. Buckle. (2004) "Developing Community Resilience as a Foundation for Effective Disaster Recovery," *The Australian Journal of Emergency Management* 19(4): 6–15.

Colorado United (2014). "News and Top Issues," Retrieved November 2014 from: https://sites.google.com/a/state.co.us/coloradounited/about-1/updates.

Comerio, M. C. (1998) *Disaster Hits Home: New Policy for Urban Housing Recovery*, Berkeley CA: University of California Press.

Comerio, M. C. (2014) "Disaster Recovery and Community Renewal: Housing Approaches," *Cityscape* 16(2): 51–68.

Conlin, M. (2012) "Sandy Refugees Say Life in Tent Feels Like Prison," *Thomson Reuters Foundation*. Retrieved December 2017 from: www.reuters.com/article/us-storm-sandy-tentcity/sandy-refugees-say-life-in-tent-city-feels-like-prison-idUSBRE8A90BV20121110.

Crisp, J., T. Morris, T., and H. Refstie. (2012) "Displacement in Urban Areas: New Challenges, New Partnerships," *Disasters* 36(S1): S23–S42.

Cutter, S.L., L. Barnes, M. Berry, C. Burton, E. Evans, E. Tate, and J. Webb. (2008) "A Place-Based Model for Understanding Community Resilience to Natural Disasters," *Global Environmental Change* 18(4): 598–606.

Cutter, S. L., C. G. Burton, and C. T. Emrich. (2010) "Disaster Resilience Indicators for Benchmarking Baseline Conditions," *Journal of Homeland Security and Emergency Management* 7(1): 1–22.

Dash, N., B. Morrow, J. Mainster, and L. Cunningham. (2007) "Lasting Effects of Hurricane Andrew on a Working-Class Community," *Natural Hazards Review* 8(1): 13–21.

Derickson, K. (2014) *Urban and Regional Policies*. Retrieved February 2015 from: http://blogs.lse.ac.uk/usappblog/2014/07/07/after-hurricane-katrina-devastated-black-neighborhoods-created-an-opportunity-for-redevelopment-that-focused-on-gentrification/.

De Wet, C. (2006) "Risk, Complexity and Local Initiative in Involuntary Resettlement Outcomes," 180–202, in *Towards Improving Outcomes in Development Induced Involuntary Resettlement Projects*, Oxford: Berghahn Books.

Dow, K. and S. L. Cutter. (2000) "Public Orders and Personal Opinions: Household Strategies for Hurricane Assessment," *Global Environmental Change Part B: Environmental Hazards* 2(4): 143–155.

Eadie, C. (1998) "Earthquake Case Study: Loma Prieta in Santa Cruz and Watsonville, California," 281–310, in J. Schwab, K. C. Topping, C. D. Eadie, R. E. Deyle, and R. A. Smith (eds.), *Planning for Post-Disaster Recovery and Reconstruction*, Planning Advisory Service Report 483/484. Chicago: American Planning Association.

Edwards, F. L. (2010) "At Home in Silicon Valley: One End to the Katrina Diaspora," 89–108, in J. D. Rivera and D. S. Miller (eds.), *How Ethnically Marginalized Americans Cope with Catastrophic Disasters: Studies in Suffering and Resiliency*, Lewiston NY: The Edwin Mellen Press.

Egeland, B., E. Carlson, and L. Sroufe. (1993) "Resilience as Process," *Development and Psychopathology* 5(4): 517–528.

Esnard, A-M. and A. Sapat. (2011) "Disasters, Diasporas and Host Communities: Insights in the Aftermath of the Haiti Earthquake," *Journal of Disaster Research* 6(3): 331–342.

Esnard, A-M. and A. Sapat. (2014) *Displaced by Disasters: Recovery and Resilience in a Globalizing World*, New York: Routledge Press.

Esnard, A-M., A. Sapat, and D. Mitsova. (2011) "An Index of Relative Displacement Risk to Hurricanes," *Natural Hazards* 59(2): 833–859.

FEMA (Federal Emergency Management Agency). (2011) *A Whole Community Approach to Emergency Management: Principles, Themes, and Pathways for Action*. Retrieved November 2014 from www.fema.gov/media-library/assets/documents/23781.

Florida Department of Community Affairs. (2010) *Post-Disaster Recovery Planning: A Guide for Florida Communities*. Retrieved from: www.dca.state.fl.us/fdcp/dcp/PDRP/Files/PDRPGuide.pdf.

Florida Department of Emergency Management. (2010) *Polk County Post-disaster Redevelopment Plan: Case Study*. Retrieved November 2014 from: www.floridadisaster.org/Recovery/IndividualAssistance/pdredevelopmentplan/tools.htm.

Florida Department of Emergency Management. (2012). "State Disaster Housing Planning: Florida Disaster Strategy for Housing Planning and Operations" Florida Division of Emergency Management, Tallahassee FL.

Gerber, B. J. (2010). "Management of Evacuee Ingress during Disasters: Identifying the Determinants of Local Government Capacity and Preparedness," *Risk, Hazards and Crisis in Public Policy* 1(3): 115–142.

Gibbs, L. I. and C. F. Holloway. (2013) *Hurricane Sandy After Action Report and Recommendations to Mayor Michael R. Bloomberg.* Retrieved December 2017 from: www.nyc.gov/html/recovery/downloads/pdf/sandy_aar_5.2.13.pdf.

Godschalk, D. (2003) "Urban Hazard Mitigation: Creating Resilient Cities," *Natural Hazards Review* 4(3): 136–143.

Gongol, B. (2005) "The Hurricane Katrina Diaspora," www.gongol.com/research/disasters/katrinadiaspora/.

Gordon, J. (1978) *Structures.* Harmondsworth UK: Penguin Books.

Gotham, K. F. and M. Greenberg. (2014) *Crisis Cities: Disaster and Redevelopment in New York and New Orleans,* Oxford: Oxford University Press.

Greenberg, M. (2014) "The Disaster inside the Disaster: Hurricane Sandy and Post-Crisis Redevelopment," *New Labor Forum* 23(1): 44–52.

Groen, J. A. and A. E. Polivka. (2008) "Hurricane Katrina Evacuees: Who They Are, Where They Are and How Are They Faring," *Monthly Labor Review* March 2008: 32–51.

Groen, J. A. and A. E. Polivka. (2010) "Going Home after Hurricane Katrina: Determinants of Return Migration and Changes in Affected Areas," *Demography* 47(4): 821–844.

Gunderson, L. (2000) "Ecological Resilience—in Theory and Application," *Annual Review of Ecology and Systematics,*" 31(1): 425–429.

Hartwig, R. P. (2002) *Florida Case Study: Economic Impacts of Business Closures in Hurricane Prone States,* New York: Insurance Information Institute.

Holling, C. (1973) "Resilience and Stability of Ecological Systems," *Annual Review of Ecology and Systematics,* 4(1): 1–23.

Hori, M., M. J. Schafer, and D. J. Bowman. (2009) "Displacement Dynamics in Southern Louisiana after Hurricanes Katrina and Rita," *Population Research and Policy Review* 28(1): 45–65.

IDMC (Internal Displacement Monitoring Centre). (2013) *Global Estimates 2012: People Displaced by Disasters,* Geneva, Switzerland: Author.

IDMC (Internal Displacement Monitoring Centre). (2014) *Global Estimates 2014: People Displaced by Disasters,* Geneva, Switzerland: Author. Retrieved January 2015 from: http://reliefweb.int/sites/reliefweb.int/files/resources/201409-global-estimates.pdf.

IDMC (Internal Displacement Monitoring Centre). (2017) *On the Grid: Global Internal Displacement in 2016,* Geneva, Switzerland: Author. Retrieved December 2017 from: www.internal-displacement.org/global-report/grid2017/pdfs/2017-GRID-part-1.pdf.

IFRC (International Federation of Red Cross and Red Crescent Societies). (2012) *World Disasters Report 2012, Focus on Forced Migration and Displacement,* Geneva, Switzerland: Author.

IFRC (International Federation of Red Cross and Red Crescent Societies). (2014) Retrieved August 2014 from: www.ifrc.org/en/what-we-do/disaster-management/about-disasters/what-is-a-disaster/what-is-vulnerability/.

King, D. (2006) "Planning for Hazard Resilient Communities," 289–299, in D. Paton and D. Johnston (eds.), *Disaster Resilience: An Integrated Approach,* Springfield IL: Charles C. Thomas Publisher.

Levine, J., A.-M. Esnard, and A. Sapat. (2007) "Population Displacement and Housing Dilemmas Due to Catastrophic Hurricanes," *Journal of Planning Literature* 22(1): 3–15.

Lin, Y. (2009) *Development of Algorithms to Estimate Post-Disaster Population Dislocation—A Research-Based Approach,* Doctoral dissertation, Texas A&M University. Retrieved November 2014 from: http://oaktrust.library.tamu.edu/bitstream/handle/1969.1/ETD-TAMU-2009-08-3266/LIN-DISSERTATION.pdf?sequence=2.

Lindell, M. K., S. D. Brody, and W. E. Highfield. (2016). "Financing Housing Recovery through Hazard Insurance: The Case of the National Flood Insurance Program," 49–66, in A. Sapat and A.-M. Esnard (eds.), *Coming Home After Disaster: Multiple Dimensions of Housing Recovery,* Boca Raton FL: CRC Press.

Lindell, M. K., J. E. Kang, and C. S. Prater (2011) "The Logistics of Household Evacuation in Hurricane Lili," *Natural Hazards* 58(3), 1093–1109.

Longstaff, P. (2005) "Security, Resilience, and Communication in Unpredictable Environments such as Terrorism, Natural Disasters, and Complex Technology," Cambridge MA: Harvard University Center for Information Policy Research.

Malone, B. (2012) "Top Ten Facts and Figures about Hurricane Sandy—New York's Worst Storm in Decades," *Irish Central,* retrieved November 2014 from: www.irishcentral.com/news/top-ten-facts-and-figures-about-hurricane-sandy-new-yorks-worst-storm-in-decades-video-176757161-237784861.html.

Maiolo, J. R., J. C. Whitehead, M. McGee, L. King, J. Johnson, and H. Stone (eds.). (2001) *Facing our Future: Hurricane Floyd and Recovery in the Coastal Plain*, Wilmington NC: Coastal Carolina Press.

Masten, A., K. Best, and N. Garmezy. (1990) "Resilience and Development: Contributions from the Study of Children who Overcome Adversity," *Development and Psychopathology* 2(4): 425–444.

Mayunga, J. S. (2007) *Understanding and Applying the Concept of Disaster Resilience: A Capital-based Approach*, A draft working paper prepared for the Summer Academy for Social Vulnerability and Resilience Building, July 22–28, 2007, Munich, Germany.

Meyer, M. A. (2013) "Internal Environmental Displacement: A Growing Challenge to the United States Welfare State," *Oñati Socio-Legal Series* 3(2): 326–345.

Miles, S. D. and S. E. Chang. (2006) "Modeling Community Recovery from Earthquakes," *Earthquake Spectra* 22(2): 439–458.

Miller, L. M. (2012a) "Katrina Evacuee Reception in Rural East Texas," 104–118, in L. Weber and L. Peek (eds.), *Displaced: Life in the Katrina Diaspora*, Austin TX: University of Texas Press.

Miller, L. M. (2012b) "Receiving Communities," 25–30, in L. Weber and L. Peek (eds.), *Displaced: Life in the Katrina Diaspora*, Austin TX: University of Texas Press.

Mitchell, C. M., A.-M. Esnard, and A. Sapat. (2012) "Hurricane Events and the Displacement Process in the United States," *Natural Hazards Review* 13(2): 150–161.

Morrow, B. H. (2005) *Recovery: What's Different, What's the Same? Presentation at National Academies of Science Disaster Roundtable, Lessons Learned Between Hurricanes: From Hugo to Charley, Frances, Ivan and Jean.* Washington DC: The National Academies.

Morrow-Jones, H. A. and C. R. Morrow-Jones. (1991) "Mobility Due to Natural Disaster: Theoretical Considerations and Preliminary Analyses," *Disasters* 15(2): 126–132.

Myers, C.A., T. Slack, and J. Singelmann. (2008) "Social Vulnerability and Migration in the Wake of Disaster: The Case of Hurricanes Katrina and Rita," *Population and Environment* 29(6): 271–291.

NOAA (National Oceanic and Atmospheric Administration) (2011) "NWS Central Region Service Assessment Joplin, Missouri, Tornado— May 22, 2011," U.S. Department of Commerce, NOAA National Weather Service Central Region Headquarters. Retrieved November 2014 from: www.weather.gov/media/publications/assessments/Joplin_tornado.pdf.

Norris, F. H., S. P. Stevens, B. Pfefferbaum, K. F. Wyche, and R. L. Pfefferbaum. (2008) "Community Resilience as a Metaphor, Theory, Set of Capacities, and Strategy for Disaster Readiness," *American Journal of Community Psychology* 41(1–2), 127–50.

Oliver-Smith, A. (2006) "Disasters and Forced Migration in the 21st Century," *Social Science Research Council, Understanding Katrina: Perspectives from the Social Sciences.* http://understandingkatrina.ssrc.org/Oliver-Smith/.

Oliver-Smith, A. (2009) *Sea-Level Rise and the Vulnerability of Coastal Peoples*, Number 7, Bonn, Germany: United Nations University Institute for Environment and Human Security.

Oliver-Smith, A. (2012) "Debating Environmental Migration: Society, Nature and Population Displacement in Climate Change," *Journal of International Development* 24(8): 1058–1070.

Oliver-Smith, A. and A. de Sherbinin. (2014) "Resettlement in the Twenty First Century," *Forced Migration Review* (45): 23–25.

Olshansky, R. B. (2006) "Planning after Hurricane Katrina," *Journal of the American Planning Association* 72(2): 147–153.

Olshansky, R. B., L. Hopkins, and L. Johnson. (2012) "Disaster and Recovery: Processes Compressed in Time," *Natural Hazards Review* 13(3): 173–178.

Olshansky, R. B. and L. A. Johnson. (2010) *Clear as Mud: Planning for the Rebuilding of New Orleans*, Chicago IL: American Planning Association.

Olshansky, R. B., L. A. Johnson, J. Home, and B. Nee. (2008) "Planning for the Rebuilding of New Orleans," *Journal of the American Planning Association* 74(3): 273–287.

Olshansky, R. B., L. Johnson, and K. Topping. (2003) *Rebuilding Communities Following Disasters: Lessons from Kobe and Los Angeles.* Abstract presented at the Third Workshop for Comparative Study on Urban Earthquake Disaster Management, Kobe, Japan, January 31.

Pais, J. F. and J. R. Elliott. (2008) "Places as Recovery Machines: Vulnerability and Neighborhood Change after Major Hurricanes," *Social Forces* 86(4): 1415–1453.

Patel, S. S., M. B. Rogers, R. Amlôt, and G. J. Rubin. (2017) "What Do We Mean by 'Community Resilience'? A Systematic Literature Review of How It Is Defined in the Literature," *PLOS Currents Disasters.* Retrieved January 2018 from: http://currents.plos.org/disasters/article/what-do-we-mean-by-community-resilience-a-systematic-literature-review-of-how-it-is-defined-in-the-literature/.

Peacock, W. G. (2009) *Advancing Coastal Community Resilience: A Brief Project Overview*. Presented at the CARRI Workshop, Broomfield CO, July 14–15 2009.

Peacock, W. G. and C. Girard. (1997) "Ethnicity and Segregation," 191–205, in W. G. Peacock, B. H. Morrow, and H. Gladwin (eds.), *Hurricane Andrew: Ethnicity, Gender and the Sociology of Disaster*, New York: Routledge.

Peacock, W. G., B. H. Morrow, and H. Gladwin (eds.). (1997) *Hurricane Andrew: Ethnicity, Gender, and the Sociology of Disasters*, New York: Routledge.

Pfefferbaum, B., D. Reissman, R. Pfefferbaum, R. Klomp, and R. Gurwitch. (2005) "Building Resilience to Mass Trauma Events," 347–358, in L. Doll, S. Bonzo, J. Mercy, and D. Sleet (eds.), *Handbook on Injury and Violence Prevention Interventions*, New York: Kluwer Academic Publishers.

Plyer, A., J. Bonaguro, and K. Hodges. (2010) "Using Administrative Data to Estimate Population Displacement and Resettlement Following a Catastrophic U.S. Disaster," *Population and Environment* 31(1–3): 150–175.

Quarantelli, E. L. (1980) *Evacuation Behavior and Problems: Findings and Implications from the Research Literature*. Columbus, OH: Disaster Research Center, Ohio State University. Retrieved from University of Delaware Library, http://udspace.udel.edu/handle/19716/1283.

Sanders, S., S. L. Bowie, and Y. D. Bowie. (2004) "Lessons Learned on Forced Relocation of Older Adults: The Impact of Hurricane Andrew on Health, Mental Health and Social Support of Public Housing Residents," *Journal of Gerontological Social Work* 40(4): 23–35.

Sapat, A. (2012) *Multiple Dimensions of Societal Resilience: Directions for Future Research*, Proceedings of the 2010 International Workshop on Societal Resilience, Department of Homeland Security, Washington DC.

Sapat, A. and A.-M. Esnard. (2012) "Displacement and Disaster Recovery: Transnational Governance and Sociolegal Issues Following the 2010 Haiti Earthquake," *Risk, Hazards and Crisis in Public Policy* 3(1): 1–24.

Sapat A. and A.-M. Esnard (eds.). (2017) *Coming Home After Disaster: Multiple Dimensions of Housing Recovery*, Boca Raton FL: CRC Press.

Sapat, A., C. M. Mitchell, Y. Li, and A-M. Esnard. (2011) "Policy Learning: Katrina, Ike and Post-Disaster Housing," *International Journal of Mass Emergencies and Disasters* 29(1): 26–56.

Schneider, A. and H. Ingram. (1993) "Social Construction of Target Populations: Implications for Politics and Policy," *The American Political Science Review* 87(2): 334–347.

Schwirtz, M. (2012) "Housing Nightmare Looms in Wake of Storm," *The New York Times* (November 4).

Singh, D. (2012) *Disaster Prevention Key to Stopping Climate Displacement*, Geneva, Switzerland: UN International Strategy for Disaster Reduction. Retrieved November 2014 from: http://reliefweb.int/report/kenya/disaster-prevention-key-stopping-climate-displacement.

SIRR (Special Initiative for Rebuilding and Resiliency). (2013) *PlaNYC: A Stronger, More Resilient New York*. Retrieved November 2014 from: www.nyc.gov/html/planyc2030/html/home/home.shtml.

Smith, G. (2011). *Planning for Post-Disaste Recovery: A Review of the United States Disaster Assitance Framework*, Washington DC: Island Press.

Smith, G. (2017). "Pre- and Post-Disaster Conditions, their Implications, and the Role of Planning for Housing Recovery," 277–292, in A. Sapat and A.-M. Esnard (eds.), *Coming Home After Disaster: Multiple Dimensions of Housing Recovery*, Boca Raton FL: CRC Press.

Smith, S. K. and C. McCarty. (1996) "Demographic Effects of Natural Disasters: A Case Study of Hurricane Andrew," *Demography* 33(2): 265–275.

Sorenson, J. and B. Vogt. (2006) "Interactive Emergency Evacuation Handbook," Retrieved November 2014 from http://orise.orau.gov/csepp/documents/planning/evacuation-guidebook/index.htm.

Soria, C. (2013) "Families Displaced by Superstorm Sandy are Running out of Time," *Gotham Gazette*. Retrieved from: www.gothamgazette.com/index.php/housing/4230-families-displaced-by-superstorm-sandy-running-out-of-time.

.US Department of Homeland Security. (2008) *National Incident Management System*. Retrieved November 2014 from: www.fema.gov/pdf/emergency/nims/NIMS_core.pdf.

US Senate. (2009) *Far from Home: Deficiencies in Federal Disaster Housing Assistance after Hurricanes Katrina and Rita and Recommendations for Improvement*, Senate Print 111–7 prepared by the Ad Hoc Subcommittee on Disaster Recovery, Committee on Homeland Security and Governmental Affairs, U.S. Senate, GPO, Washington, DC.

Van Zandt, S., W. G. Peacock, W. Highfield, and Y. Xiao. (2009) *Housing Inequalities and Social Vulnerability: Findings from 2008's Hurricane Ike*, Paper presented at the Association of Collegiate Schools of Planning Annual Meeting, Crystal City, Virginia.

Waller, M. (2001) "Resilience in Ecosystemic Context: Evolution of the Concept," *American Journal of Orthopsychiatry* 71(3): 290–297.

Weber, L. and L. Peek (eds.). (2012a) *Displaced: Life in the Katrina Diaspora*, Austin TX: University of Texas Press.

Weber, L. and L. Peek. (2012b). "Documenting Displacement: An Introduction," 1–20, in L. Weber and L. Peek (eds.), *Displaced: Life in the Katrina Diaspora*, Austin TX: University of Texas Press.

Welsh, M. G. and A.-M. Esnard. (2009) "Closing Gaps in Local Housing Recovery Planning for Disadvantaged Displaced Households," *Cityscape: A Journal of Policy Development and Research* 11(3): 87–104.

Wu, H. C., M. K. Lindell, and C. S. Prater. (2012) "Logistics of Hurricane Evacuation in Hurricanes Katrina and Rita," *Transportation Research Part F: Traffic Psychology and Behaviour* 15(4), 445–461.

Wu, H. C., M. K. Lindell, C. S. Prater, and S.-K. Huang. (2013) "Logistics of Hurricane Evacuation in Hurricane Ike," 127–140, in J. Cheung and H. Song (eds.), *Logistics: Perspectives, Approaches and Challenge*, Hauppauge NY: Nova Science Publishers.

Xiao, Y. and S. Van Zandt. (2012) "Building Community Resiliency: Spatial Links between Household and Business Post-Disaster Return," *Urban Studies* 49(11): 2523–2542.

Zensinger, L. W. (2013) "Observations by a Practitioner," *Public Administration Review* 73(4): 623–624.

Zhang, Y. and W. G. Peacock. (2010) "Planning for Housing Recovery?" *Journal of the American Planning Association* 76(1): 5–24.

18

BUSINESS AND ECONOMIC IMPACTS AND RECOVERY

Yu Xiao

Introduction

How businesses and local economies respond to and recover from natural disasters is arguably the least understood area in disaster management. An even more intriguing question is: in a market economy, why should we spend taxpayers' money to help private for-profit businesses recover? Shouldn't we let their fates be decided by the market's "invisible hand"? This chapter provides insights on these questions by reviewing literature on the business and economic impact of disasters, reasons both for and against public assistance to businesses, and some suggestions for practice.

Paradox of Business Vulnerability and Economic Resiliency

Business Vulnerability

Businesses are vulnerable to disasters. An earthquake can turn commercial buildings into piles of rubble. After a hurricane, you may see gas stations with collapsed roofs, office towers with broken windows, and foundations of beachside souvenir shops that have been completed washed away by the surge. Even if it suffered only several inches of flood damage, the corner store may be closed for weeks and the neighborhood grocery store may never return. According to the Federal Emergency Management Agency (FEMA), about 40 percent of small businesses can fail after a disaster due to damage (FEMA n.d.).

Business vulnerability comes from vulnerabilities in all aspects of its operation. In the daily operation, a business utilizes capital, labor, and supplies to produce goods and services and deliver them to customers. If any one aspect of the production chain stops functioning, the entire operation halts. Physical damage to capital—such as building, equipment, furnishings, and inventory—can force business relocation and cause business failure (Tierney 1997; Wasileski et al. 2010). A few large sample surveys of businesses in the 1990s conducted by the Disaster Research Center at the University of Delaware suggest that a business can suffer losses due to supplier problems even if there is no direct physical damage to the business itself. For instance, interruptions to utility lifeline services (i.e. water, electricity, gas, sewage, mail, phone, and internet) and key material suppliers can keep the business's doors closed (Mayer et al. 2008; Tierney 1997; Tierney et al. 1996; Zhang et al. 2009). The business can also encounter operational problems due to injury, dislocation, or commuting problems of employees (Zhang et al. 2009). The disaster can permanently change the socio-demographic composition of community resulting from the relocation of original residents and the influx of newcomers (Graham 2007; Xiao 2008). This loss or change in the customer base can be detrimental to any business (Alesch et al. 2001; Mayer et al. 2008; Tierney 1997).

Comparable to the social vulnerability factors for households, certain types of businesses are more likely to suffer damage and fail. Businesses that are small, engaged in retail sales, or renting their premises are more likely to suffer losses (Chang and Falit-Baiamonte 2002) and close (Wasileski et al. 2010) due to the disaster. Minority-owned, woman-owned, and locally-owned businesses are more vulnerable due to the owners' resource constraints (Aldrich and Auster 1986; Tigges and Green 1994; Webb et al. 2002). Pre-disaster financial conditions and previous disaster experiences may also affect business disaster impact and its ability to bounce back (Drabek 1994; Durkin 1984).

Economic Resiliency

Contrary to the common belief of long-lasting negative economic impact after disasters, research shows that local, regional, and national economies are quite resilient to natural disasters. Studies of earthquakes, hurricanes, floods, and tornadoes in the United States conclude that natural disasters had only short-term effects but no—or very minimal—long-term effects on local economic indicators (Brady and Perkins 1991; Friesema et al. 1979; Kroll et al. 1991; Wright et al. 1979; Xiao and Feser 2014). In a study of the 1993 Midwest flood, Xiao (2011) found that although the flood's negative impacts on agriculture seemed to be long-lasting, its impact on the county economy as a whole was quite negligible in the long run. Nor did the September 11 terrorist attack, a manmade disaster, affect the long-term prospects of the housing market of New York City despite the negative effects felt in the two months following the attack (Bram et al. 2002). A few studies show that stricken counties could ultimately be much better off in housing (Dacy and Kunreuther 1969) and labor market conditions (Ewing and Kruse 2002; Ewing et al. 2005) *ex post*. Most studies at the regional or national level demonstrate that disasters have either no effect (Hewings et al. 2000) or only temporary negative economic impacts (Albala-Bertrand 1993; Horwich 2000). Worthington and Valadkhani (2004) showed that Australia's capital market returns were impacted by bushfires on the day of, and the day following, an event. Economic impact was felt during the two to five days after a cyclone, and there were no immediate or long-term effects observed after other storms and floods. In a cross-national study, Cunado and Ferreira (2014) found significant positive effects of moderate floods on per capita Growth Domestic Product (GDP) growth in developing countries. The case of lingering negative economic impact after disasters is rare. Chang (2000, 2010) reported that although Kobe City, Japan regained its population within ten years after the 1995 earthquake, its port activities never recovered. However, she acknowledged that these long-run losses and structural change were not new creations of the quake but reflections of pre-disaster trends.

Reconcile the Paradox

There are several explanations for the paradox of vulnerable individual businesses but resilient aggregate economy. The first is the economic resiliency of scale. Although a disaster can impact a large geographic area, its destructive power dissipates over space from the epicenter, hurricane landfall point, flood inundation area, or site of a terrorist attack. Empirical research tends to focus on businesses located in the areas hit hard by disasters and finds many of them vulnerable. As one moves up to more aggregate geographic levels, relative damage proportionate to the economy's size diminishes. Economic resiliency can come from being large in geographic scale. The larger the unit of analysis, the less damage and negative effect of disasters one would observe.

The second explanation is the benefits of creative destruction. Local, regional, and national economies are composed of many individual businesses. It is true that the stronger the individual business, the stronger the aggregate economy. However, the well-being of the aggregate economy does not depend on the health of every individual business. Instead, some level of turnover at the micro level is good for the macro economy. There is evidence that businesses with weak financial situations prior to the event are less likely to come back afterwards (Alesch et al. 2001; Webb et al. 2002). A disaster acts like a catalyst that makes what is already bound to happen, simply happen faster. It frees labor and resources from the weak or dying businesses to be absorbed into more productive pursuits. Disasters also bring opportunities for technology upgrades, which improve overall economic productivity (Skidmore and Toya 2002).

Third, post-disaster consumer demand shifts across business sectors as well as over space. Households may cut back consumption of luxury goods to finance replenishments and repairs after the disaster. Consequently, businesses in recovery-related sectors, such as roof and plumbing repairs, furnishing, hardware, electric appliances, and plant nurseries, are likely to see increased demand while repairs are underway. Those that sell luxury goods, such as high-end furniture or jewelry, are likely to experience lower consumer demand. At the aggregate level, the winners and losers balance out (Scanlon 1988). In some cases, the winners may outweigh the losers and result in an income boom in the aggregate economy. For instance, Horwich (2000) showed that the real GDP of Japan had a higher growth rate in the year of the Kobe earthquake than in any other year since 1990. Population can relocate from heavily-damaged areas to surrounding ones after a major disaster, at least temporarily (Chang 2010; Xiao and Nilawar 2013). This displacement may cause business failures in the out-migration areas and create new business opportunities in the in-migration areas, while these two effects balance out in the aggregate. Moreover, goods and services can flow from adjacent (undamaged) areas causing a positive effect there that offsets the negative effect in the disaster impact area (Zhang et al. 2009). However, this only happens if those in the impact area have savings, insurance, credit (eligibility for commercial loans), or political capital (state/national/international aid) that can pay for the reconstruction.

And lastly, recovery outcome depends on how you define and measure "recovery". The multiple ways to define recovery were pointed out by Quarantelli (1999) and operationalized by Chang (2010). There are at least three ways to measure recovery: returning to pre-disaster level, attaining the pre-disaster trend/path as if the disaster had never occurred, and establishing a new normalcy. Using different definitions may lead to different conclusions about recovery. In the case of 2005 Hurricane Katrina, the New Orleans-Metairie-Kenner Metropolitan Statistical Area (referred as New Orleans Metro hereafter), with more than 55,000 fewer jobs in 2013 compared to 2004, had not recovered to its pre-disaster level eight years after the event (Figure 18.1). However, the New Orleans Metro enjoyed a relatively lower unemployment rate despite the short-lived high unemployment spike during Katrina (Figure 18.1). In the 1990s, the unemployment rate of the New Orleans Metro surpassed the national level; it was on par with the national level in the first half of 2000s; and since 2006, it has been much lower than the national level. Because the unemployment rate is calculated as the percentage of unemployed workers in the labor force, a lower unemployment rate indicates a higher percentage of the labor force being actually employed. Although operating on a smaller size of workforce than pre-Katrina, the New Orleans Metro economy has recovered in terms of establishing a new equilibrium (aka, "a new normal"). Whether this recovery outcome is acceptable depends on your point of view. For the local governments in the region, a smaller workforce and population may mean a decreased tax base to support public services. For the economists, the market worked quite well to adjust the demand and supply of labor. For some others, this outcome is desirable because now there are fewer people and properties in the harm's way. If the next Katrina happens, the damage will probably be much less.

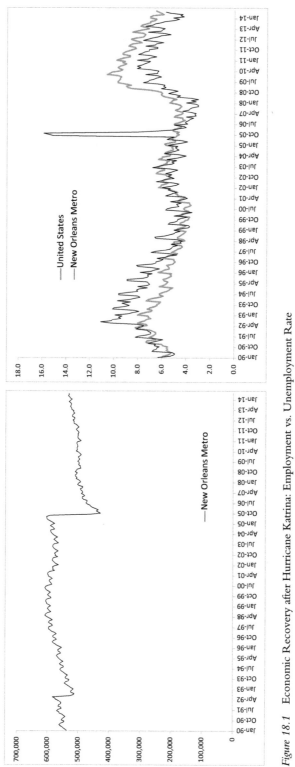

Figure 18.1   Economic Recovery after Hurricane Katrina: Employment vs. Unemployment Rate

Data source: Bureau of Labor Statistics. Data were not seasonally adjusted.

A Market Failure Approach to Public Assistance for Business Recovery

Is business recovery a target to shoot for? For most small business owners, the answer is definitely 'yes', because the business brought them income and fulfilled their dreams to provide unique products and/or services to the community, to run business in a certain kind of way, or simply to be their own boss. A family business may have been serving the community for generations and the business carries family pride, not to mention the value of the time and effort the owner has devoted to it. It is hard for the business owner to let go and accept "failure," especially if the business has been in operation for a long time and was quite successful before the disaster. It has been documented that only the weakest businesses fail right away after the disaster; many others will struggle to recover until they run out of money, resources, and energy (Alesch et al. 2001).

But should business recovery be a goal for the society in general? In a market economy, why should taxpayers subsidize the private sector for-profit businesses in their post-disaster recovery? Based on the welfare economics principles, the next section discusses the reasons for and against public assistance to businesses.

Welfare Economics and Market Failure

Welfare economics studies the aggregate welfare of a society. Market failure occurs when the allocation of goods and services decided by the free market is not societally efficient. In other words, the free market fails when there exists a possible improvement in the aggregate societal welfare (meaning making someone better-off without making any others worse off) but the free market does not achieve it.

Market failure can take the form of public goods, externalities, and imperfect information (Bartik 1990). Public goods, such as music provided by a street musician, result in inefficiency because the market cannot coerce all those who listened to the music to pay for it. Due to the free-rider problem, the musician receives less compensation than the true value of his service. Thus, the quantity of goods provided by the free market is lower than the social optimum. Externalities, such as second-hand smoke from cigarettes, are another form of market failure because the relative costs and benefits are not reflected in market prices. Since the smokers are not asked to compensate for the health damages of those others who also inhaled his/her smoke, the social costs of smoking are much higher, but such costs are not completely factored into the price of cigarettes. Imperfect information causes market failure because it prevents possible markets from forming. Due to the lack of information, potential suppliers might not enter the market because they underestimate the demand and buyers might not be able to find suppliers of goods and services that they desire.

Economists believe government should intervene in the market if and only if market failure exists. Taxes and subsidies are a double-edged sword. They can be used to correct market failure; however they can also cause allocation inefficiency (aka deadweight loss) if used inappropriately where market failure did not exist (Case and Fair 1999). Before providing public subsidies to help private-sector businesses recover, we need to ask where the market failure is in a post-disaster context.

Reasons Against Public Subsidies to Businesses

As discussed above in "Reconcile the Paradox," there are several mechanisms that explain why aggregate-level economies are resilient to disaster shocks despite the vulnerabilities of individual businesses. Creative destruction is one mechanism of adjustment in the market. The functioning market should allow businesses whose services are no longer needed to die off and new ones to materialize to capture the opportunities in emerging customer demand. Providing assistance to a no longer profitable business will not save the business from failure but delay its death and therefore delay the market adjustment. In this case, public subsidies to businesses cause allocation inefficiency.

Providing subsidies to businesses will also distort the balance of supply and demand in the market. The "invisible hand" of the market adjusts supply and demand through the pricing mechanism. Giving money to businesses will make them less sensitive to the market signal on the already shifted demand. For example, the subsidy may keep the business going at its original location although the best option for it is to move to where the customers are now located. Also, when households are spending their money for repairs and reconstruction after a disaster, providing money to a luxury goods dealer will not save it from closing its doors.

Empirical studies have found that taking loans and assistance for business recovery is counterproductive. For example, in Dahlhamer and Tierney's (1998) study of business recovery after the Northridge earthquake, they concluded that the businesses that used more aid sources were less likely to report positive recovery outcomes. Graham (2007) reported that many businesses in Manhattan were locked into their no longer profitable locations after the September 11 terrorist attack in 2001 because they took funding assistance from the Lower Manhattan Initiative, which required them to remain in their place of business for at least one year.

It should be pointed out that not finding business assistance helpful does not mean assistance in general is bad for business recovery. Many government assistance programs are ill-conceived and bad for business recovery because they stiffen business adjustment after a disaster shock. These programs should be revamped to facilitate business adjustment. More on government business disaster assistance programs will be discussed below.

Reasons Supporting Public Subsidies to Businesses

Community recovery depends on the recovery of both households and businesses due to the close ties between the two in the market economy. Businesses provide goods and services to households, and receive revenue in exchange when households fulfill their consumption needs. Businesses also provide jobs to households. On the other side, households provide essential factors of production, i.e. labor, land, capital, and entrepreneurial skills, to businesses, and in exchange, they receive monetary income, such as wages, rents, interest, and profits. Xiao and Van Zandt (2012) showed that business and household post-disaster recovery are spatially linked. They found that after Hurricane Ike, households are more likely to return when businesses in the surrounding areas have returned and household recovery also increases the chances for businesses to come back.

Is the household–business linkage in economic production sufficient on its own to justify subsidies to businesses? The answer is no, because this is how the market naturally works without any assistance, matching supply of labor, goods, and services with demand. As discussed earlier, public subsidies should be used only if there is market failure. That raises the question of where the market failures are in a post-disaster situation. In fact, there are three types of market failure. First, the benefits of business return go beyond just providing jobs to people in the community. Helping businesses recover means reducing the number of people on welfare and unemployment benefits, which saves public money. Business owners, especially owners of small locally owned businesses, are households as well. They may struggle more in recovery because both their business and residence need to be repaired. Helping their business recovery facilitates their household recovery.

Second, private sector businesses can provide public goods to assist disaster relief and recovery. After Hurricane Sandy, many businesses allowed people to use their electrical outlets to charge phones and laptops for free. Some lent their generators to help power pumps removing water from basements, helped cut down broken trees in the neighborhood, or let people take a hot shower (Brooks 2012). Businesses can also volunteer their space for community meetings, host donation drives and fundraisers, and serve as information hubs during a disaster. Businesses are also eyes on the street that keep potential looters under watch. In emergencies, businesses do not typically exclude free riders from using their services. The costs of providing these services are not completely incorporated into market prices.

Third, in the course of recovery, the market mechanism may undervalue the "sense-of-place": an intangible, location-specific asset that provides emotional security, stability, and trust to people. The "sense-of-place" is established through everyday contacts, many of which are made in coffee shops, corner stores, and grocery stores in the community. It can also involve public investment, such as historic preservation, landscape protection, and growth management (Bolton 1992). Damage to buildings, infrastructure, and community businesses reduces the "sense-of-place," so repairing the damage is valuable to people who live in the area (let's call it Place A) because it restores their "sense-of-place." But why should taxpayers living in other places (such as Place B) care about the recovery of Place A? Taking Bolton's perspective, there are three reasons. The first is "option value," where the value resides in the option for people from Place B to visit or live in Place A. The second, "pure existence value," assumes that people in Place B may never move to or visit Place A, but would like Place A to continue its existence. The last, "donor preferences," is similar to giving donations to people we don't know; taxpayers in Place B may simply prefer people in Place A to continue their "sense-of-place."

Small, locally owned businesses are more important than big-box and chain stores in creating and maintaining a "sense-of-place". What make a place unique is not the standardized chain stores like Wal-Mart and McDonald's. Instead, small, locally owned stores create memorable places because they are often unique in their decorations and lines of products, reflecting the owners' personalized tastes and styles. Rooted in the community, locally owned businesses are more psychologically invested in the community than the remotely owned and managed chain stores. They serve as "eyes" on the street to guard the safety of the community, provide opportunities for casual interactions, and places to leave your package for someone else to pick up (Jacobs 1993). In an interview with a locally owned pharmacy in a low-income community on Staten Island, New York, the owner told me that because he knows the people in the neighborhood, he would allow them to pick up an urgently needed prescription drug and pay him back later when they have money, a service never provided by chain pharmacies like CVS and Walgreens.

The capitalist market with the goal of profit maximization may lead to the replacement of small, locally owned stores by big-box and chain stores after a disaster, which destroys the valuable but intangible "sense-of-place" asset. Empirical studies have often found that small, locally-owned businesses have more trouble recovering than the chain stores because of their resource constraints (as shown by Chang and Falit-Baiamonte 2002; Wasileski et al. 2010; Xiao and Van Zandt 2012). Therefore, public subsidies to businesses are justified because they help communities recover their "sense-of-place."

Implications for Practice

Implication for Businesses

To business, there are a few things to do: make a disaster response plan, mitigate to reduce damage, and be prepared to adjust to change. In the study by Xiao and Van Zandt (2012), damage was the most significant factor to explain business return. Businesses that suffered greater damage were less likely to reopen three months after the disaster. Damage was the most significant factor to explain business continuity. Less damage means a higher chance of recovery.

For a long time, disaster researchers found that business disaster mitigation and preparedness was low and did not help damage reduction and recovery. For instance, in a study of the 2001 Nisqually earthquake, Chang and Falit-Baiamonte (2002) found no significant correlation between mitigation/preparedness and businesses' disaster losses. Neither did preparedness enhance the long-term viability of businesses affected by the Loma Prieta earthquake and Hurricane Andrew (Webb et al. 2002). This may be caused by a low level of mitigation and preparedness amongst businesses (Dahlhamer and D'Souza 1997; Mayer et al. 2008; Wasileski et al. 2010), resulting in actions too limited to make a difference (Tierney 2006). Focusing on steps designed to reduce direct property losses, instead of strategies to cope with disaster-induced community disruption, may also have contributed to the ineffectiveness of hazard mitigation and preparedness undertaken by the businesses (Alesch et al. 2001).

A recent study by Xiao and Peacock (2014) introduced some new findings. Their survey of businesses in Galveston County, Texas seven months after Hurricane Ike found that businesses with an emergency management plan implemented more mitigation and preparedness measures and therefore, suffered less severe physical damage. This research shows that disaster planning and the adoption of mitigation and preparedness measures can significantly reduce physical damage to businesses. Therefore, businesses should take mitigation and preparedness measures to enhance their resilience to disasters.

Business owners and managers should adjust and adapt to changes. In normal situations, businesses provide goods and services to customers. They need to pay attention to the market signals on customer demand. In a post-disaster context, to stay afloat, businesses must be adaptive and respond to the market signals. Business owners and managers should evaluate the disaster impacts on their customers and understand the varied recovery trajectories their customers might take. They should differentiate the stable long-term customer base from the temporary short-run demand shock. For instance, an automobile repair shop located in a low-income neighborhood might see a boost in business during the emergency response and restoration period because a large number of vehicles from the outside emergency response crews came to help with cleanup. Working in an environment with debris, these vehicles break down and require service and repairs. However, after the cleanup is over and the outside response crews leave the area, the automobile repair shop may be left with few customers because their regular customers from the low-income neighborhood did not come back. Therefore, businesses should be prudent in their post-disaster reinvestment. Business recovery should be in sync with customer recovery. For businesses serving mostly local customers, the goal of business recovery is not to return to pre-disaster condition as soon as possible, but rather making reinvestment decisions according to the pace of community, and therefore customer, recovery.

Implications for Government Assistance Programs

Government post-disaster assistance programs should facilitate business adaptation and change. Watson (2014) found that smaller but quicker loans are the most impactful for business recovery in Galveston, TX after 2008 Hurricane Ike. Galveston County was declared a federal disaster area, therefore, Small Business Administration (SBA) disaster assistance loans and economic injury loans were made available to businesses in Galveston right after Hurricane Ike. The average response time of these SBA loans was 12 days, which is quite reasonable. Of those who applied, the SBA loan approval rate was about 25 percent. Due to the heavy amount of paperwork needed for SBA loan application and not so cheerful loan approval rate, many businesses did not bother to apply for the SBA loans. Besides SBA, Department of Housing and Urban Development's Community Develop Block Grant for Disaster Recovery (CDBG-DR) and Economic Development Administration (EDA) received money from Congress to assist post-Hurricane Ike recovery. The CBDG-DR funds did not become available until 33 months after Hurricane Ike and EDA's revolving loan fund (RLF) took 19 months to set up. Meanwhile, most businesses reported that they cannot function in a deficit beyond one year; hence, it is critical to have timely allocation of assistance to businesses in need (Watson, 2014). Without prior agreement, negotiating terms and conditions of the government assistance programs could take a long time, which makes the program less effective.

References

Albala-Bertrand, J. M. (1993) *Political Economy of Large Natural Disasters*, Oxford: Clarendon Press.

Aldrich, H., and E. R. Auster. (1986) "Even Dwarfs Started Small: Liabilities of Age and Size and Their Strategic Implications," in B. M. Staw and L. L. Cummings (eds.), *Research in Organizational Behavior* (8): 165–198. Greenwich, CT: JAI Press.

Alesch, D. J., J. N. Holly, E. Mittler, and R. Nagy. (2001) *Organizations at Risk: What Happens When Small Businesses and Not-for-Profits Encounter Natural Disasters*, Fairfax, VA: Public Entity Risk Institute.

Bartik, T. J. (1990) "The Market Failure Approach to Regional Economic Development Policy," *Economic Development Quarterly* 4(4): 361–370.

Bolton, R. (1992) "Place Prosperity vs People Prosperity Revisited: An Old Issue with a New Angle," *Urban Studies* 29(2): 185–203.

Brady, R. J. and J. B. Perkins. (1991) *Macroeconomic Effects of the Loma Prieta Earthquake*, Oakland CA: Association of Bay Area Governments.

Bram, J., A. Haughwout, and J. Orr. (2002) "Has September 11 Affected New York City's Growth Potential?" *Economic Policy Review* 8(2): 81–96.

Brooks, C. (November 2, 2012) "Small Business to the Rescue After Hurricane Sandy," *Business News Daily*. Retrieved from www.businessnewsdaily.com/3359-small-business-rescue-hurricane-sandy.html on August 22, 2014.

Case, K. F. and R. C. Fair. (1999) *Principles of Macroeconomics*, Upper Saddle River, NJ: Prentice Hall.

Chang, S. E. (2000) "Disasters and Transport Systems: Loss, Recovery and Competition at the Port of Kobe after the 1995 Earthquake," *Journal of Transport Geography* 8: 53–65.

Chang, S. E. (2010) "Urban Disaster Recovery: A Measurement Framework and its Application to the 1995 Kobe Earthquake," *Disasters* 34(2): 303–327.

Chang, S. E. and A. Falit-Baiamonte. (2002) "Disaster Vulnerability of Businesses in the 2001 Nisqually Earthquake," *Global Environmental Change Part B: Environmental Hazards* 4(2–3): 59–71.

Cunado, J. and S. Ferreira. (2014) "The Macroeconomic Impacts of Natural Disasters: The Case of Floods," *Land Economics* 90(1): 149–168.

Dacy, D. C. and H. C. Kunreuther. (1969) *The Economics of Natural Disasters: Implications for Federal Policy*, New York: Free Press.

Dahlhamer, J. M. and M. J. D'Souza. (1997) "Determinants of Business Disaster Preparedness in Two U.S. Metropolitan Areas," *International Journal of Mass Emergencies and Disasters* 15(2): 265–281.

Dahlhamer, J. M. and K. J. Tierney. (1998) "Rebounding from Disruptive Events: Business Recovery following the Northridge Earthquake," *Sociological Spectrum* 18: 121–141.

Drabek, T. E. (1994) *Disaster Evacuation and the Tourist Industry*, Boulder CO: Institute of Behavioral Science, University of Colorado.

Durkin, M. E. (1984, October 6–8) *The Economic Recovery of Small Businesses after Earthquakes: The Coalinga Experience*, Paper presented at the International Conference on Natural Hazards Mitigation Research and Practice, New Delhi, India.

Ewing, B. and J. B. Kruse. (2002) "The Impact of Project Impact on the Wilmington, NC Labor Market," *Public Finance Review* 30(4): 296–309.

Ewing, B., J. B. Kruse, and M. A. Thompson. (2005) "Empirical Examination of the Corpus Christi Unemployment and Hurricane Bret," *Natural Hazards Review* 6(4): 191–196.

FEMA. (n.d.) "Protecting Your Businesses," from www.fema.gov/protecting-your-businesses on August 20, 2014.

Friesema, H. P., J. Caporaso, G. Goldstein, R. Lineberry, and R. McCleary. (1979) *Aftermath: Communities after Natural Disasters*, Beverly Hills and London: SAGE.

Graham, L. T. (2007) "Permanently Failing Organizations? Small Business Recovery After September 11, 2001," *Economic Development Quarterly* 21(4): 299–314.

Hewings, G. J. D., S. Changnon, and C. Dridi. (2000) "Testing for the Significance of Extreme Weather and Climate Events on State Economies," discussion paper. www.researchgate.net/profile/Geoffrey_Hewings/publication/237372896_Testing_for_the_Significance_of_Extreme_Weather_and_Climate_Events_on_State_Economies/links/0c96052570e00a7b24000000/Testing-for-the-Significance-of-Extreme-Weather-and-Climate-Events-on-State-Economies.pdf.

Horwich, G. (2000) "Economic Lessons of the Kobe Earthquake," *Economic Development and Cultural Change* 48(3): 521–542.

Jacobs, J. (1993) *The Death and Life of Great American Cities*, New York: The Modern Library.

Kroll, C. A., J. D. Landis, Q. Shen, and S. Stryker. (1991) "Economic Impacts of the Loma Prieta Earthquake: A Focus on Small Business," Berkeley CA: University of California Transportation Center.

Mayer, B. W., J. Moss, and K. Dale. (2008) "Disaster and Preparedness: Lessons from Hurricane Rita," *Journal of Contingencies and Crisis Management* 16(1): 14–23.

Quarantelli, E. L. (1999) *The Disaster Recovery Process: What We Know and Do Not Know from Research*, University of Delaware Disaster Research Center, Preliminary Paper #286.

Scanlon, J. (1988) "Winners and Losers: Some Thoughts about the Political Economy of Disasters," *International Journal of Mass Emergencies and Disasters* 6(1): 47–63.

Skidmore, M., and H. Toya. (2002) Do Natural Disasters Promote Long-Run Growth?" *Economic Inquiry* 40(4): 664–687.

Tierney, K. J. (1997) "Business Impacts of the Northridge Earthquake," *Journal of Contingencies and Crisis Management* 5(2): 87–97.

Tierney, K. J. (2006) "Businesses and Disasters: Vulnerability, Impacts, and Recovery," 275–296, in H. Rodriguez, E. L. Quarantelli, and R. Dynes (eds.), *Handbook of Disaster Research*, New York: Springer.

Tierney, K. J., J. M. Nigg, and J. M. Dahlhamer. (1996) "The Impact of the 1993 Midwest Floods: Business Vulnerability and Disruption in Des Moines," 214–233, in R. T. Sylves and W. L. J. Waugh (eds.), *Disaster Management in the U.S. and Canada: The Politics, Policymaking, Administration and Analysis of Emergency Management*, Springfield IL: Charles C. Thomas Publisher, LTD.

Tigges, L. M. and G. P. Green. (1994) "Small Business Success among Men and Women-Owned Firms in Rural Areas," *Rural Sociology* 59(2): 289–310.

Watson, M. (2014) *Business Recovery Financing: Galveston, TX after Hurricane Ike* (Master paper), Department of Landscape Architecture and Urban Planning, Texas A&M University.

Wasileski, G., H. Rodriguez, and W. Diaz. (2010) "Business Closure and Relocation: A Comparative Analysis of the Loma Preita Earthquake and Hurricane Andrew," *Disasters* 35(1): 102–129. doi: DOI: 10.1111/j.1467-7717.2010.01195.x.

Webb, G. R., K. J. Tierney, and J. M. Dahlhamer. (2002) "Predicting Long-term Business Recovery from Disaster: A Comparison of the Loma Prieta Earthquake and Hurricane Andrew," *Environmental Hazards* 4: 45–58.

Worthington, A. and A. Valadkhani. (2004) "Measuring the Impact of Natural Disasters on Capital Markets: An Empirical Application Using Intervention Analysis," *Applied Economics* 36(19): 2177–2186.

Wright, J. D., P. H. Rossi, and E. Weber-Burdin. (1979) *After the Clean-Up: Long-Range Effects of Natural Disasters*, Beverly Hills CA: Sage.

Xiao, Y. (2008) *Local Labor Market Adjustment and Economic Impacts after a Major Disaster: Evidence from the 1993 Midwest Flood*, PhD Dissertation, University of Illinois at Urbana-Champaign, Urbana, IL.

Xiao, Y. (2011) "Local Economic Impacts of Natural Disasters," *Journal of Regional Science* 51(4): 804–820. doi: DOI: 10.1111/j.1467-9787.2011.00717.x.

Xiao, Y. and E. Feser. (2014) "The Unemployment Impact of the 1993 U.S. Midwest Flood: A Quasi-experimental Structural Break Point Analysis," *Environmental Hazards* 13(2): 93–113.

Xiao, Y. and U. Nilawar. (2013) "Winners and Losers: Analysing Post-Disaster Spatial Economic Demand Shift," *Disasters* 37(4): 646–668.

Xiao, Y. and W. G. Peacock. (2014) "Do Hazard Mitigation and Preparedness Reduce Physical Damage to Businesses in Disasters? Critical Role of Business Disaster Planning," *Natural Hazards Review* 15(3): 04014007.

Xiao, Y. and S. Van Zandt. (2012) "Building Community Resiliency: Spatial Links between Households and Businesses in Post-Disaster Recovery," *Urban Studies* 49(11): 2523–2542.

Zhang, Y., M. K. Lindell, and C. S. Prater. (2009) "Vulnerability of Community Businesses to Environmental Disasters," *Disasters* 33: 38–57.

19

FACILITATING QUALITY DESIGN AND COMMUNITY ENGAGEMENT IN HOUSING RECOVERY

Jaimie Hicks Masterson and Katherine Barbour Jakubcin

Introduction

Poor aesthetic and functional quality of disaster housing reduces the quality of life for victims—lengthening recovery and generating community metamorphoses that perpetuate incongruous elements in the built environment. From formaldehyde-ridden FEMA trailers to metal shelters in India colloquially named "microwaves", housing following disasters has not been sufficient. The housing recovery process is a complex post-disaster challenge (see Chapter 16 by Zhang and Drake), but it is invaluable to propel individuals, families, and economies forward (Andrew et al. 2013; Sanderson 2000; Barakat 2003). We know that by establishing housing, residents can reestablish routines and get "back to normal," decreasing recovery time (Masterson et. al, 2014; Peacock et al. 2007; Quarantelli 1982). For most, a home not only provides shelter, but it is the main source of investment (Zhang and Peacock 2009). Although it is important to meet basic needs and functionality during disaster recovery, oftentimes homes are culturally inappropriate, do not include community input, and create neighborhoods that are unwanted by both residents and the surrounding communities. Whether it is the construction of emergency shelters, temporary shelters, temporary housing, or permanent housing, architects and landscape architects can design for such scenarios. Pre-event planning can help communities streamline housing recovery with the ultimate goal of meeting victims' basic needs sooner and creating spaces that allow victims to cope and return to normal. This chapter evaluates the evolution of housing structures following a disaster and investigates current gaps in housing recovery as one transitions from temporary housing to permanent housing. We evaluate the issues and obstacles in design decisions and construction, as well as best practices revealed in the literature.

Evolution of Housing Following Disasters

Quarantelli (1982) suggests that there are four categories of housing following a disaster—emergency sheltering, temporary sheltering, temporary housing, and permanent housing. *Emergency sheltering* typically occurs immediately after a disaster and is meant to be a short-term emergency solution, usually lasting a day. It often takes place within schools, churches, or other venues and the primary focus is keeping people safe. *Temporary sheltering* is still considered a short-term solution and is not intended to be primary housing. Churches with showering, feeding, and sleeping facilities often fulfill this need. Usually during this time, disaster victims are applying for temporary housing solutions through state or federal programs. *Temporary housing* is more private and facilitates that transition into daily routines. It is during the temporary housing phase that the household begins to recover and reestablish a sense of normalcy in their lives but in nonpreferred locations or structures (Johnson 2007). Trailers such as those seen in the aftermath of Hurricane Katrina and other temporary structures are considered temporary housing. During this time, individuals and families are applying for permanent housing assistance to make needed repairs or to replace damaged homes. The last type is *permanent housing*, which reestablishes household routines in preferred locations and structures. The speed of housing recovery depends on the proximity to the former place of residence, the neighborhood structure and support system to rebuild, and guidance through procedures and processes that lead to permanent housing (Johnson 2007).

Post-disaster housing has taken many forms across the globe. By assessing disaster housing case studies, three rehousing themes have emerged across disasters. The first focuses on the speed of construction and the functionality of space for disaster victims. The second theme addresses the involvement of architects in designing aesthetically pleasing and/or sustainable housing for victims. The final theme is the degree of disaster victims' participation in the design and construction of their own homes and their communities following a disaster.

Speed and Function

There is time compression following a disaster due to a push to re-house victims as quickly as possible to get back to normal. Typically, the desired speed of construction to meet basic functional needs has yielded mass fabrication of housing, usually led by national governments or other top-down mechanisms. The following is a series of examples that describe the benefits and drawbacks to a mass-manufactured housing approach.

In 1999, Turkey's Marmara region was affected by two earthquakes that left thousands of victims without homes. To handle the massive housing need, the Prime Minister's office and the Ministry of Public Works and Settlements provided 32,000 prefabricated temporary housing units. The temporary housing was not sited on disaster victims' private property, but instead on government land or vacant privately rented lots. Twenty-five prefabricated home manufacturers were selected to construct the housing, which was completed within 6 months with a 98.5 percent occupancy rate. There was little or no participation from disaster victims regarding physical and financial needs, reconstruction, and the design of housing. Although the process rapidly rehoused disaster victims, there were many complaints. One common criticism was the location, which was far from the city and available work. Moreover, the design of the homes did not meet the needs of the disaster victims; the orientation of the buildings were not culturally appropriate, and many residents altered the homes with additions and accessory units afterward (Davidson et al. 2007). Also, the housing ignored the need for seismic-resistant construction practices. This housing strategy is not unique, and even in man-made disasters, top-down rehousing solutions have persisted. Following Lebanon's war in 1991, standardization of homes and technologically oriented solutions (i.e., mass housing with pre-fabricated solutions) took place. It gave little consideration to the community's socioeconomic, cultural, and developmental issues (El-Masri and Kellett 2001).

The strategy of mass-producing temporary housing to rapidly meet post-disaster needs has also been adopted in the United States. The Federal Emergency Management Agency (FEMA) was responsible for temporary housing during the 2005 hurricane season, during which its trailers were extensively used (AHSDR 2009). Because of the scale of destruction following Hurricane Katrina, an accelerated bidding process took place and 150,000 trailers were contracted, built, and sent to affected areas. Each trailer cost roughly $59,000 over an 18-month period—$14,000 for the trailer, $12,000 for hauling and installation, $5,400 for maintenance, $750 for site costs, $25,000 for pad construction and site prep—costing the federal government $5.5 billion (AHSDR 2009). Nevertheless, contractors hit contract ceilings quickly and many demanded increased funds from FEMA (AHSDR 2009). Unfortunately, many disaster victims were on waitlists or had to wait for utility hookups to be installed, further slowing recovery (Committee on Homeland Security and Governmental Affairs, 2007). Nearly 81 percent of all trailers were placed on private property of disaster victims, allowing ease of access to repair damaged homes. Many residents, particularly along the Gulf Coast, were unable to place trailers on their private property due to their location in floodplains and high velocity surge zones. In other cases, neighboring communities did not want to host group trailer sites due to the cost of providing services such as fire, police, water, etc., among other reasons. There was no participatory involvement of disaster victims and the primary focus was the speed at which victims could be housed—a traditional top-down approach. Shortly after trailer installation, unhealthy formaldehyde levels were detected, adding to the mounting problems of the federal government's inability to adequately house disaster victims (AHSDR 2009). FEMA also acknowledged an inability to sufficiently assist disabled, special needs, and low-income assistance households during the housing recovery process (Committee on Homeland Security and Governmental Affairs 2007). These deficiencies slowed recovery and left hundreds of thousands of victims without their basic needs met.

These are just a couple of examples of a top-down approach to temporary housing recovery, demonstrating that this approach "often ignore[s] the complexity of the built environment, the local conditions and the users' needs and potential" (El-Masri and Kellett 2001: 536; United Nations 1982). After decades of seeing this expensive, homogeneous recovery housing, architects and designers began considering more sensitive approaches with design solutions intended to accommodate household needs and improve disaster victims' quality of life (El-Masri and Kellett 2001; United Nations 1982).

Developing Disaster-Resistant Designs

Over the last two decades there has been a surge in design competitions and designer involvement to help solve some of the housing challenges that disasters pose. For example, in 1998 FEMA sponsored a Wind Summit to foster the creation of a hurricane-resistant home. Although this competition did not address the complexities of disaster recovery and its attendant time compression, it does point to an interest in including designers in the discussion of high-quality structures for disasters. The result was a permanent housing solution—the Hurricane Home—created by Jack Jackson and now displayed in Chesapeake, Virginia. The home is a wood frame construction that boasts structural hardening, wind-resistant doors and windows, a reinforced concrete ceiling, a safe-room, and the ability to withstand 250 mph winds. The home also provides increased energy efficiency with high insulation values, among other features (Farnsworth 2000). Costs of the home were not taken into consideration. Another design competition was held after a tornado hit Greensburg, Kansas, in 2007. The non-profit organization Greensburg GreenTown set up the "Chain of Eco-Homes Competition," in which more than 250 teams from across the world submitted designs that featured structural hardening techniques, eco-friendly elements, and cost less than $110 per square foot (GreenTown 2013; Sherriff 2010). Several of the designs have been built and others are underway (GreenTown 2013).

Other design trends have emerged to produce temporary housing that can be partially assembled and shipped to affected areas to be fully constructed. For example, after Hurricane Katrina a number of designers and architects mobilized to help the hundreds of thousands of disaster victims. Habitat for Humanity International initiated the "home in a box" program, Operation Home Delivery. Habitat for Humanity groups across the country built wood frames to be packaged and shipped to sites. The ecoMOD was a demonstration model developed by the University of Virginia and Habitat for Humanity of the Mississippi Gulf Coast. The ecoMOD is an energy-efficient and sustainable pre-fabricated home that can resist hurricane-force winds and mold. The three-bedroom home is just over 1,000 square feet and costs $65 per square foot. It takes advantage of natural ventilation and includes photovoltaic solar panels. Another example was developed in response to the 2004 Indian Ocean earthquake and tsunami. Design students at the McGill School of Architecture created BuildAid, an organization to export post-disaster temporary housing. BuildAid also sent housing to Kashmir and the Philippines to help with their housing reconstruction needs. Garrison Architects developed a temporary housing solution for urban areas. The modular pre-fabricated construction is stackable to provide multi-story and multi-family temporary housing solutions. The units can be deployed within 15 hours and come in one-, two-, and three-bedroom options with full kitchen, bathroom, living area, and outdoor space (Fredrickson 2014).

Other strategies include hiring architectural firms to design housing for long-term recovery. After the 2000 earthquake in the Cankiri region of Turkey, the ministry of Public Works and Settlements hired an architectural firm to design three housing prototypes that disaster victims could select to qualify for interest-free loans for housing reconstruction. The original three designs were one-, two-, or three-bedroom brick masonry. Unfortunately, the homes were not adaptable to site conditions. Although the homes appealed to many disaster victims prior to construction, many did not find them adequate once occupied. Some design issues included the long distance from villages and roadways, which made it difficult to commute to work. Also, the lots lacked enough space for cultural necessities, such as cattle or straw sheds. There was little participation with the future homeowners, particularly in early decision making, which ultimately led to the rejection of new settlements—which were either abandoned or extensively renovation later (Davidson et al. 2007).

Other design concepts have been explored to create transformational landscapes during disasters. Transformational designs focus not on static solutions, but on the ability of a design and space to transform in various scenarios (Bloomer 2000; Resilience Alliance 2007). This idea has been promoted by the Resilience Alliance as a solution for areas that are regularly affected by disasters, particularly coastal areas (Resilience Alliance 2007). Some ideas include the ability of a typical residential street to transform during the disaster, during short-term recovery, and long-term recovery (Rosenblatt Naderi 2009). Residential development could be designed to absorb and resist disaster impacts—for example, streets that accommodate flood water and structures that are elevated and constructed to withstand wind conditions (Rosenblatt Naderi 2009). During short-term recovery, the street could transform into a disaster relief site with the ability to quickly assemble readymade "pop-up" facilities within the first 72 hours after a disaster (Rosenblatt Naderi 2009). These disaster relief sites would address basic needs, with 18-wheeler drop-offs for clothing and perishable items, FEMA case management assistance and information, and mental health or spiritual assistance. Finally, during long-term recovery, residents could live in temporary housing on their private property while they repair their homes (Rosenblatt Naderi 2009). Such ideas have been explored conceptually by architects and designers that are looking to lessen disaster impacts, because we know that the "social, psychological, and cultural recovery process [should] be undertaken in parallel" with debris removal, rehousing, and reconstruction, bringing communities back to "normality" (Aysan and Oliver 1987: 9).

Although many designs yield creative solutions and inspire the next generation of out-of-the-box strategies, many still abide by top-down approaches. Aysan reflects that it is "difficult for architects and industrial designers to recognize their limitation to assist constructively in the shelter field" (Davis and Aysan 2013: 12). This is not to say that designers should not or cannot play a role, but, despite their efforts, some problems persist, new issues arise, and recovery is slow. Even with designers leading the effort, there continues to be a "widespread, simplistic assumption that an instant, universal shelter unit is needed and is an effective solution" (Davis and Aysan 2013: 13).

Participation of the Local Community

High-quality designs for housing during disaster recovery should be infused with local knowledge from the community (United Nations 1982). By developing pre-disaster designs *with* the community, one can achieve multiple goals—creating culturally appropriate spaces in locations that are most convenient for residents (Davis and Aysan 2013), developing local human capital by including victims in the physical rebuilding (Davis and Aysan 2013; United Nations 1982), and reducing psychological trauma and providing therapeutic healing by involving victims in the decision-making process and engaging them in positive collective action (Aysan and Oliver 1987; Choguill 1996; Davis and Aysan 2013; El-Masri and Kellett 2001; Pugh 1997; Turner 1972; United Nations 1982). Community re-building can take place when when residents are informed and included throughout the planning process (Innes and Booher 2004). Too often community planning and design has been about the "expert" handing down their knowledge from "on high" (Teaford 2000). The following examples point to trends that engage disaster victims in their own housing recovery process.

Post-disaster housing reconstruction projects can incorporate community-building techniques to effectively engage the public. For example, after the 1999 Colombia earthquake, residents of a coffee-farming community were satisfied with the outcome of their post-disaster homes because the Coffee Growers Organization (CGO) led them through a community reconstruction program (Davidson et al. 2007). Community-based organizations such as the CGO have been found to successfully address the real needs people have in recovery (Maskrey 2011). For instance, the CGO already had personnel and resources at local, regional, and national levels and had experience working with local communities. In general, the CGO was successful because it did not act like most nongovernmental organizations— which are "help-givers" and thus create disaster victims who are only "help-receivers." Instead, disaster victims were provided with a wide variety of services (loans, subsidies, technical aid, information) to choose from, which empowered the victims. They were also able to choose (a) which type of project to invest in, a home or business, (b) whether to repair the damage, rebuild entirely, or request a pre-fab home, and (c) whether to do the work themselves or hire contractors. Because the CGO did not dictate one solution, the participants were fully engaged in the recovery process (Davidson et al. 2007).

In contrast, after an earthquake struck El Salvador in 2001, the housing recovery process involved participants, but with poor results. The European Red Cross built a model community in which disaster victims had limited input on the design of the homes and were involved in the actual reconstruction, but all 300 homes were identical. Moreover, victims had to first qualify through an eligibility process and once qualified, they received training from hired masons. Because the entire project required physically demanding manual labor, most participants could not work other jobs during the project. Laborers' fatigue and illness slowed the process (Davidson et al. 2007).

A Combined Approach

A bottom-up approach incorporating public input into disaster housing recovery has been advocated since the 1980s (Aysan and Oliver 1987; Davis and Aysan 2013; El-Masri and Kellett 2001). Cockburn and Barakat (1991) argue that in order for a resident to feel a true sense of belonging, residents must be involved from the very beginning in order for their reconstructed home to have a "sense of place." Sheltering should be "place-specific" (El-Masri and Kellett 2001) and culturally appropriate, which can combat homogenization (Oliver 1987). The evolution of disaster recovery housing has moved from governments and well-intentioned architects "directing and imposing" top-down approaches to "enabling and empowering" (Oliver 1987: 230). Therefore, a combined approach in which communities are a part of the decision-making process for their home and where designers are involved to translate residents' social and cultural needs into high-quality designs can improve the quality of life for the people that will live there.

The Katrina Cottage is an example in which high-quality housing designs merged with the inclusion of residents in the design process. In 2005, following Hurricane Katrina, Andres Duany and Marianne Cusato designed 15 floor plans ranging from 500–1500 square feet to act as temporary-to-permanent structures. The temporary-to-permanent concept (or temp-to-perm) seeks to bridge the gap between temporary and permanent housing by constructing a temporary house rather quickly, with the ability to add on to it over time to create the final permanent housing structure. These larger semi-permanent houses accelerate the housing recovery process by facilitating households' return to basic daily routines (Perkes 2012).

In 2006, the Louisiana Recovery Authority began a series of outreach meetings, "Louisiana Speaks," which involved over 27,000 citizens. This program engaged citizens to develop long-term recovery plans to create more resilient communities and to provide direct input to the Katrina Cottage designs through planning and charrettes. Architectural and design specifications were published in a pattern book based on community feedback. In the end, the project produced seven different home designs ranging from $100,000 to $135,000 (National Building Museum 2009). Cultural appropriateness, cost, environmental sustainability, and structural hardening were all addressed in the final design. Unfortunately, there were a number of problems despite the efforts to create sustainable, aesthetically pleasing, and functional homes. Some persistent issues included site selection, property acquisition, the lengthy eligibility process, heirship, proof of ownership, property tax issues, and clear title with potential residents (Abt Associates Inc and Amy Jones and Associates, 2009a; NDHRAC 2010). To fully understand a broad range of barriers in housing recovery across different disasters, we reviewed the literature to determine obstacles to the transition from temporary housing to permanent housing.

Evaluating the Gaps in Temporary-to-Permanent Housing Recovery

To identify obstacles in design decisions and construction in the temporary-to-permanent housing recovery process, we conducted a content analysis of housing recovery reports, articles, and policies. A qualitative evaluation of 21 articles and reports focused on disasters from 2005 to the present in the United States. Design decisions and construction were evaluated based on the issues and obstacles that emerge and the best practices that have been identified in the literature (see Table 19.1).

Design Decisions

As previously discussed, design decisions made through architects' collaboration with future residents can yield housing the community embraces and that promotes overall disaster recovery. Design decisions encompass the functional, aesthetic, and regulatory components of housing design, including federal, state, and local standards (National Flood Insurance Program, National Building Codes, and locally specific codes and ordinances). Many physical and cultural factors must be considered when rebuilding housing. Reconstruction efforts should take into consideration residents' short-term and long-term needs. A sustainable design solution must not only work to provide a rapid transition from temporary housing to permanent housing, but should also be adaptable to households' future needs.

In the evaluation of the literature, 13 of 20 related to design decisions, which discussed the importance of community acceptance of housing designs. In the Louisiana Katrina Cottages program, many residents did not fully understand that the temporary house was a step towards the final permanent housing outcome. In another instance, the Alabama Alternative Housing Pilot Program (AHPP) failed to communicate clearly with the community, due in part to program overextension as the number of service areas became too large. Delivering accurate and timely information to the community also became difficult because plans were continuously changing. An important finding of pilot housing programs in Mississippi and Louisiana following Hurricane Katrina was that managing

Table 19.1 Content Analysis of Housing Recovery Reports, Articles, and Policies

Title	Design Decisions	Construction
Closing Gaps in Local Housing Recovery Planning for Disadvantaged Displaced Households		X
Creating Safe Harbor after Hurricane Katrina: A Case Study of the Bayou La Batre Alternative Housing Pilot Program	X	X
Developing a More Viable Disaster Housing Unit: A Case Study of Mississippi Alternative Housing Pilot Program	X	X
Emergency Housing Program Research and Recommendations	X	X
Far From Home	X	X
GAO: Hurricane Katrina Improving Federal Contracting Practices in Disaster Recovery Operations		X
Handbook of Disaster Research		X
Hurricane Katrina Improving Federal Contracting Practices in Disaster Recovery Operations: Testimony before the Committee on Government Reform		X
National Disaster Housing Strategy	X	X
National Disaster Housing Reconstruction Plan	X	X
OIG: Effectiveness and Costs of FEMA's Disaster Housing Assistance Program, Aug 2011	X	X
OIG: Unless Modified, FEMA's Temporary Housing Plan will Increase Cost by an Est. $76 million Annually, June 2013	X	X
Rapid Housing Recovery Program Research Summary	X	X
Rebuilding or Recovering? Considering Sustainability in the Context of Disaster Rehousing	X	X
Research Trends of Post-Disaster Reconstruction: The Past and the Future		X
Resourcing Challenges Post-Disaster Housing Reconstruction: A Comparative Analysis	X	X
SERRI Project	X	X
TDHCA: Community Development Block Grant Disaster Recovery Program Hurricanes Ike & Dolly Round 2		X
The Barriers to Environmental Sustainability in Post-Disaster Settings: A Case Study of Transitional Shelter Implementations in Haiti		X
The Effects of Housing Assistance Arrangements on Household Recovery: An Empirical Test of Donor-Assisted and Owner-Driven Approaches	X	
TOTAL	13	19

community expectations was vital to the perceived success or failure of a re-housing approach (Abt Associates and Amy Jones and Associates, 2009b).

Additional concerns related to fair compensation and the residual effect of post-completion values. Following Katrina, residents of other homes felt that the style of the replacement homes did not fit with the character of their neighborhoods (Abt Associates and Amy Jones and Associates 2009b; NDHRAC 2010). Many feared that the aesthetic quality and size would lower their property values. Specifically, the Katrina Cottages were thought to be too small, resembling a trailer, and akin to "outsider housing" that were "indistinguishable mobile homes" (Wilson 2013). Some proposed cottages were vetoed by adjacent property owners and others were removed after a city vote () (Abt Associates Inc & Amy Jones & Associates 2009a; NDHRAC 2010). Because temporary housing was associated with FEMA trailer camps, many residents feared similar conditions in their communities. In the Mississippi Alternative Housing Program (MAHP), a county supervisor said that if the homes had come on a flatbed rather than on wheels, residents would have been more likely to accept them (Abt Associates and Amy Jones and Associates, 2009b). The stigma of low-quality housing coupled with the permanent nature of the homes resulted in extreme measures from jurisdictions. Many jurisdictions only permitted temp-to-perm units on private residential lots if a FEMA trailer had previously been located onsite, there was evidence the owners were building a permanent structure, or local zoning codes allowed modular or manufactured homes (Abt Associates and Amy Jones and Associates 2009b; NDHRAC 2010). Such reactions "constrained the ability of households to participate in the decision-making process, including design locations and reconstruction of damaged homes" (Andrew et al. 2013: 18).

The selection of sites for temp-to-perm housing also contributes to community acceptance. Temp-to-perm construction on private sites minimally displaces residents, provides community continuity, and is cost effective. When households are able to rebuild in their previous communities, it significantly increases capacity for recovery because it "determines whether an occupant's social network, community resources, and employment opportunities remain intact during the recovery process" (Perkes 2012: 13). In reality, many mobile homes that were purchased after Katrina could not be placed on any private property that was located in a floodplain (AHSDR 2009). In such cases, commercial sites or group sites are an alternative. However, displacing residents to sites other than their own property leads to rejection of temp-to-perm housing solutions. Many group sites were not well accepted in the communities where they were located, and many host communities were reluctant to provide group sites because of the increased demand for fire and police services and rising infrastructure costs. To offset these costs FEMA provided incentives by paying these "impact fees" (AHSDR 2009). In general, obtaining properties is a lengthy process, as seen with the Katrina Cottages (NDHRAC 2010). The longer the time lapse to construct housing after a disaster, the more negative the recovery outlook is perceived—and therefore the less community acceptance there is (Abt Associates and Amy Jones and Associates 2009b). In general, a "lack of understanding and consultation with affected communities have sometimes resulted in poor site selection for resettlement or socially and culturally inappropriate housing layouts and design leading in administrative failures" (Andrew et al. 2013: 20).

Design Best Practices

Of the 13 articles that discuss design decisions, 11 discussed two best practices—construction of temp-to-perm housing (known as the "Grow Home Approach") and design compatibility with local aesthetic styles. Six of 13 articles discussed a "Grow Home Approach," which takes into consideration the different phases that residents will go through as they move through the recovery process. Residents stated that, during the Mississippi Alternative Housing Program, it was easier to begin to return to their basic daily routines when they had a larger, semi-permanent house (Perkes 2012). This is one of the main benefits of the temp-to-perm housing solution and a successful housing program—residents are established in a temporary house quickly to make the transition to permanent housing more efficient.

In addition, eight of 13 articles discuss the importance of culturally sensitive housing designs appropriate for local aesthetic standards. The selection of materials for rebuilding houses should mirror cultural norms (Chang et al. 2010), and the materials selected must take into consideration local motifs (Chang et al. 2010). Neighborhood amenities should be worked into the community design to maintain the community character and the attractiveness and desirability of the neighborhood (Abt Associates and Amy Jones and Associates 2009b). The cultural design requirements may play as much of a role in meeting long-term recovery needs as the more classically utilitarian building necessities. The Mississippi Alternative Housing Program took into consideration the style of the homes that are built in the coastal South when designing the Cottages (Abt Associates and

Amy Jones and Associates 2009b). Also, residents can work with the case managers and designers in order to make design decisions that are tailored to fit their long-term needs. Resident are empowered when they are involved in the design decisions of what their new home will look like. Perkes (2012) found that the more input a resident had into the decisions regarding their home, the more successful the recovery process was. With the Gulf Coast Community Design Studio (GCCDS), clients were pre-qualified for home details. Designers worked with clients to identify what they had liked about their damaged homes and then provided variations on the original designs (Wilson 2013). Even though the design team was individualizing floor plans for each client, the basic structural details and wall selections stayed the same (Wilson 2013). Other considerations for design choices include materials and uses for temporary and permanent housing (Abt Associates and Amy Jones and Associates 2009b). In all, a home should also be designed and structured in ways to establish a sense of community.

Construction

Housing construction was also assessed in the housing recovery literature and comes with a variety of challenges. Although the construction process may appear to begin post-disaster, it really should begin long before then because "ad hoc arrangements after a disaster seem to be unable to perform well to alleviate resource shortages in the long run" (Chang et al. 2010: 250). A significant amount of planning goes into a proper temp-to-perm housing construction process prior to a disaster and "the absence of pre-event planning and preparedness, the inadequacy of efficient and flexible institutional arrangements, and the lack of proactive engagement of the construction industry into disaster management are underlying contributions to undermining resourcing performance in a post-disaster event" (Chang et al. 2010: 250). The construction process includes permitting and inspections, the procurement of materials, and the construction of homes. Solutions like the one proposed in this program, which are intended to transition from temporary to permanent housing, must comply with zoning and building code regulations applicable for both temporary and permanent development.

Two themes emerged from the 19 construction-related articles—issues with contractors and local building codes and ordinances. Five of 19 articles discussed the variety of issues associated with contractors. Many states have laws that require cities to use a competitive bidding process to select a contractor. This bidding process often slowed the construction process and forced many pilot programs to make changes to their designs due to the cost of construction (Abt Associates and Amy Jones and Associates 2009b). Although competitive bidding can reduce costs, many programs still received high minimum bids. For instance, the Alternative Housing Pilot Program initiated in 2006 in the states of Florida, Alabama, Mississippi, Louisiana, and Texas experienced higher costs than expected, reducing the number of units that could be built (Office of Inspector General, 2011). The increased home cost was important because even reasonably priced housing is often not affordable to many low-income households because they must pay for homeowner insurance and hazard insurance in addition to the mortgage payment (Abt Associates Inc. and Amy Jones and Associates 2009a; NDHRAC 2010). Moreover, FEMA previously had only a small pool of registered Individual Assistance–Technical Assistance Contractors from which to choose—most of which were large multinational companies. This ineligibility of local construction companies stalled local business and housing recovery, resulting in limited circulation of money within the community that would otherwise have accelerated business recovery (Abrahams 2014; AHSDR 2009). Additionally, bids were high because construction materials were more difficult to obtain. This was the case in countries with limited natural resources and the cost to import materials through customs, as was the case in Haiti. Strangely enough, construction materials are also difficult to procure, not because of the lack of available materials worldwide, but instead when compared to normal orders of the North

American construction industry, disaster recovery housing was just not large enough to procure materials with contractors (Abrahams 2014).

Although the disaster recovery process has its own emerging set of regulations and mandates, these must fit within the constraints of current zoning and land use regulations of the affected area. Building codes and regulations were an obstacle cited by eight of the 19 construction-related articles. A significant number of issues arose during the attempt to construct temp-to-perm housing following Katrina. Many jurisdictions would not allow a former home site to have a temp-to-perm home on the same parcel (NDHRAC 2010). In other cases, jurisdictions would only permit temp-to-perm homes in designated mobile home or manufactured housing areas, largely prohibiting the temp-to-perm housing process to take place (Abt Associates and Amy Jones and Associates 2009b). Related regulation obstacles included manufactured homes not meeting the residential zoning minimum square footage requirements and temporary homes not meeting the municipalities' setback requirements. Local setback requirements, National Flood Insurance Program requirements, and environmental impact reviews all contributed to the lengthy process to obtain building permits and inspection for occupancy (Abt Associates and Amy Jones and Associates 2009b; OIG 2011). These sources also found that, as time passed after a disaster, jurisdictions were less lenient with zoning and land use regulation requirements.

Working with a variety of jurisdictions to design and build housing proved to be time-consuming in housing recovery. Following Katrina, memoranda of understanding (MOUs) were developed to agree upon design choices for disaster housing in each community. MOUs were tailored for each jurisdiction in the Mississippi Alternative Housing Program, based on their needs. For example, MOUs gave precise instructions regarding how the Cottages would be used and where the temp-to-perm housing could be placed (Abt Associates and Amy Jones and Associates 2009b). Nonetheless, each jurisdiction modified the designs, which ultimately slowed reconstruction (Abt Associates and Amy Jones and Associates 2009b).

As previously discussed, in areas where there was little community acceptance of designs, jurisdictions used zoning and code enforcement to limit the construction of temp-to-perm housing (Abt Associates and Amy Jones and Associates 2009b; Wilson forthcoming). Federal regulations also limited the use of temp-to-perm structures, as FEMA requirements prohibited permanent installation in Coastal High Hazard Areas and floodplains (NDHRAC 2010). After the disaster, many areas once considered "low risk" were reclassified into categories on which National Flood Insurance Program regulations precluded rebuilding.

Construction Best Practices

Three major best practice themes emerged from the literature—local contractors, pre-procurement, and sustainable development. Using one pre-determined local or regional contractor who is more in tune with local needs and cultural considerations was recommended in four of the 20 articles. Purchasing locally also helps stimulate local economies—not only bringing purchasing power back to the local economy, but creating a higher demand for more jobs, which cultivates more investment by local labor and citizens into the success of the community (Abrahams 2014). Using one local contractor to coordinate the efforts of the rebuilding will benefit all aspects of efficiency and consistency. The Alabama Alternative Housing Pilot Program chose to use one general contractor to manage construction, which was said to enhance collaboration and help reduce the chances of multiple contractors causing delays (Abt Associates and Amy Jones and Associates 2009b). Using one contractor can also cut down on the time for the bidding process (NDHRAC 2010). Using local contractors is further beneficial since they are familiar with the local permitting and inspecting regulations. For instance, the Mississippi Alternative Housing Program contracted with a local company to help ensure that the installation was coordinated with the permitting and applicant preparation.

This local company was able to deliver homes to an intermediary site where inspections and repairs could be performed prior to final site installation, thus speeding the process (Abt Associates and Amy Jones and Associates 2009b).

Pre-procurement identifies vendors, contractors, materials, supplies, and services pre-disaster that will be ready to be deployed after a disaster strikes. This practice was recommended in six of 19 articles (Woods 2006). Florida's Division of Emergency Management developed a pre-procurement database that identified supplies and services needed (Woods 2006). Pre-procurement is important in order to control the costs of materials—which historically increase significantly after a disaster. In an instance where materials have not been pre-procured, another useful tool is joint purchasing and shipping, which cuts costs, speeds procurement, and limits the total transportation (Abrahams 2014). With the vendors and contractors pre-procured, communities can also give contractors the specifications their work must meet. As seen in the Mississippi Alternative Housing Program, "uniform design standards that could be shared with housing providers and manufacturers in advance of an emergency could shorten production time and improve quality of the units" (Abt Associates and Amy Jones and Associates 2009b: 10).

Sustainable development was a best practice cited in three of the 19 articles. Sustainable houses can be environmentally, economically, and socially sustainable. One of the biggest misconceptions in reconstruction is that sustainability is unfeasible. Sustainability is possible when principles are injected into the construction phase as well as the entire housing recovery process, leading to a more resilient and robust built environment (Yang and Yi 2014). An important aspect to successfully implementing sustainable development is setting goals and involving stakeholders before the construction process (Yang and Yi 2014). The phrase "building back better" should be used in conjunction with "building back safer," which not only incorporates building more aesthetically pleasing structures but also doing so in a way that incorporates a more sustainable use of the land and resources. Incorporating sustainable development into the housing recovery process is mainly based upon pre-procuring services, as mentioned previously. Other forms of sustainable development are more holistic and include "flexible and interchangeable materials, proactive processing of waste from deconstruction, and coordinated recycling and reuse, [which] can also be new research topics that respond to the challenges of construction waste reduction and resourcing problems during post disaster reconstruction" (Yang and Yi 2014: 28). Rubble reuse programs used for nonload-bearing structures can also be a part of a sustainable housing program. An example of this is rubble reused as an aggregate for concrete blocks and in concrete slabs (Abrahams 2014). By utilizing reuse programs, communities can alleviate an additional hurdle in the recovery process.

Conclusion

Top-down housing projects with uninspired designs consistently fail to meet disaster victims' needs (Davidson et al. 2007). Moving forward, communities can increase their quality of life following a disaster by incorporating quality design into post-disaster housing programs. Quality designs come from the input of residents and communities. To facilitate the transition from temporary housing to permanent housing, it's important to consider gaps in design decisions and construction alongside the desires of community members. Policies should be established at all levels to decrease housing recovery times, particularly for those in the greatest need. These efforts must begin long before a disaster strikes and be a part of routine planning processes.

Acknowledgment

We wish to thank Bara Safarova for assistance in preparing this chapter.

References

Abrahams, D. (2014) "The Barriers to Environmental Sustainability in Post-Disaster Settings: A Case Study of Transitional Shelter Implementation in Haiti," *Disasters* 38(s1): S25–S49.

Abt Associates Inc and Amy Jones and Associates. (2009a) *Developing A More Viable Disaster Housing Unit: A Case Study of the Mississippi Alternative Housing Program*. Washington DC: Federal Emergency Management Agency.

Abt Associates and Amy Jones and Associates. (2009b) *Creating a Safe Harbor after Hurricanes: A Case Study of the Bayou La Batre Alternative Housing Pilot Program*. Washington DC: US Government Printing Office.

AHSDR—Ad Hoc Subcommittee on Disaster Recovery. (2009) *Far from Home: Deficiencies in Federal Disaster Housing Assistance After Hurricanes Katrina and Rita and Recommendations for Improvement*. United States Senate, Committee on Homeland Security and Governmental Affairs. Washington DC: US Government Printing Office.

Andrew, S. A., S. Arlikatti, L. C. Long, and J. M. Kendra. (2013) "The Effect of Housing Assistance Arrangements on Household Recovery: An Empirical Test of Donor-Assisted and Owner-Driven Approaches," *Journal of Housing and the Built Environment* 28(1): 17–34.

Aysan, Y., and Oliver, P. (1987) *Housing and Culture After Earthquakes: A Guide for Future Policy Making on Housing in Seismic Areas*. Oxford: Oxford Polytechnic.

Barakat, S. (2003) "Housing Reconstruction After Conflict and Disaster," *Humanitarian Policy Group, Network Papers* 43: 1–40.

Bloomer, K. (2000) *The Nature of Ornament: Rhythm and Metamorphosis in Architecture*. New York: Norton.

Chang, Y., S. Wilkinson, R. Potangaroa, and E. Seville (2010) "Resourcing Challenges for Post-Disaster Housing Reconstruction: A Comparative Analysis," *Building Research and Information* 38(3): 247–264.

Choguill, M. (1996) "A ladder of community participation for underdeveloped countries," *Habitat International* 20(3): 431–444.

Cockburn, C., and S. Barakat (1991) "Community Prosperity through Reconstruction Management," *Architecture and Design* 8(1): 60–65.

Committee on Homeland Security and Governmental Affairs (2007) *Beyond Trailers: Creating a More Flexible, Efficient, and Cost-Effective Federal Disaster Housing Program*. United States Senate, Senate Hearing 110–302. Washington DC: US Government Printing Office.

Davidson, C. H., C. Johnson, G. Lizarralde, N. Dikman, and A. Sliwinski (2007) "Truths and Myths about Community Participation in Post-Disaster Housing Projects," *Habitat International* 31(1): 100–115.

Davis, I. and Y. Aysan (2013) "Disasters and the Small Dwelling-Process, Realism and Knowledge: Towards an Agenda for the IDNDR" 8–22 in Y. Aysan and I. Davis (Eds.) *Disasters and the Small Dwelling: Perspectives for the UN IDNDR*. New York: Routledge.

El-Masri, S. and P. Kellett (2001) "Post-War Reconstruction. Participatory Approaches to Rebuilding the Damaged Villages of Lebanon: A Case Study of al-Burjain," *Habitat International* 25(4): 535–557.

Farnsworth, C. B. (2000) "Building for Disaster Mitigation," *Home Energy Magazine*, January–February: 28–33.

Fothergill, A. and L. Peek. (2015) *Children of Katrina*. Austin TX: University of Texas Press.

Fredrickson, T. (2014, June 25) "Urban Post-Disaster Housing Protoype for NYC by Garrison Architects," in *DesignBoom Architecture*, June 25. www.designboom.com/architecture/urban-post-disaster-housing-prototype-nyc-garrison-architects-06-25-2014/.

GreenTown (2013) *Silo Eco-Home*. Retrieved March 2014, from Greensburg GreenTown: www.greensburg greentown.org/silo-eco-home/.

Innes, J. E. and D. E. Booher. (2004) "Reframing Public Participation: Strategies for the 21st Century," *Planning Theory and Practice* 5(4): 419–436.

Johnson, C. (2007) "Strategic Planning for Post-Disaster Temporary Housing," *Disasters* 31(4): 435–458.

Maskrey, A. (2011) "Revisiting Community-Based Disaster Risk Management," *Environmental Hazards* 10(1): 42–52.

Masterson, J. H., W. Peacock, S. Van Zandt, H. Grover, L. Schwarz and J. Cooper (2014) *Planning for Community Resilience*. Washington DC: Island Press.

National Building Museum (2009) "Transcript: The Louisiana Cottages and Carpet Cottages Project," *Community in the Aftermath*. Washington DC: US Department of Housing and Urban Development.

NDHRAC—Natural Disaster Housing Reconstruction Advisory Committee (2010) *Natural Disaster Housing Reconstruction Plan*. As required by HB2450, 81st Legislative Session. Washington DC: US Department of Homeland Security.

Oliver, P. (1987) *Dwellings: The House Across the World*. Oxford: Phaidon.

Peacock, W. G., N. Dash, and Y. Zhang (2007) "Sheltering and Housing Recovery Following Disaster," 258–274, in H. Rodriguez, E. Quarantelli, and R. Dynes (Eds.) *Handbook of Disaster Research*. New York: Springer-Verlag.

Perkes, D. (2012) *SERRI Project: Prototype Design for Temporary Disaster Housing*. Oak Ridge TN: Oak Ridge National Laboratory.

Pugh, C. (1997) "Changing Roles of Self-Help in Housing and Urban Policies 1950–1996: Experience in Developing Countries," *Third World Planning Review* 19(1): 91–106.

Quarantelli, E. L. (1982) "General and Particular Observations on Sheltering and Housing in American Disasters," *Disasters* 6(4): 277–281.

Resilience Alliance (2007) *Urban Resilience Research Prospectus*. Canberra Australia: CSIRO.

Rosenblatt Naderi, J. (2009) "Post Hurricane Mitigation and Sustainable Design in Key West," 61–69, in K. R. Brooks (Ed.) *CELA*. Tucson AZ: Council of Educators in Landscape Architecture.

Sanderson, H. (2000) *Person-Centered Planning: Key Features and Approaches*. New York: Joseph Rowntree Foundation.

Sherriff, R. (2010) *Linking Disaster Resistance and Energy Efficiency*. Washington DC: Center for Housing Policy.

Teaford, J. (2000) "Urban Renewal and Its Aftermath," *Housing Policy Debate* 11(2): 443–465.

Turner, J. (1972) "Housing as a Verb," 148–175, in J. Turner and R. Fichter (Eds.) *Freedom to Build: Dweller Control of the Housing Process*. New York: Collier Macmillan.

United Nations (1982) *Shelter after Disaster: Guidelines for Assistance*. New York: United Nations.

Wilson, B. (2013). *Rebuilding or Recovery? Considering Sustainability in the Context of Disaster Rehousing*. Unpublished Manuscript. Austin TX: University of Texas at Austin Center for Sustainable Development.

Woods, W. T. (2006) *Hurricane Katrina Improving Federal Contracting Practices in Disaster Recovery Operations: Testimony before the Committee on Government Reform, US House of Representatives*. Washington DC: US Government Accountability Office.

Yang, J. and H. Yi. (2014) "Research Trends of Post Disaster Reconstruction: The Past and the Future," *Habitat International* 42: 21–29.

Zhang, Y., and Peacock, W. G. (2009) "Planning for Housing Recovery? Lessons Learned from Hurricane Andrew," *Journal of the American Planning Association* 76(1): 5–24.

PART V

Contributions of Research to Practice

20

INFLUENCES OF RESEARCH ON PRACTICE

Kenneth C. Topping

Introduction

In what ways have practitioners influenced research and how are they interpreting hazards research? That is the thematic question addressed in Part V of this book, *Contributions of Research to Practice*, which examines the interface between research and practice in the area of disaster planning. Some answers are provided in the following five chapters.

This chapter, *Influences of Research on Practice*, provides an interpretive evaluation of how research has been translated into practice within the context of disaster planning practice. Special attention is given to hazard mitigation and disaster recovery—those areas with which urban planning professionals are most directly involved, in contrast with preparedness and response which are more the realms of emergency managers. In concert with the other chapters in Part V, it addresses key questions that have been asked or have yet to be asked within differing hazard mitigation and disaster recovery contexts:

- How does research influence mitigation and recovery practice, especially at state and local levels?
- In what ways has research either enhanced or failed practice?
- What are some remaining knowledge gaps?
- What are the opportunities and constraints for enhanced communications between researchers and practitioners?

Zhenghong Tang in Chapter 21 *Incorporating Hazard Mitigation into the Local Comprehensive Planning Process*, applies such questions to hazards planning, exploring the benefits of initiatives to integrate local hazard mitigation plans prepared under the Disaster Mitigation Act of 2000 into comprehensive plans, as opposed to preparing such plans as stand-alone documents. He argues that local capacity in hazard risk reduction and community resilience building can be maximized by adopting the local hazard mitigation plan as an integrated part of comprehensive land use plans, functional plans, area plans, capital improvement programs, zoning regulations, and municipal ordinances.

Gavin Smith in Chapter 22 *The Role of States in Disaster Recovery: An Analysis of Engagement, Collaboration, and Capacity Building*, describes the widely varying recovery policies and practices of differing states. Noting experiences of certain states with *ad hoc* post-disaster recovery organizations, he observes that states must work within a scarcity of practical guidance for post-disaster recovery.

The intent of his chapter is to better understand (1) the roles in recovery of state actors, including governors and mid-level managers; (2) the influence of pre- and post-disaster recovery planning at the state level; and (3) identification of state-level lessons learned, how such lessons are shared with other states, and the manner by which such lessons are incorporated into a state's policies for longer term institutional benefit.

Laurie Johnson in Chapter 23 *Recovery Planning with U.S. Cities* , examines planning for long-term recovery at the city level, identifying factors distinguishing recovery planning from normal city planning. Among these are (1) pressures during post-disaster recovery to balance speed vs. deliberation, (2) tensions between opportunities for transformation of the urban environment during reconstruction vs. return to preexisting patterns, (3) requirements for large amounts of funding needed for implementation, and (4) creating the management capacity within local governments to sustain recovery over long periods of time. Noting the relatively recent evolution of pre-event recovery planning, she provides observations regarding needs for further research that takes such factors into account.

John Cooper in Chapter 24 *Reflections on Engaging Socially Vulnerable Populations in Disaster Planning*, addresses the ongoing lack of attention to meaningful public involvement in disaster planning, particularly within disadvantaged communities. Despite recent emphasis on "whole community" planning, many emergency managers have yet to fully embrace meaningful public participation, especially with socially vulnerable groups for whom disaster preparedness capacity can provide access to resources critical to survival and recovery. Even when emergency managers believe in inclusive planning, they may lack the time, capability, or incentives to make it a priority. Using the *Emergency Preparedness Demonstration*, a FEMA-funded program, as a model of participatory planning for disaster in disadvantaged communities, he focuses on critical variables such as leadership skills, effective approaches to increasing awareness and preparedness, capitalizing on community knowledge, and building community connections with disaster agencies.

The following discussion in this chapter addresses the significant relationships between research and practice that have emerged in the past quarter century. Drawing from the literature as well as findings of other Part V authors, it (1) identifies who is involved with mitigation and recovery at various federal, state, local, and private sector levels; (2) outlines key influences on the relationships between research and practice; (3) assesses ways in which research has influenced persons responsible for mitigation and recovery in meeting their responsibilities; (4) summarizes ways research has enhanced or failed practice; and (5) identifies remaining knowledge gaps.

Who Has Mitigation and Recovery Responsibilities?

Within the large and highly decentralized U.S. federal system, hazard mitigation and disaster recovery activities are conducted in coordination with the other key components of disaster management, preparedness, and response. Together, these operate at a variety of levels—federal, state, and local—within the federal system, influenced by a wide range of legislation and administrative directives.

Key among these laws is the *Robert T. Stafford Disaster Relief and Emergency Assistance Act* of 1988, a.k.a. the *Stafford Act*, the nation's basic, comprehensive disaster law under which emergency assistance activities are coordinated among various levels of government both in a bottom-up and top-down manner. Under Stafford Act provisions, a disaster-stricken community requests assistance from its state if locally available resources are insufficient to handle the size and intensity of the impacts. If the state considers its resources to be insufficient, it requests federal assistance. In response to a state request, a presidential disaster declaration (PDD) authorizes deployment of national emergency resources through a top-down coordination system using states as intermediaries between the federal and local agencies (Stafford Act, 1988).

Federal Responsibilities

Within this complex, multi-tiered system, federal employees charged with disaster-related responsibilities work in a variety of agencies. Key agencies for hazard mitigation and disaster recovery include the Federal Emergency Management Agency (FEMA), Department of Housing and Urban Development (HUD), Department of Transportation (DOT), and Small Business Administration (SBA).

FEMA employees are responsible for implementing the Stafford Act, which authorizes major mitigation, preparedness, response, and recovery programs jointly administered by the FEMA and counterpart state organizations. Specific Stafford Act programs include the Individual and Household Assistance Program that provides basic disaster relief and emergency assistance, the Public Assistance Program that provides grants for financing for state and local infrastructure restoration, and the Hazard Mitigation Grant Program that provides grants to mitigate hazards posing risks of future disaster losses (Stafford Act, 1988). These programs can be particularly important for local communities after a disaster.

FEMA employees also are responsible for implementing two other important, formative pieces of federal legislation. One is the *National Flood Insurance Act* of 1968, which established the National Flood Insurance Program (NFIP). This program promotes disaster recovery through private companies' sale of flood insurance, which is re-insured by the federal government. The NFIP also has provided floodplain mapping and incentives for communities to mitigate flood hazards (NFIA 1968).

The other important legislation administered by FEMA employees is the *Disaster Mitigation Act* of 2000, which amended the Stafford Act by requiring states to prepare hazard mitigation plans as a precondition for eligibility for hazard mitigation grants and post-disaster assistance (DMA 2000). DMA 2000 represented a national initiative toward building local capacity for hazard mitigation planning on an unprecedented scale. By July 2009, more than 19,000 local jurisdictions had FEMA-approved local hazard mitigation plans (Schwab and Topping 2010, p. 18).

Federal employees in the other mitigation and recovery-relevant agencies are responsible for implementing emergency provisions within laws authorizing other basic functions, such as housing, transportation, or small business assistance. For example, HUD administers post-disaster recovery provisions of the *Housing and Community Development Act* of 1974, under which billions of dollars of Community Development Block Grant program funds are made available to states and localities basically for low- and moderate-income housing (HCDA 1974).

Similarly, U.S. DOT employees administer the *Federal-Aid Highway Act* of 1956, which supports engineering and construction of basic links in the nationwide freeway and highway transport network (FHWA 1956). After a disaster, DOT provides grants for rebuilding damaged or destroyed segments of the national network within states covered by PDDs. Operated jointly under cooperative agreements with states, the Federal-Aid Highway Act disaster assistance program provides a major stimulus to overall physical and economic recovery in areas devastated by disasters.

SBA employees implement the *Small Business Act* of 1953, which was established to provide low-interest rate loans and other services for small businesses. After a disaster, SBA employees are authorized to make low-interest rate loans to small businesses and to homeowners in areas covered by PDDs (SBA 1953). Such assistance can be an important resource expediting business recovery for small businesses impacted by disasters.

In addition to these primary agencies, many other federal agencies deal with mitigation and recovery issues under laws dating back a similar period. Examples include: the *Flood Control & Coastal Emergency Act* of 1955, administered by the U.S. Army Corps of Engineers (FCCEA 1955); the *Public Works Act* of 1976, administered by the U.S. Department of Commerce (PWA 1976); and *Emergency Planning and Community Right-to-Know Act* (EPCRA) of 1986, which amended the *Comprehensive Environmental Response, Compensation, & Liability Act* (CERCLA or Superfund) of 1980, both administered by EPA (EPCRA 1986; CERCLA 1980).

Also important within the federal disaster management system is the emergence after the 9/11 New York World Trade Center disaster of 2001 and the Hurricane Katrina disaster in 2005 of a series of federal administrative directives issued through the Department of Homeland Security (DHS). These were instituted to more effectively guide vertical and horizontal coordination of multiple statutory disaster-related responsibilities. Key examples of administrative directives related specifically to mitigation and recovery include:

- National Response Plan (NRP), 2004, an inter-agency coordination protocol which introduced Emergency Support Function (ESF) #14–Long-Term Recovery Planning. The NRP was later superseded by the National Response Framework (NRF).
- National Preparedness Goal and Presidential Preparedness Policy Directive–8 (PPD-8), 2011, creating a framework for preparedness, including mitigation and recovery.
- National Disaster Recovery Framework (NDRF), 2011, which incorporated ESF #14 by establishing a federal-state-local framework for recovery, as well as the "whole community" concept which emphasizes engagement of all sectors of the community in disaster preparedness.
- National Mitigation Framework (NMF), 2013, establishing a federal-state-local mitigation framework.

Federal employees involved with disaster programs must function in a highly complex institutional environment, interacting with many state and local counterparts. Incremental layering of federal administrative directives over a fragmented statutory base has tended to reinforce a strong top-down emphasis for integration of disaster-related planning systems at the federal, state, and local levels. Federal employees who administer mitigation and recovery laws and systems are responsible for activities involving billions of dollars in grants and loans. Such responsibilities require careful attention to vertical federal-state and state-local coordination.

State and Local Responsibilities

States play a potentially important, yet highly variable, role in facilitating disaster mitigation and recovery. Those responsible for disaster-related activities at the state level are employed in a wide variety of departments such as firefighting, law enforcement, housing, transportation, medical and social services, planning, water, and natural resources, as well as emergency management departments tasked with coordination.

States adopt and implement disaster management systems in compliance with federal requirements, usually for financial reasons. States do not wish to reject or ignore disaster management programs that make substantial amounts of federal revenue available. In order to qualify for federal funds, states adopt emergency operations plans, hazard mitigation plans, emergency operations center protocols, mutual aid agreements, and similar measures. However, differences in state history, geography, land area, population size, economic resources, and political culture have led to considerable variation among states regarding the incorporation of mitigation and recovery research into practice.

For example, a planning practice commonly accepted for many years has been adoption of a comprehensive plan as a framework for local planning and development. Yet, as of 2010, less than half of the 50 states had passed laws mandating local adoption of comprehensive plans. Moreover, only 11 states had adopted laws requiring inclusion of a hazards element in these state-mandated comprehensive plans (Schwab and Topping 2010: 24–27).

Local government personnel responsible for mitigation and recovery functions comprise a wide range of professionals, including city managers, county administrative officers, risk managers, emergency managers, fire and police personnel, public works engineers, urban planners, social service workers, as well as operators of special districts for water, wastewater, flood control, and fire. As of

2010, there were approximately 88,000 local government organizations in the U.S., including cities, counties, townships, special districts, and school districts (Schwab and Topping 2010: 18). Local government organizations vary widely in size and staffing, ranging from cities with millions of residents and tens of thousands of professional staff on the one hand, to rural townships with a few hundred residents and a single employee such as a town clerk, on the other.

Private Sector Responsibilities

Persons responsible for mitigation and recovery are also found in the private sector within international, national, regional, and local business organizations, as well as non-profit organizations such as economic development, community service, environmental, and other stakeholder groups. Among the larger firms, corporate contingency planners and risk managers pay closest attention to mitigation and recovery issues.

One private sector component important to advancement of research into practice is the network of consulting firms. These range from large international engineering companies to small, specialized firms with only a few persons. This particular web of expertise, which penetrates all layers of government and the business community, can be an important conveyor of new research findings emanating from both academic and commercial research.

An additional private sector element important to the implementation of research findings is the variety of professional associations with members within various federal, state, and local agencies responsible for disaster management. Such associations are an especially important conduit for transmission of new information on best mitigation and recovery practices emerging from research, as discussed further below.

Tensions between Research and Practice

The dramatic increase in federally declared disasters over the past quarter century (FEMA 2014) has stimulated research on a wide array of hazard mitigation and recovery topics. Research is routinely conducted in academic, government, and private sector settings as part of an ongoing cycle by which the state of hazard mitigation and recovery practice is assessed, potentially desirable changes to practice are identified, research findings and recommendations are disseminated, and practice is monitored for adjustments reflecting possible advances emerging from research.

Communications Gaps

Substantial communications gaps between researchers and practitioners, however, constitute barriers to advancements in application of research findings to practice which help reinforce tensions between the two groups. Although both groups are burdened by specific financial and social constraints, communications gaps are further exacerbated by a variety of underlying factors.

1   *Size and Scope of Practitioner Audiences.* Perhaps most daunting is the size and variety of practitioner audiences targeted by researchers for dissemination of research findings. In light of the array of disciplines reflected at various governmental levels as well as the private sector, effective outreach is a substantial challenge.

2   *Limited Personal Contacts.* Infrequent direct personal contact maintains communications gaps between researchers and practitioners. This is reinforced by existence of separate associations dealing with the same disciplines—one for researchers, another for practitioners (e.g., Association of Collegiate Schools of Planning and American Planning Association). When direct contacts occur, communications are sometimes hampered by tensions arising from misunderstanding or mistrust.

3 *Transmission Issues.* Researchers tend to rely on academic journals as a primary medium of dissemination of research findings and recommendations for practice. The actual audience for such journals is primarily other researchers. Practitioners are less inclined to read academic journals for various reasons such as the use of abstract language, reliance on statistics, or insufficient acknowledgment of practical concerns.

4 *Reception Issues.* Practitioners rely more on professional trade journals as a primary medium by which to access research findings. By giving preference to distilled information put forward in practice-oriented magazines, practitioners miss out on a vast array of research findings as well as a variety of conceptual and technical issues reflected in the research literature.

5 *Limited Practitioner Interest.* Among state and local government practitioners, city and regional planners represent an important influence structuring not only land use but also the associated economic, physical, environmental, and social fabric of American cities. Yet planners have tended to see hazard mitigation and disaster recovery as subjects for which other professionals, such as engineers, fire professionals, and emergency managers, were primarily responsible—a view that has significantly diminished but not disappeared. Although planning practitioners have necessarily become increasingly involved with mitigation and recovery in recent decades due to the growing numbers of federally declared disasters (FEMA 2014), they may still tend to see these as secondary areas of responsibility, or at best, duties to be exercised on an ad hoc, emergency basis, as further discussed below. This view may have tended to reflect perspectives found within some academic schools of urban planning despite evolving knowledge.

Such factors have tended to work in combination to impede progress toward translation of hazard mitigation and recovery research outcomes to practice.

Influences Facilitating Research Applications

Efforts to improve communication between researchers and planning practitioners have grown in recent decades. Key influences bringing research and practice closer together have included federally funded studies, legislation prompted by disasters, and outreach by research centers and professional associations promoting researcher–practitioner contact. Such elements have interacted in various combinations, rather than working separately.

Mitigation and Recovery Studies

A common source for dissemination of research findings potentially applicable to practice has been a series of major studies funded by government and foundations. For example, federally sponsored research programs have comprised an important stimulus for the conduct and dissemination of hazard mitigation and recovery research coordinated through entities such as the National Science Foundation, National Academy of Sciences, National Institute of Building Sciences, and National Institute of Science and Technology, as well as federal operating departments such as FEMA. Hazard mitigation and disaster recovery studies sponsored by such institutions have grown extensively in the past quarter century.

A seminal work in the area of recovery was *Reconstruction Following Disaster*, which provided the first systematic, comparative assessments of post-disaster restoration and reconstruction processes in several North American cities. This study identified various time phases related to near- and longer-term processes experienced during disaster recovery (Haas et al. 1977). A subsequent work by the consulting firm of William Spangle and Associates, *Pre-Earthquake Planning for Post-Earthquake Rebuilding*, explored further the post-disaster time phases, and added the theme of pre-event planning for post-earthquake disaster recovery. The latter was distinguished by inclusion of practice-oriented materials such as a model recovery earthquake ordinance by which local jurisdictions could adapt pre- and post-event recovery planning principles into local regulations that would help organize reconstruction (William Spangle and Associates 1986).

From the practice side, these studies were accompanied by the City of Los Angeles *Recovery and Reconstruction Plan*, a pre-event recovery plan which made direct use of ideas presented by the Spangle Report, applying them to specific responsibilities of city departments. The pre-event recovery plan, which ultimately helped guide rebuilding after the Northridge Earthquake (City of Los Angeles 1994), appears to have been successful in accelerating the rate of housing reconstruction and the incorporation of hazard mitigation into the recovery process (Wu and Lindell 2004).

The theme of pre-event recovery planning introduced by the Spangle Report and the Los Angeles recovery plan was subsequently applied to disasters caused by multiple hazards in a FEMA-sponsored study by the American Planning Association, *Planning for Post-Disaster Recovery and Reconstruction* (Schwab et al. 1998). The latter expanded on practice-oriented materials, including a more generic, annotated, multi-hazard model recovery ordinance for local adaptation (Topping, Chapter 5, in Schwab et al. 1998). An extensive update with online elements, *Planning for Post-Disaster Recovery: Next Generation*, includes a revision of the previous model recovery ordinance (Schwab 2014).

Paralleling the preceding studies was the widely acknowledged NSF-funded study, *Disasters by Design: A Reassessment of Natural Hazards in the United States*. This summarized major research findings on natural hazards from 20 years of research, based on a combination of well-focused academic research and practitioner feedback, including findings capturing more than two decades of NFIP practice outcomes (Mileti 1999).

Also in the practitioner community, FEMA undertook *Project Impact* during the latter part of the 1990s. This program was a bold initiative that funded multi-hazard mitigation projects in communities across the nation, helping to lay the practical foundation for later passage of DMA 2000.

Project Impact also generated attention to the efficacy of mitigation programs, exemplified by the 2005 Multihazard Mitigation Council study, *Natural Hazard Mitigation Saves*. This important study identified benefit-cost outcomes for FEMA mitigation grants approved during the period 1993–2003, confirming mitigation as a worthwhile investment that had an overall loss avoidance ratio of four dollars saved for every dollar invested (Multihazard Mitigation Council 2005; Rose et al. 2007). This overall loss avoidance ratio has been recalculated as six dollars saved for every dollar invested in a recent study by the National Institute of Building Sciences (2017).

Meanwhile, attention has turned to the relationships of hazard mitigation with other elements of local planning, in response to nearly a decade of FEMA-sponsored local hazard mitigation plans following passage of DMA 2000 (Schwab and Topping 2010, Ch. 2). This issue has received increasing emphasis within the context of state hazard mitigation plans.

Legislation Influences

Legislation has been an important impetus for advancement of hazard mitigation and disaster recovery practice at all levels of government, often reinforced by nationally funded studies. Moreover, legislation reflects a policy response to problems arising from negative experiences in large-scale disasters. Passage of the mitigation and recovery laws cited previously has often been preceded by a significant disaster that triggered public questions from key constituencies regarding needed solutions. Congressional hearings often have presented opportunities for quick, focused communication of research findings in a way that captures policy makers' attention and subsequently is translated to practice through new regulations.

An early example of legislation prompting reform of hazard mitigation and recovery practices was the *National Flood Insurance Reform Act* of 1994, which amended the National Flood Insurance Act of 1968. Spurred by Hurricane Andrew, which hit Florida two years before, the National Flood

Insurance Reform Act captured lessons learned from practice under NFIP. Significant changes in this round of reforms included: 1) establishment of Flood Mitigation Assistance (FMA) grants for states and localities, together with 2) the requirement for local governments to adopt flood hazard mitigation plans as a precondition for eligibility of localities to receive such grants (NFIRA 1994). Such changes provided a model for later broadening of planning-grant linkages under DMA 2000 through addition of a requirement for adoption of local multi-hazard mitigation plans as a precondition for local eligibility for mitigation grants.

Insurance aspects of NFIP administration have been the subject of subsequent rounds of flood insurance reform legislation, such as the *Biggert-Waters Act* of 2012, passed not long after Hurricane Irene hit the Northeast in 2011 but before SuperStorm Sandy hit New York and New Jersey in 2012. Biggert-Waters was intended, among other things, to place NFIP on a more financially sound basis by eliminating certain types of policyholder subsidies. However, the *Homeowner Flood Insurance Affordability Act* of 2014, which was passed in the aftermath of Sandy, eliminated some of the more stringent Biggert-Waters insurance financing reforms (Schwab 2014).

Institutional Outreach

A third, and perhaps most important, method for conveying research findings and recommendations potentially advancing practice is outreach by research centers and professional associations promoting communications between researchers and practitioners. A familiar example of the latter is the NSF-funded Natural Hazards Center (NHC) at Boulder, Colorado, which sponsors the annual Natural Hazards Workshop, bringing together disaster management researchers and practitioners in both formal conference sessions and informal face-to-face contact settings.

Also conducting outreach to advance hazard mitigation and recovery practice through systematic communication of research outcomes are such professional associations as the American Planning Association, the American Society of Civil Engineers, the Earthquake Engineering Research Institute, and others. Although largely composed of practicing professionals, these organizations sponsor both academic journals and professional magazines, reaching out to both researchers and practitioners for purposes of improving communications and, possibly, greater mutual understanding. Outreach services include professional magazines by which research can be interpreted to practice, and conferences similar to the NHC's Natural Hazards Workshops. Particularly valuable for convenience of participants are ongoing one- and two-day training workshops, and one- or two-hour web-based seminars.

How Research Has Enhanced or Failed Practice

The preceding discussion has outlined both the means and effectiveness of transmitting research findings and recommendations related to hazard mitigation and recovery practices to practitioner audiences. In this context we explore below ways in which research has either enhanced or failed practice.

Planning practitioners may not be consciously aware of the influence of research on what they do, yet it can be observed on the basis of experience that research continuously and routinely enhances practice by:

1   Correcting factual inaccuracies in the information practitioners have used as a basis for action;
2   Drawing practitioners' attention to new information that provides a foundation for new mitigation and recovery practices;
3   Providing opportunities for integration of mitigation and recovery study findings with professional goals, values, and practices;

4   Enlarging contacts between researchers and practitioners;
5   Stimulating a feedback flow of practitioner responses fueling new research.

In the broadest sense, research has the capacity to refresh practice by infusing new ideas and life into practice, redirecting it towards enhanced goals and values emerging from the research literature.

Observations about how research may have failed practice are somewhat more conjectural and perhaps more subtle. A perception held by some practicing planners is that researchers make minimal efforts to reach out to them to ascertain fundamental reasons underlying their existing practices. Without direct consultation and inquiry, it may be difficult for researchers to truly understand factors influencing translation of emerging hazards and recovery research outcomes into practice, especially if the research, for whatever reason, reflects erroneous assumptions or interpretations of practice.

Additionally, there exists among practicing planners a limited awareness regarding hazard mitigation and disaster recovery as subjects relevant to their professional responsibilities. This may be explainable by the inherent "newness" of these emerging professional areas. As previously suggested, planners appear to be prone to see these as subjects for which emergency managers are responsible. However, studies appear to be missing on the levels of planner awareness of hazards and recovery challenges in the framework of day-to-day professional practice. Awareness of their potential responsibilities in these areas may be growing, but planners may still see these as functions to be exercised on an ad hoc, emergency basis rather than as part of their regular duties.

Failures of Research

Thus, it is reasonable to assert that there are two distinct ways in which mitigation and recovery research, especially that conducted within the context of schools of urban and regional planning, may have failed planning practice. First, as noted above, there is an apparent absence of studies determining levels of interest and awareness among planners regarding potential hazard mitigation and disaster recovery responsibilities within the profession, as well as research into ways to boost such levels of interest if truly found to be lacking.

Second, and perhaps more fundamental, academic schools of urban planning appear to lack a nationally shared academic standard for inclusion of core hazard mitigation and disaster recovery knowledge as required course content in planning curricula. Indeed, only a handful of planning programs in the U.S. appear to provide substantive content on mitigation and recovery at either the graduate or undergraduate levels.

To the extent that academic schools of city planning are responsible for preparing students for professional life, then they need to get out ahead of the growing demand in society for planning practitioners with mitigation and recovery knowledge. By examining levels of professional interest in these subjects, reaching out directly to practitioners, and strengthening core curriculum content, researchers in schools of planning have an opportunity to strengthen the contribution of planners to mitigation and recovery practice.

Planning Practice Comes to the Rescue

As discussed previously, planning practitioners have in the past tended to see hazard mitigation and disaster recovery as subjects for which other practitioners such as engineers, fire professionals, and emergency managers were primarily responsible. In some cases, this view reflected perspectives found within some academic schools of urban planning. With proliferation of larger and more intense disasters, however, levels of interest in mitigation and recovery issues has become greater both among planners and within planning schools.

As evidence of this, the American Planning Association Board in 2014 adopted a substantial document titled *Hazard Mitigation Policy Guide*, which also addresses recovery (APA 2014). This is significant for practitioners in that mitigation and recovery concepts contained within this guide will ultimately work their way into professional certification examinations administered through the affiliated American Institute of Certified Planners. Thus, academic planning departments are challenged to bring core planning curriculum more closely in line with the leading edges of best practices in mitigation and recovery as seen by professional planners. For planning practitioners, this represents an opportunity to guide the research community into useful directions for refinement of both research and practice.

Why is This Important?

City and regional planners are a basic part of state and local government and related consulting services which routinely structure the physical, economic, social, and environmental fabric of our nation's communities. To the extent that they are continuously informed by evolving knowledge derived from both experience and research, opportunities will be enhanced for creating safer, more resilient urban and rural settings designed to minimize natural and human-caused hazards, risk, and vulnerability, and thus reduce long-term disaster losses. Consequently, it is imperative that the years ahead see more consistent inclusion of experience-based hazard mitigation and recovery knowledge in required core curricula for graduate planning schools.

Remaining Knowledge Gaps

Layered among the outcomes of these counterinfluences between researchers and practitioners remain sizeable knowledge gaps on substantive issues of potential interest to both groups. Among areas needing closer examination by both research and practice are: (1) stronger linkages between mitigation, recovery, and comprehensive planning; (2) climate action and adaptation planning; (3) improved understanding of evolving concepts of resilience and sustainability; (4) reversing decay in aging infrastructure; (5) assessing long-term recovery outcomes (especially failure); and (6) broadening stakeholder outreach, communications, and informed engagement in mitigation and recovery.

Conclusions

It has often been observed that researchers and practitioners tend to inhabit "different worlds." There remains a wide gulf in understandings and perspectives between them, reflecting differences in work contexts and daily tasks as well as varying professional motivations. Quite naturally, researchers tend to hold more respect for theoretical aspects of the phenomena they are researching, whereas practitioners are more likely to focus on details of day-to-day planning processes far removed from theory. Practitioners may be reluctant to accept researcher recommendations calling for changes in practice for a variety of reasons, including added cost, disruption of routines, philosophical differences, and/or institutional inertia.

However, any lack of progress in closing communications and knowledge gaps between the two groups can be characterized figuratively as a "two-way street." On the one hand, hazard mitigation and recovery research may impede advancement of practice if it is does not sufficiently recognize practical needs of planners. On the other, planning practitioners may impede translation of research into practice if they view mitigation and recovery as primarily being the responsibility of emergency managers, not planners.

Therefore, both sets of parties need to change their outlooks and approaches so they can be more effective in dealing with the other. Researchers could be more successful in influencing practitioners with their findings by sharpening their sensitivity to practice. Likewise, practitioners could be more effective in dealing with hazard mitigation and recovery challenges by accepting these subjects as relevant to their core professional responsibilities and improving their receptivity to related research findings.

References

American Planning Association. (2014) *Hazard Mitigation Policy Guide*. Chicago IL: Author. Accessed 22 December 2017 at www.planning.org/policy/guides/adopted/hazardmitigation.htm.

Biggert-Waters Flood Insurance Reform Act of 2012. Public Law 112–141, 42 U.S.C. Ch. 50.

City of Los Angeles, Emergency Operations Organization. (1994) *Recovery and Reconstruction Plan*. Los Angeles CA: Author.

Comprehensive Environmental Response, Compensation, & Liability Act (CERCLA or Superfund). 42 U.S.C. §9601 et seq. (1980)

Disaster Mitigation Act of 2000, Public Law 106–390; Mitigation Planning (44 CFR Part 201—Section 201.4–201.7).

Emergency Planning and Community Right-to-Know Act (EPCRA) of 1986. 42 U.S.C. Ch. 116.

Federal-Aid Highway Act of 1956, Section 125, U.S. Code, Title 23: Emergency Repairs.

Federal Emergency Management Agency (2004) *National Response Plan*. Washington DC: Author.

Federal Emergency Management Agency (2011) *National Disaster Recovery Framework*. Washington DC: Author.

Federal Emergency Management Agency (2013) *National Mitigation Framework*. Washington DC: Author.

Federal Emergency Management Agency (2014) *Federal Declared Disasters by Year or State*. Accessed 22 December 2017 at www.fema.gov/news/disaster_totals_annual.fema.

Federal Emergency Management Agency (FEMA). *State and Local Mitigation Planning How-to Guides*. Various dates for ten guides from 2001 to 2003, available online, in print, and as CDROM. Washington DC: Author. www.fema.gov/media-library/collections/6.

Flood Control & Coastal Emergency Act of 1955, Public Law 84–99. 33 U.S.C. 70.

Flood Insurance Reform Act of 2004, Public Law 108–264, 42 U.S.C. Ch. 50.

Florida Department of Community Affairs. (2010) *Post-Disaster Redevelopment Planning: A Guide for Florida Communities*. Tallahassee FL: Author.

Haas, J.E., R. Kates, and M. Bowden (1977) *Reconstruction Following Disaster*. Cambridge MA and London, England: The MIT Press.

Homeowner Flood Insurance Affordability Act of 2014, Public Law No. 113–89, 42 U.S.C. Ch. 50.

Housing and Community Development Act of 1974, Public Law 93–383, 42 U.S.C. Ch. 69 § 5301.

Mileti, D.S. (1999) *Disasters by Design: A Reassessment of Natural Hazards in the United States*. Washington DC: National Academies Press.

Multihazard Mitigation Council. (2005) *Natural Hazard Mitigation Saves*. Washington DC: National Institute of Building Sciences.

Multihazard Mitigation Council. (2017) *Natural Hazard Mitigation Saves 2017 Interim Report: An Independent Study*. Washington DC: National Institute of Building Sciences.

National Dam Safety Act of 2006, Public Law 109–460, 44 CFR §1724.55.

National Flood Insurance Act of 1968, Public Law 90–448, 42 U.S.C. Ch. 50.

National Institute of Building Sciences. (2017) *Natural Hazard Mitigation Saves: 2017 Interim Report*.

National Flood Insurance Reform Act of 1994, Public Law 103–325, 42 U.S.C. Ch. 50.

Presidential Preparedness Policy Directive/PPD-8: National Preparedness. Accessed 22 December 2017 at www.dhs.gov/presidential-policy-directive-8-national-preparedness.

Public Works Act of 1976, Public Law 94–369, 90 Stat. 2359.

Robert T. Stafford Disaster Relief and Emergency Assistance Act of 1988, Public Law 93–288, 42 U.S.C. §5121 et seq.

Rose, A. et al. (2007) "Benefit-Cost Analysis of FEMA Hazard Mitigation Grants," *Natural Hazards Review* 8(4): 97–111.

Schwab, J.C., and K.C. Topping. (2010) Chapter 1, "Hazard Mitigation: An Essential Role for Planners," Chapter 2, "Hazard Mitigation and the Disaster Mitigation Act," Chapter 3, "Integrating Hazard Mitigation Throughout the Comprehensive Plan," in J. Schwab (ed.), *Hazard Mitigation: Integrating Best Practices into Planning*, Federal Emergency Management Agency and American Planning Association, Planners Advisory Service (PAS) Report 560. Chicago: American Planning Association.

Schwab, J.C. (2014). *Planning for Post-Disaster Recovery: Next Generation*. Federal Emergency Management Agency (FEMA) and American Planning Association (APA), Planners Advisory Service (PAS) Report 576. Chicago: American Planning Association.

Schwab, J.C. et al. (1998) *Planning for Post-Disaster Recovery and Reconstruction*. Federal Emergency Management Agency (FEMA) and American Planning Association (APA), Planners Advisory Service (PAS) Report 483/484. Chicago: American Planning Association.

Small Business Act of 1953, 15 U.S.C. Ch. 14A.

Smith, G. and D. Wenger. (2007) "Sustainable Disaster Recovery: Operationalizing an Existing Agenda," 234–257, in H. Rodríguez, E.L. Quarantelli, and R.R. Dynes (eds.) *Handbook of Disaster Research*. New York, NY: Springer.

Topping, K.C. (2010) "Using National Financial Incentives to Build Local Resiliency: The U.S. Disaster Mitigation Act," *Journal of Disaster Research* 5(2): 164–171.

Topping, K.C. (2011) "Strengthening Resilience Through Mitigation Planning," *Natural Hazards Observer* 36(2): 15–16.

Topping, K.C. (1999) "The Planner's Tool Kit – Model Recovery Ordinance" Chapter 5 in Schwab et al. *Planning for Post-Disaster Recovery and Reconstruction*. FEMA–American Planning Association. Planners Advisory Service Report Number 483/484. Chicago: American Planning Association.

Water Resources Development Act of 1974, Public Law 93–251, 88 Stat. 20.

William Spangle and Associates. (1986) *Pre-Earthquake Planning for Post-Earthquake Rebuilding*. Portola Valley CA: Author.

Wu, J.Y. and Lindell, M.K. (2004). "Housing Reconstruction after Two Major Earthquakes: The 1994 Northridge Earthquake in the United States and the 1999 Chi-Chi Earthquake in Taiwan," *Disasters* 28(1): 63–81.

21

INCORPORATING HAZARD MITIGATION INTO THE LOCAL COMPREHENSIVE PLANNING PROCESS

Zhenghong Tang

Introduction

Local jurisdictions play a critical role in hazard mitigation and long-term sustainability. In a local comprehensive planning system, hazard mitigation policies and practices are important in building community resilience and sustainability (FEMA 2013a, 2013b; Godschalk et al. 1998; Schwab 2010; Tang 2008; Tang et al. 2012). This chapter summarizes two categories of local hazard mitigation plans—a stand-alone hazard mitigation plan and an integrated hazard mitigation plan. Although a stand-alone plan is typically developed and administered by emergency managers, integrating hazard planning into a local comprehensive planning framework is increasingly recognized as an effective means of maximizing local capacity in hazard risk reduction and community resilience building. This chapter summarizes the critical steps needed to integrate hazard mitigation into a local comprehensive planning framework comprising local comprehensive land use plans, function plans, area plans, capital improvement programs, zoning regulations and municipal ordinances, and administrative procedures.

The Disaster Mitigation Act of 2000 and Local Hazard Planning

The 1988 Stafford Act required state governments to prepare post-disaster mitigation plans after receiving federal disaster assistance. At the time the legislation was passed, federal hazard mitigation planning focused on post-disaster issues, which typically exemplified a crisis-response model rather than a proactive risk management model. Hazard mitigation planning was conducted at the state level for some specific hazards (e.g., floods, earthquakes) in specific geographic areas (e.g., flood zone areas, faults). However, there were no federal government mandates for local jurisdictions to plan for all possible hazards at the community level.

Twelve years later, the Disaster Mitigation Act of 2000 (DMA 2000) promulgated a set of requirements for state, tribal, and local governments to develop hazard mitigation plans. The extension of hazard mitigation planning requirements to the local level was a dramatic change. Under DMA 2000, local government entities are required to develop and submit local hazard mitigation plans in order to be eligible to receive federal funds for disaster planning and recovery. Local jurisdictions that do not have hazard mitigation plans are ineligible for the federal Hazard Mitigation Grant Program, the Pre-Disaster Mitigation Program, and the Flood Mitigation Assistance Program. Since 2000, over 10,000 local jurisdictions have developed local hazard mitigation plans, and additional jurisdictions are interested in participating in local hazard planning processes.

DMA 2000 pays more attention to annual pre-disaster mitigation grant programs than post-disaster programs, and it asks localities to consider multiple hazards rather than a single hazard in their planning processes. DMA 2000 also encourages a collaborative integrated planning approach to reduce the costs of natural and man-made disasters in terms of casualties, property damage, and economic disruption. The responsibility for hazard mitigation is placed on local governments to assess risks, implement loss reduction measures, and assure the safety and security of critical facilities that must survive a disaster. Both natural hazards (e.g., earthquakes, landslides, avalanches, volcanoes, tornados, hurricanes, floods, fires) and man-made hazards (e.g., fires, explosive devices, firearms, structural collapses, transportation events, industrial events) should be considered in local hazard planning.

Hazard Mitigation in a Local Planning System

Local jurisdictions have the authority and responsibility to adopt proactive mitigation policies and actions that help reduce risk and create safer and more hazard-resilient communities (FEMA 2013b). Hazard mitigation can be integrated naturally into the local planning framework to reduce the long-term risks to people and property from hazards (Fu et al. 2013a, 2013b; Fu and Tang 2013; Godschalk et al. 1998; Schwab 2010). Hazard mitigation in local planning systems is widely viewed by various levels of governments and jurisdictions as a critical step in building community resilience and sustainability (Burby 2005, 2006; Godschalk et al. 1998; Smith et al. 2013; Tang 2008; Tang et al. 2008, 2010, 2011a, 2011b). This paper details two approaches that can be used to incorporate hazard mitigation principles and practices at the local level: the stand-alone hazard plan and the integrated hazard plan (Figure 21.1).

The stand-alone hazard plan focuses only on hazard issues. Stand-alone plans include local all-hazards plans, local emergency management plans, and specific types of hazard plans. A local all-hazards plan identifies all possible hazards, assesses the community's vulnerability and risk from these hazards, and identifies community mitigation strategies. Local emergency management plans

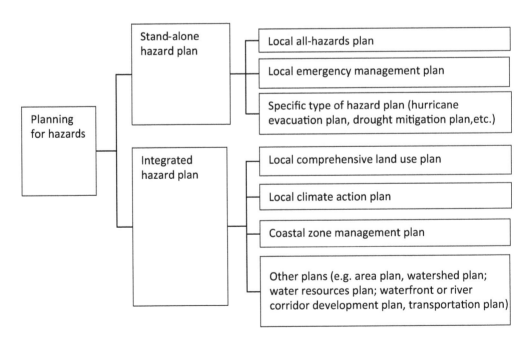

Figure 21.1   Stand-Alone Hazard Plans and Integrated Hazard Plans

mainly focus on community emergency preparedness and response. The specific type of hazard plan emphasizes both a specific type of hazard and a specific phase of hazard management (e.g., mitigation, response, recovery, or preparedness). Examples include a local hurricane response plan, a tornado recovery plan, or a local drought preparedness plan. Stand-alone hazard plans can include a thorough examination of specific hazards and extensive details on specific hazards or emergency management phases. They tend to lead to high-quality planning documents in hazard mitigation, response, recovery, or preparedness (Tang et al. 2008). However, stand-alone hazard plans require a separate planning process within local land use planning. Specific resources and personnel are necessary to develop and implement these plans. Stand-alone hazard plans have only superficial linkages to other community issues (e.g., land use, housing, transportation) and lack depth in treatment of hazards within an existing planning framework (e.g., zoning regulations, design standards, site review requirements).

Integrated hazard planning, on the other hand, incorporates hazard mitigation principles and practices into an existing planning framework, such as local comprehensive land use plans (Tang 2008; Tang et al. 2008), local climate action plans (Tang et al. 2013), local coastal zone management plans (Tang et al. 2011a), and other types of plans (e.g., watershed plans, water resources plans, transportation plans). These planning efforts are not specifically for hazard mitigation, but their planning processes and platforms offer tremendous opportunities to incorporate hazard mitigation and adaptations into these plans. Integrated hazard planning provides valuable opportunities to build local capacity for resilient communities. The integrated hazard planning model is increasingly viewed as an effective approach to collaborate with community stakeholders and engage in community development strategies (FEMA 2013a). This chapter specifically explains the steps for integrating hazard mitigation principles and practices into local comprehensive planning frameworks.

Opportunities for Integrating Hazard Mitigation into Local Planning Frameworks

In local land use planning frameworks, multiple planning opportunities can be considered in the integration of hazard mitigation with local land use development. Schwab (2010) notes that hazard mitigation can be fully integrated with four types of local plans—local comprehensive plans, area plans (e.g., subarea, neighborhood), functional plans (e.g., sewer and water, stormwater management), and operational plans (e.g., emergency operations plans). The Federal Emergency Management Agency (FEMA 2013a) suggests a full-extent integration structure that includes a local comprehensive land use plan, area plans, function plans, zoning ordinances and municipal codes, site review criteria, capital improvement programs, stakeholder and public engagement plans, and other aspects of local planning activities.

This study proposes three different layers of integration of hazard mitigation principles and practices into local land use planning and implementation. The first layer is the local comprehensive plan, the second layer comprises function plans and area plans, and the third layer includes the capital improvement program, zoning regulations and municipal ordinances, and administration procedures (Figure 21.2).

The Local Comprehensive Land Use Plan

This plan serves as a blueprint for future development and provides a proactive platform in preparing for potential community challenges. Over 20 states in the United States have mandated local governments to develop comprehensive land use plans and most have also required hazard elements within these plans (Schwab 2010). Local comprehensive land use plans play an integral role in promoting hazard resilient development (Schwab 2010). The major reasons for integrating hazard mitigation principles and practices into local comprehensive planning can be summarized in terms of four principles—consistency, holism, comprehensiveness, and collaboration.

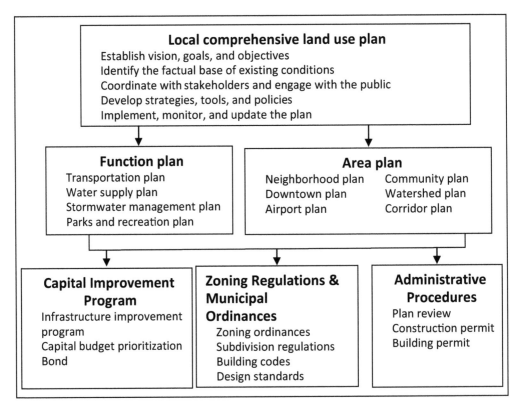

Figure 21.2 Opportunities for Integrating Hazard Mitigation into the Local Planning Framework

Consistency is important because local planning systems normally require consistency in their land use policies. Local comprehensive land use plans should be internally consistent and consistent with other planning documents (e.g., functional plans, area plans, capital improvement programs, regulations and ordinances, and administrative procedures) in translating a hazard-resilient vision into all of the community's planning activities. The policies and strategies outlined in local comprehensive plans can leverage available financial resources and capital improvement programs to improve hazard mitigation and adaptation practices.

Holism is important because almost all critical community sectors (e.g., land use, housing, economic development, transportation and infrastructure, open space, environment, conservation) are naturally involved in a local comprehensive planning framework. Holistic comprehensive planning creates strategic opportunities to engage decision makers and citizens to find feasible solutions for hazard risk reduction. Decision makers and citizens can think holistically about the challenges facing a community, how to allocate the available resources, and how to engage stakeholders toward desired community goals.

Comprehensiveness is important in identifying hazardous areas, balancing future development, guiding development towards less vulnerable areas, setting development standards for hazard reduction, and educating the population through public participation (FEMA 2013a). A comprehensive planning process also can provide opportunities to incorporate hazard risks into the stages that occur after plan adoption—implementation, monitoring, evaluation, and updating.

A collaborative framework is important because a joint planning process can identify and resolve stakeholders' (e.g., planners, emergency managers, developers, engineers, and citizens) conflicting goals at an early stage, leading to the development of more feasible solutions in pre-disaster planning. Stakeholders are engaged in this collaborative framework to reduce the risk of potential hazards by avoiding uncoordinated planning. This framework provides an ideal channel to integrate mitigation policies, strategies, codes, standards, programs, and administrative procedures to achieve the community goal of hazard resilience.

Function Plans

Function plans provide an avenue to integrate best hazard management practices into specific functions or services, such as transportation, water supply, stormwater management, or parks and recreation. This can avoid the problem of different community functions exacerbating hazard vulnerability, such as constructing city offices and other facilities in flood plains. Function plans help build a larger capacity of local resiliency beyond land use elements to other municipal services.

Area Plans

Area plans address specific geographic areas such as neighborhoods, central business districts, watersheds, and transportation corridors that have high levels of hazard exposure. Area plans must be consistent with local comprehensive plans in terms of land use and regulations. They can help translate jurisdictional vision and goals to local communities, neighborhoods, or special districts.

Capital Improvement Programs

These programs can prioritize capital expenditures in order to implement hazard mitigation measures, as well as guide these expenditures to be consistent with resiliency goals and strategies (FEMA 2013a). Many capital improvement programs (e.g., fire stations, utility lines, wastewater treatment plans) serve vital functions for community operations and need to be protected from hazard risks. Capital improvements can encourage hazard mitigation components in new projects and promote safer development patterns without decreasing tax revenues. Capital improvement programs also stimulate private investment and support new development in safe areas through the review of potential impacts of proposed project improvements on hazard vulnerability.

Zoning Regulations and Municipal Ordinances

Zoning ordinances and municipal regulations govern land use development and guide redevelopment location, type, density, and patterns to areas with less risk. Subdivision requirements can mandate that construction and land parcels be located away from known hazards and encourage adoption of best practices in hazard mitigation. Zoning ordinances can restrict development in known hazard areas and encourage growth in safe locations (Schwab 2010). Subdivision requirements and zoning ordinances can avoid exacerbation of hazard risks in new development and construction practices (FEMA 2013a). Moreover, building codes and design standards can prescribe that new construction and renewable projects be more hazard resistant (Burby and May 1999).

Administrative Procedures

Administrative procedures include site plan review and construction and building permitting. Site plan review provides a process whereby community planners examine development plans before construction in order to avoid proposed development in an area with known hazards and ensure consistency with the safe growth goal in a local comprehensive plan. The plan review procedure normally includes specific standards for evaluating risks and has certain requirements for safe growth best practices. These administrative procedures essentially double-check the hazard mitigation requirements as a condition for permitting the development (FEMA 2013a). They also establish requirements for daily planning decisions to be consistent with the long-term goal of hazard resilience.

Steps for Promoting Integrated Hazard Planning

There are five steps involved in promoting the integration of hazard mitigation into the local comprehensive planning framework: (1) engage stakeholders and the general public in local integrated hazard planning, (2) set community goals and objectives for hazard resilience, (3) assess community hazard conditions, (4) develop hazard mitigation and adaptation strategies, and (5) monitor implementation progress and update plans.

Engage Stakeholders in Local Integrated Hazard Planning

Stakeholder coordination and public engagement is a crucial step to integrate hazard mitigation into local comprehensive plans. The local comprehensive land use planning process provides a platform to bridge different constituencies and maximize the expertise of multiple technical professions. All stakeholders should be included who either have the expertise to develop the plan or have the authority to implement it. The key stakeholders in community hazard planning include community leaders and decision makers such as the mayor, county manager, city council members, county commissioners, planning commission members, city attorney, and finance director. Other stakeholders are departmental managers and experts such as emergency managers, planners, public works managers, transportation engineers, natural resources managers, economic development experts, public utility managers, police chief, fire chief, and emergency medical services director. It is also important to include stakeholders such as community leaders from partner agencies such as school administrators, neighborhood association leaders, and representatives from downtown districts, neighboring jurisdictions, metropolitan planning organizations, chambers of commerce, or special districts. In addition, the planning process should include technical experts (builders, developers, engineers, consultants, environmental professionals, Geographical Information Systems specialists), as well as representatives from development groups and non-governmental organizations (NGOs). Finally, meetings should be open to the general public so that everyone can attend who will be, or thinks they will be, affected by the plan.

Three different tiers of involvement can be identified in a local integrated hazard planning process (FEMA 2013b). The first tier comprises the core leadership team including community planners, emergency managers, GIS specialists, public works staff, engineering staff, elected and appointed officials, and floodplain administrators. The second tier includes other stakeholders such as regional planning organizations, city councils, boards of commissioners, planning commission members, and state and federal partners. The third tier is the general public. Stakeholder engagement and public participation increase community capabilities for reducing long-term hazard risks through improving institutional capacity (e.g., regulations, policies, administrative procedures), technical capacity (e.g., shared hazard datasets, more accurate hazard maps), financial capacity (e.g., prioritized hazard mitigation programs or projects, bonds), and education and outreach capacity (e.g., community hazard awareness programs).

Set Community Goals and Objectives for Hazard Resilience

Local community leaders, professional experts, and citizens can shape their communities' long-term goals and specific objectives. For example, community plans can include the following general goals:

(1) protect public safety and prevent injury and loss of life; (2) reduce property damage and economic loss; (3) protect community resources and assets from hazard impacts; and (4) build community capacity for hazard resilience. These goals are broad policy statements as well as community strategic visions. The objectives defining the goals should be more specific and represent the current needs and values of the community. The objectives also need to be more measurable so that community stakeholders and citizens can track progress toward meeting them.

The vision, goals, and objectives need to be agreed upon by the leadership team, elected officials, key stakeholders, and the public. A review of existing plans and other policy documents will identify, and ideally eliminate, conflicts with the proposed hazard mitigation goals and objectives. In addition, the proposed hazard-resilience goals and objectives must be incorporated into local comprehensive plan elements such as economic development, growth management, environmental preservation, historic preservation, parks and recreation, land use and zoning, and transportation. Lastly, the proposed goals and objectives need to reflect community realities and conditions.

The public's involvement in developing the community's goals and objectives is a critical factor to ensure its support for the planning process and the final planning products. Consensus-building between key stakeholders in government and the general public is important in order to maximize overall agreement with the proposed goals and objectives.

Assess Community Hazard Conditions

In order to build a more resilient community, a necessary step is to identify and profile the community's current hazard vulnerability. Three stages of assessment are crucial to analyze hazard conditions at the community level: hazard identification, vulnerability assessment, and risk analysis (Deyle et al. 1998; UNISDR 2011; Wisner et al. 2004). The section below illustrates the three stages of community hazard analysis.

Hazard Identification

Hazard identification examines the geographic extent of natural and technological/manmade hazards in terms of the probability of different intensities throughout the community. Hazard identification systematically documents the type, frequency of previous occurrence, magnitude and potential intensity, location, speed of onset, probable spatial extent, duration, seasonal pattern, and availability of protective responses such as warning and evacuation. In the first stage of hazard identification, planners need to collect first-hand information from multiple sources, typically via telephone or face-to-face interviews with various local, regional, and national hazard experts. Other information can be gathered from a review of local or regional historical hazard records, existing plans and reports, local newspapers, government documents, and online public-access information.

Once a hazard has been identified, it should be analyzed first for geographic extent and then for the probabilities of different event magnitudes (e.g., the recurrence interval for different levels of flooding). When considering hazard probabilities, planners should recognize that some hazards occur seasonally (e.g., flooding, hurricanes), whereas others lack any periodic character (e.g., dam failure, terrorism). With regard to magnitude, the intensity of each hazard is significantly different with respect to damage or loss of life. A matrix is helpful to establish the relationship between the experienced frequency of a hazard event and the potential impacts that may result.

In order to ensure an effective consensus-building process, hazard identification can begin with those hazards that are more frequent and also have a greater adverse impact on life and property. Based on the different probabilities and potential impacts, hazard identification can label hazard probability as high (once a year), moderate (once every two to ten years), and low (once every ten to fifty years). Similarly, hazard impact can be classified as high (major loss of life, significant property damage and economic impact), moderate (few lives lost, some injuries, and some property damage), or low (no loss of life, few injuries, and little damage). When hazard identification is completed, planners can produce a matrix of all hazards' probability of occurrence and impact intensity. Additionally, they can develop a series of maps for estimating the geographic extent of the hazards.

Vulnerability Assessment

Vulnerability assessment involves the evaluation of the community's susceptibility to physical (casualties and damage) and social (psychosocial, sociodemographic, socioeconomic, and political) impacts (Lindell and Prater 2003). In turn, vulnerability can be assessed for different population segments (especially low-income households), economic sectors (manufacturing and retail), infrastructure types (water, sewer, electric power, fuel, telecommunications, and transportation), and buildings (residential, commercial, industrial). It is also important to consider the vulnerability of other important community assets such as environmental and historical resources. Vulnerability assessment extends the results of the hazard identification by providing a detailed inventory of the people and property that are likely to be affected by a disaster. It is important to examine differences in hazard vulnerability among population segments because a disaster can cause significantly different degrees of impact to groups of people that differ in their vulnerabilities (Bolin and Stanford 1999). In a vulnerability assessment, an inventory with maps can illustrate the population segments, buildings, and economic sectors at risk.

Two approaches have been used for vulnerability assessment. The first approach is typified by the HAZUS model (www.fema.gov/hazus). The HAZUS model has been developed by FEMA to identify the boundaries of high-risk locations and estimate the impacts of earthquakes, floods, hurricanes, and tsunamis. Specifically, planners can use HAZUS in combination with building inventory data and Census data to estimate the physical damage to buildings, infrastructure, and critical facilities; economic losses such as the costs of building repair, business interruption, and job loss; and social impacts such as the number of displaced households and requirements for accommodations in public shelters.

A second approach to vulnerability assessment is the Social Vulnerability Index (SoVI—Cutter et al. 2003), which uses data from the US Census to assess the distribution of vulnerable groups within geographical areas such as Census divisions (blocks, block groups, and tracts) and higher levels of aggregation (cities, counties, and states). SoVI's statistical analysis of the Census data yields broad factors (e.g., wealth, race, and poverty, Hispanic immigrants, age, gender, and race—Schmidtlein et al. 2008, Table II) that can then be weighted and combined into the overall index of social vulnerability for each geographical area. This allows planners to identify the areas within their jurisdictions that have the highest concentrations of population segments that are likely to experience the greatest adverse effects of a disaster.

Each of these two approaches—SoVI and HAZUS—has its own strengths and weaknesses. A HAZUS analysis focuses primarily on identifying the geographical areas of greatest physical (casualties and damage) and economic impact, and only secondarily on an assessment of other social impacts—especially the distribution of impacts over population segments. Conversely, a SoVI analysis focuses primarily on identifying the geographical areas that are likely to need the most assistance *if a disaster strikes that area*. SoVI does not assess hazard exposure (the probabilities of disasters of different intensities) or physical (casualties and damage) or economic impacts. Despite their differences, both types of vulnerability analysis are geo-references, so their results can be depicted in maps that show planners which geographic areas are likely to need special attention.

Risk Analysis

Risk analysis can quantitatively assess the probability distribution of event magnitudes at a given location (or area) within a given time period. This information can then be combined with physical and social vulnerability analyses to produce a spatiotemporal distribution of the physical and social impacts of a particular hazard. By incorporating information about the probability of an event of a given magnitude within a specific time period, risk analysts can provide a more precise, detailed, site-specific identification of the geographic areas, structure types, and population segments that need to be addresses in their hazard mitigation and disaster recovery plans.

Risk analysis needs a sophisticated quantitative analysis to assess impacts at multiple scales, which can require a large amount of data. For example, detailed analyses of earthquake recurrence interval data and accurate geological maps are useful for seismic risk analyses. However, some risk analyses suffer from inadequate data due to limited historical records. This, in turn can produce significant levels of uncertainty in estimates of the probabilities of the occurrence and consequences of extreme events.

Many risk analyses involve experts from different disciplines but local community citizens, planning or environmental agencies, NGOs, and media may also be able to contribute, as well. Although professional and scientific experts should be primarily responsible for determining hazard probabilities and consequences, the other stakeholders should be involved to be sure the risk analyses address their information needs. Furthermore, policy decision-makers need to fully consider expert opinions and public interest, and then propose risk mitigation initiatives such as property acquisition or conservation tools.

Develop Hazard Mitigation and Adaptation Strategies

Hazard mitigation and adaptation strategies commonly divided into structural approaches and non-structural approaches (Godschalk et al. 1999). A structural approach uses engineered protection works, as in the case of dams and levees to reduce flood impacts. A non-structural approach uses non-engineering approaches such as land use practices and building construction practices to reduce hazard risks. A local comprehensive land use plan is a perfect example of a non-structural strategy because it addresses many areas in which a community can reduce its hazard vulnerability. In general, a local comprehensive land use plan comprises multiple elements including land use, transportation, housing, environmental, and natural resources conservation, open space and recreation, historical preservation, community development, and public facilities and services. In addition, it also includes a description of the community's condition; the community vision and overall goals, policies and strategies; and implementation procedures. Hazard mitigation principles and practices can be integrated into many of these elements (May and Deyle 1998; Olshansky and Kartez 1998). A non-structural approach can enhance the community ecosystem capacity to reduce impacts from natural hazards. This study summarized the previous literature (Berke et al. 2009; Brody et al. 2009; FEMA 2013a; Godschalk et al. 1998, 1999; Peacock 2003; Peacock and Husein 2011; Tang et al. 2011a, b, 2013) and proposed 15 categories of strategies that can be used to incorporate hazard risk reduction into local comprehensive planning system. The detailed hazard mitigation and adaptation strategies for local comprehensive land use plans are summarized in Table 21.1. (1) Land use development regulations; (2) Zoning ordinances and municipal codes; (3) Building codes and design standards; (4) Regulations and limitations for shoreline, waterfront, riverfront development; (5) Natural system protection; (6) Education and public awareness programs; (7) Local incentive programs; (8) Federal incentive programs; (9) Property acquisition programs; (10) Financial tools; (11) Critical public and private facilities policies; (12) Public-private sector initiatives; (13) Hiring professionals; (14) Public and stakeholder engagement; and (15) Structural and infrastructure practices.

Table 21.1 Mitigation and Adaptation Strategies in Local Comprehensive Land Use Plans

Category	Strategies for hazard mitigation and adaptation
Land use development regulations	Hazard overlay regulations; Planned unit development; Project review and permitting; Stormwater management regulations.
Zoning ordinances and municipal codes	Zoning ordinances; Subdivision ordinances; Agricultural or open space zoning; Performance zoning; Hazard setback ordinances; Form-based zoning.
Building codes and design standards	Building codes; Wind hazard resistance for new homes; Flood hazard resistance for new home construction; Retrofit for existing buildings; Special utility codes.
Regulations and limitations for shoreline, waterfront, riverfront development	Limitation of shoreline development to water-dependent uses; Restrictions on shoreline armoring; Restrictions on dredging/filling; Dune protection; Vegetation protection.
Natural system protection	Watershed management; Wetland restoration and protection; Habitat protection/restoration; Protected areas; Erosion and sedimentation control; Open space preservation; Stream corridor restoration.
Education and public awareness programs	Public education for hazard mitigation; Citizen involvement in hazard mitigation planning; Seminar on hazard mitigation practices for developers and builders; Hazard disclosure; Hazard zone signage; Electronic information channels (webpage, social media); Radio or television spots.
Local incentive programs	Transfer of development rights; Density bonuses; Clustered development.
Federal incentive programs	National Flood Insurance Program; FEMA Community Rating System; Hazard Mitigation Grant Program; Pre-Disaster Mitigation Program; Flood Mitigation Assistance Program.
Property acquisition programs	Fee simple purchases of undeveloped lands; Acquisition of developments and easements; Relocation of existing structures out of hazardous areas.
Financial tools	Tax increment financing for implementing mitigation measures; Lower tax rates; Special tax assessment; Impact fees or special assessments; Capital Improvement Plans; Capital Infrastructure Programs.
Critical public and private facilities policies	Requirements for locating public facilities and infrastructure; Requirements for locating critical private facilities and infrastructure; Using municipal service areas to limit development.
Public–private sector initiatives	Land trusts; Public–private partnerships.
Hiring professionals	Identify suitable building sites; Develop special building techniques.
Public and stakeholder engagement	Engage key decision makers from federal, state, regional, and local agencies; Convene a resilient community advisory committee; Identify champions; Input public and stakeholder comments in plans.
Structural and infrastructure practices	Acquisitions and elevations of structures in hazard prone areas; Utility underground; Structural retrofits; Floodwalls and retaining walls; Detention and retention structures; Culverts; Safe rooms; Levees; Dams; Dykes.

Monitor Implementation Progress and Update Plans

To ensure implementation effectiveness, community planners must monitor, evaluate, and update their plans to reflect dynamic changes in local development conditions and implementation progress toward mitigation goals. The key responsible agencies need to be invested in implementing the planned actions identified in local comprehensive plans. The successful implementation of the hazard mitigation elements in a local comprehensive plan depends on several factors, including financial commitment, technical assistance, materials, and staff commitment. Planners can use a Gantt chart to coordinate efforts among stakeholders by listing the partners that are responsible for each task and checking off completed milestones to track progress toward implementation. This monitoring process can help promote continued engagement in the implementation process.

Long-term monitoring and continual updates of the plans can help community planners re-evaluate priorities and re-allocate resources when conditions change in the community. As changes in conditions occur, adjustment of mitigation actions is necessary to reflect the new realities. FEMA (2013b) suggested a set of criteria to consider when considering the need for these adjustments—life safety, property protection, technical feasibility, political support, legal authority, environmental impacts, social impacts, administrative capability, local champions, community objectives, and budget. The public should be involved in plan maintenance and updates.

Discussion and Policy Implications

In recent decades, many studies have advocated the integration of hazard mitigation into the local comprehensive planning system (FEMA 2013a; Godschalk 2009; Schwab 2010; Tang 2008). Although the value of integrated hazard planning has been increasingly recognized by decision makers and the general public, the experience and lessons learned from the academic literature and professional practice need to be considered.

Process vs. Product

The *process* of hazard planning is as equally important to the final *product*—the hazard plan itself (FEMA 2013b). Effective integration of hazard mitigation into a local comprehensive plan should reflect a well-organized planning document that has been produced by a fully engaged planning process. In the past, hazard planners have paid more attention to the product (the plan) rather than the planning process itself, but a balance between the two should be considered necessary. DMA 2000 requires that the planning process be clearly documented in local hazard plans. Integrated hazard planning needs to adjust its focus on both "planning for lands" and "planning for people." Traditionally, the focus has been predominately on physical hazard planning rather than the social planning process. The final plan itself is usually viewed as "a hard product" and the planning process is always treated as "a routine working procedure." The priority given to the plan can be seen in the fact that there are no criteria for evaluating planning process quality that are equivalent to the criteria for assessing plan quality. A fully engaged planning process with community planners can improve final product quality (Lyles et al. 2014). More research is needed to learn how to effectively move from a sound "planning process" to a high quality "planning product."

Planned Agenda vs. Implementation Actions

A high-quality hazard mitigation plan does not mean that its provisions will be well implemented. The mitigation policies and strategies listed in local plans are not a sufficient condition for implementing these mitigation practices in the community. The effectiveness of any implementation action is subject to financial conditions, institutional capacity, and political structures, so disconnects are often found between the initially planned actions and actual implementation. Although recent research has greatly improved our understanding of mitigation policies adopted in local plans (Berke et al. 2006; Brody and Highfield 2005; Brody et al. 2009, 2010, 2011; et al. 2013), little is known about how to smoothly convert planned actions into effective implementation practices. One aspect of this problem is that few tracking systems exist to compare how well-planned items were successfully implemented. Godschalk et al. (1989) analyzed mitigation practices (acquisition of development rights, disclosures, and educational seminars) and found that some have relatively higher adoption rates than others. Peacock and Husein (2011) conducted surveys to detect the adoption and implementation ratios of hazard mitigation policies and strategies by coastal jurisdictions in Texas. Ge and Lindell's (2016) study of planners from counties in the five U.S. Pacific states found that they viewed effectiveness (a desirable attribute) as positively correlated with economic costs and other impediments (which are undesirable attributes). Consequently, these planners must make trade-offs among these attributes and choose the most appropriate tool when formulating community plans. These studies help fill in the gap between the planned agenda and the actual implemented actions.

Crisis Management (Planning for Emergency) vs. Risk Management (Planning for Resiliency)

Crisis management is still a crucial focus of local hazard planning systems. Many stand-alone hazard plans (e.g., hurricane emergency response plans) are suitable for responding to emergencies but cannot prevent the emergencies from occurring in the first place. By recognizing the limitations of the crisis management model, planners are increasingly moving from a short-term crisis response approach to a long-term risk management approach. Land use planners have started to pay more attention to long-term hazard risk reduction and community resilience building. Although significant progress has been made, a large gap still exists between crisis management and risk management. Emergency managers still lead mitigation planning efforts in many communities despite the fact that planners have stronger qualifications in community development and incorporating hazard mitigation into local comprehensive plans. Consequently, only limited efforts have been made to integrate planning tools, such as local comprehensive land use plans and development ordinances, with local all-hazard plans. Local land use planners and community development leaders still have limited participation and leadership in local hazard planning because many of them regard these efforts as falling outside their professional purview (FEMA 2013a). Nonetheless, both emergency managers and community planners share the responsibility to determine what shared long-term values and feasible risk management strategies work best for their communities (Schwab 2010). Partnerships between emergency managers and land use planners should be promoted by establishing a shared mission and responsibility.

Local vs. Regional Plans

Most hazards span jurisdictional boundaries, so many hazard planning efforts require the cooperation of multiple jurisdictions at a regional scale. At the same time, many smaller jurisdictions have limited resources and technical capabilities to conduct hazard planning (Tang et al. 2011a, 2011b). These small jurisdictions lack planning staff and experts in community risk assessment and mitigation planning. A regional planning approach is perfectly suitable for multiple jurisdictions in order to share resources, reduce planning costs, and avoid duplication of planning efforts. A regional platform offers a feasible framework to engage multiple jurisdictions in community hazard mitigation planning. Regional organizations, such as Metropolitan Planning Organizations, Councils of Government, or Natural Resources Districts, can coordinate land development planning to achieve more hazard resiliency goals at a larger scale and utilize resources more efficiently. A regional coordination platform enables comprehensive mitigation strategies to be implemented as joint planning efforts.

Although the benefits of regional integrated hazard planning are well recognized, challenges still exist in using a regional approach to mitigate hazard risks. Under a regional framework, some local jurisdictions may lose control of the process because of varied capacities, priorities, political will, and/or institutional experiences. A strong governance mechanism is required to balance diverse interests and priorities and deal with the large cross-boundary datasets needed for planning.

Safe Growth vs. Smart Growth

The safe growth audit (Godschalk 2009; Schwab 2010) is a proactive planning approach to balance land development, public safety, and long-term hazard risk management at the community level. It provides a systematic way to fully engage planning policies and tools in hazard-resilient development. The safe growth concept incorporates hazard mitigation principles and practices into local plans, policies, ordinances, and procedures, which can reduce long-term risks from hazards. Safe growth is different from smart growth but the two are not necessarily mutually exclusive; in fact, they are essentially consistent in their principles and practices. Smart growth principles emphasize compact development and open space preservation—strategies that can be "safe" in helping to reduce physical vulnerability to hazards. However, the concepts of safe growth and smart growth (as well as New Urbanism) are still disconnected in planning practices, so there are cases in which smart growth and New Urbanism can produce concentrated growth in hazard-prone areas (Berke et al 2009). Consequently, hazards researchers have recommended a more proactive land use planning approach to integrate hazard mitigation techniques in New Urban project site designs (Stevens et al. 2010a, 2010b).

Conclusions

This chapter reviewed the influence of DMA 2000 on local hazard mitigation planning and compared two categories of hazard mitigation plans: the stand-alone hazard mitigation plan and the integrated hazard mitigation plan. Five steps were proposed to promote the integration of hazard mitigation into the local comprehensive planning framework: (1) Engage stakeholders and the general public in local integrated hazard planning; (2) Set community goals and objectives for hazard resilience; (3) Assess community hazard conditions; (4) Develop hazard mitigation and adaptation strategies; and (5) Monitor implementation progress and update the plan.

The experiences and lessons from recent research indicate that integration of hazard mitigation into the local comprehensive planning framework can increase public awareness and leverage resources to achieve shared hazard-resilient goals. The partnerships established in an integrated planning process offer a collaborative platform to expand community capacities with feasible solutions in hazard risk management. A proactive planning approach can shift the balance of land use projects and capital improvement programs toward more resilient development. Through joint planning efforts, local planners will be able to create hazard-resistant communities that provide current and future generations with a full range of choices and opportunities.

Recent research has highlighted a number of obstacles that impede the integration of hazard mitigation into local comprehensive plans (FEMA 2013a; Tang et al. 2012; Tang et al. 2013). *Motivational barriers* arise from low awareness of hazards, low motivation for hazard mitigation, and low political will to implement mitigation solutions. *Institutional barriers* include a localized vision or perspective regarding hazard risks, low priority for hazard mitigation in planning agendas, weak incentives for integrated planning, and insufficient frameworks for intergovernmental collaboration. *Resource barriers* include the lack of technical experts in community risk assessment and lack of financial resources.

To address these obstacles, this chapter makes the following suggestions for integrated hazard mitigation planning. First, as much emphasis should be placed on the process of hazard planning as on the final product of the hazard plan itself. Second, planners need more effective roadmaps to translate adopted plans to implemented practices. Third, regional planning coordination should be adopted as a useful approach to engage small communities and spur the cooperation of multiple jurisdictions in dealing with cross-boundary hazards. Finally, safe growth principles and practices need to be essentially fused into the early stage of community development planning and site-design.

References

Berke, P. R., M. Backhurst, L. Laurian, J. Crawford, and J. Dixon. (2006) "What Makes Plan Implementation Successful? An Evaluation of Local Plans and Implementation Practices in New Zealand," *Environment and Planning B: Planning and Design* 33(4): 581–600.

Berke, P. R., Y. Song, and M. Stevens. (2009) "Integrating Hazard Mitigation into New Urban and Conventional Developments," *Journal of Planning Education and Research* 28(4): 441–455.

Bolin, R. and L. Stanford. (1999) *The Northridge Earthquake: Vulnerability and Disasters*, New York: Routledge.

Brody, S. D. and W. E. Highfield. (2005) "Does Planning Work? Testing the Implementation of Local Environmental Planning in Florida," *Journal of the American Planning Association* 71(2): 159–175.

Brody, S. D., Z. Sahran, S. P. Bernhardt, and J. E. Kang. (2009) "Evaluating Local Flood Mitigation Strategies in Texas and Florida," *Built Environment* 35(4): 492–515.

Brody, S. D., J. E. Kang, S. Bernhardt. (2010) "Identifying Factors Influencing Flood Mitigation at the Local Level in Texas and Florida: The Role of Organizational Capacity," *Natural Hazards* 52(1): 167–184.

Brody. S. D., W. E. Highfield, and J. E. Kang. (2011) *Rising Waters: The Causes and Consequences of Flooding in the United States*, New York: Cambridge University Press.

Burby, R. J. (2005) "Have State Comprehensive Planning Mandates Reduced Insured Losses from Natural Disasters?" *Natural Hazards Review* 6(2): 67–81.

Burby, R. J. (2006) "Hurricane Katrina and the Paradoxes of Government Disaster Policy: Bringing about Wise Governmental Decisions for Hazardous Areas," *The Annals of the American Academy of Political and Social Science* 604(1): 171–191.

Burby, R. and P. May. (1999) "Making Building Codes and Effective Tool for Earthquake Hazard Mitigation," *Environmental Hazards* 1(1): 27–37.

Cutter, S. L., B. J. Boruff, and W. L. Shirley. (2003) "Social Vulnerability to Environmental Hazards," *Social Science Quarterly* 84(2): 242–261.

Deyle, R. E., S. P. French, R. B. Olshansky, and R. G. Paterson. (1998) "Hazard Assessment: The Factual Basis for Planning and Mitigation," 119–166, in R. J. Burby (ed.), *Cooperating with Nature: Confronting Natural Hazards with Land-Use Planning for Sustainable Communities*, Washington DC: Joseph Henry Press.

Federal Emergency Management Agency (FEMA). (2013a) *Integrating Hazard Mitigation into Local Planning: Case Studies and Tools for Community Officials*, Washington DC: Author.

Federal Emergency Management Agency (FEMA). (2013b) *Local Mitigation Planning Handbook*, Washington DC: Author.

Fu, X. and Z. Tang. (2013) "Planning for Drought-Resilient Communities: An Evaluation of Local Comprehensive Plans in the Fastest Growing Counties in the U.S.," *Cities* 32: 60–69.

Fu X., M. Svoboda, Z. Tang, Z. Dai, and J. Wu. (2013a) "An Overview of U.S. State Drought Plans: Crisis or Risk Management?" *Natural Hazards* 69(3): 1607–1627.

Fu, X., Z. Tang, J. Wu, and K. McMillan. (2013b) "Drought Planning Research in the United States: An Overview and Outlook," *International Journal of Disaster Risk Science* 4(2): 51–58.

Ge, Y. and M. K. Lindell. (2016) "County Planners' Perceptions of Land Use Planning Tools for Environmental Hazard Mitigation: A Survey in the U.S. Pacific States," *Environment and Planning B: Planning and Design* 43(4): 716–736.

Godschalk, D. R. (2009) "Safe Growth Audits," 2–7, In *APA Zoning Practice*, No. 10, Chicago IL: American Planning Association. Available at: www.planning.org/zoningpractice/open/pdf/oct09.pdf.

Godschalk, D. R., T. Beatley, P. R. Berke, D. Brower, and E. J. Kaiser. (1999) *Natural Hazard Mitigation: Recasting Disaster Policy and Planning*, Washington DC: Island Press.

Godschalk, D. R., D. J. Brower, and T. Beatley. (1989) *Catastrophic Coastal Storms: Hazard Mitigation and Development Management*, Durham NC: Duke University Press.

Godschalk, D., E. J. Kaiser, and P. R. Berke. (1998) "Integrating Hazard Mitigation and Local Land Use Planning," 85–118, in R. Burby (ed.) *Cooperating with Nature: Confronting Natural Hazards with Land-Use Planning for Sustainable Communities*, Washington DC: Joseph Henry Press.

Lindell, M. K. and C. S. Prater. (2003). "Assessing Community Impacts of Natural Disasters," *Natural Hazards Review* 4(4): 176–185.

May, P. J. and R. E. Deyle. (1998) "Governing Land Use in Hazardous Areas with a Patchwork System," 57–84, in R. Burby (ed.) *Cooperating with Nature: Confronting Natural Hazards with Land-Use Planning for Sustainable Communities*, Washington DC: Joseph Henry Press.

Lyles, W., P. R. Berke, and G. Smith. (2014) "Do Planners Matter? Examining Factors Driving Incorporation of Land Use Approaches into Hazard Mitigation Plans," *Journal of Environmental Planning and Management* 57(5): 792–811.

Olshansky, R. B. and J. D. Kartez. (1998) "Managing Land Use to Build Resilience," 167–202, in R. J. Burby (ed.), *Cooperating with Nature: Confronting Natural Hazards with Land-Use Planning for Sustainable Communities*, Washington DC: Joseph Henry Press.

Peacock, W. G. (2003) "Hurricane Mitigation Status and Factors Influencing Mitigation Status among Florida's Single-Family Homeowners," *Natural Hazards Review* 4(3): 149–158.

Peacock, W. G. and R. Husein. (2011) *The Adoption and Implementation of Hazard Mitigation Policies and Strategies by Coastal Jurisdictions in Texas: The Planning Survey Results*, Report submitted to the Texas General Land Office and the National Oceanic and Atmospheric Administration. College Station TX: Texas A&M University Hazard Reduction and Recovery Center.

Schmidtlein, M. C., R. C. Deutsch, W. W. Piegorsch, and S. L. Cutter. (2008) "A Sensitivity Analysis of the Social Vulnerability Index," *Risk Analysis* 28(4): 1099–1114.

Schwab, J. C. (ed.). (2010) *Hazard Mitigation: Integrating Best Practices into Planning*, Planning Advisory Service (PAS) Report 560, Chicago IL: American Planning Association.

Stevens, M., P. R. Berke, and Y. Song. (2010a) "Creating Disaster-resilient Communities: Evaluating the Promise and Performance of New Urbanism," *Landscape and Urban Planning* 94(2): 105–115.

Stevens, M., P. R. Berke, and Y. Song. (2010b) High-Density Developments in High-Risk Locations: Why do New Urbanist Developments Locate Built Structures inside the Floodplain?" *Natural Hazards Review* 53(3): 605–629.

Smith, G., W. Lyles, and P. R. Berke. (2013) "The Role of the State in Building Local Capacity and Commitment for Hazard Mitigation Planning," *International Journal of Mass Emergencies and Disasters* 31(2): 178–203.

Tang, Z. (2008) "Evaluating the Capacities of Local Jurisdictions' Coastal Zone Land Use Planning in California," *Ocean and Coastal Management* 51(7): 544–555.

Tang, Z., M. Lindell, C. Prater, and S. D. Brody. (2008) "Measuring Tsunami Hazard Planning Capacity on the U.S. Pacific Coast," *Natural Hazards Review* 9(2): 91–100.

Tang, Z., S. D. Brody, C. Quinn, L. Chang, and T. Wei. (2010) "Moving from Agenda to Action: Evaluating Local Climate Change Action Plans," *Journal of Environmental Planning and Management* 53(1): 43–62.

Tang, Z., M. K. Lindell, C. S. Prater, T. Wei, and C. Hussey. (2011a) "Examining Local Coastal Zone Management Capacity in U.S. Pacific Coastal Counties," *Coastal Management* 39(2): 105–132.

Tang, Z., S. D. Brody, R. Li, C. Quinn, and N. Zhao. (2011b) "Examining Locally-Driven Climate Change Policy Efforts in Three Pacific States," *Ocean and Coastal Management* 54(5): 415–426.

Tang, Z., N. Zhao, Z. Lei. (2012) "Can Planners Take the Leadership in Local Environment Management?" *Journal of Environmental Assessment Policy and Management* 14(2): 1–24.

Tang, Z., Z. Dai, X. Fu, and X. Li. (2013) "Content Analysis for the U.S. Coastal States' Climate Action Plans in Managing the Risks of Extreme Climate Events and Disasters," *Ocean and Coastal Management* 80: 46–54.

United Nations International Strategy for Disaster Reduction (UNISDR). (2011) *Global Assessment Report on Disaster Risk Reduction: Revealing Risk, Redefining Development*, Geneva, Switzerland: United Nations International Strategy for Disaster Reduction.

Wisner B., P. Blaikie, T. Cannon, and I. Davis. (2004) *At Risk: Natural Hazards, People's Vulnerability and Disaster*, 2nd ed., London: Routledge.

22

THE ROLE OF STATES IN DISASTER RECOVERY

An Analysis of Engagement, Collaboration, and Capacity Building

Gavin Smith

Introduction

Federal, state, and local disaster recovery planning and policy have suffered from the lack of a national recovery strategy (GAO 2016; Mitchell 2006; Smith 2011; Smith and Wenger 2006; Smith et al. 2018; Topping 2009), weak state recovery plans (Sandler and Smith 2013; Smith and Flatt 2011), and a scarcity of local plans focused on disaster recovery (Berke et al. 2014). For years, disaster recovery remained largely marginalized within the federal policy milieu, buried within the National Disaster Response Framework (Smith 2011: 40–42). A common refrain among researchers and practitioners over the past several decades is that disaster recovery is the least understood phase of emergency management (Berke et al. 1993; Rubin 1991, 2009; Smith and Wenger 2006). Until recently, recovery research lagged behind studies addressing response, preparedness, and hazard mitigation. Disaster recovery research experienced a dramatic uptick following Hurricane Katrina, remained somewhat constant for nearly a decade, and increased again following Hurricane Sandy. Much of this research, which consisted extensively of case studies, has been focused on communities (Cutter et al. 2014; Friesema et al. 1979; Geipel 1982; Haas et al. 1977; Oliver-Smith and Goldman 1988; Olshansky et al. 2006, 2008; Peacock et al. 1997; Rubin 1982; Rubin et al.1985; Seidman 2013; Wright et al.1979). There are very few cases in which states were the primary unit of analysis (Sandler and Smith 2013; Smith 2014a, 2014b; Smith et al. 2018). The degree to which there is another resurgence in research following Hurricanes Harvey, Irma, and Maria, including a focus on states, remains to be seen.

Two significant gaps remain in our understanding of disaster recovery: the effective translation of the growing body of research findings to practice and the role that states play in this process (Sandler and Smith 2013; Smith 2008, 2011: 5, 91–100, 104–106; Smith et al. 2018). The importance of tackling these two issues became increasingly clear when Hurricane Katrina brought national and international attention to the long-standing problems in our nation's approach to disaster recovery (CREW 2007; GAO 2008; Smith 2011: 5, 321). Heightened Congressional awareness of these weaknesses resulted in the passage of the Post Katrina Emergency Management Reform Act (PKEMRA), which required FEMA to increase its commitment to assist state and local governments' recovery from disasters and enhance their pre-event capacity to plan for future events through the creation of a national recovery strategy—something that did not exist prior to Katrina (Smith 2011). The resulting National Disaster Recovery Framework (NDRF) remains an emerging federal policy that provides substantial, albeit unrealized, potential to address fundamental, vexing problems (Johnson and Olshansky 2016: 47–56; Smith and Wenger 2006; Smith et al. 2017). Given the modest improvements in disaster recovery policy, it is important to continue to advance a national dialogue that strives to improve the level of pre- and post-disaster engagement, collaboration, leadership, and capacity building. Although states are uniquely positioned to address these issues, there remains a limited research and practice-based literature addressing the roles states play in recovery, the quality of state plans, and the degree to which national and state policy enables state-based agencies and organizations, including the Office of the Governor, to effectively confront what we know to be major challenges (Sandler and Smith 2013; Smith and Sandler 2012; Smith et al. 2018).

Role of States in Disaster Recovery

States, which have been described as a "linchpin" between federal and local actors (Sandler and Smith 2013; Smith and Flatt 2011) undertake a number of important activities, including: 1) translating and/or administering federal programs, policies, and planning mandates; 2) creating state programs and grants that address local needs that are not met by federal assistance; 3) initiating capacity-building initiatives such as training, education, and outreach programs; and 4) assisting local governments plan for post-disaster recovery, although the latter two activities remain underemphasized, with some notable exceptions (Durham and Suiter 1991; Florida Department of Community Affairs 2010; Smith 2011: 273; Smith et al. 2018).

The roles that states play in pre- and post-disaster recovery planning and operations vary significantly due to their experience, their level of capacity and commitment to the process, the engagement and leadership exhibited by the governor, and the local needs that emerge following a disaster (Smith 2011: 43–49; Smith et al. 2018). In this chapter, problems surrounding capacity and commitment are framed in the context of state-level planning, engagement, collaboration, and leadership.

The Influence of Federal Recovery Policy and State Leadership: Why Some States Take a More Active Role in Recovery than Others

The federal commitment of large sums of post-disaster resources without a concomitant emphasis on pre-event capacity building at the state and local level is a serious problem (GAO 2010, 2016; Smith 2011; Topping 2009). The federal approach to recovery, including the politicization of the disaster declaration process and the disaster assistance that follows, limits the proactive commitment of state-level resources (May and Williams 1986; Platt 1999; Smith et al. 2018). This is because states receive large infusions of funding after disasters regardless of their own levels of pre-event recovery readiness. Indeed, states that have failed to establish adequate capacity before a major disaster must divert resources needed at the local level to build the state's capacity to seek Congressional funding, create new state organizations and associated policies, or administer post-disaster federal grant programs (Smith 2011: 49; Smith et al. 2013, 2018). States receive very limited pre-disaster federal assistance targeting recovery capabilities such as planning. Nor do many states provide adequate resources to agencies tasked with disaster recovery for them to proactively address the significant problems identified throughout this chapter.

In some ways, federal disaster recovery policy can hinder, rather than promote, proactive state and local action. FEMA suggests that a key aim of its overall recovery strategy is to reduce disaster losses and associated payouts after an event occurs, even though they have little control over the actions taken by states and local governments. Ray Burby (2006) refers to this dilemma as the "local government paradox" wherein subnational governments have greater control over local actions such as land use planning, but view disaster management as less salient relative to other competing priorities.

This is particularly true in the case of disaster recovery policy. In a study of more than 2,000 state and local individuals, including governors, mayors, planners, developers, and others, Rossi et al. (1982) found that when asked about varied "disaster philosophies" (e.g., structural mitigation, non-structural mitigation, free-market compulsory insurance, and post-disaster assistance), respondents preferred post-disaster assistance and structural mitigation measures.

The preferred approaches noted in the survey have the additional long-term effect of furthering the safe growth paradox whereby disaster relief and an overemphasis on structural measures such as sea walls, levees, and dams increase development pressures in adjacent areas, leading to potentially greater losses and federal payouts when the design parameters of the protective measures are exceeded (Burby 2006; Montz and Tobin 2008; Tobin 1995). The passage of PKEMRA, and the ensuing development of the NDRF, neither establish clear planning requirements or standards nor hold state and local governments accountable for pre-event action (or inaction) that increase their hazard vulnerability and adversely impacts their capacity to recover after a disaster strikes (Smith 2011; Smith et al. 2017).

State Capacity and Commitment

May and Williams (1986) suggest in their analysis of disaster management policy that commitment is a "willingness to work toward a goal or solve a problem," whereas capacity is the "ability to reach the goal as reflected by available resources, and by political, managerial, and technical competence" (28). The pre- and post-event conditions described in the previous section of this chapter influence a state's willingness to invest the time and resources needed to build local disaster recovery capacity through various activities, including state-initiated grant programs that fund the development of local plans; the creation of training, education, and outreach programs; and the hiring of state staff needed to sustain these efforts over time (Smith et al. 2013, 2018). As part of a 7-year study assessing the quality of state and local hazard mitigation plans, researchers evaluated the capacity and commitment of states to help local governments with a range of hazard mitigation activities including planning. According to Smith et al.:

> The findings suggest that states maintain a wide variation in state capacity and commitment to support local hazard mitigation activities, including that which is influenced by disaster-based funding. They also tend to emphasize building local governments' capacities to gain access to project funding rather than focusing on helping them identify and establish a comprehensive, proactive, and sustained risk reduction strategy grounded in land use policy. In addition, state land use policies are not well integrated into state hazard mitigation plans and capacity building initiatives. Finally, state mitigation officials believe that most local governments do not possess the capacity or commitment necessary to develop sound hazard mitigation plans or administer hazard mitigation grants.
>
> *(2013: 178)*

While state officials lamented the limited capacity of local government to administer what largely amounts to post-disaster hazard mitigation grants or their inability to proactively address key land use issues through planning, their commitment to address recognized local shortfalls in capacity remains insufficient (Smith et al. 2013).

Important disaster recovery parallels can be drawn from these findings to include the disproportionate effect of federal assistance strategies (e.g., emphasizing access to post-disaster funding versus a more comprehensive, proactive approach to pre-event planning) on the actions taken by subnational levels of government, including states. The findings of the study are disconcerting as the hazard mitigation plans were developed more than five years after the passage of the Disaster Mitigation

Act of 2000 (DMA 2000). The intent of DMA 2000 is to help states and local governments build the capacity needed to reduce risk and speed the implementation of identified risk-reduction projects in pre- and post-disaster settings through the development of hazard mitigation plans. Pre- and post-disaster funds, which are provided to state and local governments to help achieve this aim, are contingent on the ability of these governments to comply with established planning standards. The 7-year study of state and local hazard mitigation plans also found that even though they met established national standards, the plans were of modest to poor quality (Berke et al. 2012, 2014; Lyles et al. 2014a, 2014b; Smith et al. 2013). Thus the establishment of proactive planning is not enough. Rather, the creation of high standards, and holding states and local governments more accountable for achieving them is vital, and should be considered during the ongoing operationalization of NDRF policy and guidance. At this time, pre-disaster federal funding is not widely available to develop recovery plans and plans are not required in order to receive post-disaster assistance. In a more recent and promising development, FEMA is proposing the creation of a "disaster deductible program" in which states would be required to spend a predetermined amount of funds to reduce risk and enhance state-wide levels of resilience in order to access post-disaster funding used to repair damaged infrastructure (GPO 2017).

Other factors that shape a state's level of capacity and commitment to disaster recovery are experience and lessons drawn from past events, including high profile disasters that have struck elsewhere; an awareness of the hazard risks prevalent in a given area; and the presence of policy advocates (Birkland 1997, 2006: 129–156; Rubin 2012; Smith 2011: 45). States that have experienced more major disasters, or multiple disasters in close succession, are more likely to take an active role in recovery—including committing state resources such as matching federal funds, creating state-level recovery programs, and hiring additional recovery staff (Smith et al. 2013). In California, for instance, the combination of engineers, earth scientists, and elected officials has proven highly effective in encouraging the development of strong earthquake legislation following a number of major disasters (Olson 2003; Geschwind 2001).

The State of Florida has developed the Post-Disaster Redevelopment Planning Initiative, which provides funding and guidance on the development of pre-disaster redevelopment plans in designated counties. Six communities—five counties and one city—were chosen as the sites for case studies from which to draw lessons to be shared across the state and to formulate a disaster recovery guide titled *Post-Disaster Redevelopment Planning: A Guide for Florida Communities*. According to Julie Dennis, State of Florida, Community Program Manager,

> The 2004 and 2005 hurricane season saw twelve named storms making landfall in Florida, resulting in seven major presidential disaster declarations and billions of dollars in damage to local communities. The long-term recovery process that ensued underscored the value of planning for long-term recovery prior to an event. While at the time, state statute required post-disaster redevelopment plans for coastal communities and encouraged them for inland communities, there were no set criteria for what should be included in these plans. The Statewide Post-Disaster Redevelopment Planning Initiative (PDRP) sought to develop the criteria and provide guidance to local governments on how to organize and plan for long-term recovery under blue skies.
>
> *(Julie Dennis, Personal communication July 29, 2014)*

It is a worthwhile research endeavor to assess the degree to which the local disaster recovery plans developed under the state PDRP program were implemented following Hurricane Irma, which struck the state in 2017, and how this affected disaster recovery outcomes in these communities.

The State of Oregon has also created a program to help communities develop pre-disaster recovery plans. In this case, the 2004 Sumatra earthquake and accompanying tsunami, as well as Hurricane

Katrina, precipitated state action (Smith 2011: 335). The Oregon Natural Hazards Workgroup at the University of Oregon's Community Service Center convened a forum of experts from the Cascadia Region Earthquake Workgroup, the US Geological Survey, and Oregon Emergency Management to prepare for recovery should an earthquake or tsunami strike the Oregon coast. One of the first pilot communities they worked with was Cannon Beach, an area highly vulnerable to earthquakes and tsunamis. The effort has since expanded to the rest of the Oregon coast, focusing on capacity building efforts and collaborative problem solving (Smith 2011: 335).

A state's institutional culture can also influence the roles and responsibilities assumed by state agencies and organizations, including leadership provided by the Governor's Office. This can be seen in a state's commitment to identify gaps in federal assistance and develop strategies to address them (Smith and Birkland 2016; Smith et al. 2018). The manner in which this is achieved may reflect broader political philosophies held by governors and is subject to change across administrations (Smith 2014a; Smith et al. 2018). Other important factors to consider include the planning culture in a state and how this may influence the quality of a state's pre-disaster recovery plan (or willingness to develop and implement a post-disaster recovery plan); the relationship between states and the federal government, including the robustness of the FEMA-state "partnership"; the connectivity of the Governor to influential members of Congressional Appropriation Committees and the President; and leadership (Boin and t'Hart 2003; Kapucu and Van Wart 2008; May 1985; May and Williams 1986; Mitchell 2006; Olshansky et al. 2012: 65; Smith 2011; Smith et al. 2018).

Key Dimensions of Disaster Recovery

State agencies, including the Governor's Office, are part of a much larger network of stakeholders involved in pre- and post-disaster recovery activities. This loosely coupled system, which is subject to change over time, is referred to as the disaster recovery assistance network (Smith 2011; Smith and Birkland 2016; Smith et al. 2017, 2018). Members of this network provide or influence the provision of three types of resources: funding, policy, and technical assistance. States play a vital role in this network and can significantly influence the following dimensions of disaster recovery: (1) the degree to which resources address local needs, (2) the timing of resource delivery, and (3) the level of horizontal and vertical integration across the network.

Resources and Local Needs

The resources delivered or influenced by members of the disaster recovery assistance network vary in the degree to which they address local needs and the prescriptiveness of rules governing their administration. For instance, highly prescriptive federal rules and the overreliance on these programs can disproportionately affect the trajectory of recovery at the local level and dominate the actions taken by state agencies, particularly those agencies that assume a more passive role (Smith 2011: Smith et al. 2017, 2018). States can address these shortfalls in a number of ways, including through the development of state recovery organizations; the creation of state-level programs; the hiring of temporary, permanent, and contractor-based staff; and the active search for alternative resources (Sandler and Smith 2013; Smith et al. 2017, 2018).

The hypothetical network shown in Figure 22.1 highlights a number of important factors, including the breadth of stakeholders involved, a set of stakeholders that we know less about compared to others (e.g., the zone of uncertainty), and the overall complexity of the network considering the potential number of interactive effects spanning financial expenditures, policy development and implementation, and the delivery of technical assistance by organizations over time. The degree to which the key factors encapsulated in the network are subject to change through state-level planning are described later in this chapter.

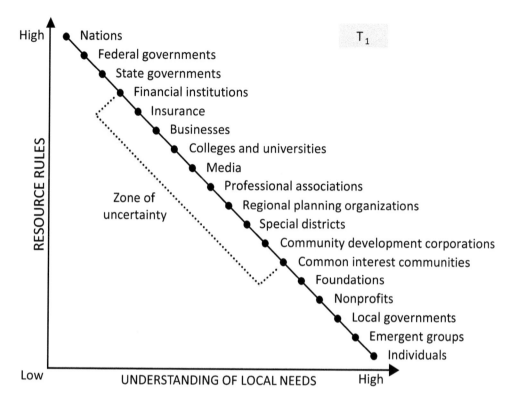

Figure 22.1 Resource Rules and Understanding of Local Needs

The stakeholder group nodes provide a means to synthesize a much larger set of agencies, organizations, groups, and individuals. For instance, the node denoting federal governments may include the Federal Emergency Management Agency, the Department of Housing and Urban Development, the National Oceanic and Atmospheric Administration, the Department of Health and Human Services, the US Army Corps of Engineers, and many others.

The nodes, including the agencies and divisions located within them, operate in different ways due to variation in an array of organizational cultures, norms, and laws. For instance, program goals often vary among divisions within agencies. This can lead to contradictory policies in pre- and post-disaster settings (Smith 2011: 16, 37). This problem is evident in FEMA's Public Assistance (PA) and Hazard Mitigation Grant Program (HMGP). Public Assistance, which funds costly infrastructure repair, has historically been driven by cost containment and speed of delivery whereas HMGP is focused on risk reduction and can take years to administer post-disaster. Differing goals in these two programs have, for instance, made it difficult to incorporate risk reduction into PA projects. This one example highlights how these conditions affect the manner in which organizations distribute resources, including the degree to which the resources they manage address local needs, the timing of their resource distribution, and the degree to which organizations coordinate with other stakeholders.

State agencies and their associated roles in recovery are also diverse, including the rules governing the degree to which state programs address disaster recovery-related issues and local needs. Typical state government agencies involved in disaster recovery include emergency management and homeland security, public safety, health and human services, finance, state planning, natural resources, community development, transportation, and agriculture. Although the authority and responsibility for emergency management rests with the governor, the state emergency management agency leads daily operations, including, in most cases, disaster recovery (Durham and Suiter 1991: 101–102; Smith et al. 2018). In larger events such as those discussed in the North Carolina case study, an organization tasked with coordinating disaster recovery efforts may be created.

State emergency management agencies can be found in a variety of different locations—the Governor's Office, a division within a civilian department (e.g., Emergency Management or Homeland Security), in the Adjutant General, or Public Safety. The locations of these agencies reflects the history of emergency management in the United States, which evolved from a response and preparedness-oriented institution focused on civil defense and the associated threat of nuclear attack (Kreps 1990), and more recently a shift towards homeland security in a post-9/11 world (Rubin 2012; Sylves 2008). According to NEMA (2012: 3), state emergency management agencies in 2012 were distributed accordingly: Public Safety (12), Military/Adjutant General (18), Governor's Office (9), Emergency Management/Homeland Security (8), and other (5).

Agency location (and its associated organizational culture) can also affect a state's approach to disaster recovery operations. This applies to planning and policymaking as well as the degree to which these actions focus on coordinating the network of organizations and the manner in which the collective array of resources they possess addresses local needs. In a study of state emergency management agencies, the National Governor's Association suggested that recovery-related activities are better managed by policy-oriented agencies whereas response activities are best managed by those that are more tactical, like the state police. The study went on to conclude

> some offices are strong and some are weak for a variety of historical, turf, political, and conceptual reasons. Clearly, there is not a state "model" to follow; rather it is the governor's understanding, concern, and support, coupled with the state director's coordination and strategy-building skills, that determine the strength of the organizational framework.
>
> *(Durham and Suiter 1991: 103–104)*

Although representative of an older era, the claim remains relevant. The complexities of disaster recovery, including the ability to work together collaboratively with a number of state agencies as well as other members of the larger disaster recovery assistance network can prove challenging for state emergency management agencies. In the 2012 Biennial Report from the National Emergency Management Association, states reported flat or declining budgets in an era of increasing disaster costs (NEMA 2012: 2). Following Hurricanes Katrina and Ike, the Government Accountability Office reported that state agencies tasked with emergency management still struggle with the coordination of the larger network of stakeholders and their associated resources. The study also found that a failure to develop sound recovery plans was partly to blame for this shortfall (GAO 2008, 2010).

Timing of Assistance

Haas et al.'s (1977) conclusions about the timing of disaster recovery emphasized the post-disaster reconstruction process but paid little attention to the effect of pre-event planning on recovery trajectories, the speed at which recovery occurred, or how varied organizations coordinated the timing of their assistance strategies to meet collaboratively defined local goals and needs (Smith 2011). Rather, they viewed recovery as a post-disaster phenomenon tied to the physical reconstruction of the built environment, a process that they contended was largely linear in nature, with clearly defined tasks that occurred sequentially over time. More recent research suggests that this orderly process is an oversimplification of many temporal elements, including the variation in the rates of recovery among different population subgroups and spatially defined areas, some of which never return to pre-event conditions (Alesch et al. 2009; Berke and Beatley 1997: 177; Cutter et al. 2014;

Peacock et al. 1997; Rubin and Popkin 1990). Factors affecting these differential outcomes and rates of recovery include the severity of the event and associated impacts on jurisdictions, neighborhoods, groups, and individuals; levels of federal, state, and local officials' pre-event experience, capacity, and commitment to act; levels of social vulnerability; and access to pre- and post-disaster resources (Smith 2011: 19–20).

Additional research likens reconstruction to the process of urban redevelopment, operating under a compressed timeframe (Olshansky et al. 2012; Johnson and Olshansky 2016: 10–11). Members of the disaster recovery assistance network, including states, must confront the challenges inherent in striking a balance between speed and deliberation, another concept advanced by Olshansky (2006). A common metric of success at multiple levels of government involves the speed at which recovery occurs, not necessarily its quality, including the degree to which local needs are met (Smith 2011: 16). Intense pressure is placed on state and local officials to quickly recover rather than adopting a deliberative, inclusive problem-solving approach, which can entail a more time-consuming process— particularly when employed after a disaster. Ideally, this process is furthered through the development of a pre-disaster recovery plan, which necessitates allocating the time required to develop collaboratively derived policies and programs in anticipation of disaster recovery challenges spanning all members of the disaster recovery assistance network. States have an obligation to work proactively with local governments to develop sound pre-disaster recovery plans that are inclusive. This relationship remains highly varied across the U.S. due to variations in the capacity of states to provide meaningful guidance or their commitment to take the time required to create state-level goals that inform local actions. Additional challenges include the ability to hire and train a sufficient number of competent staff that are able to work with local governments to build local capacity, to invest the financial resources necessary to implement planning policies and projects, and to procure and manage the data needed to develop robust local plans (Sandler and Smith 2013; Smith and Flatt 2011).

Smith (2011) suggests that the temporal aspects of recovery should be viewed in terms of pre-event planning and post-event actions to include the effects of how and when resources are provided by differing members of the disaster recovery assistance network, including state officials. The interactive effects of funding, policy, and technical assistance distributed by varied actors can significantly affect recovery trajectories that are not limited to physical reconstruction, but also include complex institutional arrangements affecting social, environmental, and economic outcomes (Smith and Birkland 2012; Smith and Wenger 2006; Smith et al. 2017).

Horizontal and Vertical Integration

The horizontal and vertical integration framework provides a useful way to describe interorganizational coordination among stakeholders at the local level and across federal, state, and local levels. The horizontal and vertical integration framework has been applied to a number of policymaking contexts, including disaster recovery (Berke and Beatley 1997; Berke et al. 1993; Smith 2011; Smith et al. 2018). Figure 22.2 depicts a commonly applied typology whereby the strength of horizontal and vertical integration is described across four community types. Type 1 communities are characterized by strong horizontal and vertical integration. That is, federal, state, and local relationships are strong, based on a clear understanding of federal recovery policies, the state's roles in recovery (translator of federal rules, capacity-builder, etc.), and the degree to which federal and state assistance addresses local needs and complements local recovery goals. Strong horizontal integration is present as well, indicated by close working relationships across local actors such as non-profits, small businesses, community groups, and others. Close local associations help to identify and address local needs that may not be met by federal or state assistance programs and empowers these groups to speak with a collective voice regarding local needs to state agencies that may develop state level programs or transmit these needs to federal agency partners.

Horizontal \ Vertical	Strong	Weak
Strong	Type 1	Type 2
Weak	Type 3	Type 4

Figure 22.2   Horizontal and Vertical Integration Typology

Type 2 communities possess strong horizontal integration, but weak vertical connectivity to state and federal organizations. A prototypical Type 2 community is the small rural community that maintains strong horizontal connectivity across local organizations, thereby understanding local conditions, but may not regularly interact with state and federal agencies until a disaster strikes. The lack of strong vertical integration may hinder a Type 2 community's ability to convey local needs to appropriate providers of external assistance or understand the often confusing array of post-disaster recovery programs. The lack of pre-event capacity to administer state and federal grant programs can overwhelm these communities, resulting in much of their time and limited staff resources being dedicated to learning about programmatic rules and eligibility rather than developing local disaster recovery plans designed to improve the linkage between external aid and local needs.

Type 3 communities are characterized by strong vertical integration, but weak horizontal integration. Strong external relationships among federal, state, and local governments materialize in a good understanding of federal and state recovery programs and policies. Weak horizontal integration can result in a lack of collaborative problem solving, including that achieved through local recovery planning. Type 4 communities face the greatest difficulty with recovery. Weak horizontal and vertical integration means that these localities do not have a good awareness of federal and state recovery programs. Nor do they possess an understanding of needs predicated on close working relationships among local organizations.

Horizontal integration is typically applied to the actions taken at the local level, spanning local governments, community-based non-profits, small businesses, and groups that may emerge post-disaster to address specific local needs. In this chapter, horizontal integration is used to assess the level of coordination across state-level agencies and organizations, which has been done in a number of other studies (Sandler and Smith 2013; Smith and Flatt 2011; Smith et al. 2018). Examples of mechanisms used to achieve the aim of improved state-level coordination across agencies include the formation of state recovery committees and task forces following major disasters as well as the development of state disaster recovery plans. Vertical integration remains focused on local, state, and federal linkages, with special emphasis placed on the role of states as a "lynchpin" connecting local and federal organizations (Sandler and Smith 2013; Smith et al. 2018).

Planning and the Modification of Key Dimensions of Disaster Recovery

State disaster recovery plans can provide tools to address the three dimensions of recovery (e.g., resources and local needs, timing of assistance, and horizontal and vertical integration) (Smith 2011: 26–30; Smith and Wenger 2006: 242–244). Specific ways in which states can improve local disaster recovery planning processes and outcomes include: (1) creating a state vision for recovery that local governments can strive to achieve; (2) helping local governments comply with state goals and policies, including their associated rules; (3) serving as an intermediary between federal policies and local actions; (4) evaluating local disaster recovery plans; and (5) building local capacity through state-level training and outreach programs (Smith 2011; Smith and Wenger 2006: 242–245; Smith et al. 2018; Waugh and Sylves 1996). State recovery efforts can also foster intergovernmental coordination (e.g., horizontal and vertical integration) and assist communities to recover from disaster impacts in a more resilient manner (May and Williams 1986; Peacock et al. 2009; Smith and Birkland 2012; Smith and Wenger 2006; Smith et al. 2017, 2018;).

Researchers have found that states can advance state goals and assist local governments create locally grounded solutions if certain conditions are met (Sandler and Smith 2013). These include well-constructed state policies that engender local commitment and strong implementation strategies, adequate funding to support planning efforts, and creating the conditions that enable state agencies to assist local governments over time to build and sustain local capacity (Burby et al. 1997; Sandler and Smith 2013; Smith et al. 2018). Additional research shows that developing intergovernmental planning programs may benefit from an approach that blends the positive effects of a command-and-control approach and cooperative approach that engenders not only participation, but also commitment across diverse subnational actors (Burby and May 2009).

Good planning processes are characterized by collaborative efforts that provide a meaningful participatory problem-solving vehicle for diverse stakeholders (Innes and Booher 1999, 2003, 2004). Applied to disaster recovery, plans can help to connect resources that span the broader assistance network with the needs that exist at the local level. Returning to a modified version of Figure 22.1, the diagonal line shifts to a vertical orientation, whereby rules governing assistance across the network may remain the same, but through the active sharing of information between resource recipients and providers, recipients gain an appreciation of programmatic constraints (and potential opportunities) and providers gain a better understanding of local needs (Figure 22.3). This requires the active and ongoing involvement of individuals, local government officials, and others at the community level. It also requires the active involvement of states, which are tasked with translating this information from local governments to federal agencies following a disaster (i.e., enhancing vertical integration), and in some cases, developing state-level programs designed to address unmet local needs.

Good planning also facilitates learning and associated changes in policy through what planning scholar John Friedmann (1987) calls social learning. Understood in the context of the disaster recover assistance network, resource providers not only recognize local needs and conditions, but also ultimately alter the rules governing these programs, thereby further shifting the location of the nodes in the network to reflect these changes (Figure 22.4). Through ongoing dialogue, members of the disaster recovery assistance network explore, and ultimately undertake changes to the rules governing disaster recovery resource management. If done collectively, this can lead to the optimal allocation of resources, striking a balance between rules and local needs across the disaster recovery assistance network.

Altering the rules governing the allocation of disaster recovery resources is not a simple process; it may require changes in Congressionally or legislatively promulgated authorities such as those found in the Post-Katrina Emergency Reform Act or state policy described in this chapter's case study. In other cases, change may result from differing interpretations of enabling rules. For instance, FEMA has been unwilling to mandate that DMA 2000-compliant hazard mitigation plans require applying land use practices to reduce risk even though this technique is widely recognized as among the most effective, proactive means to accomplish the Act's central aim.

Altering the level of prescriptiveness of disaster recovery programs may also require changes in organizational culture among non-governmental actors such as foundation boards that manage post-disaster assistance, or among private sector actors, such as insurance providers. Change requires actors' willingness to modify their program rules to complement those provided by others. This may include actors that have a history of distributing recovery resources in a manner that reflects their

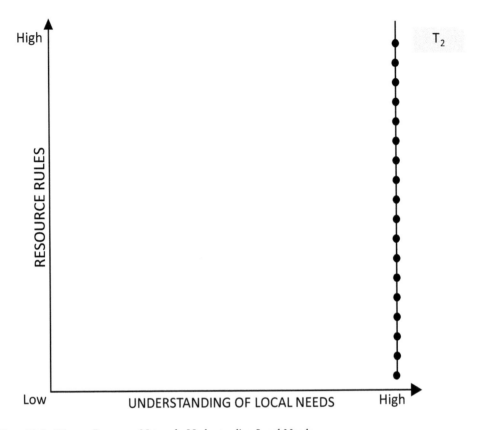

Figure 22.3 Disaster Recovery Network: Understanding Local Needs

frustrations with others whom they view as overly bureaucratic or acting in ways that are counter to their organizational missions (Smith 2011: 127).

The disaster literature shows that policies can change following major disasters but remain unaltered in other instances, even in the face of clear evidence that change is warranted (Birkland 1997, 2006). Public policy scholars have shown that policies can remain constant for decades, followed by significant episodic change—a process known as "punctuated equilibrium" (Baumgartner and Jones 1993). Sabatier and Jenkins-Smith note in their Advocacy Coalition Framework that "alterations require changes in the distribution of political resources of subsystem actors arising from shocks exogenous to the subsystem" (1993: 42). Olson et al. (1999) applied this framework to the earthquake hazard mitigation policymaking process and found that change can occur through the use of pre-established policymaking venues such as local task forces. They also found that the need to adopt more rigorous risk reduction codes and standards became increasingly salient in the post-disaster environment and subsystem actors advocating for change became more impactful. Furthermore, the decisions surrounding these measures were tied to the political economy and associated costs of the repair or retrofitting of structures in the city of Oakland, California, not necessarily geotechnical or engineering-based solutions. Finally, policy change was closely connected to resources, which they defined as funding, recognizing that change takes time and the policy subsystems must include an intergovernmental dimension that included existing and new partnerships.

Planning can also improve the timing of pre- and post-disaster assistance by identifying the resources maintained by the disaster recovery assistance network, and developing a collaboratively

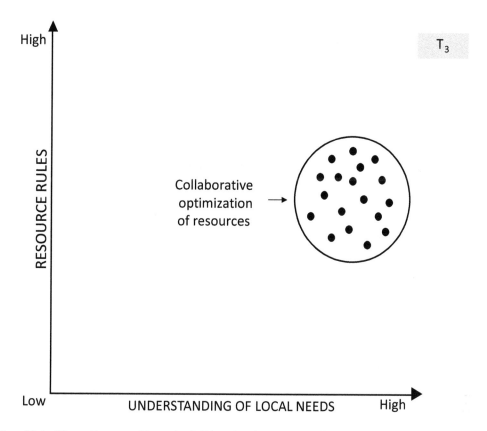

Figure 22.4   Disaster Recovery Network: Collaborative Optimization of Resources

derived strategy to better coordinate their use over time (i.e., pre- and post-disaster) in accordance with higher order goals established by a cohesive network. Altering the rules governing resource distribution is not an easy task and change benefits from the development of a well-constructed resource distribution strategy before a disaster occurs that allows for the thoughtful timing of resource delivery.

States can play an important role in the process of pre-disaster recovery planning by bringing together stakeholders to develop a state recovery plan that focuses on the temporal coordination of assistance across members of the assistance network. The development of vertically and horizontally integrated coalitions helps to address the time-consuming task of modifying resource rules and distribution strategies that, as the public policy literature shows, can take years to accomplish. As Smith et al. (2018) note, state coalitions can also stand ready to advocate for significant policy change following a major disaster by collecting supporting data (often included in plans), garnering political support (often led by governors), and offering clear policy alternatives (also included in plans). Although waiting to develop disaster recovery plans in the post-disaster environment can take time, research has shown that the additional time required to educate stakeholders and build consensus across the network following a disaster can reduce conflict and actually speed the overall recovery effort (Chandrasekhar et al. 2014). Once the stakeholders agreed to the plan, acting collectively can be expedited (Johnson and Olshansky 2016: 13).

The ability to slow the process of post-disaster recovery in order to engage disaster recovery assistance networks requires a strong political will among elected officials (including local and state levels of government), as well as an ability to see the value of post-disaster recovery planning and convey its benefits to the public and those in positions of power. Although the development of pre-disaster recovery plans is more common among states than local governments, the quality of state plans appears to reflect powerful forces driving recovery actions, namely a strong focus on the administration of post-disaster federal assistance rather than the development of a comprehensive approach to recovery that involves the larger assistance network (Sandler and Smith 2013). States should do more to address these shortfalls, as encouraged through the NDRF and as suggested by hazards scholars, to include developing plans in pre- and post-disaster timeframes that are of high quality.

State Recovery Plan Quality

Plan quality principles have been developed to assess plans across widely recognized standards of practice (Baer 1997; Berke and Godschalk 2009; Kaiser et al. 1995). This technique allows researchers to pinpoint defining characteristics and evaluate strengths and deficiencies that help determine a plan's usefulness. Initial studies of criteria for assessing plan quality emphasized the fact base, goals, and policies, whereas later studies included vision, internal consistency, plan implementation, monitoring and evaluation, organizational clarity, coordination with other plans, and compliance with government mandates (Berke and French 1994; Berke and Godschalk 2009). Many of these studies have applied plan quality principles to assess state and local disaster management plans, including those addressing hazard mitigation (Berke et al. 2012; Brody 2003; Godschalk et. 1999; Smith et al. 2013) and disaster recovery (Berke et. al 2014; Smith and Flatt 2011; Sandler and Smith 2012).

Figure 22.5 applies the dimensions of plan quality described by Berke and Godschalk (2009) and modified by Sandler and Smith to state disaster recovery plans. The dimensions include (1) vision and issue identification, (2) fact base, (3) goals, (4) policies, (5) implementation, (6) evaluation and monitoring, (7) internal consistency, (8) interdependent actions, (9) participation, and (10) organizational clarity (Sandler and Smith 2013). The application of plan quality principles to a small sample of state recovery plans has shown that the plans do not effectively address many of the important issues tied to a state's roles in recovery nor plan quality principles (Sandler and Smith 2013). The study of state plans in Florida, California, Mississippi, and North Carolina found that plans scored poorly across most dimensions of plan quality. This is discouraging, considering that the study selected states that were widely considered to have strong emergency management programs, as evidenced by their professional accreditation status and significant disaster experience (Sandler and Smith 2013). It is also important to note that this study was completed in 2013 and many of the plans were written in 2010.

State recovery plans were characterized as a collection of documents, guides, and administrative rules rather than plans as defined by the plan quality principles' evaluative criteria (Sandler and Smith 2013). No plan scored higher than 5 on a 10-point scale. Plans did not articulate a vision or define a set of goals but, rather, articulated grant management authorities, state agency responsibilities, and the process by which disasters will be declared. Recovery plans were largely devoid of policies designed to balance competing interests, guide the coordinated actions of the disaster recovery assistance network, and serve as a larger decision-making tool before and after a disaster. Plans did not identify an underlying fact base or describe how data supported the development of policies. The plans' implementation elements were among the weakest of all principles and remained focused on grant administration rather than local planning and capacity-building.

A brief review of North Carolina's plan provides additional insights. The Comprehensive Recovery Plan, developed by the North Carolina Division of Emergency Management in 2010 emphasized how agencies administer federal and state programs across short-, intermediate-, and long-term operations (Sandler and Smith 2013). The plan is best described as an assemblage of several documents developed over time. Nested within the Comprehensive Recovery Plan is the North Carolina Disaster Recovery Guide, which was prepared by the Office of State Budget and Management and the Statewide Disaster Recovery Leadership Team. The Recovery Guide describes

Vision and Issue Identification	A vision statement defines the principal themes and intent of the plan.
Fact Base	The fact base is an analysis of current and projected conditions within the study area (i.e. the state). The fact base should include characteristics about relevant plans, policies, and programs; fiscal, legal, and administrative capabilities to address recovery (e.g., capability assessment); population and other social characteristics; current and projected growth rates; the economy; the environment; and a hazard identification and vulnerability assessment.
Goals	Goals are statements of future desired conditions that are tied to the overall vision. Goals are instrumental in setting a direction to guide policies and actions described within the plan.
Policies	Policies are statements intended to guide public and private decisions and should achieve identified goals. Policies should be specific and tied to definitive actions.
Implementation	Implementation is the commitment to carry out policy-driven actions. This includes the identification of resources, responsible organizations, and the timing of assistance.
Evaluation and Monitoring	Evaluation and monitoring are defined as the steps necessary to track changes in the fact base, assess the progress of recovery according to predetermined benchmarks, and update the plan over time.
Internal Consistency	Internal consistency is the degree to which vision, issues, goals, policies, and implementation are clearly linked and mutually reinforcing.
Interdependent Actions	Interdependent actions refer to the vertical and horizontal integration of organizations involved in recovery. With the state as the unit of analysis, vertical integration is the coordination between public sector organizations (local, regional, state, and federal); horizontal integration is the coordination across state agencies and departments.
Participation	Participation is measured by the level of engagement and involvement of the network of recovery stakeholders in the preparation and implementation of the recovery plan.
Organizational Clarity	Organizational clarity is defined by the overall legibility of the plan. The clarity is influenced by the degree to which the plan is logical and consistent and whether it includes visual aids such as charts and diagrams to clearly convey information.

Figure 22.5 Plan Quality Principles Applied to State Disaster Recovery Plans

the roles, responsibilities, and types of assistance provided by federal, state, and community organizations involved in recovery (Smith and Flatt 2011: 15).

In 1999, as shown in the subsequent Hurricane Floyd case study, the state developed some of the most extensive disaster recovery programs in the nation, and the Governor argued that sustainability should be an overriding focus of state recovery policies. Yet the quality scores for the plan (which was developed 10 years after Hurricane Floyd struck) were very low, including the points tied to vision, goals, policies, and organizational clarity. The North Carolina plan lacked a clear vision and a set of goals designed to achieve it. The policy section scored low in large part because it did not address policies beyond federal and state programs. North Carolina relies upon multiple recovery documents that should work together to achieve higher-order goals specified in the recovery plan. Examples include administrative plans tied to FEMA programs that must be developed in order to receive post-disaster federal assistance. However, these documents are not referenced, and it is not clear that they function together as part of a larger disaster recovery decision-making process (Sandler and Smith 2013).

An important element of plan quality involves the interconnectivity of principles (Berke and Godschalk 2009). A weak link in any of the individual plan quality principles reduces the effectiveness of the plan. For instance, plans that contain strong policies without an accompanying implementation strategy are likely to remain unrealized. Although North Carolina had a disaster recovery committee at the time and this committee was tasked with the plan's creation, the plan did not describe the planning process. Nor were key members of the larger disaster recovery assistance network—non-profit, private sector, and quasi-governmental organizations—involved in the preparation of the disaster recovery plan. Although North Carolina created procedures to inform local governments and individuals about programs available after disasters, there was very limited information provided about the processes used to develop local pre-disaster recovery plans. Each of these findings and their potential implications will be discussed next in the context of North Carolina's disaster recovery experiences, including the failure to incorporate the lessons learned from Hurricane Floyd into the state recovery plan.

Disaster Recovery Case Study: North Carolina Following Hurricanes Fran and Floyd

The State of North Carolina case study emphasizes three areas: (1) pre- and post-disaster conditions, (2) post-disaster recovery processes, and (3) disaster recovery outcomes. Framing the overall discussion is the exploration of how planning practices were applied and the manner in which they may have taken into account and/or influenced the three dimensions of disaster recovery. Specific pre-event conditions explored include vulnerability, recovery capacity, disaster experience, the types of plans and policies in place, and political influence and power. Post-disaster conditions include a description of the nature of the disasters that struck the state such as event type, magnitude, speed of onset, scope (geographical extent) of impacts, duration, and post-disaster decision-making processes.

The discussion of post-disaster recovery processes consists of a review of the level of collaboration across the disaster recovery assistance network and the manner in which post-disaster resources were distributed. Examples include the timing of assistance, the degree to which assistance met local needs, the equitable distribution of assistance, the drawing of lessons learned from past events, the degree to which hazard mitigation measures were incorporated into recovery, and the degree to which an emphasis was placed on capacity-building measures. Disaster recovery outcomes comprise the degree to which state goals were attained, the speed and quality of recovery, and the institutionalization of state recovery policies in the state disaster recovery plan.

North Carolina: Observations and Lessons

North Carolina has a long hurricane history, with the 1990s being a particularly active decade as Emily (1993), Bertha and Fran (1996), and Dennis, Floyd, and Irene (1999) struck the state (Barnes 2001). Compounding the effects of a high frequency of extreme events include the ditching and draining of protective wetlands, the construction of elevated roadways with inadequately sized culverts and bridge openings, rapid development on highly dynamic barrier islands, and the preponderance of low-income individuals living in riverine floodplains (Riggs 2001: 37–44). Additional factors that led to increased vulnerability were a declining agriculture-based economy and local governments with limited capacity, as evidenced by dwindling budgets and reduced staff (Delia 2001).

Hurricane Fran, which struck in 1996, was at the time the costliest disaster in North Carolina's history, totaling 5 billion dollars in damages (Barnes 2001: 203). At that time, the most recent disaster of comparable force was Hurricane Hazel, which impacted the same area in 1954. Since Hazel, significant coastal development had occurred, greatly increasing the amount of property at risk.

In the interim, the nation's approach to disaster recovery had also changed. In 1979, FEMA was established, which led to the consolidation of many federal disaster relief programs within one agency (Rubin 2012; Sylves 2008). In 1988, with the passage of the Stafford Disaster Relief and Emergency Assistance Act, disaster recovery programs were further codified, and new ones created, including the Hazard Mitigation Grant Program (HMGP).

The magnitude of Fran and the scale of federal assistance that followed overwhelmed the capacity of the North Carolina Division of Emergency Management (NCEM) and local governments. This led Governor James B. Hunt to recruit a North Carolina native to serve as the new director of the state agency. Eric Tolbert, who originally worked in NCEM, had moved to Florida following Hurricane Andrew to lead disaster response efforts in that state. Shortly after he returned to North Carolina, Tolbert increased the size of the Division by hiring temporary state employees using post-disaster funding to help better manage the influx of federal dollars (Smith 2014a: 201–202). The Governor also created the Hurricane Fran Disaster Recovery Task Force. The task force, comprising state agency leaders, identified local needs through listening sessions in the affected area, developed new state recovery policies based on local feedback, and developed a congressional relief package request that sought assistance beyond that available under the Stafford Act.

Three years later, Hurricane Floyd struck, causing more than $6 billion in damage and killing 52 people (Barnes 2001: 254, 256). The slow-moving storm did not possess the strong winds associated with Fran, which was a Category 3 hurricane when it made landfall. Floyd's destructive force was due to heavy rainfall, dumping as much as 24 inches of rain in eastern North Carolina on land already saturated by Hurricane Dennis, which had arrived 11 days earlier. A tropical depression and frontal system added even more rain, which cumulatively totaled more than 40 inches (Riggs 2001: 30–31). The resulting damage to mid-sized cities and small, rural communities throughout eastern North Carolina was immense. More than 45,000 homes and 24 wastewater treatment plans were flooded. Two hundred and thirty-five Red Cross shelters were opened, providing accommodations for more than 50,000 residents (Smith 2014a).

Floyd flooded many of the same communities that were still recovering from Hurricane Fran, which struck three years earlier. Given the magnitude of the flooding and the more active role of the state, the disaster recovery assistance that followed was substantially larger than for Fran. Governor Hunt, serving in his third term, was highly influential in the statehouse and in Washington DC. Not only did the state actively pursue supplemental funding appropriated by Congress, they also committed to perhaps the most significant allocation of state disaster recovery assistance in U.S. history, which led to the creation of twenty-two state programs designed to address gaps in federal assistance (Smith 2011: 57; Smith 2014a; Smith et al. 2018). The Governor also created the North Carolina Redevelopment Center, which was tasked with identifying unmet needs, procuring congressional appropriations, and developing new state recovery programs (Smith et al. 2018).

The Paradox of Federal and State Funding

The state undertook a blend of administrative negotiation and congressional lobbying to modify federal recovery policy, procure additional recovery funds, and create new state programs (Smith et al. 2018). The use of political influence, post-disaster data, a dramatic increase in state staffing, and the lack of a pre- or post-disaster recovery plan also influenced the trajectory of recovery at the community level. Governor Hunt successfully lobbied the state legislature to use $836 million from the state's "rainy day fund" and existing state agency budgets to pay for the cost of the new programs (Smith et al. 2018).

The governor also advocated for the adoption of sustainable development "principles" to guide the recovery process, yet no state recovery plan was created to clearly link these principles to tangible projects and programs (Smith et al. 2018). The State did require local governments to develop hazard mitigation plans in order to be eligible to receive post-disaster hazard mitigation funding. In an interesting twist on national policy formulation, similar language was incorporated into the Disaster Mitigation Act of 2000's enabling rules based on North Carolina's policy requirement (Smith 2014a: 204–205). The state also encouraged communities to consider land use in their hazard mitigation plans as a viable strategy to reduce future losses. While this produced mixed results, an evaluation of North Carolina's plans found that they did place greater emphasis on risk reduction policies when compared with the State of Florida's plans, which tended to be more focused on the development of plans as a means to access post-disaster mitigation funds that would pay for risk reduction projects (Berke et al. 2012).

In North Carolina, the local disaster recovery process was driven by a large assemblage of state recovery programs. These programs, while providing assistance not available from the federal government, had the added effect of further overwhelming local officials by increasing an administrative burden that ultimately exceeded the capacity of local governments and available consulting firms that were hired to do much of the work (Smith 2011: 57–58; Smith et al. 2018). Although some of the state funding was used to hire housing counselors, they worked for the state, not local governments.

Increasing State Capacity: Staffing the Redevelopment Center and North Carolina Division of Emergency Management

At the state level, a commitment was made to increase the size of staff in the North Carolina Division of Emergency Management (NCDEM) and the newly created Redevelopment Center. The Governor's Task Force, created after Hurricane Fran, had a staff of 5, whereas the Floyd Redevelopment Center hired a larger managerial and administrative staff, including a number of housing counselors. In NCDEM, the hazard mitigation section grew from a staff of 11, focused on the administration of Fran-related grant programs, to a staff of 50—including individuals focused on grants management, hazard mitigation planning, risk assessment, and floodplain management (Smith 2014a: 218; Smith et al. 2013). Most of these positions were filled by temporary employees, who often sought employment in the private sector once they learned highly marketable skills, leaving NCDEM with a loss of institutional knowledge and shifting levels of competency (Smith et al. 2013). This required the regular hiring and training of new staff to fill the void.

Post-Disaster Hazard Mitigation Programs

A key aspect of recovery following Fran and Floyd involved implementing hazard mitigation projects and policies. Following Hurricane Fran, approximately 1,150 homes scheduled for acquisition and 400 homes slated to be elevated were in various stages of completion when Floyd struck. After Floyd, the state chose to focus entirely on the acquisition of flood-prone homes. Eventually, the state approved more than 5,000 homes to be acquired and demolished, and the land converted to open space. The administration of post-disaster hazard mitigation programs is a highly technical and cumbersome process. In the case of an acquisition, or "buyout," the process involved assessing community and individual interest in the voluntary program, determining each property's eligibility (e.g., cost effectiveness, technical feasibility, and environmental impact), conducting title searches and closing the sale, demolishing the structure, and clearing the site.

The magnitude of the demand among local communities, and the clear message from the Governor that it was incumbent on the state to identify ways to speed the process, led to two Floyd-specific changes in federal policy that significantly streamlined the program. The first change was the development of a negotiated agreement with FEMA about what constituted a "cost-effective" project and the second change was the blending of HMGP and supplemental Community Development Block

Grant (CDBG) eligibility criteria that were used to acquire flood-prone homes. Based on lessons learned after Fran, and supported by data collected after Floyd, it was found that homes located in the floodplain that received over 2 feet of water were cost-effective using FEMA's benefit–cost analysis module, and therefore eligible for buyout under the HMGP.

State officials were able to ascertain that approximately 10,000 properties were cost-effective. The technique involved estimating ground elevation, flood extent, and building foundation types prevalent in the area. This data was mapped using a geographic information system, and the resulting maps were used to help bolster the state's case for additional congressionally appropriated funding beyond that available under the HMGP. In addition, a negotiated agreement with FEMA was eventually approved that established a proxy measure of cost-effectiveness, thereby eliminating the requirement to conduct a much more time-consuming benefit–cost analysis on each structure. This agreement also had the effect of substantially reducing the amount of supplemental funds available for other states receiving Hurricane Floyd disaster declarations (due to the creation of higher eligibility standards) and highlights the reality that, in major disasters, states often compete with one another for available federal dollars (Smith 2011: 266; Smith et al. 2018).

Federal programs, including those with similar goals, such as risk reduction and resilience, may have differing eligibility criteria, which can confuse local officials and hinder the timely expenditure of these funds. Following major disasters, congressionally appropriated CDBG funds are typically made available to states, which set priorities for the use of these funds in accordance with federal guidance. In North Carolina, the state purposefully established similar criteria for both CDBG and HMGP funds, thereby easing confusion at the local level and speeding the implementation of both programs (Smith et al. 2018).

The State Acquisition and Relocation Fund (SARF), which proved instrumental in encouraging participation in the hazard mitigation housing acquisition program, was created by the state in response to local officials' concerns about purchasing predominantly low-income properties. SARF provided up to $75,000 to participating homeowners in addition to the pre-disaster fair market value available through HMGP. Without the additional funding, many grant recipients whose homes were repeatedly flooded, or in general disrepair due to the lack of general maintenance, faced limited replacement housing options as they were ill-equipped to purchase a replacement home outside the floodplain. At the same time, local officials were concerned about the loss of existing tax base. As a result, a stipulation was added to the rules governing the program that homeowners who accepted the SARF money were required to use the funds to acquire a replacement home in the same county. In practice, the SARF program did help to incentivize participation among low-income homeowners, but it proved difficult to track compliance with the purchase of replacement homes.

Additional State Recovery Programs

The state created a number of additional programs to address long-standing problems that became evident after the storm. The Housing Repair and Replacement program provided local governments with funding to build new neighborhoods located outside the floodplain. The intent of the state program was to help offset the loss of available affordable housing that was either damaged or purchased and demolished as part of the large-scale buyout efforts. However, the program was not widely used and much of the funding allocated for this effort was reprogrammed into other state initiatives (Smith et al. 2018).

Another concern expressed by local governments following Hurricane Floyd was that many of the Flood Insurance Rate Maps (FIRMs) in the state were significantly outdated and development that relied on the accuracy of the data led to homes and public infrastructure being built in a manner that did not reflect the true nature of the flood hazard (Smith 2014a: 207). In an effort to resolve this problem, over $20 million of the state's redevelopment funding was used to update the FIRMs, beginning in the eastern part of the state and moving west. This process has continued over time, with the state assuming responsibility for floodplain mapping, the only state in the nation to do so. These maps, once created, are reviewed and approved by FEMA as required under the National Flood Insurance Program. The resulting North Carolina Floodplain Mapping Program has continued through funding from the state and a range of other sources. Most recently, the state digitized and geo-referenced all structures larger than 1,000 square feet in size that are located in the floodplain and has documented their first floor elevations. This digital information is unprecedented in the US and provides a level of detail found in no other state.

The flooding from Floyd not only exposed the inadequacies of the maps used to regulate flood-plain development, but also demonstrated the magnitude of inappropriate land uses in this largely rural area. Two of the most egregious examples include hog farms and automobile junkyards located in the floodplains of eastern North Carolina. Originally identified as potential buyouts under the state's HMGP criteria, FEMA deemed them ineligible. As a result, the state set aside funds to acquire the properties and turn the purchased land into open space. While some hog farms and junkyards were acquired and converted into more appropriate uses such as public parks and greenways, the program proved less successful than intended, as few applicants sought funding (Smith 2014a; Smith et al. 2018).

In an effort to assist cash-strapped jurisdictions with limited financial reserves, the state set aside funds to cover the non-federal cost share of disaster recovery programs. This North Carolina policy remains unusual because only four other states provided the entirety of the non-federal share in 2012 (NEMA 2012: 6–7). The same holds true for the Public Assistance Program, typically the largest federal payout among FEMA recovery programs, which is used to fund the costs associated with debris management, personnel costs, and infrastructure repair following a federally declared disaster.

Following Hurricane Floyd, the state sought to codify many of the programs and institute a typology of disaster assistance for events of various sizes or levels of damage, including those that did not meet the criteria for a federal disaster declaration. The resulting tiered state disaster declaration process triggers scalable levels of commitment based on the magnitude and associated damages caused by an event to include state programs targeting gaps in federal assistance (Smith 2014a: 218). A Type 1 disaster is triggered when a local state of emergency is declared, but no presidential disaster declaration has been made and preliminary damage assessments equal or exceed the criteria set for the Small Business Administration Loan Program. A Type 1 disaster may also be declared if the following criteria have been met—the jurisdiction has identified a minimum of $10,000 in uninsurable losses, uninsured losses exceed one percent of a jurisdiction's annual operating budget, the jurisdiction has an approved hazard mitigation plan, and it participates in the National Flood Insurance Program. A Type 1 disaster triggers two state programs: Individual Assistance and Public Assistance. These programs, which assist with the repair of damaged housing and infrastructure, mirror traditional FEMA Individual Assistance and Public Assistance grant eligibility criteria.

The Governor may declare a Type II disaster following a presidential disaster declaration. A Type II declaration triggers Type I programs as well as the State Acquisition and Relocation Fund and supplemental repair and replacement housing grants. A Type III disaster declaration criteria requires that the president has declared a major disaster, damage is of a sufficient magnitude to warrant a change in the federal cost share from 75 percent to 90 percent, and the governor calls a special session of the legislature to consider additional state funding to address unmet local needs (in Hurricane Floyd, the cost share for the Public Assistance program was reduced to a 90 federal/10 non-federal split). Eligible programs in a Type III disaster include those associated with Type II disasters as well as those authorized by the General Assembly (Smith and Sandler 2012: 60–61).

Fluctuations in State Capacity and Commitment: Lessons Learned and the Role of Planning

More recent assessments of state-level capacity and commitment in North Carolina highlights how state capacity can change over time. In 2014, the rainy day fund used to pay for the 22 state programs was gone, most of the staff involved in Floyd recovery efforts have left in pursuit of other job opportunities, temporary positions funded with federal dollars have been eliminated, and the state legislature and governor largely ignored the risks associated with a changing climate despite studies that warn of the potentially serious consequences of inaction (Smith 2014a). More recently, following Hurricane Matthew, which struck in 2016, the state elected a new governor (who recognizes the effects of a changing climate on flood hazard vulnerability) several months after the storm, the state legislature appropriated funds to support state recovery programs, and the state hired additional staff housed in the Governor's Office and the NCDEM. Some of these hires have been used to fill positions needed to administer post-disaster CDBG funds that were eliminated in the previous administration. The fluctuating capacity of the state is indicative of a significant, endemic problem across the US, namely the failure of states and the federal government to invest the appropriate resources in the pre-event timeframe in order to build and sustain the capacity needed to be prepared for the multitude of issues associated with disaster recovery (Smith 2011: 13, 40; Smith et al. 2013).

One closely related challenge remains the development of procedures to collect, archive, and share disaster recovery lessons, both internally (i.e., within organizations and the disaster recovery assistance networks in which they operate) and externally (i.e., across states and other networks). Good plans provide a valuable means to share institutional knowledge and identify the resources (i.e., funding, data, and staff) needed to implement policies over time. Many of the policies North Carolina developed after Floyd remain, codified in the tiered disaster declaration process. North Carolina crafted an improved state disaster recovery plan after Hurricane Katrina struck the Gulf Coast, which served as an impetus for many state and federal agencies to rethink their past approaches to disaster recovery and adhere to the tenets of the NDRF (Smith 2014a: 218). The 10-year gap between Floyd and the development of the state recovery plan hindered lessons learned, as most of the state officials involved in leading recovery efforts following Hurricane Floyd had left for other jobs. In 2017, the state began the process of updating its disaster recovery plan to reflect advances in the NDRF and to incorporate lessons from Hurricane Matthew and yet officials struggled to locate important background information undergirding the multiple programs developed and past experience garnered post-Floyd (Smith et al. 2018).

Lessons from the North Carolina Case Study

The North Carolina case study highlights a number of important themes, including 1) large-scale disasters in the U.S. trigger considerable amounts of post-disaster assistance and attention; 2) the manner in which states use the "window of opportunity" to garner and effectively coordinate the distribution and use of these resources varies significantly over time; and 3) the transfer of lessons from one event to another within a state, including the institutionalization of policies and the hiring of personnel needed to sustain the implementation of these policies, requires a significant commitment of political, financial, technical, and administrative resources that is difficult to achieve in practice.

North Carolina's ability to develop state-level recovery programs that address gaps in federal assistance is unique and is the result of an uncommon confluence of factors. These include a charismatic governor who was able to influence state legislators and members of Congress to provide resources that are significantly broader than usual; the availability of substantial state funding used to create an unprecedented array of state programs addressing local needs not met by federal aid; and the significant, albeit temporary hiring of additional state staff (Smith et al. 2013, 2014a, 2018). Considering the unique interaction of events and stakeholders, one might question the transferability of these lessons to other states. Viewed in a different light, the case study not only highlights how difficult it is for states to actively insert themselves into the federally dominated recovery process, but also uncovers subtle ways in which states of various capacities can play an important role in recovery to include identifying specific areas in which state capacity needs to be increased and maintained over time (Smith et al. 2018).

Following major disasters, the salience of disaster recovery policy increases dramatically. Much of the disaster recovery literature has focused on the degree to which major disasters affect federal policy change (Birkland 1997, 2006; Platt 1999; Rubin 2012). Much less attention has been placed on how extreme events shape state disaster recovery policy (Smith et al. 2018). North Carolina established a disaster recovery office for both Fran and Floyd; in the latter case, the state created 22 programs and established a tiered disaster declaration process to determine when these state programs would be available in the future. Although the Redevelopment Center remained in place for more than 10 years, the NCDEM staffing level decreased to pre-Fran levels with the exception of the North Carolina Floodplain Mapping Initiative (Smith 2014a).

North Carolina did not have a disaster recovery plan in place before or following Hurricane Floyd. Initial efforts to develop a post-Floyd recovery plan in the immediate aftermath of the storm were discontinued and not renewed until more than 10 years later. An important role of state plans is to identify the means to help local governments build capacity to achieve an agreed upon vision and associated set of goals. In North Carolina, the development of new state programs further overwhelmed local governments. Nor were the 22 state programs tied to a "sustainable recovery" vision initially voiced by the governor, as state officials failed to develop a broader, inclusive plan for how all of the new programs fit together across the larger disaster recovery assistance network, the timing of that assistance, and the integration of federal and state programmatic objectives (Smith 2011: 57).

The current state recovery plan, although a step in the right direction, ignores many of the key plan quality principles and the three key dimensions of recovery discussed in this chapter. Thus, an important question remains—do high quality plans lead to superior recovery outcomes? Some case study evidence suggests that this is the case (Schwab et al. 1998, 2014). It is much less clear about the direct nexus between good state recovery plans and outcomes at the state and local level. Recovery plans should help build state and local capacity, coordinate the distribution of assistance in a way that meets local needs, and facilitate good horizontal and vertical integration between federal, state, and local units of government.

The Promise of the National Disaster Recovery Framework: Building State Capacity and Commitment

The continued development of the NDRF stands to help states develop better plans, if in fact key dimensions of recovery are addressed as part of clear federal guidance (Smith et al. 2018). GAO studies in 2010 and 2016 identified problems that are similar to those described in this chapter. Two specific problems include: 1) delivering vague guidance regarding the appropriate time to employ FEMA's long-term recovery assistance, which leads to confusion among federal agencies and state officials, and 2) providing disaster recovery assistance before state and local government officials are ready to receive it, and once provided, the assistance ending before many long-term recovery needs for planning and coordination are met. The emphasis on post-disaster recovery planning and the lack of sustained engagement has led to the creation of disaster recovery plans that set unrealistic expectations, or outlined project ideas that did not have clear implementation mechanisms, such as funding (Smith 2011: 322). The 2010 GAO study, which pre-dates the creation FEMA's

Community Planning and Capacity Building (CPCB) team, which is responsible for coordinating disaster recovery planning efforts with states, highlights problems that still exist.

Additional timing-related problems include a disproportionate focus on post-disaster aid. Pre-disaster recovery planning assistance and capacity building still remain limited, but are improving unevenly across FEMA regions. Conversations with state and federal officials, including those involved in more recent events such as Hurricanes Harvey and Maria, and a 2010 GAO report suggest that CPCB officials are not always welcome in the post-disaster environment. Reasons cited included efforts to engage with state and local officials at inappropriate times, seeking information from overwhelmed state and local officials, and conveying information about federal programs that they didn't manage.

In order for the NDRF to be successful and address identified limitations at the state level, five actions should be taken.

1   Assist states in building capacity and commitment to address the three dimensions of recovery at the local level.
2   Ensure that states commit the time and resources needed to adequately foster local capacity and commitment to recovery planning.
3   Improve the timing of assistance to include more investments in pre-event planning for post-disaster recovery.
4   Ensure that state leadership, particularly governors, play an active role in disaster recovery policy and planning.
5   Hold state and local governments more accountable for disaster recovery actions and outcomes as plans are developed and capacity is built over time.

Significant changes are required to address disincentives to develop and sustain strong state disaster recovery capabilities. May and Williams (1986) characterize the political dilemma associated with disaster policymaking as intergovernmental implementation nested within the concept of shared governance where states and local governments have important roles to play, and yet the roles assumed by states can be hampered by an excessive level of dependence on federal programs. Given that federal disaster recovery policy and associated grants-in-aid programs are often shaped by the latest major disaster, the notion of shared governance remains substantially skewed towards a federal orientation. Further compounding this unequal distribution of capacity and commitment is that federal programs are created with limited involvement of subnational governments and states receive substantial assistance regardless of their level of pre-event preparedness for post-disaster recovery. Perhaps the disaster deductible concept proposed by FEMA can begin to shift this focus as part of a larger effort to hold states more accountable while investing more resources before disasters to build the capacity of states (and local governments) to act.

The federal government has produced disaster management policy, such as the DMA 2000 and PKEMRA, that are intended to proactively build state and local capacity tied to national goals of reducing future losses and improving disaster recovery outcomes respectively. However, these efforts tend to be less politically salient at subnational levels and are less likely to receive significant and sustained attention as predicted by May and Williams (1986: 3). This is evident in state and local mitigation plans that are of low quality (Berke et al. 2012; Lyles et al. 2014a; Smith et al. 2013) and in the dearth of local disaster recovery plans (Berke et al. 2014). Furthermore, the low quality of hazard mitigation plans suggests that they are a means to an end (i.e., access to federal funding) rather than a process used to comprehensively reduce future hazard-related losses. The passage of PKEMRA provides a vehicle to improve disaster recovery outcomes by investing more resources before a disaster occurs and gradually holding communities more accountable over time. Unfortunately, the reality remains that PKEMRA has not sufficiently addressed this substantial problem (Smith et al. 2017).

Future Research: Do State Recovery Plans Matter?

The ability to implement the policies proposed in this chapter, as well as to create a more robust and enduring national recovery strategy, should be predicated on a better understanding of states' roles in disaster recovery (Sandler and Smith 2013; Smith 2011; Smith et al. 2018).

A number of important questions raised in this chapter merit additional research.

1 Why do some states, including agency officials and governors, take a more active role in recovery than others?
2 How can state capacity and commitment be fostered and sustained over time?
3 What is the quality of state disaster recovery plans across the U.S.?
4 Based on an analysis of state recovery plans, is there a correlation between the quality of state recovery plans and tangible recovery outcomes?

These questions suggest that a multi-method approach is needed to tease out complex procedural issues through a well-designed set of case studies, interviews of key stakeholders across disaster recovery assistance networks, and a national survey of those involved in disaster recovery planning. This approach could benefit from an empirical analysis of the three dimensions of disaster recovery described in this chapter, to include the degree to which planning can transform the relationship between resource rules and local needs, the timing of assistance, and horizontal and vertical integration. Additional methods include an evaluation of state recovery plans using plan quality principles coupled with an approach that unpacks the correlation between plan quality and recovery outcomes. Such studies stand to benefit from a pre- and post-testing of state recovery plans to include those developed before and after NDRF guidance is codified. Ideally, the pre- and post-testing of plans would include longitudinal assessments of select states (i.e., continued study between pre- and post-testing) chosen from the findings of the national survey and plan quality analysis. Although evaluating the disaster recovery process is difficult and time consuming to achieve due to the lengthy nature of on-site review and participant observation procedures, it would help to further clarify the still murky process by which states operate before and after disasters strike.

References

Alesch, D. L., L. A. Arendt, and J. N. Holly. (2009) *Managing for Long-Term Community Recovery in the Aftermath of Disaster*, Fairfax VA: Public Entity Risk Institute.

Baer, W. C. (1997) "General Evaluation Criteria: An Approach to Making Better Plans," *Journal of the American Planning Association* 63(3): 329–345.

Barnes, J. (2001) *North Carolina's Hurricane History*, Chapel Hill NC: The University of North Carolina Press.

Baumgartner, F. R. and B. D. Jones. (1993) *Agendas and Instabilities in American Politics*, Chicago IL: University of Chicago Press.

Berke, P. R. and T. Beatley. (1997) *After the Hurricane: Linking Recovery to Sustainable Development in the Caribbean*, Baltimore MD: Johns Hopkins University Press.

Berke, P. R., J. Cooper, M. Aminto, J. Horney, and S. Grabich. (2014) "Adaptive Planning for Disaster Recovery and Resiliency: An Evaluation of 87 Local Recovery Plans in Eight States," *Journal of the American Planning Association* 80(4): 310–323

Berke, P.R. and S. P. French. (1994) "The Influence of State Planning Mandates on Local Plan Quality," *Journal of Planning Education and Research* 13(4): 237–250.

Berke, P. R. and D. R. Godschalk. (2009) "Searching for the Good Plan: A Meta-analysis of Plan Quality Studies," *Journal of Planning Literature* 23(3): 227–240.

Berke, P. R., J. Kartez and D. Wenger. (1993) "Recovery After Disasters: Achieving Sustainable Development, Mitigation and Equity," *Disasters* 17(2): 93–109.

Berke, P., W. Lyles, and G. Smith. (2014). "Impacts of Federal and State Mitigation Policies on Local Land Use Policy," *Journal of the Planning Education and Research* 34(1): 60–76.

Berke, P. R., G. Smith, and W. Lyles. (2012) "Planning for Resiliency: Evaluation of State Hazard Mitigation Plans Under the Disaster Mitigation Act," *Natural Hazards Review* 13(2): 139–150.

Birkland, T. A. (1997) *After Disaster, Agenda Setting, Public Policy, and Focusing Events*, Washington DC: Georgetown University Press.

Birkland, T. A. (2006) *Lessons of Disaster: Policy Change After Catastrophic Events*, Washington DC: Georgetown University Press:

Boin A. and 't Hart, P. (2003) "Public Leadership in Times of Crisis: Mission Impossible?" *Public Administration Review* 63(5): 544–553.

Brody, S. (2003) "Are We Learning to Make Better Plans? A Longitudinal Analysis of Plan Quality Associated with Natural Hazards," *Journal of Planning Education and Research* 23(2): 191–201.

Burby, R. J. (2006) "Hurricane Katrina and the Paradoxes of Government Disaster Policy: Bringing About Wise Government Decisions in Hazardous Areas," *The Annals of the American Academy of Political and Social Science* 64(1): 171–192.

Burby, R. J. and P. J. May. (2009) "Command or Cooperate? Rethinking Traditional Central Governments' Hazard Mitigation Policies," 21–33, in Urbano Fra Paleo (ed.), *Building Safer Communities: Risk Governance, Spatial Planning and Responses to Natural Hazards*, Amsterdam: IOS Press.

Burby, R. J., P. J. May, P. R. Berke, L. C. Dalton, S. P. French, and E. J. Kaiser. (1997) *Making Governments Plan: State Experiments in Managing Land Use*, Baltimore MD: Johns Hopkins University Press.

Chandrasekhar, D., Y. Zhang, and Y. Xiao. (2014) "Nontraditional Participation in Disaster Recovery Planning: Cases from China, India, and the United States," *Journal of the American Planning Association* 80(4): 373–384.

CREW—Citizens for Responsibility and Ethics in Washington. (2007) *The Best Laid Plans: The Story of How the Government Ignored its Own Gulf Coast Hurricane Plan*, Washington DC: Citizens for Responsibility and Ethics in Washington.

Cutter, S., C. T. Emrich, J. T. Mitchell, W. W. Piegorsch, M. M. Smith, and L. Weber. (2014) *Hurricane Katrina and the Forgotten Coast of Mississippi*. New York: Cambridge University Press.

Delia, A. A. (2001) "Population and Economic Changes in Eastern North Carolina Before and After Hurricane Floyd," 199–204, in J. R. Maiolo, J. C. Whitehead, M. McGee, L. King, J. Johnson, and H. Stone (eds.), *Facing Our Future: Hurricane Floyd and Recovery in the Coastal Plain*, Greenville NC: Coastal Carolina Press.

Durham, T. and L. Suiter. (1991) "Perspectives and Roles of the State and Federal Governments," 101–127, in T. Drabek and G. Hoetmer (eds.) *Emergency Management: Principles and Practice for Local Government*, Washington DC: International City Management Association.

Florida Department of Community Affairs, Division of Community Planning. (2010) *Post-Disaster Redevelopment Planning: A Guide for Florida Communities*, Tallahassee FL: Author.

Friedmann, J. (1987) *Planning in the Public Domain: From Knowledge to Action*, Princeton NJ: Princeton University Press.

Friesema, H.P, J. Caporaso, G. Goldstein, R. Lineberry, and R. McCleary. (1979) *Aftermath: Communities After Natural Disasters*, Beverly Hills CA: Sage Publications.

Geipel, R. (1982) *Disaster and Reconstruction: The Friuli (Italy) Earthquake of 1976*, London: Allen and Unwin.

Geschwind, C-H. (2001) *California Earthquakes: Science, Risk, and the Politics of Hazard Mitigation*, Baltimore MD: Johns Hopkins University Press.

GAO—Government Accountability Office. (September 2008) *Disaster Recovery: Past Experiences Offer Insights for Recovering from Hurricanes Ike and Gustav and Other Recent Natural Disasters*, GAO-08-1120, Washington DC: Government Accountability Office.

GAO—Government Accountability Office. (March 2010) *Disaster Recovery: FEMA's Long-term Assistance Was Helpful to State and Local Governments but Had Some Limitations*, Report to Congressional Requesters, GAO-10-404, Washington DC: Author.

GAO—Government Accountability Office. (2016) *Disaster Recovery: FEMA Needs to Assess its Effectiveness in Implementing the National Disaster Recovery Framework*, Washington DC: Author.

GPO—Government Printing Office. (2017) *Establishing a Deductible for FEMA's Public Assistance Program* 82(8): 4064–4097, Washington, DC: Government Printing Office.

Haas, J. E., R. W. Kates, and M. J. Bowden. (1977) *Reconstruction Following Disaster*, Cambridge MA: The MIT Press.

Innes, J. and D. Booher. (1999) "Consensus Building and Complex Adaptive Systems: A Framework for Evaluating Collaborative Planning," *Journal of the American Planning Association* 65(4): 412–423.

Innes, J. and D. Booher. (2003) "Collaborative Policymaking: Governance through Dialogue," 33–59, in M. A. Hajer and H. Wagenarr (eds.), *Deliberative Policy Analysis: Understanding Governance in a Network Society*, New York: Cambridge University Press.

Innes, J. and D. Booher. (2004) "Reframing Public Participation: Strategies for the 21st Century," *Planning Theory and Practice* 5(4): 419–436.

Johnson, L. A. and R. B. Olshansky. (2016) *After Great Disasters: How Six Countries Managed Community Recovery*, Cambridge MA: Lincoln Institute of Land Policy.

Kaiser, E. J., Godschalk, D. R., and Chapin, S. Jr. (1995) *Urban Land Use Planning*, 4th Edition, Urbana IL: University of Illinois.

Kapucu, N. and M. Van Wart. (2008) "Making Matters Worse: An Anatomy of Leadership Failures in Managing Catastrophic Events," *Administration and Society* 40(7): 711–740.

Kreps, G. A. (1990) "The Federal Emergency Management System in the United States: Past and Present," *International Journal of Mass Emergencies and Disasters* 8(3): 277.

Lyles, W., Berke, P., and Smith, G. (2014a) "A Comparison of Local Hazard Mitigation Plan Quality in Six States, USA," *Landscape and Urban Planning* 122: 89–99.

Lyles, W., Berke, P., and Smith, G. (2014b). "Do Planners Matter? Examining Factors Driving Incorporation of Land Use Approaches into Hazard Mitigation Plans," *Journal of Environmental Planning and Management* 57(2): 792–811.

May, P. J. (1985), *Recovering from Catastrophes: Federal Disaster Relief Policies and Politics*, Westport CT: Greenwood Press.

May, P. J. and W. Williams. (1986) *Disaster Policy Implementation: Managing Programs Under Shared Governance*, New York: Plenum Press.

Mitchell, J. K. (2006) "The Primacy of Partnership: Scoping a New National Disaster Recovery Policy," *ANNALS of the American Academy of Political and Social Science* 604(1): 228–255.

Montz, B. E. and G. A. Tobin (2008). "Livin' Large with Levees: Lessons Learned and Lost," *Natural Hazards Review* 9(3): 150–157.

NEMA—National Emergency Management Association. (2012) *NEMA Biennial Report*, Lexington KY: National Emergency Management Association.

Oliver-Smith, A. and R. Goldman. (1988) "Planning Goals and Urban Realities: Post-Disaster Reconstruction in a Third World City," *City and Society* 2(2): 67–79.

Olshansky, R. B., L. A. Johnson, and K. C. Topping. (2006) "Rebuilding Communities Following Disaster: Lessons from Kobe and Los Angeles," *Built Environment* 32(4): 354–374.

Olshansky, R. B., L. A. Johnson, J. Horne, and B. Nee. (2008) "Planning for the Rebuilding of New Orleans," *Journal of the American Planning Association* 74(3): 273–287.

Olshansky, R. B., L. D. Hopkins, and L. A. Johnson. (2012) "Disaster and Recovery: Processes Compressed in Time," *Natural Hazards Review* 13(3): 173–178.

Olson, R. A. (2003) "Legislative Politics and Seismic Safety: California's Early Years and the 'Field Act,' 1925–1933," *Earthquake Spectra* 19(1): 111–131.

Olson, R. S., R. A. Olson, and V. T. Gawronski. (1999) *Some Buildings Just Can't Dance: Politics, Life Safety, and Disaster*, Stamford CT: JAI Press.

Peacock, W. G., B. H. Morrow, and H. Gladwin. (1997) *Hurricane Andrew: Ethnicity, Gender, and the Sociology of Disasters*, New York: Routledge.

Platt, R. (1999) *Disasters and Democracy: The Politics of Extreme Natural Events*, Washington DC: Island Press.

Riggs, S. (2001) "Anatomy of a Flood," 29–45, in J. R. Maiolo, J. C. Whitehead, M. McGee, L. King, J. Johnson, and H. Stone (eds.), *Facing Our Future: Hurricane Floyd and Recovery in the Coastal Plain*, Greenville NC: Coastal Carolina Press.

Rossi, P. H., J. D. Wright, and E. Weber-Burdin. (1982) *Natural Hazards and Public Choice: The State and Local Politics of Hazard Mitigation*, New York: Academic Press.

Rubin, C. (1991) "Recovery from Disaster," 224–259, in T. E. Drabek and G. J. Hoetmer (eds.), *Emergency Management: Principles and Practice for Local Governments*, Washington DC: International City Management Association.

Rubin, C. (2009) "Long-Term Recovery from Disasters: The Neglected Component of Emergency Management," *Journal of Homeland Security and Emergency Management* 6(1): 1–17.

Rubin, C. (2012) *Emergency Management: The American Experience 1900–2010*, Boca Raton FL: CRC Press.

Rubin, C. and R. Popkin. (1990) *Disaster Recovery after Hurricane Hugo in South Carolina*, Boulder CO: University of Colorado Institute of Behavioral Science.

Rubin, C., M. Saperstein, and D. G. Barbee. (1985) *Community Recovery from a Major Natural Disaster*, Boulder CO: University of Colorado, Institute of Behavioral Science.

Sabatier, P. A., and H. C. Jenkins-Smith. (1993) *Policy Change and Learning: An Advocacy Coalition Approach*. Boulder CO: Westview Press.

Sandler, D. and G. Smith. (2013) "Assessing the Quality of State Disaster Recovery Plans: Implications for Policy and Practice," *Journal of Emergency Management* 11(4): 281–291.

Schwab, J., A. Boyd, J. B. Hokanson, L. A. Johnson, K. Topping. (2014) *Planning for Post-Disaster Recovery: Next Generation*, Chicago IL: American Planning Association.

Schwab, J., K. C. Topping, C. C. Eadie, R. E. Deyle, and R. Smith. (1998) *Planning for Post-Disaster Recovery and Reconstruction*, Chicago IL: American Planning Association.

Seidman, K. F. (2013) *Coming Home to New Orleans: Neighborhood Rebuilding After Katrina*, New York: Oxford University Press.

Smith, G. (2008) "Recovery from Catastrophic Disasters," Paper presented at the 33rd Annual Hazards Research and Applications Workshop, Broomfield, CO, July 12–15.

Smith, G. (2011) *Planning for Post-Disaster Recovery: A Review of the United States Disaster Assistance Framework*, Washington DC: Island Press.

Smith, G. (2014a) "Applying Hurricane Recovery Lessons in the U.S. to Climate Change Adaptation: Hurricanes Fran and Floyd in North Carolina, USA," 193–229, in B. Glavovic and G. Smith (eds.), *Adapting to Climate Change: Lessons from Natural Hazards Planning*, New York: Springer.

Smith, G. (2014b) "Disaster Recovery in Coastal Mississippi, USA: Lesson Drawing from Hurricanes Camille and Katrina," 339–368, in B. Glavovic and G. Smith (eds.), *Adapting to Climate Change: Lessons from Natural Hazards Planning*, New York: Springer.

Smith, G. and T. Birkland. (2012) "Building a Theory of Recovery: Institutional Dimensions," *International Journal of Mass Emergencies and Disasters* 30(2): 147–170.

Smith, G. and V. Flatt. (2011) *Assessing the Disaster Recovery Planning Capacity of the State of North Carolina*, Research Brief, Durham, NC: Institute for Homeland Security Solutions.

Smith, G., W. Lyles, and P. R. Berke. (2013) "The Role of the State in Building Local Capacity and Commitment for Hazard Mitigation Planning," *International Journal of Mass Emergencies and Disasters* 31(2): 178–203.

Smith, G., A. Martin, and D. Wenger. (2017) "Disaster Recovery in an Era of Climate Change: The Unrealized Promise of Institutional Resilience," 595–619, in H. Rodriguez, W. Donner, and Joseph E. Trainor (eds.), *Handbook of Disaster Research*, 2nd ed., New York: Springer.

Smith, G., L. Sabbag, and A. Rohmer. (2018) "A Comparative Analysis of the Roles Governors Play in Disaster Recovery," *Risk, Hazards, & Crisis in Public Policy* 9(2): 205–243.

Smith, G. and D. Sandler. (2012) *State Disaster Recovery Guide*, Chapel Hill NC: Department of Homeland Security Coastal Hazards Center of Excellence.

Smith, G. and D. Wenger. (2006) "Sustainable Disaster Recovery: Operationalizing an Existing Framework," 234–257, in H. Rodriguez, E. Quarantelli, and R. Dynes (eds.), *Handbook of Disaster Research*, New York: Springer.

Sylves, R. (2008) *Disaster Policy and Politics: Emergency Management and Homeland Security*, Washington DC: CQ Press.

Tobin, G. A. (1995) "The Levee Love Affair: A Stormy Relationship?" *JAWRA Journal of the American Water Resources Association* 31(3): 359–367.

Topping, K. (2009) "Toward a National Disaster Recovery Act of 2009," *Natural Hazards Observer*. Invited Comment. Vol. XXXIII, No. 3.

Waugh, W. and R. Sylves. (1996) "The Intergovernmental Relations of Emergency Management," 46–48, in R. T. Sylves and W. L. Waugh (eds.), *Disaster Management in the U.S. and Canada*, 2nd ed., Springfield IL: Charles C. Thomas.

Wright, J. D., P.H. Rossi, S.R. Wright, and E. Weber-Burdin. (1979) *After the Clean-Up: Long Range Effects of Natural Disasters*, Beverly Hills CA: Sage.

23

RECOVERY PLANNING WITH U.S. CITIES

Laurie A. Johnson

Introduction

When it comes to disasters, planning is a core function of local disaster management. Planning for emergency response is essential to the creation of a common operational picture and for helping to set priorities, objectives, strategies, and tactics (DHS 2008). Planning for disaster recovery, however, is not nearly as common, standardized, or well-researched as emergency response and other aspects of disaster management. In particular, the sub-field of long-term recovery has been cited as an outlier, or "orphan," when it comes to concerted policy attention, pre-disaster planning, and post-disaster planning and implementation (Abramson et al. 2011; Smith and Wenger 2007).

This chapter focuses on plans and planning for both pre- and post-disaster recovery at the local level. As noted in the National Disaster Recovery Framework (NDRF), "local, State and tribal governments have primary responsibility for the recovery of their communities and play the lead role in planning for and managing all aspects of community recovery" (FEMA 2016: 6). First, some of the key triggers for city planning for long-term recovery in the United States are considered. Next, the body of research on local-level recovery planning processes and plans is reviewed. Then an analysis of a combination of recovery research and practical observations posits how recovery planning differs from normal city planning, with the goal of identifying future recovery research needs and informing future long-term recovery planning policy and practice.

Triggers for City-Level Planning for Long-Term Recovery

All planning begins with a decision that some construction and shaping of recommended future policies and actions is needed. For long-term recovery planning in the United States, there are few regulatory triggers for such decisions, pre- or post-disaster. Instead, recovery planning has largely been a function performed by communities after they were impacted by significantly damaging disasters. It has also been undertaken, in far fewer instances, by communities that engaged in pre-disaster recovery planning because they faced significant threats.

A handful of cities are known to have developed pre-disaster recovery plans in the 1990s and earlier—notably Los Angeles, California, which is discussed later in the chapter (City of Los Angeles EOO 1994). Many of these pre-disaster recovery planning efforts were motivated by evidence of a large-scale risk, such as earthquakes or hurricanes, and political recognition of the potentially catastrophic impact that would necessitate a sustained organizational commitment to long-term recovery of the affected communities (Spangle Associates 1997; Spangle 1987). For a limited few, pre-disaster recovery planning was initiated in order to codify lessons learned in their actual recovery experience, such as the City of Watsonville, California, which in the years following the 1989 Loma Prieta earthquake developed a pre-disaster recovery plan that focused on involving community-based organizations (CBOs) in the city's governmental structure following a future disaster (City of Watsonville 1992).

Other communities across the United States have developed pre-disaster recovery plans as part of emergency management, continuity of operations, or risk management planning efforts. One such impetus comes from the Emergency Management Accreditation Program (EMAP) established in 1997, which is a non-profit voluntary assessment and accreditation process for state and local emergency management programs. It requires that an accredited Emergency Management program has the following elements: prevention, preparedness, mitigation, response, and recovery (EMAP 2013). It defines the recovery process as "development, coordination, and execution of plans or strategies for the restoration of impacted communities and government operations and services through individual, private sector, nongovernmental and public assistance" (EMAP 2013: 3). It then calls for the recovery plan to "address short-term and long-term recovery priorities and provide guidance for restoration of critical functions, services, vital resources, facilities, programs, and infrastructure post-disaster" (EMAP 2013: 8).

A national movement towards pre-disaster recovery planning began in the aftermath of the unusually active hurricane seasons of 2004 and 2005, as governments at all levels across the country became concerned about their own long-term recovery capabilities. In 2007, the state of Florida launched a three-year recovery planning initiative to develop a post-disaster redevelopment plan (PDRP) guidebook for coastal counties and cities in the state (State of Florida 2010). The guidebook is based on the consensus input and support of the local governments, planners, emergency responders, business organizations and other stakeholders participating in the six pilot community efforts. An addendum to the guidebook incorporates sea level rise data and concerns into the post-disaster redevelopment planning framework using southeast Florida—specifically Palm Beach County—as the pilot community (State of Florida 2015).

With the 2011 introduction of the NDRF, the federal government also signaled its intent to encourage the formalization of state and local pre-disaster recovery planning (FEMA 2011a). The second edition of the NDRF was published in 2016 (FEMA 2016). The NDRF specifies local government as having "the primary role of planning and managing all aspects of the community's recovery" and also recommends appointing a Local Disaster Recovery Manager to coordinate development, training and exercise of the jurisdiction's disaster recovery plan (FEMA 2016: 14, 17) It identifies core principles to guide recovery efforts that include pre-disaster recovery planning, partnerships and inclusiveness, unity of effort, resilience and sustainability, and individual and family empowerment. The NDRF defines eight core capabilities for recovery: Planning; Public Information and Warning; Operational Coordination; Economic Recovery; Health and Social Services; Housing; Infrastructure Systems; and Natural and Cultural Resources. Although no mandates or standards for pre-disaster recovery planning have been set as of this writing, FEMA has been adding recovery planning expertise to all of its regional offices and developing both state and local recovery planning guides and training offerings. State and local jurisdictions across the country are implementing recovery planning efforts that also incorporate the core principles, management structure, and recovery core capabilities defined in the NDRF.

One of the first post-disaster recovery plans created by a modern U.S. city was in Santa Cruz, California after the 1989 Loma Prieta earthquake (Arnold 1999). Vision Santa Cruz focused on rebuilding the city's downtown and emphasized the desired outcomes as opposed to the process (William Spangle and Associates 1991). Santa Cruz was also one of the first U.S. cities to have the assistance of the Urban Land Institute (ULI), which conducted a one-week planning charrette in the city a few months after the earthquake. Since that time, the ULI and other professional organizations, planning consultants, and governmental agencies have provided similar post-disaster planning assistance to disaster-stricken communities across the U.S, in some instances on a pro-bono basis.

Over the past 20 years, the number of U.S. cities conducting recovery planning in the post-disaster period has increased exponentially. This trend is arguably rooted in two federal-level triggers:

- FEMA's provision of federally funded technical assistance for post-disaster planning in communities that receive Presidential Disaster Declarations (PDDs), and
- Department of Housing and Urban Development (HUD) action planning requirements for Community Development Block Grant-Disaster Recovery (CDBG-DR) funds awarded through disaster-specific Congressional appropriations.

FEMA Post-Disaster Planning Technical Assistance

Almost a decade before the NDRF, the 2004 National Response Plan included Emergency Support Function (ESF)-14—Long-Term Community Recovery Planning (LTCR) with the purpose of coordinating federal agency resources to support long-term recovery of states and communities and to mitigate future risks (DHS 2004). FEMA was designated as the lead agency for ESF-14, which was activated on a trial basis in 2004 in Florida after Hurricanes Charley, Frances, and Jeanne and in Utica, Illinois after a tornado. Between 2004 and 2011, FEMA's LTCR teams worked with more than 180 communities across 23 states, two Indian tribal governments, and the Commonwealth of Puerto Rico, producing some 90 community post-disaster recovery plans, strategies, or documents and providing assistance to 11 states to organize for recovery (FEMA 2011b).

For example, following the 2005 hurricanes, FEMA applied ESF-14 broadly along the Gulf Coast. In Louisiana, FEMA and the state government launched an ESF-14 LTCR process, working with each affected Louisiana parish to consult with the community and prepare a recovery plan. At its peak, more than 350 planners and technical specialists were working as part of the ESF-14 LTCR process across the 19 Louisiana parishes affected by Hurricanes Katrina and Rita (FEMA 2011b). The Louisiana Recovery Authority (LRA) then linked the allocation of a portion of federal CDBG "disaster recovery" funds to parishes with LRA-accepted, long-term recovery plans—most of which were developed through the ESF-14 LTCR process (LRA 2006).

Similarly, in Iowa following tornados, torrential rain, and flooding in the summer of 2008, the State's Rebuild Iowa Office (RIO) also established a Community and Regional Recovery Planning team that partnered with the FEMA-LTCR team and local Councils of Government to assist 10 severely impacted communities in their long term recovery efforts (Rebuild Iowa Office 2011). Support ranged from a single technical specialist to teams of up to eight specialists and planners assisting with the development of recovery plans and strategies (FEMA 2011b).

With the introduction of the NDRF in 2011, the FEMA-LTCR program was superseded by the NDRF's Community Planning and Capacity Building (CPCB) Recovery Support Function (RSF), for which FEMA was the designated lead federal agency (FEMA 2011a). The mission of the CPCB RSF was to support and build the capacity and resources of local, State and tribal governments to plan for, manage, and implement disaster recovery activities. The CPCB RSF not only promoted local ownership of recovery and a long-term recovery planning process, it also emphasized a holistic recovery management process and coordination among federal agencies and non-governmental partners to support states and communities.

The first large-scale implementation of the CPCB-RSF occurred in New York and New Jersey following Hurricane Sandy's landfall in late October 2012. FEMA and the CPCB RSF coordinated assistance from over 15 federal agencies, as well as a wide range of professional and CBOs, to assist states and communities in organizing, planning, and building recovery management capability across the affected area (Hurricane Sandy Rebuilding Task Force 2014). For example, in New York the CPCB RSF and its partners assisted the State in developing and launching the New York Rising Community Reconstruction Program (New York Rising). In doing so, it developed case study guidance and recovery planning-related documents, trained the state contractors, and developed Geographic Information System (GIS) data and community conditions information for severely impacted communities (Hurricane Sandy Rebuilding Task Force 2014). In New Jersey, CPCB RSF and partners have helped impacted communities develop and implement recovery planning processes and also train and fund the hiring of Local Disaster Recovery Managers in several communities.

With the second edition of the NDRF, the CPCB RSF is replaced by the recovery core capability of Planning which emphasizes pre-disaster recovery planning, community-based planning and post-disaster recovery planning (FEMA 2016). The framework calls for community planning efforts to reflect and involve the whole community and be supported by voluntary, faith-based community organizations; businesses; and local, regional/metropolitan, state, tribal, territorial, insular area, and federal governments. FEMA CPCB program efforts continue under the new framework.

HUD CDBG-DR Planning Requirements

Since 1992, following Hurricane Andrew, Congress has appropriated disaster recovery (DR) grants through the Community Development Block Grant (CDBG) program (McCarty et al. 2005). CDBG-DR grants are noncompetitive, nonrecurring grants awarded to state and local governments by a formula that considers disaster recovery needs unmet by other federal disaster assistance programs (HUD 2014). As of March 2018, HUD was managing over $55 billion in CDBG-DR funds for disasters dating back to the 2001 World Trade Center disaster (HUD 2018).

Before receiving a grant, an eligible government must develop and submit an action plan that describes the needs, strategies, and projected uses of the CDBG-DR funds. In recent disasters, a large portion of the CDBG-DR funds has been directed to disaster-impacted state governments, which then submit the CDBG-DR action plan. In some cases, however, HUD entitlement cities such as New York City post-Sandy have been direct recipients and thus prepared their own CDBG-DR action plans.

Many post-disaster recovery planning processes and plans have resulted from the CDBG-DR planning requirements. As an example, following catastrophic flooding on the Red River of the North in 1997, the City of Grand Forks, North Dakota, was awarded $171.6 million in one of the largest CDBG-DR grants ever made to a single jurisdiction at the time (Natural Hazards Center 1999). The states of North Dakota and Minnesota managed the CDBG-DR funds for the smaller and less affected local jurisdictions in the disaster. For Grand Forks, HUD also committed $1 million of on-site technical assistance and utilized its National CDBG Technical Assistance Contract to lead the effort (HUD Technical Assistance Team 1997). Until 1997, this contract had been used primarily to provide a more limited technical assistance to many communities participating in the HUD Empowerment Zone/Enterprise Communities Initiative.

The City of Grand Forks used the HUD-funded team and CDBG-DR funds to undertake a number of recovery planning efforts (Johnson 2014). A citywide Short-Term Recovery Action Plan integrated and documented in one place all the work that had already been done in the previous weeks of the disaster, including the City's initial plan for using CDBG-DR funds, its FEMA Hazard Mitigation Grant Program (HMGP) plan for the voluntary acquisition program to purchase substantially damaged residences in the 100-year floodplain, and plans for constructing a permanent flood hazard mitigation project along the Red River (City of Grand Forks 1997). Following this, plans for a new housing development, downtown redevelopment, and a greenway park along the Red River were also developed with the HUD technical assistance and CDBG-DR funds.

Following Hurricane Sandy, Congress approved a Sandy supplemental aid package on January 28, 2013 and HUD directed an initial allocation of $1.7 billion in CDBG-DR funds to the State of New York (Cuomo 2013). The State's proposed action plan was approved by HUD in late April 2014, at which time the Governor unveiled the Community Reconstruction Zone program (now called the NY Rising Community Reconstruction program), which uses CDBG-DR funds to provide additional rebuilding and revitalization assistance to communities severely impacted by Hurricanes Sandy and Irene and Tropical Storm Lee; communities impacted by the 2013 flooding were added later (New York State 2018). This program enabled communities to identify resilient and innovative reconstruction projects and other needed actions by developing community-driven recovery plans that consider current damage, future threats, and economic opportunities. Communities successfully completing a recovery plan are then eligible for state funding to support the implementation of projects and activities identified in the plans.

As part of its effort, each NY Rising community was asked to establish a Planning Committee composed of local residents and business leaders, as well as municipal representatives and elected officials with non-voting status, to lead the plan development. In addition, the State provided each community with a planning team to help prepare the plan. Consultants have been hired through a process administered by New York State Homes and Community Renewal's Office of Community Renewal and the Housing Trust Fund Corporation. Planning experts from the New York Department of State and Department of Transportation were also assigned to each community to provide technical assistance and help oversee the planning consultants. The magnitude of this effort can be seen in the 66 completed Community Reconstruction Plans representing more than 124 communities participating in the program (New York State 2018).

Studies of Local Recovery Planning, Pre- and Post-Disaster

Although much of the recovery literature supports the need for recovery planning, few studies contain direct guidance for urban planners who suddenly find themselves with significant post-disaster responsibilities (Olshansky and Chang 2009). A challenge in the recovery literature is that researchers define the problem in different ways. Community-level recovery can be viewed as a problem of urban systems (Alesch et al. 2009); physical design and reconstruction (Arnold 1993; Spangle Associates 2002); management (Inam 2005; Johnson 2014; Rubin 1985); urban planning (Schwab 1998; William Spangle and Associates et al. 1980); finance (Comerio 1998); or social and institutional organization (Smith 2011; Tierney 2007). Furthermore, the content and timing of recovery planning is much less well defined than other recovery activities, such as housing, public facilities, and infrastructure reconstruction (William Spangle and Associates 1991).

Scope and Focus of Recovery Plans and Planning

Most researchers cited in this section have made recommendations for recovery plan contents and the recovery planning process and these are synthesized and referenced in Table 23.1 (Johnson 2009). Analyzing the compilation, there are many consensus items and a few, but crucial, potential conflicts that are reflective of the lack of consensus in the field of disaster recovery research. The first disagreement is about the core content of the plan and whether it should be focused on general principles, which in many cases are desired physical and socioeconomic outcomes, or on action-oriented procedures. The second disagreement is about scope and goals; some researchers see plans as similar to comprehensive or master plans whereas others see them as procedural plans. These conflicts can be explained as the difference between *physically oriented* and *process-oriented* recovery plans, which can also be resolved by ensuring that recovery plans address both the desired physical outcomes of a city's recovery as well as the management structure, policies, and procedures that it wants to establish.

Table 23.1 Compilation of Research Recommendations for Local Disaster Recovery Plan Contents and Recovery Planning Processes

---

Local Disaster Recovery Plan Contents

- Clear overall vision, goals, and/or philosophy for recovery and reconstruction
- Establish the recovery management organization, with clear roles and relationships, and including all that can provide specific or assigned types of assistance
- Focused on general principles while maintaining flexibility
- Action-oriented procedures that clearly establish relationships and priorities, include detailed information, traceable milestones, and articulate operational tasks with assigned responsibilities
- Finances linking identified needs to funds
- Human resources and technical assistance needs
- Detailed implementation plan that is phased (e.g. short-, medium-, and long-term), focusing time and resources, and setting clear policies and expectations for each phase
- Policies, strategies, and responsibilities assigned to ensure that long-term recovery is linked with comprehensive planning, growth management, hazard mitigation, and emergency planning
- Long-term policies for redevelopment, infrastructure, densities, nonconforming uses, and future land use patterns
- Policies for preserving local communities and local heritage.

Recovery Planning Processes

- Strong local leadership
- Cooperation among local, state, and national officials
- Focused on reducing the unknowns associated with disasters and their consequences
- Use of modern technologies, data collection and sharing, sound information, empirical data, and valid knowledge; not untested assumptions
- Involve the community and stakeholders, including elected officials, citizens, and community organizations, partly for education and also to build consensus. Expect resistance
- Use of previous plans and pre-existing planning institutions
- Distinct from the activity of disaster management, but involving response agencies in the process.

Source: Johnson (2009).

Thus, in some sense, research suggests that recovery plans should be a hybrid of urban planning and emergency operations planning which is consistent with the planning examples coming from FEMA and other state and local agencies as they work to incorporate the core principles and operational functions outlined in the NDRF (Fairfax County 2012; FEMA 2011b; Rebuild Iowa Office 2010). However, the differences between an emergency operations plan and a pre-disaster recovery plan should not be overstated because the FEMA (2010) *Comprehensive Planning Guide (CPG) 101* advocates a collaborative planning process that differs from other planning processes only in the set of government agencies, NGOs, and private sector organizations that should be involved.

As previously noted, another issue of inconsistency in the literature surrounds the process of recovery planning and when it is best done. Key policy guidance, practitioner planning guidance, and most researchers strongly advocate pre-disaster recovery planning (FEMA 2011b; Schwab 1998; William Spangle and Associates et al. 1980). Although the potential payback from pre-disaster recovery planning might seem obvious, there can still be some significant challenges, such as engaging elected officials and the public in the process. Similarly, those communities that initiate a post-disaster recovery planning effort face some of the same difficulties that any political process normally does, but the literature raises some distinct advantages and disadvantages to initiating such discussions post-disaster.

Reconstruction Following Disaster undertook one of the first comprehensive evaluations of the plans and planning following disasters and also documented some of the dilemmas of planning and the timing of plans (Haas et al. 1977). As the authors astutely recognized, any new post-disaster "studies, plans and designs compete with the old . . . There is already a plan for reconstruction, indelibly stamped in the perception of each resident—the plan of the pre-disaster city" (Haas et al. 1977: 268). This is the "first plan," but rebuilding back exactly as the community was before is impossible and may deny opportunities for risk reduction or other community betterments which might otherwise be attained in the rebuilding (Burby 1998; Olshansky et al. 2006; Schwab 1998; Smith and Wenger 2007). Thus, there is almost inevitably a "second plan" that may come from previous plans developed before the disaster or from new plans made following the disaster.

Value of Recovery Plans and Planning

Many researchers have noted the value of pre-disaster plans and pre-existing planning institutions in helping facilitate recovery (Olshansky et al. 2006; Ota 2007; Rubin 1985; Spangle Associates 1997; Schwab 1998; William Spangle and Associates et al. 1980). Communities with up-to-date general or comprehensive plans, redevelopment plans, or other subarea plans are described as having the best foundation for (post-disaster) recovery planning. Disaster-impacted communities may find that they are able to implement plans, or parts of plans, that were not possible before the disaster and turn adversity into opportunity.

Pre-disaster plans are also important in recovery because they represent consensus policies about the future and demonstrate that the community has an active planning process, including well-established community organizations, lines of communication, and a variety of planning documents and tools (Olshansky et al. 2006). Many researchers have acknowledged the role of previous plans and information management in improving both the speed and quality of post-disaster decisions (Haas et al. 1977; Olshansky and Chang 2009; Ota 2007). Post-disaster planning decisions must be based on the best available information and are easiest to accomplish if plans and policies have been adopted before the disaster.

In 1998, the American Planning Association developed the first-ever nationally focused recovery planning guide for practicing planners entitled *Planning for Post-Disaster Recovery and Reconstruction*, which made several arguments for the value of pre-disaster recovery planning (Schwab 1998).

- Pre-disaster plans reduce the chances of making short-term decisions following a disaster that may limit future options.
- Pre-disaster plans help public officials to respond to the pressures of the moment, and reduce the risk that they might make promises that compromise opportunities for achieving a safer community.
- The planning process itself is valuable. Planners play an important role in building consensus around a vision before a disaster, and then in making key rebuilding decisions after the disaster.
- Pre-disaster plans help save critical time and better position a community to access additional post-disaster funds as they become available, such as for hazard mitigation or for infrastructure improvement.

The 1998 APA guide also included a model recovery and reconstruction ordinance for local governments to adopt before a disaster. The ordinance was modeled after the City of Los Angeles plan, calling for the establishment of a recovery organization to prepare a recovery plan and define post-disaster authorities, as well as the adoption of the plan and ordinance by a governing body. It also defined a variety of regulatory powers and procedures that a city might use following a disaster, including damage assessment, development moratoria, expedited permit processing, handling non-conforming uses, and demolition of damaged historic buildings. More recently Schwab (2014) updated the 1998 APA guide to recognize the significant body of planning literature addressing post-disaster recovery that was developed over the prior decades as well as a host of new laws, programs, and conditions influencing post-disaster recovery, such as the emergence of the NDRF, local hazard mitigation, and the increasing policy focus on disaster resilience.

Effectiveness of Recovery Plans and Planning Approaches

There are two common pre-disaster recovery planning approaches used by local governments. The first is a standalone plan and the second is an element of the community's comprehensive plan, but there are benefits to both (Berke and Campanella 2006). A standalone recovery plan "can be easier to revise, has more technical sophistication, is less demanding of coordination, and is simpler to implement" (Berke and Campanella 2006: 194). However, an integrated plan or a plan that has elements integrated into other plans, such as the comprehensive plan or emergency operations plan, can bring more resources together for implementation, broaden the understanding of the integrative nature of recovery issues with other local issues (e.g., transportation, housing, land use, environment), and provide access to a wider slate of planning and regulatory tools.

Although systematic assessments of traditional urban planning processes and planning content are well established (Baer 1997; Berke and Godschalk 2009), systematic local recovery plan assessments have only been given limited attention. In part, this is due to the infrequent nature of disasters, which hinders systematic study to develop quantitative data as well as qualitative indicators of recovery, such as plans, processes, key actors, and institutions (Miles and Chang 2006).

The Northridge earthquake, which struck the Los Angeles region in 1994, provided a well-timed opportunity for the first known assessment of the effectiveness of a pre-disaster recovery plan (Olshansky et al. 2006; Spangle Associates 1997). Starting in the late 1980s, the City of Los Angeles undertook the nation's first comprehensive, local pre-disaster recovery and reconstruction planning process which resulted in the preparation of the draft Los Angeles Recovery and Reconstruction Plan in 1994, just prior to the Northridge earthquake (City of Los Angeles EOO 1994). The draft plan was approved by the city's Emergency Operations Board on January 22—just five days after the Northridge earthquake—and later revised to incorporate lessons from the Northridge earthquake (City of Los Angeles 1995). The plan is very process-oriented, with policies and actions for physical elements of recovery (e.g. residential, commercial, and industrial rehabilitation, public sector services, economic recovery, and land use) as well as key governmental functions (e.g., organization and authority, vital records, inter-jurisdictional relationships) (City of Los Angeles EOO 1994).

Post-earthquake assessments credit both Los Angeles' pre-disaster recovery plan and planning process as major factors in the City's positive inter-organizational and multi-governmental relationships, as well as its overall ability to manage the post-Northridge recovery (Olshansky et al. 2006; Spangle Associates 1997; Tierney 1995). This observation is consistent with earlier observations that communities with effective leadership generally have adopted plans before the disaster, reflecting the consensus of community networks (Rubin 1985); and with observations that local governments with well-established planning functions tend to be the most effective at managing reconstruction (William Spangle and Associates et al. 1980). They also described the plan as having a hybrid quality, with action statements "falling somewhere between a checklist of items and descriptions of implementing programs" (Spangle Associates 1997: 20). They concluded that this hybrid quality makes sense since the plan covers both emergency response and normal municipal operations. Finally, Wu and Lindell's (2004) comparison of the 1994 Northridge earthquake and the 1999 Chi-Chi Earthquake in Taiwan concluded that the Los Angeles pre-disaster recovery plan accelerated housing recovery and allowed local officials to be more effective in integrating hazard mitigation into the recovery process.

Berke et al. (2014) studied the plan quality of the pre-disaster recovery plans developed by 87 local governments in eight states. They found that that the core principles used to evaluate traditional city plan quality were not fully suited for pre-disaster recovery plan evaluation. In particular, they called for more adaptive plan quality principles to suit the complexity and uncertainty associated with pre-disaster recovery planning, given the potentially large variations in timing, location, and severity of impacts on social, natural, and built environment systems from different hazards.

A similar systematic evaluation of local post-disaster recovery plans in the U.S. has not yet been conducted, but it is plausible to assume that, even with the disaster's footprint revealed, there are still considerable complexities and uncertainties of long-term recovery that require flexible and adaptive policies, strategies, and organizational structures. This is also consistent with comparative research observations of post-disaster planning for long-term recovery (Alesch et al. 2009; Haas et al. 1977; Inam 2005; Olshansky et al. 2012; Olshansky et al. 2006; Rubin 1985).

For cities, the work of long-term recovery looks in many ways a lot like normal urban life, governance, development, and renewal. However, from the perspective of city building, and thus urban planning, what is uniquely different after a disaster is that all these activities—which previously took years and even generations to accomplish—must happen concurrently and at a considerably faster pace (Olshansky et al. 2012). This "time compression" of activities also varies considerably both spatially and temporally across the disaster-impacted community, which creates a sort of "warping" that can cause processes of physical construction, the supply of financial resources, and restoration of neighborhood social and economic networks, among others, to happen unevenly and unnaturally. As a result, certain urban activities get out of sync compared to normal times; things get rebuilt in the wrong order; and some apparently lower priority recovery actions can get completed before higher priorities. Olshansky et al. (2012) posit that the concepts of time compression and differential time compression are useful in adapting ideas about urban development and planning in normal times to the specific situation of post-disaster recovery planning.

Time compression is also an explanation for the dilemma of speed versus deliberation in post-disaster planning, which has been raised by many researchers over the years starting with Rubin (1985). Haas et al. (1977) recommended that post-disaster planners make decisions as soon as possible so as to reduce uncertainty among private decision makers. They caution that "over-ambitious and detached planning . . . takes too long, raises too many issues simultaneously, and produces massive resistance and counter-attack" (Haas et al. 1977: 67). But, moving too quickly can cause problems as well. Planning should aim to evoke the most appropriate, but not necessarily most rapid, actions to address problems (Quarantelli 1982). One concern is the challenging pace at which post-disaster planning must often happen. As Olshansky and Chang (2009: 208) state, post-disaster planning is "a high speed version of normal flow of the most basic information upon which planning normally depends." This means that none of the participants quite knows what anyone else is doing, or how their own activity fits into the big picture (because no one yet understands the big picture). In many ways, this is an extension of the information challenges in emergency response and is commonly referred to as a "planning fog of war" within that context (Olshansky and Johnson 2010). Under these circumstances, citizen involvement in planning—essential for urban planning—is particularly challenging to accomplish. However, many researchers caution that consensus is critical to successful planning especially in complex multi-stakeholder settings (Birch and Wachter 2006; Innes and Booher 2010; Olshansky et al. 2006).

Several post-disaster recovery research studies echo those of Berke and his colleagues (Berke et al. 2014; Berke and Godschalk 2009), calling for adaptation as an essential element of recovery planning (Alesch et al. 2009; Inam 2005). As Alesch and his colleagues observe "[a]daptation doesn't mean abandoning the plan and starting again from scratch; it requires working with the reality of the situation to increase the probability of reaching the intended goal" (2009: 174–175). This is quite similar to Inam's (2005: 180) contention

[o]ne way of coping with uncertainty, complexity and change. . . is to recognize that all planning projects are policy experiments, and to plan them incrementally and adaptively by disaggregating problems and formulating response through processes of decision making that join learning with action.

These statements express much the same idea as Lindell and Perry's observation (2007: 118) "emergency responders should be trained to implement the most likely responses to disaster demands but they should also be encouraged to improvise" (see also Kreps, 1991).

Observing the complexities of recovery following large-scale urban disasters, researchers have emphasized that recovery involves multiple plans by multiple actors—one form of adaptive policy development and implementation (Olshansky et al. 2012; Ota 2007). Each of the citywide plans produced in New Orleans has served a purpose for both its authors and audiences (Olshansky and Johnson 2010). Alesch et al. (2009: 174) agree with the notion of planning as a continuum, describing effective post-disaster plans as providing for "branching out into different directions at various junctures if such revision is called for."

Olshansky and Chang (2009) also note a gap in the recovery and reconstruction literature regarding planning implementation challenges. Similarly, the Natural Hazards Center (2005) recommends that recovery plans provide a complete picture for holistic recovery and that every goal in the plan has an accompanying implementation strategy that includes action items, lead agency/entity and its deliverables, partnerships that will make the actions effective, methods for obtaining technical expertise and advice, local regulations needed, and funding methods. This document further recommends that the plan contain an overall budget, schedule, funding details, a monitoring process, and a process for public review and comment.

Overall, most of the commentary on post-disaster recovery planning is consistent with the literature on urban plans and urban planning processes in general. Speaking of the latter, Spangle Associates (1997: 23) observes "[p]lans are pictures of a desired future; they provide a vision that can make people willing to sacrifice short-term pleasures or gains for a better future." Plans also provide information about intentions and future actions of various actors and communicate information from specific actors to appropriate audiences and, in this way, affect behavior (Hopkins 2001). Thus, recovery plans are demonstrations of leadership, making persuasive arguments for various parties for various purposes, particularly among city, state, and federal officials (Mammen 2011; Olshansky and Johnson 2010; Rubin 1985). In particular, recovery actors often create plans in order to more effectively obtain and apply external sources of funds to meet recovery goals (Johnson and Olshansky 2017).

Plans are also the result of thinking about multiple alternatives before taking an action, and planning scholars have long recognized that the benefits of planning come both from the process as well as the plans themselves (Innes and Booher 1999, 2010). Even when a group does not agree on a plan, the planning process itself can substantively inform and modify the plans of individuals and organizations that participate in the process. Finally, plans serve as signals of collaboration to other recovery actors to jointly participate in recovery and rebuilding, which is a collective-action process (Olshansky et al. 2009).

Distinguishing Recovery Planning from Other City Planning Activities

Recovery planning typically has many of the same elements as traditional city plans such as a vision statement, goals and objectives, identified needs or problems, assumptions and method of reasoning, specific proposals, stakeholder input and deliberation, implementation devices, and criteria for monitoring and assessing outcomes (Baer 1997). However, post-disaster time compression creates some unique challenges for governments and managers of long-term recovery that distinguish recovery planning for normal city planning activities and that should be considered in recovery plan design and guidance (Olshansky et al. 2012). The following four extraordinary recovery characteristics are proposed and illustrated with examples of approaches taken in local recovery plans and planning processes (initiated both pre- and post-disaster in the United States) to manage them.

Balancing the Considerable Tension Between Speed and Deliberation

As noted earlier, a central issue in post-disaster recovery is the tension between speed and deliberation—between rebuilding as quickly as possible and considering how to improve on what existed before the disaster. Governments at all levels feel the burden of having to choose between speed and deliberation and, thus, the degree to which the public and stakeholders are involved in post-disaster efforts to both plan for and facilitate long-term recovery (Alesch et al. 2009; Rubin 1985). The balance between these two goals can affect how both the planning process is conducted and the plan is implemented. Acting too fast can result in poorly targeted, poorly prepared, and/or poorly designed recovery policies and programs, whereas taking too long can prolong human suffering and exacerbate the economic losses resulting from the loss of capital services (MDF and JRF 2012). Despite warnings from Haas et al. (1977) to avoid slowing down to plan, the historic record is full of examples of carefully considered post-disaster recovery plans that led to significant improvements. Current research only provides limited anecdotal guidance regarding appropriate responses to this tension. As noted by Rubin (1985), the ability to make this tradeoff strategically and purposefully is rare.

Olshansky et al. (2012) identified three effective methods for compressing both the local recovery planning and implementation processes post-disaster. They are:

- Iteration (e.g., making decisions and plans, and then acting on them in stages as more information becomes available);
- Decentralization (e.g., setting standards for planning and plan implementation as guides for simultaneous decision making and actions undertaken by multiple agencies); and,
- Increasing capacity (e.g., providing technical assistance and other external resources).

Many post-disaster planning efforts, particularly following large-scale disasters, offer examples of one or more of these time-compressed approaches. Also, post-disaster planning observations have shown that, particularly in large disasters, there is no "one" plan or blueprint for the recovery, but rather a series of plans developed over time to influence various decisions and persuade the ecosystem of recovery actors to act.

As an example of iteration and increasing capacity, Grand Forks' short-term recovery action plan described the plans that had been made up to that point as well as the organizational structure, goals, short-term objectives, and actions that would be taken to implement these plans and begin developing updated and more detailed plans in the first six months of recovery (City of Grand Forks 1997). Area-specific plans as well as a long-term, citywide vision were developed over the next several years that relied heavily on outside planning and other technical assistance to reach completion (City of Grand Forks 2011; Johnson 2014).

The FEMA ESF#14, CPCB and Planning processes, as previously discussed, are also examples of increasing capacity for recovery planning. In combination with these programs, as well as with federal funding support, many states—including Louisiana, Iowa, and New York—have developed post-disaster recovery planning frameworks and programs that allowed for both increasing capacity and decentralization of planning at the local level. The Louisiana Speaks (LRA 2006) and New York Rising (New York State 2018) efforts both established specific requirements and guidelines for community recovery planning efforts, and also tied some local recovery funding opportunities to the planning processes. The Iowa Community and Regional Recovery Planning team coordinated its efforts to ensure consistency in the long-term recovery planning assistance it provided to ten severely impacted communities (Rebuild Iowa Office 2011).

The Unified New Orleans Plan process initiated after Hurricane Katrina also used the increasing capacity and decentralization approaches with its two-tiered planning process (Olshansky and Johnson 2010). A citywide planning team had two key charges, the first of which was to assess the more systemic, citywide recovery needs, such as infrastructure recovery, and second of which was to unify the previous and ongoing planning efforts into one comprehensive Citywide Strategic Recovery and Rebuilding Plan (UNOP 2007). Another group of planning consultants worked at the district level, constructing District Recovery Plans for each of the city's 13 planning districts (administrative areas delineated by the New Orleans City Planning Commission during the 1980s). Both the citywide and district teams followed a similar three-phase structure: (1) conducting a comprehensive recovery assessment, (2) developing and selecting recovery scenario preferences, and (3) constructing the recovery plans and prioritized list of recovery projects. Throughout the process, the planning teams maintained a top-down and bottom-up interaction that, coupled with the broad citizen input, helped establish the recovery scenario preferences and principles for the plans (Johnson and Rabalais 2007).

Pre-disaster designs for effective post-disaster planning processes are more limited. The Fairfax County, Virginia, pre-disaster recovery plan defines a post-disaster recovery planning cycle that is modeled on the action planning cycle of the Incident Command System, which is linked to county-specific RSFs informed by the NDRF (Fairfax County 2012). Following a disaster, the plan specifies that the county's recovery organization will develop a recovery action plan (RAP) based upon input from each of the activated RSF branches to guide actions, tactics, and mission assignments of the recovery operations. It will also develop community recovery plans with substantial involvement of stakeholders and the public to provide strategic integration of policies, programs, and projects into a holistic community context (Fairfax County 2012: iv–7). Ideally, the community recovery plan would be developed ahead of the RAP, but might not be possible due to post-disaster time compression. Thus, the county's plan offers guidance on how plan consistency and integration can be achieved before, during, and after the development and completion of the community recovery plan.

Balancing Unprecedented Opportunity for Transformation, or Betterment, With Considerable Pressures for Restoration

Disasters cause unusually large and simultaneous depletions or losses of capital services (e.g., housing, bridges, roads, schools) compared with normal life cycles and replacement rates (Olshansky et al. 2012). This concentrated destruction of physical assets after a disaster can also reduce the normal constraints and opportunity costs of changing, adapting, and transforming physical, social, and economic systems. However, this opportunity often competes with the desires of residents to quickly restore the familiar. Haas et al. (1977) describe this tension as one of competing plans—competition between the planners' ideas for improvement and the "plan" of the pre-disaster city in residents' minds. Transparency, accountability, and equity both in the planning process and plan content and dissemination can help manage this tension (Olshansky et al. 2012).

The visioning process led by Joplin's Citizen Advisory Recovery Team (CART) offers guidance on how transformation and restoration interests can be blended and balanced in a post-disaster recovery planning process (Cage 2013; Joplin Area CART 2014). An international example, the post-earthquake restoration plan for Kobe, Japan, illustrates how boundaries between transformation and restoration can quite literally be drawn (City of Kobe 1995). The plan identified priority recovery areas in which transformation would be considered in greater detail as part of recovery implementation actions (e.g., redevelopment, land readjustment, and housing reconstruction policies).

This distinction gave clarity and confidence to residents, businesses, and other stakeholders outside these areas to move ahead with restoration to pre-existing or new conditions as individually desired (Olshansky et al. 2006).

Demonstrating Need and Justification for Unusually Large Amounts of Funding Needed for Plan Implementation

One of the most difficult problems inherent to the compressed time environment following disasters is obtaining and providing funds quickly, transparently, and with accountability and equity (Alesch et al. 2009). Also, access to various funds comes at different rates during recovery and this may require new policy and organizational approaches for managing and monitoring their use. The Unified New Orleans Plan citywide recovery and rebuilding plan developed a policy framework to define recovery strategies and estimate the $14 billion in funding required to implement them in different parts of the city over different time phases (UNOP 2007). The framework established three policy areas defined by the two overarching issues affecting the city's recovery—varying rates of repopulation and differing levels of future flood risk. The framework was also designed as an implementation tool to allow future planners and decision makers to monitor and evaluate progress, adjust funding to meet the changing demands, and determine how strategies should change.

The increased flow of funds in compressed time also means that rates and forms of communication must increase, both horizontally among agencies and organizations in the community and vertically among spatial scopes and levels of government (Chandrasekhar and Olshansky 2007; Smith 2011). The Fairfax County, Virginia, pre-disaster recovery plan proposes the post-disaster establishment of a Recovery Policy Advisory Board to advise the Recovery Coordinator and Recovery Agency and ensure accountability and representation of the public's interests in overarching policy guidance and general prioritization of recovery activities (Fairfax County 2012). It also identifies a Finance and Administration section (based upon the Incident Command System structure) within the Recovery Agency to track and coordinate recovery financing both across the county government and with recovery program funders and distributors. As previously noted, the NDRF also recommends appointing a Local Disaster Recovery Manager with post-disaster responsibilities to lead a local recovery organization and serve as the key point of contact with state and federal recovery partners as well as local political bodies and the public to keep them appropriately involved in decision-making and to ensure effective and consistent communication (FEMA 2011a).

Demonstrating Scalable and Sustained Capacity to Manage Long-Term Recovery

Local government is only one of the important recovery actors, but it has critical management and leadership roles in motivating the ecosystem of actors necessary to successfully achieve community recovery and rebuilding (Johnson 2014). Local government is the level of government that most citizens interact with on a daily basis regarding the regulation of land use, building construction, redevelopment, and the provisions of basic urban services. However, large-scale disasters by definition exceed local capacities to meet demands and there are many observations of state- and even national-level interventions in normal local government functions during post-disaster recovery.

Pre-disaster response and recovery planning efforts to identify the roles and responsibilities of various departments and agencies in the City of Los Angeles were identified as major factors in the city's positive inter-organizational and multi-governmental relationships, as well as its overall ability to manage recovery following the 1994 Northridge earthquake (Spangle Associates 1997). Some of the measures undertaken by the City of Grand Forks, North Dakota, were the designation of staff leadership for recovery, the creation of a short-term recovery action plan and weekly action planning sessions, the administration of recovery staffing and information management needs assessments, and the preparation of a communications plan; these efforts are all credited with demonstrating the city's resource capacity and needs to manage recovery to the satisfaction of key funding partners such as HUD in the aftermath of the 1997 floods (City of Grand Forks 2006; Johnson 2014).

Interpretations for Research and Practice

Local governments have the primary role of planning and managing community recovery. Although there has been an increase in both pre- and post-disaster recovery planning by U.S. cities, there are not yet established triggers, mandates, or standards for recovery planning during either phase. The NDRF, introduced in 2011, defined the federal government's management structure for recovery and outlines a set of core principles and capabilities to support disaster recovery. State and local recovery planning efforts that incorporate the core principles, management structure, and core capabilities defined in the NDRF are also happening across the country.

Assessments of the quality and content of local recovery planning processes and planning content have only received limited attention to date. Berke and his colleagues' (2014) systematic study of the quality of local governments' pre-disaster recovery plans found that that the core principles used to evaluate traditional city plan quality were not fully suited for recovery planning. Given the potentially large variations in timing, location, and severity of impacts on social, natural, and built environment systems from different hazards, they called for more adaptive plan quality principles to suit the complexity and uncertainty associated with recovery. Although a similar systematic evaluation of local post-disaster recovery plans throughout the United States has not yet been completed, it is plausible to assume that even after a disaster's footprint has been revealed there will still be considerable complexities and uncertainties about long-term recovery that require flexible and adaptive policies, strategies, and organizational structures.

The phenomenon of post-disaster time compression defines some unique conditions that distinguish recovery planning from normal city planning activities and that should be carefully considered in recovery policy development and local recovery plan design and guidance, both pre- and post-event. They are:

- Balancing the considerable tension between speed and deliberation;
- Balancing the unprecedented opportunity for transformation, or betterment, with considerable pressures for restoration;
- Demonstrating the need for unusually large amounts of funding needed for plan implementation; and,
- Demonstrating a scalable and sustained capacity to manage long-term recovery.

More systematic studies of recovery plans, processes, key actors, and institutions are also needed to advance long-term recovery policy, planning, and implementation.

References

Abramson, D., D. Culp, J. Sury, and L. Johnson. (2011). *Planning for Long-Term Recovery Before Disaster Strikes: Case Studies of 4 US Cities*, Final Project Report, New York: Columbia University, Mailman School of Public Health, National Center for Disaster Preparedness. http://academiccommons.columbia.edu/item/ac:152838.

Alesch, D. J., L. A. Arendt, and J. N. Holly. (2009) *Managing for Long-Term Community Recovery in the Aftermath of Disaster*, Fairfax VA: Public Entity Risk Institute.

Arnold, C. (1993) *Reconstruction After Earthquakes: Issues, Urban Design, and Case Studies*, Report to the National Science Foundation, San Mateo CA: Building Systems Development, Inc.

Arnold, C. (1999) "Earthquake as Opportunity: The Reconstruction of Pacific Garden Mall, Santa Cruz, After the Loma Prieta Earthquake of 1989," *Earthquake Engineering Research Institute, Learning from Earthquakes Series, Lessons Learned Over Time* 1: 1–40.

Baer, W. (1997) "General Plan Evaluation Criteria: An Approach to Making Better Plans," *Journal of the American Planning Association* 63(3): 329–344.

Berke, P. and T. J. Campanella. (2006) "Planning for Post Disaster Resiliency," *The Annals of the American Academy of Political and Social Science* 604(1): 192–207.

Berke, P., J. Cooper, M. Aminto, S. Grabich, and J. Horney. (2014) "Adaptive Planning for Disaster Recovery and Resiliency: An Evaluation of 87 Local Recovery Plans in Eight States," *Journal of the American Planning Association* 80(4): 310–323.

Berke, P. and D. R. Godschalk. (2009) "Searching for the Good Plan: A Meta-Analysis of Plan Quality Studies," *Journal of Planning Literature* 23(3): 227–240.

Birch, E. L. and S. M. Wachter (eds.). (2006) *Rebuilding Urban Places After Disaster*, Philadelphia PA: University of Pennsylvania Press.

Burby, R. (ed.). (1998) *Cooperating with Nature: Confronting Natural Hazards with Land-Use Planning for Sustainable Communities*, Washington DC: Joseph Henry Press.

Cage, J. (ed.). (2013) *Joplin Pays It Forward: Community Leaders Share Our Recovery Lessons*. Accessed 11 June 2018 at www.joplinmo.org/joplinpaysitforward.

Chandrasekhar, D. and R. B. Olshansky. (2007) "Managing Development after Catastrophic Disaster: A Study of Organizations That Coordinated Post-Disaster Recovery in Aceh and Louisiana," presented at the Annual Conference of the Association of Collegiate Schools of Planning in Milwaukee WI.

City of Grand Forks, North Dakota. (1997) "The First Season of Recovery, Grand Forks' Flood Recovery Action Plan, Action Plan Period: June 1 through November 1, 1997." City Council formal adoption, July 7, 1997.

City of Grand Forks, North Dakota. (2006) "Grand Forks Flood Disaster and Recovery Lessons Learned." www.grandforksgov.com/home/showdocument?id=528.

City of Kobe, Japan. (1995) "Kobe City Restoration Plan (abridged version)." June, City of Kobe: Secretariat of the Earthquake Restoration Headquarters

City of Los Angeles. (1995) "In the Wake of the Quake, A Prepared City Responds. A Report to the Los Angeles City Council." Mayor and Ad Hoc Committee on Earthquake Recovery.

City of Los Angeles, Emergency Operations Organization. (1994) "Recovery and Reconstruction Plan." http://eird.org/cd/recovery-planning/docs/2-planning-process-scenario/Los-angles-recovery-and-reconstruction-plan.pdf.

City of Watsonville Planning Department. (1992) "Community Based Disaster Plan." City of Watsonville Planning Department.

Comerio, M. (1998) *Disaster Hits Home: New Policy for Urban Housing Recovery*, Berkeley CA: University of California Press.

Cuomo, A. (2013) "Governor Cuomo Announces Community Reconstruction Zones Funded by Federal Supplemental Disaster Aid to Guide Local Rebuilding Process." www.governor.ny.gov/press/04262013cuomo-reconstruction-federal-disaster-aid.

DHS—Department of Homeland Security. (2004) *National Response Plan*. Washington DC: Author.

DHS—Department of Homeland Security. (2008) *National Incident Management System*, Washington DC: Author. www.fema.gov/pdf/emergency/nims/NIMS_core.pdf.

EMAP—Emergency Management Accreditation Program. (2013) *Emergency Management Standard*, Lexington KY: Author. www.emaponline.org/index.php/root/for-programs/23-2013-emergency-management-standard.

Fairfax County, Virginia. (2012) Fairfax County Pre-Disaster Recovery Plan. Fairfax VA: Author.

FEMA—Federal Emergency Management Agency. (2010) *Developing and Maintaining Emergency Operations Plans—Comprehensive Planning Guide (CPG) 101*, Version 2.0, Washington DC: Author. www.fema.gov/media-library/assets/documents/25975.

FEMA—Federal Emergency Management Agency. (2011a) *National Disaster Recovery Framework: Strengthening Disaster Recovery for the Nation*, Washington DC: Author. www.fema.gov/pdf/recoveryframework/ndrf.pdf.

FEMA—Federal Emergency Management Agency. (2011b) *Lessons in Community Recovery: Seven Years of Emergency Support Function #14 Long-Term Community Recovery From 2004 to 2011*, Washington, DC: Author. www.fema.gov/pdf/rebuild/ltrc/2011_report.pdf.

FEMA—Federal Emergency Management Agency. (2016) *National Disaster Recovery Framework*, 2nd edition. Washington DC: Author. www.fema.gov/media-library/assets/documents/117794.

Haas, J. E., R. Kates, and M. Bowden (eds.). (1977) *Reconstruction Following Disaster*, Cambridge MA and London, England: The MIT Press.

Hopkins, L. D. (2001) *Urban Development: The Logic of Making Plans*, Washington DC: Island Press.

HUD Technical Assistance Team. (1997) *Team Evaluation of Post-Flood Technical Assistance Provided to the City of Grand Forks, North Dakota, A Report for the U.S. Department of Housing and Urban Development*, Washington DC: Author.

HUD—Department of Housing and Urban Development. (2018) *CDBG Disaster Recovery Assistance*, Washington DC: Author. www.hudexchange.info/programs/cdbg-dr/cdbg-dr-grantee-contact-information/#all-disasters.

Hurricane Sandy Rebuilding Task Force. (2014) *Hurricane Sandy Rebuilding Strategy: Progress Update—Community Planning and Capacity Building*. Washington DC: U.S. Department of Housing and Urban Development. portal.hud.gov/hudportal/documents/huddoc?id=cpcb-0614.pdf.

Inam, A. (2005) *Planning for the Unplanned: Recovering from Crises in Megacities*, New York: Routledge.

Innes, J. E. and D. E. Booher. (1999) "Consensus Building and Complex Adaptive Systems: A Framework for Evaluating Collaborative Planning," *Journal of the American Planning Association* 65(4): 412–423.

Innes, J. E. and D. E. Booher. (2010) *Planning with Complexity: An Introduction to Collaborative Rationality for Public Policy*, New York: Routledge.

Johnson, L. A. (2009) *Developing a Management Framework for Local Disaster Recovery: A Study of the U.S. Disaster Recovery Management System and the Management Processes and Outcomes of Disaster Recovery in 3 U.S. Cities*, PhD Dissertation, School of Informatics, Kyoto University.

Johnson, L. A. (2014) "Developing a Local Recovery Management Framework: Report on the Post-Disaster Strategies and Approaches Taken by Three Local Governments in the U.S. Following Major Disasters," *International Journal of Mass Emergencies and Disasters* 32(2): 242–274.

Johnson, L. A. and R. B. Olshansky. (2017) *After Great Disasters: An In-Depth Analysis of How Six Countries Managed Community Recovery*, Cambridge MA: Lincoln Institute of Land Policy.

Johnson, L. A. and R. Rabalais. (2007) "Planning for Post-Disaster Rebuilding: An Update from New Orleans—An Invited Comment," *Natural Hazards Observer* 31(5): 1–3.

Joplin Area CART. (2014) "The Citizens Advisory Recovery Team (CART) for the Recovery of Joplin, Missouri," Joplin MO: Author.

Kreps, G.A. (1991). "Organizing for Emergency Management," 30–54 in T. S. Drabek and G. J. Hoetmer (eds.). *Emergency Management: Principles and Practice for Local Government*. Washington DC: International City/County Management Association.

Lindell, M. K. and R. W. Perry. (2007) "Planning and Preparedness," 113–141 in K. J. Tierney and W. F. Waugh, Jr. (eds.) *Emergency Management: Principles and Practice for Local Government*, 2nd ed. Washington DC: International City/County Management Association.

LRA—Louisiana Recovery Authority. (2006) *Louisiana Speaks. Long-Term Community Recovery Planning, Parish Recovery Planning Tool*. Baton Rouge LA: Author.

Mammen, D. (2011) *Creating Recovery: Values and Approaches in New York After 9/11*, Tokyo: Fuji Technology Press Ltd.

McCarty, M., L. Perl, and B. Foote. (2005) *The Role of HUD Housing Programs in Response to Disasters*, RL33078, Washington DC: Congressional Research Service, The Library of Congress.

MDF—Multi Donor Fund for Aceh and Nias and JRF—Java Reconstruction Fund. (2012) *Effective Post-Disaster Reconstruction of Infrastructure: Experiences from Aceh and Nias*. Working Paper 3. Lessons Learned from Post-Disaster Reconstruction in Indonesia. Jakarta, Indonesia: MDF-JRF Secretariat, The World Bank Office.

Miles, S. and S. E. Chang. (2006) "Modeling Community Recovery from Earthquakes," *Earthquake Spectra* 22(2): 439–458.

Natural Hazards Center. (2005) *Holistic Disaster Recovery, Ideas for Building Local Sustainability After a Natural Disaster*, Boulder CO: University of Colorado Natural Hazards Center.

Natural Hazards Center. (1999) *An Assessment of Recovery Assistance Provided After the 1997 Floods in the Red River Basin: Impacts on Basin-Wide Resilience*. Ottawa, Canada: International Red River Basin Task Force, International Joint Commission.

New York State. (2018) *New York Rising Community Reconstruction Program*, NY Rising Communities. http://stormrecovery.ny.gov/community-reconstruction-program.

Olshansky, R. B. and S. E. Chang. (2009) "Planning for Disaster Recovery: Emerging Research Needs and Challenges," *Progress in Planning* 72(4): 200–209.

Olshansky, R. B. and L. A. Johnson. (2010) *Clear as Mud: Planning for the Rebuilding of New Orleans*, Chicago IL: American Planning Association.

Olshansky, R. B., L. D. Hopkins, D. Chandrasekhar, and K. Iuchi. (2009) "Disaster Recovery: Explaining Relationships among Actions, Decisions, Plans, Organizations, and People," in *Proceedings of 2009 NSF Engineering Research and Innovation Conference*. Honolulu, HI.

Olshansky, R. B., L. D. Hopkins, and L. A. Johnson. (2012) "Disaster and Recovery: Processes Compressed in Time," *Natural Hazards Review* 13(3): 173–178. doi:10.1061/(ASCE)NH.1527-6996.0000077.

Olshansky, R. B., L. A. Johnson, and K. C. Topping. (2006) "Rebuilding Communities Following Disaster: Lessons from Kobe and Los Angeles," *Built Environment* 32(4): 354–374.

Ota, T. (2007) "The Kobe City Restoration Plan: Factors That Facilitated Its Rapid Formulation," *The Second International Conference on Urban Disaster Reduction Proceedings*, Taipei, Taiwan.

Quarantelli, E. L. (1982) "Ten Research Derived Principles of Disaster Planning," *Disaster Management* 2(1): 23–25.

Rebuild Iowa Office. (2010) *Iowa Disaster Recovery Framework.* publications.iowa.gov/10060/1/2010-11-16_Iowa_Disaster_Recovery_Framework.pdf.

Rebuild Iowa Office. (2011) *Rebuild Iowa Office Quarterly Report & Economic Recovery Strategy: April 2011.* rio.urban.uiowa.edu/sites/rio/files/2011.04_RIO_Quarterly_Report.pdf.

Rubin, C. (1985) *Community Recovery from a Major Natural Disaster*, Monograph No. 41. Boulder CO: University of Colorado Institute of Behavioral Science, Program on Environment and Behavior.

Schwab, J. C. (1998) *Planning for Post-Disaster Recovery and Reconstruction*, Planning Advisory Service Report 483/484, Chicago IL: American Planning Association.

Schwab, J.C. (2014). *Planning for Post-Disaster Recovery: Next Generation*, Planning Advisory Service Report 576. Chicago, IL: American Planning Association.

Smith, G. (2011) *Planning for Post-Disaster Recovery: A Review of the United States Disaster Assistance Framework*, Washington DC: Island Press.

Smith, G. and D. Wenger. (2007) "Sustainable Disaster Recovery: Operationalizing an Existing Agenda," 234–257, in H. Rodriguez, E. L. Quarantelli, and R. R. Dynes (eds.), *Handbook of Disaster Research*, New York: Springer.

Spangle Associates. (1997) *Evaluation of Use of the Los Angeles Recovery and Reconstruction Plan after the Northridge Earthquake*, Portola Valley CA: Author.

Spangle Associates. (2002) *Redevelopment After Earthquakes*, Portola Valley CA: Author.

Spangle, W. (ed.). (1987) *Pre-Earthquake Planning for Post-Earthquake Rebuilding*, Pasadena CA: Southern California Earthquake Preparedness Project.

State of Florida. (2015) "Post Disaster Redevelopment Planning Addendum: Addressing Adaptation During Long-Term Recovery," Tallahassee FL: State of Florida Department of Economic Opportunity and Florida Division of Emergency Management. www.floridajobs.org/docs/default-source/2015-community-development/community-planning/pdr/pdrpsealeveriseaddendum.pdf?sfvrsn=2.

State of Florida. (2010) "Post Disaster Redevelopment Planning," Tallahassee FL: State of Florida Department of Community Affairs (DCA) and Division of Emergency Management. www.floridadisaster.org/globalassets/importedpdfs/post-disaster-redevelopment-planning-workshop.pdf.

Tierney, K. (1995) *Social Aspects of the Northridge Earthquake*, Preliminary Paper 225, Newark DE: University of Delaware Disaster Research Center.

Tierney, K. (2007) "From the Margins to the Mainstream? Disaster Research at the Crossroads," *Annual Review of Sociology* 33: 503–525.

UNOP—Unified New Orleans Plan. (2007) "Citywide Strategic Recovery and Rebuilding Plan, Final Draft, Appendix E: Preliminary Citywide Financial Assessment."

William Spangle and Associates. (1991) *Rebuilding After Earthquakes: Lessons from Planners*, Portola Valley CA: Author.

William Spangle and Associates, Earth Science Associates, H. J. Degenkolb & Associates, G. S. Duggar, and N. Williams. (1980) *Land Use Planning After Earthquakes*, Portola Valley, CA: Author.

Wu, J. Y. and M. K. Lindell. (2004). "Housing Reconstruction After Two Major Earthquakes: The 1994 Northridge Earthquake in the United States and the 1999 Chi-Chi Earthquake in Taiwan," *Disasters* 28(1): 63–81.

24

REFLECTIONS ON ENGAGING SOCIALLY VULNERABLE POPULATIONS IN DISASTER PLANNING

John T. Cooper, Jr.

Introduction

The aftermath of recent disasters such as Hurricanes Katrina (2005) and Ike (2008) underscores the importance of accounting for the most vulnerable segments of the public (e.g., the elderly, poor, and disabled) in local disaster plans. In fact, even decades prior to Katrina and Ike, disaster researchers and emergency management professionals were aware of the disproportionate impacts of disasters on certain population segments. For example, some of the deadliest natural disasters in American history occurred in the low-lying areas of South Carolina (1893), Louisiana (1893), and on multiple occasions in Mississippi, where the vast majority of victims were poor and black (Steinberg 2000). As early as the 1960s, researchers were documenting cases where Blacks died at higher rates than Whites (Bates et al. 1963). Others observed that the disadvantaged face disproportionately low access to opportunities and disproportionately high exposure to risks, which are a consequence of the socioeconomic system (Bolin and Bolton 1986) and these inequities result in disadvantaged people experiencing elevated probabilities of loss, injury, death, and reduced ability to recover from hazards or disasters (Blaikie et al. 1994; Boyce 2000; Mileti 1999)

One explanation for the disproportionately higher impact of disasters on poor versus wealthy populations is that the wealthy often choose to locate in flood-prone areas to enjoy environmental, recreational, or aesthetic amenities. Thus exposure to flood hazards is usually voluntary for the wealthy. The poor on the other hand, tend to live in old, low-quality housing that is less able to withstand extreme events (Dolbeare 1999). Their housing also tends to be located in low-lying areas where land is cheap and they lack sufficient resources to either mitigate disasters where they are or move elsewhere; they are place-bound by virtue of their poverty (Bolin and Bolton 1986; Denton and Massey 1993; Dolbeare 1999; Peacock and Girard 1997; Sundet and Mermelstein 1988).

Acknowledging the disproportionate impact of disasters on the disadvantaged leads to questions about what can be done to reduce those impacts. To this end, research and professional experience point to the importance of involving the public in disaster planning as a method of increasing the extent to which individuals are able to prepare for, survive, and recover from disasters (Berke et al. 2010). In keeping with this notion, the earliest versions of federal programs offering guidance on how to prepare local disaster plans encouraged planners to involve the public early in the process as a way to learn about citizens' conditions and concerns (Godschalk et al. 1999). Historically, the disproportionate impacts of disasters have actually been compounded by government emergency management practices that overlook or discount disadvantaged persons, rendering them more vulnerable to future losses of life and property from disasters while protecting the status quo (Bolin and Bolton 1986). In short, one can find many examples—both before and after Katrina—of socially vulnerable disaster survivors who feel they have been victims of insufficient planning, unfair treatment, misinformation, and/or inadequate assistance from federal and state relief workers (Peacock et al. 1997; Highfield et al. 2014). Since Katrina however, there has been an increasing focus on accounting for disadvantaged populations in disaster planning and practice, but there is still room for improvement.

The purpose of this chapter is to discuss how the public, especially disadvantaged populations, could be more meaningfully engaged in local disaster planning programs. Specifically, the discussion will focus on the methods used during the Emergency Preparedness Demonstration (EPD), a 2009 FEMA-funded effort aimed at identifying ways to increase disaster awareness and preparedness in disadvantaged communities. Later, the chapter will discuss two key differences in the EPD process compared to the process prescribed by FEMA's *Local Mitigation Planning Handbook*, which was released in March 2013 and is based on FEMA's *Whole Community Approach to Emergency Management*. The *Local Mitigation Guidebook* is also considered a guidance document for communities developing plans in compliance with the Disaster Mitigation Act of 2000 (DMA 2000).

Historical Context

Before diving into a discussion of how federal guidance documents address socially vulnerable populations, it is useful to provide some recent context to further illuminate the social forces that shape the documents we have today.

Thirteen years before Katrina, Hurricane Andrew made landfall near Homestead Florida in the early morning hours of August 24, 1992. This event is significant because it is likely the first time the notion of a disproportionate impact of disasters on certain populations seeped into the collective consciousness of the US public. This was followed by the 1993 Midwest flood, and Hurricanes Fran and Floyd, which devastated rural eastern North Carolina in 1996. Not long afterwards, Peacock et al. (1997) cast more light on the subject with the publication of a series of studies looking back on Hurricane Andrew. The result was a detailed examination of how race, ethnicity, class, and gender interact in a disaster context. Then in 1999, six years before Katrina, Hurricane Floyd provided another compelling example of how historic societal inequities can result in greater hazard vulnerability for some populations.

At about 3 a.m. on September 16, 1999, Hurricane Floyd made landfall at Cape Fear, North Carolina, as a Category 2 hurricane with 105 mph winds. The heavy rains associated with the storm caused the Tar River to crest at 24 feet and flood the town of Princeville. Princeville, North Carolina, is an African-American community started by newly freed slaves just after the Civil War. Princeville was originally called Freedom Hill for a nearby hill where Union soldiers first announced that a Union victory had made slaves free, and it became the nation's first black incorporated town. It is also situated on the low, swampy side of the Tar River on land that nobody else wanted—the only place where newly freed slaves and their descendants could form their own community with little resistance. For the former slave masters and other potential employers in communities nearby, having a separate black community in Princeville settled the dilemma of retaining a ready labor supply while keeping former slaves at a respectable social (segregated) distance. Unfortunately, prior to Hurricane Floyd, the Tar River overflowed its banks at least seven times (1865, 1889, 1919, 1924, 1940, and 1958). Though none of Princeville's residents lost their lives, they lost homes, pets, and family treasures (Sturgis 2005).

The impact of Hurricane Floyd on Princeville was the second time the plight of socially vulnerable populations caught the nation's attention. A few days after Princeville flooded, President Clinton stood at the edge of the flood water and promised the federal government would do whatever necessary to help Princeville and surrounding communities recover. During February (Black History Month) of the following year, President Clinton signed an executive order establishing the President's Council on the Future of Princeville, North Carolina ("Council on Princeville"). The Council on Princeville comprised designees from 15 federal agencies including FEMA. The Council on Princeville was charged with developing recommendations for the President on further federal agency and legislative actions that could be undertaken to address the future of Princeville, taking into consideration the "unique historic and cultural importance of Princeville in American history" and the "views and recommendations of the relevant State and local governments, the private sector, citizens, community groups, and non-profit organizations, on actions that they all could take to enhance the future of Princeville and its citizens" (White House 2000: 1) among other things.

After the President established the Council on Princeville, and with the momentum of positive national public sentiment, members of the US Congress led by Representative Eva Clayton (the first African-American to represent to represent North Carolina since 1901), whose district included Princeville, began working to get national disaster policy to focus more on reducing the disproportionate impact of disasters on disadvantaged communities. However, by the start of the 2000 hurricane season, hope for more proactive approaches to mitigating the impacts of disasters by federal agencies began to fade as campaigns for the presidential and congressional elections overshadowed their endeavor. Then early in President Bush's administration, he signaled intent to scale back spending on FEMA programs by announcing the end of one of those programs (Project Impact) on the same day the mayor of Seattle praised the program for sponsoring the earthquake mitigation efforts that prevented further damage from the magnitude 6.8 Nisqually earthquake (Block and Cooper 2006: 68; Congressional Research Service 2009). Later that year, the tragic events of September 11, 2001, diverted the investment of more national resources to preventing the next terrorist attack rather than preparing for environmental disasters.

Although Congress did not take specific steps before 2001 to establish federal disaster programs that would ameliorate the extent to which disasters disproportionately affect socially vulnerable populations, it had passed DMA 2000, which required state and local governments to have FEMA-approved mitigation plans in place prior to receiving federal disaster support. In 2002, FEMA produced a *How-to Planning Guide* for DMA 2000 that emphasized the importance of hazard mitigation planning and calling for partnerships to bring together the skills, expertise, and experience of a broad range of groups to achieve a common vision for their communities and ensure that the most appropriate and equitable mitigation projects would be undertaken (FEMA 2003).

The DMA 2000 *How-to Planning Guide*, unlike FEMA's previous planning guidance documents, devoted more attention to encouraging local planners to be collaborative in the design and execution of planning programs, for example by ensuring local planning teams are diverse and equitable. The issue of community collaboration remained unresolved, however, and in 2004, as part of its appropriation to assist communities after Hurricane Isabel, Congress set aside funds to establish the *Emergency Preparedness Demonstration* (EPD) *Program*.

On December 22, 2004, FEMA announced the availability of $1.5 million for an eligible organization to conduct post-disaster critiques and evaluations in disadvantaged communities affected by Hurricane Isabel, provide an assessment of emergency preparedness awareness in these communities, and develop recommendations for new methods to improve outreach to these and similar communities (Federal Register 2004). In May 2005, two months before Hurricane Katrina made landfall, FEMA entered into a cooperative agreement with MDC Inc., a non-profit organization based in Durham, North Carolina, and the University of North Carolina, Chapel Hill, to design and manage the EPD. There were three primary goals of the program.

1 To learn why disadvantaged communities are typically less prepared for disasters and which strategies have and have not worked to prepare these communities for disasters;

2 To raise awareness among disadvantaged residents about their vulnerability in future disasters, and to test ways that minority and disadvantaged communities can be engaged to help with disaster planning and preparedness; and

3 To reduce community and household vulnerability to harm from disasters while positioning the community to undertake comprehensive and equitable disaster recovery in the future.

In September, 2005, as the Bush Administration and FEMA faced heavy public criticism for the federal response to Hurricane Katrina, staff in FEMA's Division of Individual and Community Preparedness were working with skilled community organizers and researchers in six states (DE, MD, NC, PA, VA, WV) and the District of Colombia to understand the barriers to increased disaster awareness and preparedness among socially vulnerable groups. It is worth mentioning that in 2004, the year before Katrina, President Bush issued an executive order requiring federal departments and agencies to consider the unique needs of both agency employees with disabilities and individuals with disabilities whom the agency serves during their emergency preparedness planning (White House 2004). After Hurricane Katrina, however, it must have been clear to even the most casual observer that the federal government still had room for improvement in complying with the order. Even the Council on Princeville failed to meaningfully engage displaced citizens of the town.

Although the Council on Princeville produced an exemplary set of potential strategies for the future of Princeville—with hundreds of pages of data, analysis, designs, and illustrations—it did not address Princeville's lack of capacity to implement the proposed strategies. As a result, despite an ambitious plan and millions of dollars in donations from around the country (how relief money was spent is another issue) after Hurricane Floyd, the town had accomplished very little by the 10-year anniversary of Hurricane Floyd. In fact, on September 28, 2009, days after the 10-year anniversary, a former town manager pleaded guilty to 13 counts of obtaining property by false pretense and one count of conspiracy to commit false pretense after a probe into the town's handling of FEMA disaster funds (WRAL.com 2009). Unfortunately, the next significant attempt to improve federal policy and practice did not come until 2011, when President Obama issued Presidential Policy Directive (PPD) 8, which called for strengthening the security and resilience of the United States in a systematic way by engaging communities, families, and individuals in the *Whole Community Approach*.

Paradigm Shift

FEMA's *Whole Community Approach* is a means by which residents, emergency management practitioners, organizational and community leaders, and government officials can collectively understand and assess the needs of their respective communities and determine the best ways to organize and strengthen their assets, capacities, and interests (FEMA 2011). In addition, FEMA Administrator Craig Fugate hosted a series of national dialogues to foster collective learning from communities' experiences across the country and extract lessons from past programs such as the previously mentioned Project Impact. These national dialogues resulted in the set of three principles, summarized below, which form the foundation of the *Whole Community Approach*.

1 *Understand and meet the actual needs of the whole community.* Community engagement leads to a deeper understanding of the unique and diverse needs of a population, including its demographics, values, norms, community structures, networks, and relationships.

2 *Engage and empower all parts of the community.* All members of the community are part of the emergency management team—including those who may not traditionally have been directly involved in emergency management—and engage in an authentic dialogue.

3 *Strengthen what works well in communities on a daily basis.* A *Whole Community Approach* requires finding ways to support and strengthen existing structures and relationships that are present in

the daily lives of individuals, families, businesses, and organizations before an incident occurs. These structures and relationships can then be leveraged and empowered to act effectively during and after a disaster strikes.

By defining and encouraging communities to adopt the *Whole Community Approach*, the Obama administration took a positive step towards reducing the disproportionate impact of disasters on socially vulnerable populations. The downside is that the principles and strategic themes of the approach are still too broad; they lack a step-by-step implementation guide or toolkit for emergency managers who know very little about organizing and facilitating change. Thus, FEMA released an updated (and improved) edition of the *Mitigation Handbook* in 2013, with an even stronger focus on public engagement, as well as cross references to sections in the Code of Federal Regulations that buttress the importance of commitment to public involvement and technical resources (e.g., worksheets, tool, links to online help).

Most public planning programs—whether they be rational, incremental or collaborative in their theoretical foundations—involve a set of activities that include forming a planning team, identifying and analyzing community problems, defining specific goals and objectives, evaluating alternative strategies for solving the problems, implementing promising (e.g., technically feasible, cost-efficient, and equitable) strategies, and reviewing the effectiveness of strategy implementation over time. Likewise, FEMA's *Mitigation Handbook* lists nine tasks. Tasks 1–3 describe the process, resources, and people needed to complete the remaining tasks. Tasks 4–8 discuss the types of analyses and decisions necessary to complete the mitigation plan. Task 9 provides suggestions and resources for implementing the plan to reduce risk. Still, although the *Mitigation Handbook* is good at providing end users with step-by-step instructions, it offers few examples of the most appropriate approaches for identifying the unique conditions, concerns, and capacities of socially vulnerable populations and engaging those populations in conversations about appropriate strategies to address the challenges they face. It is important to note that, although many local planners believe in the efficacy of inclusive planning, too often they lack the experience necessary to design and implement inclusive planning programs that meet culturally diverse needs. Other planners may possess the experience but lack the time and resources. Still others simply may not believe that accounting for the disadvantaged distinctly is an ethical obligation (Godschalk et al. 1999; Phillips 1993). Finally, those who do have the experience, commitment, time, and resources are sometimes unable to overcome the mistrust that marginalized populations have of local officials.

Informing a Community of Practice

For almost 40 years the standard protocol for managing disasters at FEMA and across the US at state and local levels is best characterized as a top-down approach. When disaster strikes a community, formal emergency response, relief, and recovery efforts are often oblivious of the way the community operated before the disaster and discount the value of local norms, social networks, and community capacity afterwards. At the same time, people from that community who know and care about each other willingly risk their own lives and give generously to help their neighbors. They are also frustrated sometimes when the formal system's bureaucracy ties their hands. For example, one can find many stories of volunteers who answer the call but often have to stand by until someone at the top decides what to do. Emergency managers can be equally frustrated when citizens "self-deploy" and inadvertently exacerbate the impact of a disaster. On the other hand, Hurricanes Katrina and Harvey have taught many that neither side of this equation is likely to be successful without the other and that the key to community resilience is an integrated system where people with badges and fiduciary responsibilities partner with individual citizens and citizen groups to share information, identify populations with special needs and gaps in current plans, and then find ways to fill those gaps by maximizing formal and informal resources—before another disaster strikes. In addition, since the passing of DMA 2000, FEMA has been doing more to promote a *Whole Community Approach* to increasing community resilience, including investing in the EPD Program discussed above.

The creation of the EPD Program was an acknowledgement of FEMA's lack of understanding of the barriers to increased disaster readiness among socially vulnerable populations and a sincere intent to provide the field of disaster planning with more insight on how to design and manage meaningful public engagement programs for marginalized or mistrustful populations. To ensure the legitimacy of the EPD, FEMA entered into a cooperative agreement with MDC (www.mdcinc.org/), a North Carolina based nonprofit, with a national reputation and nearly 50 years of experience helping communities become more resilient. Over the course of its history, MDC's work focused on identifying ways to help communities maximize their effectiveness and efficiency, by providing communities with the resources and guidance necessary to work in an inclusive way and connecting communities to opportunities to leverage their assets as they grow. The primary principles and methods of MDC's work—steeped in evidence-based theories of inclusion and focused on creating new leaders, new relationships, and tangible outcomes—were used in carrying out the EPD Program. For example, when initiating the work of community transformation, in the MDC approach it is critical to start by building the right team.

The value of building the right team for the work is based on MDC's decades of experience in the design and execution of community-centered projects. That experience taught MDC that a single citizen, no matter how committed, will likely have a difficult if not impossible challenge working alone to increase community resilience. By the same token, a group of people, no matter how well intentioned, will struggle to gain traction unless they begin by inviting a diverse set of stakeholders to participate in the process. In the EPD Program, this diverse set of stakeholders was called the EPD Taskforce. One might note that at the outset at least, MDC's process is similar to the process FEMA prescribes in the *Mitigation Handbook*. Specifically, this *Handbook* calls for planners to determine which agency or individual will lead the process and what resources (human, technical, and financial) are on hand to complete the process. The *Mitigation Handbook* notes that agencies such as the local emergency management agency (LEMA) in particular have an interest and responsibility to be included in the process. Likewise, in designing the EPD process MDC was mindful of the importance of involving local emergency managers in the work and actively targeted places with LEMA staffers who were supportive of the goals of the program. However, there is a subtle difference in the role of local emergency planners in the EPD versus what is called for in the *Mitigation Handbook*.

According to the *Mitigation Handbook*, after identifying a lead agency within local government, planners can focus on assembling a core team of planning partners that have the authority to implement the strategies identified in the process. Specifically, the *Handbook* suggests it is important to distinguish between those who should serve as members of the planning team and other individual or group stakeholders (e.g., businesses, private organizations, and citizens) who will be, or who think they will be, affected by a program. Further, unlike planning team members, other stakeholders may not be involved in all stages of the planning process. Instead, they may inform the planning team on a specific topic or provide input from different points of view in the community. Local planners may also choose to create an outreach strategy for how and when to involve other stakeholders in the planning process in lieu of inviting socially vulnerable stakeholders to join the planning team. To be clear, the *Mitigation Handbook* calls on local planners to provide stakeholders the opportunity to be on the planning team or otherwise involved in the planning process, presumably from the outset. Nevertheless, regardless of one's interpretation of what constitutes a "stakeholder," one thing seems clear—in the *Mitigation Handbook*, the choice of when and how to involve key stakeholders is made at the top. By contrast, the EPD Program sought to involve private individual and group stakeholders in process design, meeting planning, and data collection and analysis *from the outset*.

As noted above, when designing the EPD, MDC targeted places with emergency planners who supported the goals of the program. One overarching goal of the program was to engage citizens as co-leaders in determining hazard risks and deciding the most appropriate strategies for mitigating risks. Therefore, in targeting pilot sites for the EPD project, MDC sought places with supportive local emergency managers, that also had at least one community-based organization (CBO) with a track record of working with socially vulnerable populations and be willing and able to provide leadership to achieving program goals.

Considering CBOs as lead planners (not just planning team representatives) was important in the EPD Program, as these organizations are increasingly called upon to help marginalized people and communities recover from disasters. As a result, CBOs have a primary interest in identifying and mitigating potential threats to daily operations so they may continue to provide a safety net for marginalized populations before, during and after disasters. The extent to which a CBO is prepared to respond when things go awry determines not only whether the organization itself can survive and recover, but it could presumably affect the extent to which the people and places they serve day-to-day in normal times will survive and adequately recover. Therefore, the work of building a Taskforce for each EPD Program community began with efforts to gauge the interest and capacity of local CBOs to achieve project goals, while simultaneously engaging the local emergency management agency. To be clear, the EPD Program would not proceed without the support of the LEMA, because that office's expertise and legal responsibility to coordinate disaster planning activities meant that efforts to plan would be informal without it. In any case, LEMA designees and a trusted CBO formed the lead or core planning team (2–4 members) of each pilot site taskforce.

With a core planning team in place, each EPD pilot site recruited local volunteers with a stake in the outcome of the process to form the broader task a taskforce (8–12 members). Members of the broader taskforce were drawn from representatives of key constituencies with disaster-coping experience, a concern for raising community disaster readiness, and a willingness to commit to following through on EPD Program goals. In every pilot site, the EPD taskforce included individuals from socially vulnerable communities who had lived through a disaster and possessed the desire, knowledge, skill, or some other attribute that could enhance the work of the team. Members also represented institutions connected to disadvantaged communities (e.g., community centers, civic clubs, faith-based institutions, environmental justice advocates, etc.) and formal public institutions or agencies that play a role in disaster readiness for socially vulnerable groups (e.g., school systems, hospitals, health centers). Just as important were representatives of voluntary institutions, networks or associations trusted in the community that could play a role in disaster planning or implementing strategies (e.g., environmental justice advocates, food pantries, or homeless shelters). In short, the EPD engagement strategy was more bottom-up by design than the one recommended in the *Mitigation Handbook*.

In addition to naming CBOs as co-leaders (and in some case the fiscal agent for planning grants) alongside the local office of emergency management in a more bottom-up design, there is another key distinction between the EPD design and the one described in FEMA's *Mitigation Handbook*. Specifically, the *Handbook* suggests local mitigation planning teams identify a strong "advocate" or "local champion" on the team to serve as the chairperson. In addition, the role of the chairperson is to help enlist the support and participation of local officials and community leaders and oversee or help manage the mitigation planning process. While this chairperson does not need to be a professional planner or subject matter expert, they must be able to communicate the purpose and importance of planning, convene the planning team, and facilitate the completion of tasks required for the mitigation plan to be finished on schedule. By contrast, the designers of the EPD believed such responsibility is too much for an unpaid volunteer. Instead and in keeping with basic tenets of community building outlined above, each EPD taskforce had a pre-assigned *community coach* to guide the process in each site.

A community coach is a technical consultant who can provide a community with ways to make use of resources—such as specialized knowledge, practices, expertise—by adapting these resources to supplement existing community assets and capacity, integrating these resources with local wisdom, and supporting the creation of new ways of seeing and doing (Emery and Hubbell 2011). In the EPD Program, community coaches served as technical advisers, facilitators, and catalysts for change who exercised a diverse mix of skills (communication, consensus building, mediating, visioning, technical competence, advocacy) needed to motivate collective action (Berke et al. 2010). EPD coaches had the responsibility of assessing participant needs, designing an appropriate curriculum to meet those needs, and delivering the curriculum in an effective, engaging manner so participants were properly equipped to go about the work of change. Although there was no taskforce chairperson in EPD sites, each EPD coach worked with a site-based coordinator—usually a paid staffer from the CBO, but never a volunteer—to ensure the process stayed on track. Each site-based coordinator was responsible for staying in contact with task force members, keeping them enthusiastic, listening to and addressing concerns. Each CBO's coordinator was also responsible for securing meeting logistics (location, date, time, food, and drink, etc.) and ensuring that task force members knew of the location, date, and time and had received meeting materials in advance. Coordinators kept members abreast of information and decisions when they missed meetings. In short, providing each taskforce with a coach and a paid coordinator, meant the volunteer taskforce members could also focus their full time, energy, and creativity simply on showing up and participating in ways most meaningful and productive for them.

EPD Process

After solidifying the local taskforce, each EPD site began a conversation about the history of disaster in their communities, things that had worked or not worked with regard to the handling of response relief and recovery in the past and things worth doing in the future to ensure the community did not repeat past mistakes. The steps in the process are summarized in Figure 24.1.

In Step 1 of the process, the coach met with the core planning team (e.g., representatives of CBO and LEMA) to clarify the purposes of the EPD Program and the roles and responsibilities of all parties. At this time the core team set a preliminary schedule of meetings and identified neutral locations. They also discussed who should be invited to serve on the task force. The second step in the process was for the core team to host an open information meeting intended to provide an overview of the project to potential taskforce members and to discuss roles and responsibilities of members. Specifically, each potential EPD planning team member was asked to commit to sharing what has and hasn't worked in disaster preparedness in their area, which included helping researchers gather data about disasters and community preparedness as necessary. Each prospective taskforce member also had to commit to the process of discovering new ideas for ensuring that socially vulnerable populations in their communities, and the organizations that serve them, would be better prepared for future disasters. Perhaps most important, prospective taskforce members had to commit to being accountable for ensuring new ideas were put into practice.

In addition to showing up and holding themselves accountable, taskforce members had to be open to learning about ways to analyze data and assess disaster readiness—especially for disadvantaged communities—with tools and technical support from subject matter experts when desired or appropriate. To this end, Step 3 in the process was to initiate an effort to collect data on factors that cause economically disadvantaged people to be more vulnerable to disasters. Unfortunately, even when socially vulnerable groups believe in the efficacy of individual and household preparedness, they are often not able to act due to their poverty or other constraints. Either way, the consequences of not understanding or taking steps to prepare for disasters are far more severe on socially vulnerable

Step 1 Core Connections
Process coach meets core taskforce leaders (e.g., CBO & local emergency management) to:
- Clarify the purposes of the EPD, roles, and responsibilities of all parties
- Set preliminary schedule of meetings, locations, and outcomes
- Discuss who should be invited to serve on the broader taskforce

Step 2 Participant Orientation
Informational meeting for potential taskforce members to learn about EPD and membership requirements including:
- Learning from the Past
- Community Research
- Strategy Planning
- Ensuring Accountability

Step 3 Vulnerability Assessment
Collect data on factors that cause some to be more "vulnerable" to disasters including:
- Local hazard history
- Risk faced by existing structures, infrastructure, and critical facilities
- Information on future land uses and development trends

Step 4 Planning Retreat
Day-long taskforce retreat to launch work. Sessions included:
- Teambuilding: i.e., skills/knowledge each person brings to the table, strategies to ensure the work is inclusive, timely, and productive
- Review of who does what in emergency management: role of FEMA, state, local government
- Elements of an "ideal" disaster awareness and preparedness plan
- Review of EPD planning process, timetable, and locations
- Discussion of additional resources required to conduct the work

Step 5 Community Input
Documenting the experiences and knowledge of local people through culturally appropriate ways, including but not limited to public meetings, surveys, and interviews.

Step 6 Developing a Plan
Coaches guided each taskforce through a series of activities and guided questions in light of lessons learned in Step 5, leading to the creation of a plan.
- _Assessing the Current Situation_: Examination of local history to determine ways local leaders responded to disasters. What are the lessons?
- _Developing a Vision for the Future_: What would be in place if the community were prepared for disasters? What values would under gird the community's attitude towards preparedness?
- _Setting Goals and Objectives_: What specific goals are needed to turn vision into reality? What measurable object ives can be set as targets?
- _Strategy Development_: What range issues need consideration prior to strategy development. Presentation from CURS about model practices from other communities. Brainstorm additional strategy ideas. Identify research needed. Decide on strategies
- _Analyze and Involving Stakeholders_: Identify other stakeholders that need to be involved in achieving goals/object ives
- _Developing a Budget for Implementation_: What assets, resources or investments –internal and external to the community-are required to implement strategies?
- _Developing Action Plans_: Who does what and when to carry out new ideas?
- _Developing Sustainability Plans_: Who will be responsible for ongoing engagement of disadvantaged communities in local disaster planning after project work ends?

Step 7 Implementation – Putting Ideas into Practice
The new ideas will come out of the planning process and they need to be put into practice. Some will be put into place with grant money via MDC. Others will not cost. The Team will need to meet for some number of months to be sure that the ideas take root.

Figure 24.1 Emergency Planning Demonstration (EPD) Process

populations. Therefore, an aim of the EPD Program was to conduct meaningful research—valid and actionable by those who could benefit most from it. Related to helping communities conduct meaningful research, each EPD site had access to the resources and technical support necessary to conduct research and sustain the work, including a pool of experts from state universities with expertise in anthropology, public health, urban planning, and hazard vulnerability analyses for disaster planning. To be clear, subject matter experts and researchers were welcomed as "co-learners" in the process, and encouraged to involve members of the taskforce in determining what questions to ask, what data to collect, and the most appropriate ways to interpret data—all to ensure that conversations about what strategies to pursue were grounded in local social, cultural, and political contexts. Then, armed with an assessment of current conditions, each EPD taskforce began a deeper conversation on the complexities of managing disasters and what it meant for their work going forward.

In Step 4, taskforce coaches led team-building activities designed to help the group set the tone for conversations about potential strategies to address issues and opportunities revealed in Step 3. They discussed the skills and knowledge each member brought, established ground rules and other ways to ensure their interactions would be constructive and enjoyable. They also learned about the role of FEMA and state and local government in disaster management and invited subject matter experts in to talk about the elements of high-quality disaster plans. Each group also talked about what other kinds of information would be necessary to inform their conversations, in light of current conditions.

At Step 5 in the EPD Program, each taskforce took time to listen to the experiences of the broader community. This was accomplished in a variety of ways, including community meetings where attendees initially had the opportunity to share what they learned from past disasters, including areas where flooding or other hazard impacts had been more severe, or where recovery had been slow, for example. In later community meetings, participants had the chance to provide feedback on draft goals and objectives developed by the taskforce and comment on draft strategies to address goals and objectives. In some cases, the EPD taskforce groups chose to use community surveys to elicit and document information about community disaster experience, locations where the most vulnerable community members go for help, as well as barriers to increased disaster awareness and preparedness. However, a few sites chose personal interviews of community members as a way to uncover views about barriers to disaster readiness for disadvantaged communities, as well as successes and failures in past efforts. With knowledge of local history informed by community input, each taskforce was poised to plan for the future.

In Step 6, the EPD planning team began developing a final plan, which required a series of meetings (about 2 hours each) to review community research, finalize taskforce goals, and identify the most appropriate ideas for increasing local awareness and preparedness. The indigenous knowledge collected in Step 5 was cross-checked with information from the more formal or publicly available data (e.g., Census, flood maps, land use maps, etc.) gathered in Step 4 to expose and reconcile discrepancies in understanding of local vulnerability. In reconciling differences, coaches used a range of activities such as SWOT (Strength/Weakness/Opportunity/Threat) Analysis, to capture the knowledge and interests of taskforce members. Reconciling discrepancies resulted in each taskforce having a shared understanding of local vulnerability, the ways local leaders had responded to disasters in the past, lessons from those experiences and a better sense of what to do or not do going forward. With a shared understanding of current conditions and the values undergirding local attitudes towards preparedness, EPD coaches had each taskforce create a vision of what would be necessary for the community to be prepared for disasters in the future.

After creating a shared vision, EPD taskforce teams identified goals and measurable objectives to achieve their vision. Some sites chose to invite public comment on goals and objectives. Others moved on to strategy development with the help of subject matter experts who presented promising strategies used in other communities. Again, EPD coaches helped teams identify the range of issues worth considering prior to settling on specific strategies. Subject matter experts were present in strategy meetings to offer advice, clarification, and feedback when appropriate. In some instances, the work of identifying strategies required the creation of subcommittees to which other stakeholders were invited to weigh in on strategy ideas. In other instances, taskforce teams chose to host community meetings to get input on strategy ideas.

In Step 7, the final set of tasks related to drafting the plan began with determining the cost of implementing the mitigation plan and creating a budget. This step also required assigning responsibility for carrying out specific tasks in the mitigation plan and setting milestones for completing those tasks. EPD coaches encouraged the planning teams to think about how best to sustain the work (e.g. maintaining groups, shifting responsibility for implementation to municipal agencies or CBOs). Finally, an enduring reality of community-based work is that change does not happen overnight. As a result, it can be hard to maintain group efficacy. Therefore, it is important to take time to celebrate the achievement of milestones. To this end, after completing draft plans, each taskforce took time to celebrate their accomplishments. Some sites invited the public to join in the celebration by presenting the final plan in a public meeting. This was also a time to build public support for implementing the mitigation plan.

Conclusions

Involving the public in planning efforts is inherently messy. However, open discussion among equal citizens in planning and decision-making is necessary for community transformation (Habermas 1984). Likewise, one could argue that the key to community disaster readiness is related to the extent to which citizens, especially the most vulnerable ones, are involved in and accounted for in local disaster planning programs. Unfortunately, however, guidance on the design and execution of inclusive planning programs is sparse in documents such as FEMA's *Mitigation Handbook*. This chapter presented methods employed by the EPD, a pilot planning program funded by FEMA intended to increase disaster awareness and preparedness among disadvantaged communities. In short, the key differences between the EPD design and the one described in the *Mitigation Handbook* are clear. For one, the EPD is a more bottom-up approach that seeks to raise the bar for citizen engagement. By design, the EPD process required citizens have equal influence over what questions are asked, what data is collected, and how strategies are prioritized. The EPD also offers a methodology aimed at meaningful community engagement. By contrast, although the *Mitigation Handbook* promotes public engagement, that engagement is more along the lines of what Arnstein (1969) described as tokenism.

Although the *Mitigation Handbook* is clear on the value of public engagement and points out the fact that DMA 2000 requires state and local governments seek public input on multi-hazard mitigation plans as a precondition for receiving FEMA mitigation project grants, it offers little detail on the methodology of engagement, and unfortunately, local emergency managers are typically not experienced in the design and management of inclusive planning programs. By contrast, the EPD process was designed by skilled "community builders" who understood that almost every public participation endeavor will create learning experiences that are either anxiety-inducing and counterproductive, or positive and nurturing depending on how well the process is managed. Therefore, each EPD taskforce was assigned a "coach" who possessed not only subject matter knowledge, but was also culturally competent and skilled in helping citizens get better at learning from and with others who see things differently, so that together they can make strategic decisions that everyone is committed to. Among other things, taskforce coaches were mindful of details such as the importance of creating safe and respectful learning environments to allow taskforce members and public officials the comfort to test new behaviors, wrestle with new ideas, explore new relationships, and address old challenges (MDC, Inc. 2001). In short, coaches had to have some knowledge of how to mitigate a range of disasters, including the interpersonal or group conflicts that keep communities from reaching their full planning potential, and this is a skill more local emergency planners should develop.

A 2015 study of efforts to engage the public in disaster recovery planning in 87 counties across US Gulf and Atlantic states found limited public involvement in the planning process (Berke et al 2014). In addition, it discovered emergency planners in those counties were skeptical about the benefits of engaging the public in planning programs. Most emergency planners surveyed expressed a concern that planning would not guarantee representation or might be dominated by special interests. Many felt citizens lacked the expertise to contribute meaningfully. Others seemed to believe the negatives (e.g., undermined trust in authorities, heightened conflict, delayed completion, and increased cost) outweighed any positives. On the contrary, household surveys and interviews with key informants in the same study revealed the other side of the coin. Although few reported participating in the planning process, many believed public input would improve the plan and the top reason for not participating in the process according to the majority of those interviewed was they didn't know they could, which likely means they were not invited.

A lesson of the EPD was that when planners are skilled in the design and facilitation of public engagement programs, ordinary citizens can participate meaningfully in the process. Specifically, they can work collaboratively, gather and understand complex data and make wise choices in light of a myriad of scientific, economic, social or other considerations. They just need to be invited, and when they show up their knowledge—anchored in experience—needs to be appreciated.

Perhaps the most promising outcome of the EPD for emergency planners to note is that it resulted in greater public awareness and support for the work of emergency planners. In every EPD site, the LEMA was understaffed and overworked during emergencies. Also, in a few cases, emergency managers were hesitant to share any information with the public following the September 11, 2001 terrorist attacks in New York and Washington DC. In addition, the heavy public criticism of the Bush Administration and FEMA's response to Hurricane Katrina only intensified the suspicion some people in EPD sites had for emergency managers, whose uniforms with badges seemed indistinguishable from the local law enforcement they feared or mistrusted. However, in every EPD site the process resulted in new or strengthened relationships of stakeholders, local institutions, agencies, nonprofits, etc., with the LEMA. Members of the EPD taskforce at each site became advocates for local emergency managers and potential partners in times of crisis. Greater job security and less work alone should be enough for local emergency managers to think about becoming better community organizers.

References

Arnstein, S. R. (1969) "A Ladder of Citizen Participation," *Journal of the American Institute of Planners* 35(4): 216–224.

Bates, F., C. W. Fogleman, B. Parenton, R. Pittman and G. Tracy. (1963) *The Social and Psychological Consequences of Natural Disasters. NRC Disaster Study 18*, Washington DC: National Academy of Science.

Berke, P., J. Cooper, D. Salvesen, D. Spurlock, and C. Rausch. (2010) "Building Capacity for Disaster Resiliency in Six Disadvantaged Communities," *Sustainability* 3(1): 1–20.

Berke, P., J. Cooper, M. Aminto, S. Grabich, and J. Horney. (2014) "Adaptive Planning for Disaster Recovery and Resiliency: An Evaluation of 87 Local Recovery Plans in Eight States," *Journal of the American Planning Association* 80(4): 310–323.

Blaikie, P., T. Cannon, I. Davis, and B. Wisner. (1994) *At Risk: Natural Hazards, People's Vulnerability, and Disasters*, Routledge: London.

Block, R. and C. Cooper. (2006) *Disaster: Hurricane Katrina and the Failure of Homeland Security*, New York: Times Books–Henry Holt and Company.

Bolin, R. and P. Bolton. (1986) *Race, Religion, and Ethnicity in Disaster Recovery*, Boulder CO: University of Colorado, Institute of Behavioral Science.

Boyce, J. K. (2000) "Let Them Eat Risk? Wealth, Rights and Disaster Vulnerability," *Disasters* 24(3): 254–261.

Congressional Research Service (2009) *FEMA's Pre-Disaster Mitigation Program: Overview and Issues*, Washington DC: Author.

Denton, A. and D. Massey (1993) *American Apartheid: Segregation and the Making of the Underclass*, Cambridge MA: Harvard University Press.

Dolbeare, C. (1999) *Out of Reach: The Gap between Housing Costs and Income of Poor People in the United States,* Washington DC: National Low Income Housing Information Service.

Dolbeare, C. N. (1999) "Conditions and Trends in Rural Housing," 13–26, in R. J. Wiener and J. N. Belden (eds.) *Housing in Rural America,* Thousand Oaks CA: Sage.

Emery, M. and K. Hubbell (2011) *A Field Guide to Community Coaching,* Ames IA: W.K. Kellogg Foundation, the Annie E. Casey Foundation, Kellogg Action Lab at Fieldstone Alliance, and the Northwest Area Foundation.

FEMA—Federal Emergency Management Agency. (2003) *Developing the Mitigation Plan: Identifying Mitigation Actions and Implementation Strategies* FEMA 386–3, Washington DC: Author. www.fema.gov/media-library-data/20130726-1521-20490-5373/howto3.pdf.

FEMA—Federal Emergency Management Agency. (2011) *A Whole Community Approach to Emergency Management: Principles, Themes, and Pathways for Action,* FDOC 104-008-1 / December 2011, Washington DC: Author. www.fema.gov/media-library/assets/documents/23781.

FEMA—Federal Emergency Management Agency. (2013) *Local Mitigation Planning Handbook,* Washington, DC: Author. www.fema.gov/media-library/assets/documents/31598.

Federal Register (2004) *Emergency Preparedness Demonstration Program* 69 FR 76772. Washington DC: Author.

Godschalk, D. R., T. Beatley, P. Berke, D. Brower, and E. Kaiser. (1999) *Natural Hazard Mitigation: Recasting Disaster Policy and Planning,* Washington DC: Island Press.

Habermas, J. (1984) *Theory of Communicative Action,* Boston MA: Beacon Press.

Highfield, W. E., W. G. Peacock, and S. Van Zandt. (2014) "Mitigation Planning: Why Hazard Exposure, Structural Vulnerability, and Social Vulnerability Matter," *Journal of Planning Education and Research* 34(3): 287–300.

MDC, Inc. (2001) *Building Community by Design,* Chapel Hill NC: Author.

Mileti, D. (1999) *Disasters by Design: A Reassessment of Natural Hazards in the United States,* Washington DC: Joseph Henry Press.

Peacock, W. G. and Girard, C. (1997) "Ethnic and Racial Inequalities in Hurricane Damage and Insurance Settlements," 171–190, in W. G. Peacock, B. H. Morrow, and H. Gladwin (eds.) *Hurricane Andrew: Ethnicity, Gender, and the Sociology of Disasters,* New York: Routledge.

Peacock, W. G., B. H. Morrow, and H. Gladwin. (1997) *Hurricane Andrew: Ethnicity, Gender, and the Sociology of Disasters,* New York: Routledge.

Phillips, B. D. (1993) "Cultural Diversity in Disasters Situations: Sheltering, Housing, and Long Term Recovery," *International Journal of Mass Emergencies and Disasters* 11(1): 99–110.

Steinberg, T. (2000) *Acts of God: The Unnatural History of Natural Disaster in America,* New York: Oxford University Press.

Sturgis, S. (2005) "Fear and Flooding in North Carolina: A Hurricane-Harried African-American Town Lives with the Specter of Future Disaster," *Southern Exposure* 32(Winter): 34–41.

Sundet, P. and J. Mermelstein. (1988) "Community Development and the Rural Crisis: Problem-Strategy Fit," *Journal of Community Development Society* 19(2): 91–107.

White House (2000) Executive Order 13146: President's Council on the Future of Princeville, North Carolina. Accessed 20 May, 2018 at www.federalregister.gov/documents/2000/03/02/00-5209/presidents-council-on-the-future-of-princeville-north-carolina.

White House (2004) *Executive Order: Individuals with Disabilities in Emergency Preparedness,* Washington DC: Author. Accessed 29/12/2017 at http://georgewbush-whitehouse.archives.gov/news/releases/2004/07/20040722-10.html.

WRAL.com. (2009) "Former Princeville Town Manager Pleads Guilty," Accessed 29/12/2017 at www.wral.com/former-princeville-town-manager-pleads-guilty/6047230/#fjd4UeFAcjs55t0W.99.

INDEX

For Product Safety Concerns and Information please contact our EU
representative GPSR@taylorandfrancis.com
Taylor & Francis Verlag GmbH, Kaufingerstraße 24, 80331 München, Germany